Process Dynamics and Control

Process Dynamics and Control

Modeling for Control and Prediction

Brian Roffel

University of Groningen, The Netherlands

and

Ben Betlem

University of Twente, The Netherlands

John Wiley & Sons, Ltd

Email (for orders and customer service enquiries): cs-books@wiley.co.uk
Visit our Home Page on www.wileyeurope.com or www.wiley.com

Other Wiley Editorial Offices

John Wiley & Sons Inc., 111 River Street, Hoboken, NJ 07030, USA
Jossey-Bass, 989 Market Street, San Francisco, CA 94103-1741, USA
Wiley-VCH Verlag GmbH, Boschstr. 12, D-69469 Weinheim, Germany
John Wiley & Sons Australia Ltd, 42 McDougall Street, Milton, Queensland 4064, Australia
John Wiley & Sons (Asia) Pte Ltd, 2 Clementi Loop #02-01, Jin Xing Distripark, Singapore 129809
John Wiley & Sons Canada Ltd, 6045 Freemont Blvd, Mississauga, Ontario, L5R 4J3, Canada

Wiley also publishes its books in a variety of electronic formats. Some content that appears in print may not be available in electronic books.

Library of Congress Cataloging-in-Publication Data

Roffel, Brian.
Process dynamics and control : modeling for control and prediction / Brian Roffel and Ben Betlem.
p. cm.
ISBN-13: 978-0-470-01663-3
ISBN-10: 0-470-01663-9
ISBN-13: 978-0-470-01664-0
ISBN-10: 0-470-01664-7
1. Chemical process control. I. Betlem, B. H. (Ben H.) II. Title.
TP155.7.R629 2006
660'.2815–dc22

2006019140

A catalogue record for this book is available from the British Library

ISBN-13: 978-0-470-01663-3 (HB) 978-0-470-01664-0 (PB)
ISBN-10: 0-470-01663-9 (HB) 0-470-01664-7 (PB)

Typeset by SNP Best-set Typesetter Ltd., Hong Kong

This book is printed on acid-free paper responsibly manufactured from sustainable forestry in which at least two trees are planted for each one used for paper production.

CONTENTS

FOREWORD

In 1970, Brian Roffel and I started an undergraduate course on process dynamics and control, actually the first one for future chemical and material engineers in The Netherlands. Our idea was to teach something of general value, so we decided to focus on process modeling. Students received a verbal description of a particular chemical or physical process, to be transformed into a mathematical model. To our surprise, students appeared to be highly motivated; they spent much additional time in developing the equations. Some of them wanted to do it all by themselves and even refused to benefit from our advice. Maybe part of the fun was to be creative, an essential ingredient in model building to complement the systematic approach.

In fact, models are situation-dependent. Already about 50 years ago we ran into a clear-cut case at Shell, during the development of dynamic models for distillation columns. There we faced the problem of defining the response of the column pressure. Chemical engineers told us that this response is very slow: it takes many minutes before the pressure reaches a new equilibrium. However, the automation engineers did not agree as in their experience automatic pressure control is relatively fast. After some thinking, we discovered that both parties were right: the pressure response can be modeled by a large first-order time constant and a small dead time (representing the sum of smaller time constants). The large time constant dominates the open loop response, while closed loop behavior is limited by the dead time, irrespective of the value of the large time constant. Evidently, modeling requires a good insight into the purpose of the model. This book provides good guidance for this purpose.

Most books on process control restrict modeling to control applications. However, inside as well as outside industry many different process models are required, adapted to the specific requirements of the application. Consequently there exists a strong need for a comprehensive text about how to model processes in general. Fortunately, this excellent book fills in the gap by covering a wide range of methods complemented by a variety of applications. It goes all the way from ‘white’ (fundamental) to ‘black’ (empirical) box modeling, including a happy mix in the form of ‘hybrid’ models. More specifically, Ben Betlem adapted the systematic approach advocated for software development to modeling in general.

Special attention is paid to the influence of the process environment, and to techniques of model simplification. The latter can be helpful, among others, in reducing the number of model parameters to be estimated.

I wholeheartedly recommend this book, both to students and to professional engineers, and to scientists interested in modeling of processes of any kind.

John E. Rijnsdorp
Emeritus Professor of Process Dynamics and Control
University of Twente
The Netherlands

PREFACE

Process dynamics and control is an inter-disciplinary area. Three disciplines, process, control and information engineering, are of importance, as shown in Fig. 1.

- *Process engineering* offers the knowledge about an application.
  Understanding a process is always the basis of modeling and control. A rigorous dynamic process model should be developed to increase the understanding about the operation fundamentals and to test the control hypothesis. Experimental model verification is essential to be aware of all uncertainties and peculiarities of the process.
- *Control engineering* offers methods and techniques for (sub-)optimal operation at all hierarchical control and operational levels.
  For all process operational problems encountered, an appropriate or promising control method should be tested to meet the defined requirements.
- *Software engineering* offers the means for implementation.
  The simulation approach or control solution that is developed should be implemented in an appropriate way and on an appropriate hardware and software platform.

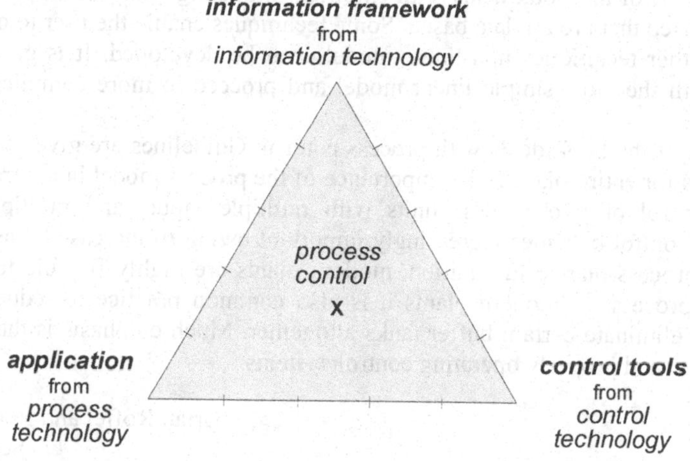

Fig. 1. Process dynamics and control area in three dimensions.

The three disciplines process, control and information technology answer questions such as: for what, why, how, and in which way.

Other disciplines are also of interest.

- *Business management* sets the production incentives and defines the coupling between the production floor and the office.
- *Human factors* study the relation between humans and automation. The contents and format of the supplied information has to meet certain standards to enable the personnel to perform their control and supervisory tasks well. This may conflict with the hierarchical structure of the control functions, which will be based on the partitioning of equipment operations. For the most part, flexibility of the automation infrastructure can solve these conflicts. In addition the degree of automation along the control hierarchy should be chosen with care.

- *Chemical analysis* supports quality control. The product quality is one of the most important operation constraints in process operation. In this respect, it should be mentioned that quality measurement is often problematic owing to its time delays and its unreliability. This can be overcome by a quality estimator based on mathematical principles.

This book will create a link between specific applications on the one hand and generalized mathematical methods used for the description of a system on the other.

The dynamic systems that will be considered are chemical and physiological systems. System behavior will be determined by using analytical mathematical solutions as well as by using simulation, for example Matlab-Simulink. Information flow diagrams will be used to reveal the model structure. These techniques will enable us to investigate the relationship between system variables and their dependencies.

This book is organized in three parts. The first part deals with physical modeling, where the model is based on laws of conservation of mass, momentum and energy and additional equations to complete the model description. In this case physical insight into the process is necessary. It is probably the best model description that can be developed, since this type of model imitates the phenomena that are present in reality. However, it can also be a very time-consuming effort.

In this first part, numerous unit operations are described and numerous examples have been worked out, to enable the reader to learn by example.

The second part of the book deals with empirical modeling. Various empirical modeling techniques are used that are all data based. Some techniques enable the user to develop linear models; with other techniques non-linear models can be developed. It is good practice to always start with the most simple linear model and proceed to more complicated methods only if required.

The last part of the book deals with process control. Guidelines are given for developing control schemes for entire plants. The importance of the process model in controller tuning is shown and control of two process units with multiple inputs and multiple outputs is demonstrated. Control becomes increasingly important owing to increased mass and energy integration in process plants. In addition, modern plants are highly flexible for the type of feed they can process. In modern plants it is also common practice to reduce the size of buffer tanks or eliminate certain buffer tanks altogether. Much emphasis is therefore placed on well-designed and properly operating control systems.

Brian Roffel and Ben H.L. Betlem
September 2006

ACKNOWLEDGEMENT

This book makes extensive use of the MATLAB® program, which is distributed by the Mathworks, Inc. We are grateful to the Mathworks for permission to include extracts of this code.

For MATLAB® product information, please contact:

The MathWorks, Inc.
3 Apple Hill Drive
Natick, MA, 01760-2098 USA
Tel: 508-647-7000
Fax: 508-647-7001
E-mail: info@mathworks.com
Web: www.mathworks.com

A user with a current MATLAB license can download trial products from the above Web site. Someone without a MATLAB license can fill out a request form on the site, and a sales rep will arrange the trial for them.

For the principal components analysis (PCA) and partial least squares regression (PLS) in Chapters 22 and 23, this book makes use of a PLS Toolbox, which is a product of Eigenvector Research, Inc. The PLS Toolbox is a collection of essential and advanced chemometric routines that work within the MATLAB® computational environment. We are grateful to Eigenvector for permission. For Eigenvector® product information, please contact:

Eigenvector Research, Inc
3905 West Eaglerock Drive
Wenatchee, Wa, 98801 USA
Tel: 509.662.9213
Fax: 509.662.9214
E-mail: bmw@eigenvector.com
Web: www.eigenvector.com

1 Introduction to Process Modeling

The form and content of dynamic process models are on the one hand determined by the application of the model, and on the other by the available knowledge. The application of the model determines the external structure of the model, whereas the available knowledge determines the internal structure. Dynamic process models can be used for simulation studies to get information about the process behavior; the models can also be used for control or optimization studies. Process knowledge may be available as physical relationships or in the form of process data.

This chapter outlines a procedure for developing a mathematical model of a dynamic physical or chemical process, determining the behavior of the system on the basis of the model, and interpreting the results. It will show that a systematic method consists of an analysis phase of the original system, a design phase and an evaluation phase. Different types of process model are also be reviewed.

1.1 Application of Process Models

A model is an image of the reality (a process or system), focused on a predetermined application. This image has its limitations, because it is usually based on incomplete knowledge of the system and therefore never represents the complete reality.

However, even from an incomplete picture of reality, we may be able to learn several things. A model can be tested under extreme circumstances, which is sometimes hard to realize for the true process or system. It is, for example, possible to investigate how a chemical plant reacts to disturbances. It is also possible to improve the dynamic behavior of a system, by changing certain design parameters. A model should therefore capture the essence of the reality that we like to investigate. Is modeling an art or a science? The scientific part is to be able to distinguish what is relevant or not in order to capture the essence.

Models are frequently used in science and technology. The concept of a model refers to entities varying from mathematical descriptions of a process to a replica of an actual system. A model is seldom a goal in itself. It provides always a tool in helping to solve a problem, which benefits from a mathematical description of the system.

Applications of models in engineering can be found in (i) *Research and Development*. This type of model is used for the interpretation of knowledge or measurements. An example is the description of chemical reaction kinetics from a laboratory set-up. Models for research purposes should preferably be based on physical principles, since they provide more insight into the coherence as to understand the importance of certain phenomena being observed.

Another application of process models is in (ii) *Process Design*. These types of model are frequently used to design and build (pilot) plants and evaluate safety issues and economical aspects.

Models are also used in (iii) *Planning and Scheduling*. These models are often simple static linear models in which the required plant capacity, product type and quality are the independent model variables.

Another important application of models is in (iv) *Process Optimization*. These models are primarily static physical models although for smaller process plants they could also be dynamic models. For debottlenecking purposes, steady-state models will suffice.

Process Dynamics and Control: Modeling for Control and Prediction. Brian Roffel and Ben Betlem.

Optimization of the operation of batch processes requires dynamic models. Optimization models can often be derived from the design models through appropriate simplification. There is, however, a shift in the degrees of freedom from design variables to control variables or process conditions.

In process operation, process models are often used for (v) *Prediction and Control*. Application of models for prediction is useful when it is difficult to measure certain product qualities, such as the properties of polymers, for example the average molecular weight. These models find also application in situations where gas chromatographs are used for composition analysis but where the process conditions are extreme or such that the gas chromatograph is prone to failure, for example because of frequent plugging of the sampling system.

Process models are also used in process control applications, especially since the development of model-based predictive control. These models are usually empirical models, they can not be too complex due to the online application of the models.

1.2 Dynamic Systems Modeling

Modeling is the procedure to formulate the dynamic effects of the system that will be considered into mathematical equations. The dynamic behavior can be characterized by the dynamic responses of the system to manipulated inputs and disturbances, taking into account the initial conditions of the system.

The outputs of the system are the dependent variables that characterize and describe the response of the system. The manipulated inputs of the system are the adjustable independent variables that are not influenced by the system. The disturbances are external changes that cannot be influenced, they do, however, have an impact on the behavior of the system. These changes usually have a random character, such as the environmental temperature, feed or composition changes, etc. The initial conditions are values that describe the state of the system at the beginning. For batch processes this is often the initial state and for continuous processes it is often the state in the operating point.

Manipulated inputs and disturbances cannot always be clearly separated. They both have an impact on the system. The former can be adjusted independently; the latter cannot be adjusted. This distinction is especially relevant for controlled systems. The inputs are then used to compensate the effects of the disturbances, such that the system is kept in a desired state or brought to a desired state.

Every book in the area of process modeling gives another definition of the term “model”. The definition that will be used here is a combination of the ideas of Eykhoff (1974) and Hangos and Cameron (2003).

A model of a system is:

- *a representation of the essential aspects of the system*
- *in a suitable (mathematical) form*
- *that can be experimentally verified*
- *in order to clarify questions about the system*

This definition incorporates the goal, the contents (the subject) as well as the form of the model. The goal is to find fitting answers to questions about the system. The subject for modeling is the representation of the essential aspects of the system. The aspects of the system should in principle be verifiable. The form of the model is determined by its application. Often an initial model that serves as a starting point, is transformed to a desirable form, to be able to make a statement about the behavior of the system.

At the extremes there are two types of models models that only contain physical-chemical relationships (the so-called "white box" or mechanistic models) and models that are entirely based on experiments (the so-called "black box" or experimental models). In the first category, only the system parameters are measured or known from the literature. In that case it is assumed that the structure of the model representation is entirely correct. In the second case, also the relationship(s) are experimentally determined. Between the two extremes there is a grey area. In many cases, some parts of the model, especially the balances, are based on physical relationships, whereas other parts are determined experimentally. This specifically holds for parameters in relationships, which often have a limited range of validity. However, the experimental parts can also refer to entire relationships, such as equations for the rate of reactions in biochemical processes.

In the sequel the most important aspects of modeling will be discussed.

1. *A model is a tool and no goal in itself.*
   The first phase in modeling is a description of the goal. This determines the boundaries of the system (which part of the system and the environment should be considered) and the level of detail (to which extent of detail should the system be modeled). The goal should be reasonably clear. A well-known rule of thumb is that the problem is already solved for 50% when the problem definition is clearly stated.
2. *Universal models are uneconomical.*
   It does usually not make sense to develop models that fit several purposes. Engineering models could be developed for design, economic studies, operation, control, safety and special cases. However, all these goals have different requirements with respect to the level of detail and have different degrees of freedom (design variables versus control variables, etc.).
3. *The complexity of the model should be in line with the goal.*
   When modeling, one should try to develop the simplest possible form of the model that is required to achieve the set goal. Limitation of the complexity is not only useful from an efficiency point of view, a too comprehensive, often unbalanced model, is undesirable and hides the true process behavior. The purpose of the model is often to provide insight. This is only possible if the formulation of the model is limited to the essential details. This is not always an easy task, since the essential phenomena are not always clear. A useful addition to the modeling steps is a sensitivity analysis, which can give an indication which relationships determine the result of the model.
4. *Hierarchy in the model.*
   Modeling is related to (experimental) observation. It is well known that during observation, the human brain uses hierarchical models. A triangle is observed by looking at the individual corners and the entire structure. This is also the starting point for modeling. The model comprises a minimum of three hierarchical levels: the system, the individual relationships and the parameters. Usually a system comprises several subsystems, each with a separate function. To understand the system, knowledge is required of the individual parts and their dependencies.
5. *Level of detail.*
   This is a difficult and important subject. Figure 1.1 shows the three dimensions in which the level of detail can be represented: time, space and function. Models can encompass a large time frame. Then only the large time constant should be considered and the remainder of the system can be described statically. If short times are important, small time constants become also important and the long-term effects can be considered as integrators (process output is an integration of the input). This will be discussed in more detail in a later chapter.

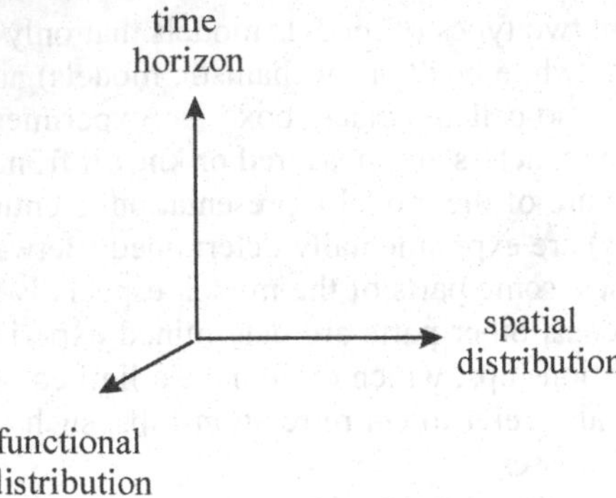

Fig. 1.1. Level of detail in three dimensions.

For the spatial description it is relevant to know whether the system can be considered to be lumped or not. Lumped means that all variables are independent of the location. An example is a thermometer, which will be discussed in the sequel. A good approximation is that the mercury has the same temperature everywhere, independent of the height and the cross sectional location.

In case of distributed systems, variables are location dependent. Most variables that change in time also change with respect to location. Examples are found in process equipment, such as heat exchangers, tubular reactors and distillation columns.

The level of detail of a function can vary considerably. The functionality of a system can be considered from the molecular level to the user level. An example is a coffee machine. At the user level it is only of interest how this system can be operated to get a cup of coffee of the required amount and quality. Knowledge about how, how fast and to what extent the coffee aroma is extracted from the powder is not required. In order to understand what really happens, knowledge at the molecular level may even be required.

Usually there is a dependency between the levels of detail of time, space and functionality. If the system is considered over a longer period, the spatial distribution and functionality will require less detail.

6. *Modeling based on network components versus modeling based on balances.*
   Mechanical, electro-mechanical and flow systems can often be modeled on the basis of elements (resistance, condenser, induction, transistor) of which the network is composed. This is often also possible for thermal systems. In many cases these elements can be described by linear relationships because they do not exceed the operating region. When combining these thermal systems with chemical systems, this network structure is not so clear anymore. The starting point in this case is usually the mass, energy and component balances. The balances can often be written as a network of differential equations and ordinary equations. But the structure is not recognizable as a network of individual components. Forcing such a system into a network of analog electrical components may violate the true situation.
7. *A model cannot explain more than it contains.*
   Modeling and simulation may enhance the insight, clarify dependencies, predict behavior, explore the system boundaries; however, they will not reveal knowledge that is unknown. A model is a reflection of all the experiments that have been performed.

8. *Modeling is a creative process with a certain degree of freedom.*
   The problem statement, the definition of the goal, the process analysis, the design and model analysis are all steps in which choices have to be made. Especially important are the assumptions. The final result will be dependent on the knowledge and attitude of the model developer.

1.3 Modeling Steps

Figure 1.2 shows the steps that are involved in the modeling process in detail. There are three main phases: system analysis, model design and model analysis. These phases can be further subdivided into smaller steps. By using the example of a thermometer, these steps will be clarified. Problems during model development are:

- What should be modeled?
- What is the desired level of detail?
- When is the model complete?
- When can a variable be ignored or simplified?

These questions are not all independent. The answers to the third and fourth question depend on the answers to the first two. One could state that during the *system analysis phase* these questions should be answered. The answers are obtained by formulating the goals of the model on the one hand and by considering the system and the environment in which it operates on the other hand. This should provide sufficient understanding as to what should be modeled.

In the *model design phase* the real model is developed and when appropriate, implemented and verified. In this phase the first question to be answered should be how the model should look like.

The starting point is the design of a basic structure that can be used to realize the goal. In case of physically based modeling, this structure is more or less fixed: differential equations with additional algebraic equations.

With these models the behavior of a variable in time can be investigated. Also other types of model are possible. Examples are so-called experimental models, or black-box models, such as fuzzy models or neural network models. The design of these models proceeds using slightly different sub-steps, which will be discussed later.

The verification and validation of the implemented model by using data is part of the design phase. The boundary between system analysis and design is not always entirely clear. When the system analyst investigates the system, he or she often thinks already in terms of modeling. During the system analysis phase it is recommended to limit oneself to the analysis of the physical and chemical phenomena that should be taken into account (the "what"), whereas during the modeling process the way in which these phenomena are accounted for is the key focus (the "how").

In the *model analysis phase* the model is used to realize the goals. Often the model behavior is determined through simulation studies, but the model can also be transformed to another form, as a result of which the model behavior can be determined. An example is transformation of the model to the frequency domain. These types of model give information on how input signals are transformed to output signals for different frequencies of which the input signal is composed. The boundary between model design and model analysis is also not always clear-cut. Implementation and transformation of the model are sometimes part of the model design.

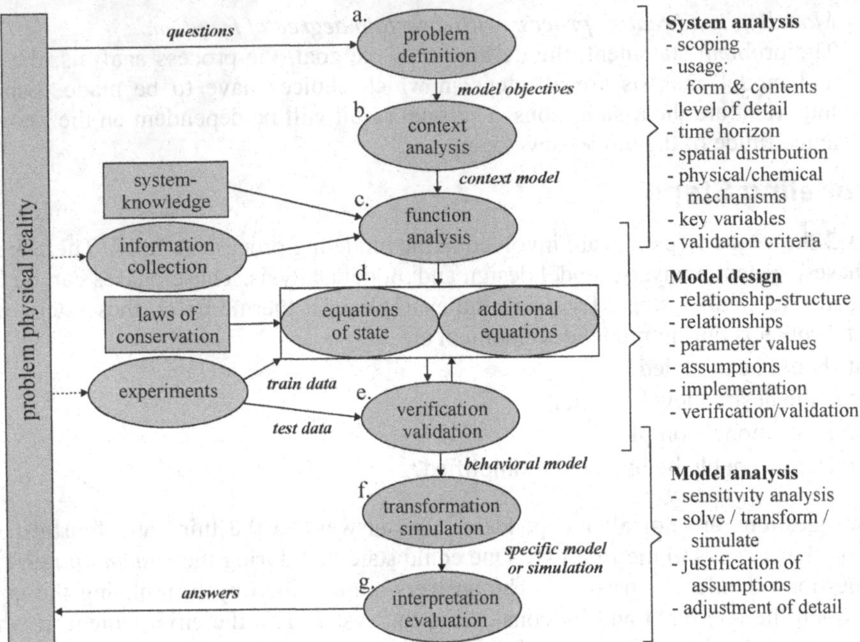

Fig. 1.2. Modeling steps.

1.3.1 System Analysis

During system analysis, the goals and the requirements of the model are formulated, the boundaries of the system are determined and the system is put into context with its environment. The primary task of a model is not to give the best possible representation of reality, but to provide answers to questions. The formulation of a clear goal is not a trivial task. The list of requirements is a summary of conditions and constraints that should be met. As mentioned before, the definition phase is the most important phase. Feedback does not happen until the evaluation phase. Then it will become clear whether the goals are met.

During the problem analysis phase, the environmental model is developed. This is an information flow diagram representing the process inputs and outputs as shown in Fig. 1.3 for the case of a simple thermometer.

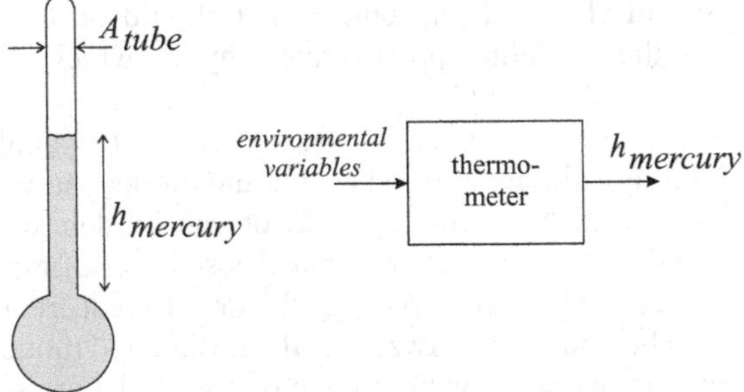

Fig. 1.3. Thermometer schematic with environmental diagram.

This representation is also called input–output model. Using the previously defined goal, the level of detail of the model with respect to time, location and functionality has to be determined. The accuracy of a dynamic model is primarily determined by its position in the time hierarchy and the corresponding frequency spectrum. The lower the time hierarchy, the higher the frequencies that should be covered. The determination of the level of detail cannot be postponed to the design phase. Parallel to the analysis of the system, it is already useful during this phase to investigate in which way the model can be verified.

During the system analysis[1] one focuses on the system itself, in order to investigate which physical-chemical phenomena take place and are relevant with respect to the modeling goal. One method that could be used is to find key mechanisms and key components, for example: evaporation (phase equilibrium), mass diffusion or transport, heat convection, conduction or radiation, and liquid flow. The key variables could be, amongst others, temperature, pressure and concentration at a certain location. Also in this case, one should take into account the required level of detail, the hierarchy within the model and the required accuracy.

The modeling steps as indicated in Fig. 1.2 will be illustrated by considering the modeling procedure for a thermometer.

Step a: Problem Definition

The starting point for the problem definition is: develop a model for a liquid thermometer as shown in Fig. 1.3. This looks like a simple problem, however, it is easy to get lost in details. Important questions that could surface are: "how does the thermometer function exactly?" and "what can be ignored?". First the problem definition and the goal should be clearly stated. What should be modeled and how?

Examples of problem definitions could be:

- What type of behavior does the height of the mercury show when the environmental temperature changes? This question could be answered qualitatively.
- How does the height change exactly? Is this change linearly dependent on the temperature? This is a quantitative question requiring a quantitative answer.
- Can the thermometer follow fast changing temperatures? This is also a quantitative question with respect to the system dynamics.
- What determines the speed of change? This is a question with respect to relative sensitivities.
- Is it possible to use alcohol instead of mercury for certain applications? This is a design question.

The goal that should be met in this case will be formulated as:

- The model should provide insight into those factors that determine the speed of change of the indication of the height of the mercury of the thermometer (see Fig. 1.4a).

The requirements for the model become then:

- *Format.* The model should be in the form of differential equations with additional equations, enabling us to perform a simulation. This is a very commonly used format.
- *Level of detail.* The thermometer will be considered as we view it, hence not at a molecular level.

The most important starting point is the assumption that the temperature in the thermometer does not have a spatial distribution and has the same temperature everywhere.
It is assumed that this assumption also holds when the thermometer is not entirely submerged in the medium of which the temperature should be measured. The impact of the volumetric expansion of the glass and the heat transfer through the glass on the operation of the

[1] If the system is a process, one could use the term "process analysis".

thermometer, have to be further investigated. These two factors have an impact on the way in which the mercury level increases or decreases when the temperature changes.

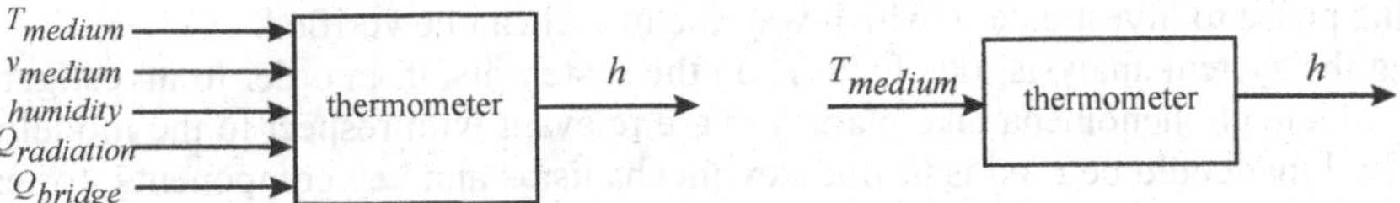

Fig. 1.4. Detailed (a) and simplified (b) context diagram of the thermometer.

Step b: Context Analysis

Let us assume that the model we are interested in is a model that will determine the height of the mercury as a function of the environmental temperature (Fig. 1.4b). Therefore, there is only one input and one output, the cause and effect. The number of inputs, however, can be disputed, since also the dynamic behavior of level changes will be modeled. Other influences from the environment could be:

- Properties of the medium in which the measurement takes place.
  The heat transfer from the medium to the glass will be determining for the rate of change of the mercury level. Heat transfer will also be determined by the state of aggregation of the medium (gas, liquid or solid), conduction and flow properties of the medium in which the temperature measurement takes place. The heat transfer coefficients for non-flowing gas, flowing liquid or boiling liquid are more than a factor of thousand apart. In a turbulent air flow, heat is transferred much faster than in a stagnant air flow. For a temperature measurement in a gas flow with fast changing temperatures, the velocity of the gas will be an additional model input variable. In the sequel it is assumed that the heat transfer conditions are constant.
- Radiation to and from the environment.
  To measure temperatures accurately is difficult, since the measurement is affected by radiation to and from objects with a different temperature, close to the actual measurement. This impact is therefore dependent on the situation.
- Heat losses through the contact point
  The way the thermometer is installed may lead to heat transport through the contact point (Q_{bridge}).

The result of the context analysis is an input–output model with assumptions about environmental effects that can be ignored. In the case of the thermometer it is assumed that only the medium temperature is relevant and that the heat transfer conditions are constant.

Step c: Function Analysis

A liquid thermometer works on the principle that the specific mass or density is a function of the temperature. The volumetric expansion of the volume is therefore a measure for the increase in temperature.

The temperature of the liquid inside the thermometer does not change instantaneously with the environmental temperature. Mercury and glass have a heat capacity and increase in temperature as a result of heat conduction through the material.

It is our experience that most modeling exercises tend to start with a description of the mercury inside the thermometer (the state of the system) in the form of a differential equation for the density or the volume of the mercury.

However, the density and the volume change instantaneously with the temperature. The density increases when the volume decreases and vice versa, since the mass of the mercury is constant:

$$\frac{dm}{dt} = \frac{d(\rho V)}{dt} = \rho\frac{dV}{dt} + V\frac{d\rho}{dt} = 0 \tag{1.1}$$

It is the temperature that determines the state of the system!

1.3.2. Model Design

During the model design phase, the behavioral model will be determined. It will represent the functional relationships between inputs and outputs, which were assigned in the input–output diagram. The model that will be used will consist of differential equations and additional equations and the assumptions under which the model is valid. A model can be formulated top down by its structure, the functions and the parameters.

The following design activities can be distinguished:

- determine the assumptions
- determine the model structure
- determine the model equations
- determine the model parameters
- model verification
- model validation

Diagrams are an efficient way to portray the model structure. The types of diagram that are suitable are 'data-flow-diagrams', which show the relationships between functions (the data flows) or function flow diagrams that portray relationships on the basis of variables (the relationships between the data are the functions). In this example, data flow diagrams, also called information flow diagrams will be used. They relate closely to the input–output models that were used in a previous phase. The functions in the model for the description of the dynamic behavior are differential equations and additional equations. These equations contain parameters.

The formulation of the assumptions is an activity that proceeds in parallel with the model structuring, the design of the equations and the determination of the parameters. The assumptions can be propositions about the way the system functions, they have to be further verified. Assumptions can also be simplifications that are made in the derivation of the equations or the determination of the parameters. In equations it often happens that one term is much smaller than another term and subsequently can be ignored. An example is the contribution of flow compared to the radiation in a heat balance. For parameters it usually concerns the assumption that a parameter value is constant (independent of temperature, pressure and composition). An example is the density or heat capacity which is usually assumed to be constant in a limited operating range.

Not all parameters in the model will be known or can be retrieved from data bases, some may have to be determined experimentally. This can happen in two different ways: the parameter in question can be measured directly or the system is "trained" using real operating data. In the last case the parameters are determined by adjusting the parameters such that the modeling error is minimal. The modeling error is the difference between the model output and the corresponding true measurement. In both cases an experimental data set of the system is required.

Verification and validation of the designed model is an essential step. Verification means that the behavior of the model should agree with the behavior of the system under study. Validation, however, refers to the absolute values the model produces. For dynamic systems it is always necessary to check whether the stationary values over the entire operating region are realistic. At this point one should decide whether the assumptions that have been made are correct. Even if the model parameters are trained based on a real situation, this check is required. However, on the other hand, the danger exists that the parameters only describe the situation for which they were trained. It could also happen that the model is overtrained. This occurs in situations where there are too many model parameters compared to the number of measurements. In that case the measurement errors are also modeled. It is obvious that the training data set and test data set should be different.

Step d: Model Relationships with Assumptions
The model that follows from the function analysis is shown in Fig. 1.5. The temperature of the mercury rises or falls as a result of the change in environmental temperature.

The level of the mercury follows from the mercury temperature. For the temperature it holds that it rises as long as there is heat transport, in other words as long as there is a temperature difference between the mercury and the environment. The heat transport is proportional to the temperature difference.

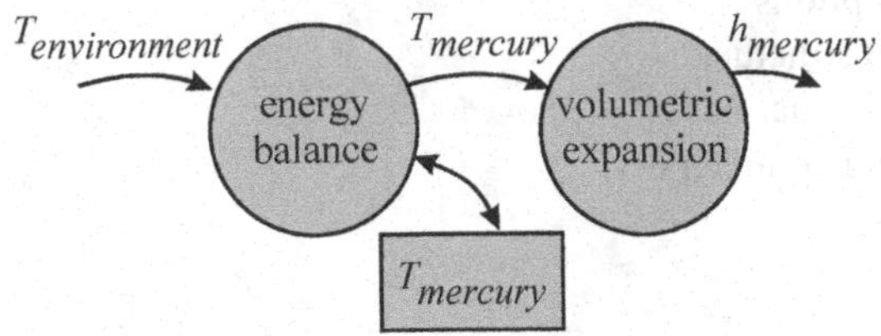

Fig. 1.5. Behavioral model of the thermometer.

The speed at which the temperature rises is determined by the heat capacity:

$$\left(c_{p,mercury}m_{mercury} + c_{p,glass}m_{glass}\right)\frac{dT_{mercury}}{dt} = UA_{glass}\left(T_{env} - T_{mercury}\right) \tag{1.2}$$

in which:

A_{glass} surface area, m^2
c_p heat capacity, J/kg.K
h height, m
m mass, kg
T temperature, K
U overall heat transfer coefficient, $J/m^2.K$
V volume, m^3
γ volumetric expansion coefficient, 1/K

The height of the mercury is a function of the temperature according to:

$$h_{mercury} = h_0 + \frac{V - V_0}{A_{tube}} = h_0 + \frac{V_0}{A_{tube}}\gamma_{mercury}\left(T_{mercury} - T_{mercury,0}\right) \tag{1.3}$$

This equation shows how the level changes with respect to the reference point (h_0,V_0,T_0) when the temperature changes with respect to that point. The reference point has the index '0'. $\gamma$is defined as:

$$\gamma_{mercury} = \frac{1}{V}\left(\frac{\partial V}{\partial T}\right)_{mercury} \tag{1.4}$$

To be able to make realistic assumptions, additional knowledge on the physical properties is required. Table 1.1 shows the properties of various materials. Ethanol has been added to the Table, since it is an alternative for the use of mercury. Water is shown as a reference. Table 1.2 shows some guidelines for the heat transfer.

Table 1.1. Physical properties at 20 °C and 1 bar.

	heat conduction coefficient λ, W/m.K	heat capacity c_p, J/kg.K	volumetric expansion coefficient γ, 1/K	density ρ, kg/m^3
mercury	10.4	0.14×10^3	1.8×10^{-4}	13.6×10^3
glass	0.8–1.2	$0.8–1.0 \times 10^3$	$0.1–0.3 \times 10^{-4}$	$2.2–3.0 \times 10^3$
ethanol	0.17	2.43×10^3	11×10^{-4}	0.8×10^3
water	0.6	4.18×10^3	2.1×10^{-4}	1.0×10^3

Source: David, R.L. (ed.) *Handbook of Physics and Chemistry*, 75th edn: CRC Press, 1994.

Table 1.2. Heat transfer coefficient at a fixed surface.

situation	heat transfer coefficient α, W/m^2.K
free convection:	
gases	3–24
liquids	120–700
boiling water	1200–240000
forced convection	
gases	12–120
viscous liquids	60–600
water	600–12000
condensation	1200–120000

Source: Bird, R.B. Stewart, W.E., and Lightfoot, E.N. *Transport Phenomena*: Wiley, 1960.

The model shown in Eqns. (1.2) and (1.3) is based on several assumptions:

- The mercury temperature can be lumped.
  The mercury inside the thermometer has the same temperature everywhere (lumped). This means that there is no temperature gradient in radial or axial direction. This assumption is realistic, since the thermal heat conduction of mercury is relatively large (about 10 times the value for glass). Whether total submersion in the medium makes any difference on the temperature can easily be verified.
- Physical properties do not depend on the temperature.
  In Eqns. (1.2) and (1.3) it is assumed that the physical properties $c_{p,mercury}$ and $c_{p,glass}$ and $\gamma_{mercury}$ do not depend on the temperature. Figure 1.6 shows that the volumetric expansion of mercury over a large range is relatively constant, the difference between 0 °C and 100 °C is only 0.5%.
- Volumetric expansion of the glass can be accounted for.
  As a result of the volumetric expansion of the glass, the surface area and the volume increase. Owing to the increased surface area for heat transfer, the heat exchange increases. However, it can be calculated that this effect is very small. The increase of the volume, however, has a significant effect on the mercury level. The volumetric expansion coefficient of mercury is 1.8×10^{-4} K^{-1}, the value for glass is in the order of $0.1–0.3 \times 10^{-4}$ K^{-1}.[2] By ignoring the volumetric expansion of glass, an error is made in the order of 10%. A possible solution is to use a modified value for mercury, which is the difference between the two expansion coefficients.

---

[2] For glass usually the linear expansion coefficient α is given, since it is a solid. When the expansion is the same in all directions, $\gamma = 3\alpha$.

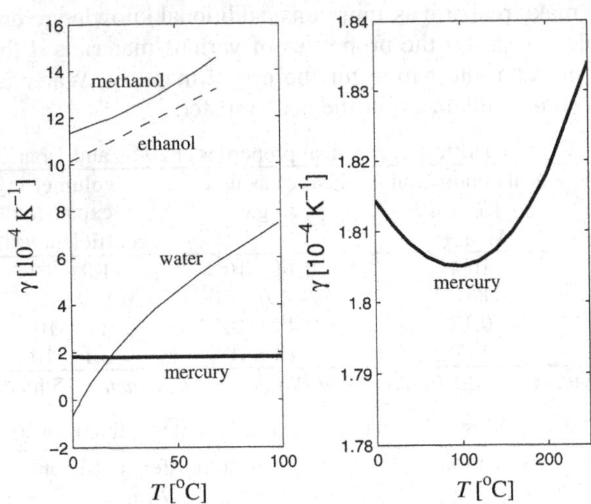

Fig. 1.6. Volumetric expansion coefficients for various liquids.

- Models for heat transfer.

  In the proposed model it is implicitly assumed that the temperatures of the glass and the mercury are the same. It is, however, questionable whether this is true. The temperature of the glass changes always before the mercury temperature, because the glass is exposed to the medium. This effect could be taken into account if required.

  In a stationary situation, the thermal resistance (inverse of the heat transfer coefficient) can be written as:

$$\frac{1}{U} = \frac{1}{\alpha_{env,glass}} + \frac{d_{glass}}{\lambda_{glass}} + \frac{1}{\alpha_{glass,mercury}} \tag{1.5}$$

in which:

U overall heat transfer coefficient, $J/m^2.K$
α heat transfer coefficient, $J/m^2.K$
d thickness, m
λ heat conduction coefficient, J/m.K

There are three resistances in series: the resistance for heat transfer from the environment to the glass, the resistance for heat transfer through the glass and the resistance for heat transfer from the glass to the mercury. The magnitude of these three individual terms will indicate whether the assumption that the mercury temperature is equal to the glass temperature is justified. Generally, the resistance between the glass and mercury will be small ($\alpha_{glas,mercury} > 200$ $J/m^2.K$), the heat conduction term also indicates a small resistance ($\lambda_{glass}/d_{glass} = 10^3$ $J/m^2.K$), but the resistance between the glass and the medium can vary considerably, depending on the turbulence in the medium ($10 < \alpha_{env,glass} < 10^4$ $J/m^2.K$).

The so-called Biot number N_B indicates to what extent convection or conduction is the limiting factor in the heat transport. N_B is the dimensionless ratio between the conductive and convective resistance:

$$N_B = \frac{\alpha d}{\lambda} \tag{1.6}$$

When the Biot number is small ($N_B < 0.1$), then the resistance of the glass is small (relatively large conduction) and the glass will therefore have a uniform temperature. Two possibilities that often happen in practice will be considered:

a. Heat transfer between the environment and the thermometer is rate limiting
If the thermometer measures the air temperature, the heat transfer from the environment to the glass will be the limiting factor for the rate at which the mercury level changes. The value for the heat transfer coefficient follows from Table 1.1: $\alpha_{env,glass} = 10$ W/m^2.K. The Biot number can then be calculated as:

$$N_B = \frac{\alpha_{env,glass} d_{glass}}{\lambda_{glass}} = \frac{10 \times 10^{-3}}{1} = 0.01 \tag{1.7}$$

In this case it may be assumed that the temperature of the glass is the same everywhere and does not depend on the location. Since the resistance between the environment and the glass is much larger then the resistance between the glass and the mercury, the glass temperature may be assumed the same as the mercury temperature. When the temperatures are the same, the heat capacities can be added. From Table 1.1 it can be seen that the heat capacity of the glass is much higher than the heat capacity of mercury. When the mass of mercury is small compared to the mass of glass, the heat capacity of the mercury can be ignored.

b. Heat transfer between the environment and the thermometer is not rate limiting
In almost all other cases, the heat conduction through the glass together with the heat transfer from the glass to the mercury will be the limiting factor for heat transfer. Assume that the thermometer is submerged in a boiling liquid. Then the heat transfer coefficient is in the order of $\alpha_{env,glass} = 1000$ W/m^2.K and N_B is a factor of one hundred larger. It is also not allowed to assume that the glass and mercury temperature are the same. For the glass and mercury temperature separate differential equations should be used. The heat transfer through the glass can then be modeled by several differential equations in series.

- Uniform mercury tube
  In the model it is assumed that the cross sectional area of the mercury tube does not depend on the height. This should be verified since it will depend on the construction of the tube.

Often, as is the case here, the description of the assumptions will be more elaborate than the mathematical formulation of the model. The most uncertain assumptions will have to be verified. In this example this is the assumption that the temperature of the mercury and the glass are the same.

Step e: Verification and Validation
In this step it should be checked whether the most important assumptions are realistic. A simple experiment is to submerge the thermometer in a bath of warm water and wait until an equilibrium situation is obtained. Subsequently place the thermometer is a bath with water at room temperature and measure the change of temperature with time (the dynamic behavior) and compare it to the model results. The experiment should be repeated by placing the thermometer from the warm bath in air at room temperature.

This simple experiment will enable us to verify whether the assumption we made about the heat transfer coefficient is correct.

Another important assumption is that the temperature of the glass and mercury are the same. If these temperatures differ, the temperature of the glass will change first and subsequently the mercury temperature will change. A consequence is that for a temperature increase, first the glass would expand and the mercury level would fall, but the longer-term effect would be that the mercury level would rise. Whether this happens in reality is easy to check.

1.3.3 Model Analysis

The developed model is the starting point for a model analysis or analysis of the behavior of the thermometer. The model will have to be solved mathematically, simulated by using a computer or be transformed to a mathematical form that is suitable for the analysis. Examples of such a suitable form are, for example, transfer functions and state models. Transfer functions are linearized input–output relationships. This form of the model will show how the output signal changes for a given change in input signal. State models are differential equations in matrix format. This form is, amongst others, suitable to perform a stability analysis.

A first step in determining the model behavior could be a sensitivity analysis. It is then investigated how the model outputs change when the model inputs or parameters change. In addition it can be determined which relationships are relevant or can be omitted. With this information the model can be further simplified or extended and one can ensure that it only captures the essence.

Step f: Transformation to a Simulation Model

The model consists of two equations:

$$\frac{c_{p,mercury} m_{mercury} + c_{p,glass} m_{glass}}{UA_{glass}} \frac{dT_{mercury}}{dt} = T_{env} - T_{mercury} \tag{1.8}$$

and

$$h_{mercury} - h_0 = \frac{V_0}{A_{tube}} \gamma_{mercury,net} \left(T_{mercury} - T_{mercury,0}\right) \tag{1.9}$$

The drawback of this model is that several values are difficult to determine, such as: $m_{mercury}$, m_{glass}, A_{glass}, U, V_0, A_{tube} and $\gamma_{mercury,net}$. The overall heat transfer coefficient U determines the dynamic behavior. This includes the heat transfer from the environment to the glass and the heat conduction through the glass. Both effects are difficult to determine accurately. For the static behavior V_0 and A_{tube} are important. An error of 10% in their values results in an error of 10% in the model. Therefore the model should be transferred to a form that is suitable for simulation and experimental verification. To do this, two new parameters are introduced: τ and K. Equation (1.8) can then be transformed to:

$$\tau \frac{dT_{mercury}}{dt} = T_{env} - T_{mercury} \tag{1.10}$$

with

$$\tau = \frac{c_{p,mercury} m_{mercury} + c_{p,glass} m_{glass}}{UA_{glass}} \tag{1.11}$$

τ is a characteristic time, called time constant. It indicates how fast the mercury and glass increase in temperature on a temperature difference with the environment. τ is an experimentally verifiable quantity and can be estimated using Eqn. (1.11). From Table 1.1 it is clear that only the heat capacity of glass is relevant. With estimated values for the volume and cross sectional area on the basis of a glass tube of 10 cm long, a diameter of 2 mm and a wall thickness of 1 mm, the following values are obtained:

Table 1.3. Values of thermometer parameters

$M_{glass} = V_{glass} \times \rho_{glass} =$	$0.5 \text{ ml} \times 2 \times 10^3 \text{ kg/m}^3 = 1 \times 10^{-3} \text{ kg}$
$c_{p,glass} =$	0.8×10^3 J/kg.K
$A_{glass} =$	1×10^{-3} m^2

Two extreme cases can be considered:

- *The heat transfer between the air and the thermometer is rate limiting.* The value for the overall heat transfer coefficient follows from Table 1.2 and is $U = 10$ W/m^2.K. Using the values before, this results in an estimated value of $\tau = 200$ s.
- *Heat transfer through the glass is rate limiting.* In this case the temperature of the glass and mercury will not be the same and the glass temperature will show a gradient. The value for the heat transfer coefficient follows from Table 1.1: $U = \lambda_{glass}/d_{glass} =$ 1 W/m.K/10^{-3} m = 10^3 W/m^2.K. Consequently, the estimated time constant $\tau = 2$ s.

From experiments it can be shown that these values are realistic. In the second case in which the heat conduction is rate limiting, a more complicated model will have to be developed, because temperature changes are fast and more model detail is required to capture this. If the thermometer is part of a larger system in which changes on a time scale of minutes are important, dynamic thermometer temperature changes are relevant in the first case, whereas in the second case these changes can be assumed to be instantaneous.

Equation (1.9) can be written as:

$$\delta h = K \delta T \tag{1.12}$$

with

$$K = \frac{V_0}{A_{tube}} \gamma_{mercury,net} \tag{1.13}$$

K indicates how the height increases with temperature. This is reflected in the thermometer scale. Many errors can be accounted for by proper scaling, such as errors in the assumption that the physical properties are constant. The scaling can be determined by calibration. For a calibration thermometer, K is a function of temperature, since small corrections are required for the expansion of mercury and glass. It is then essential that the thermometer is submerged in the medium.

The global model is more accurate than the detailed model, since the constants of the global model can be experimentally determined, whereas this is not possible for the detailed model.

Step g: Interpretation and Evaluation
Upon a step change in the temperature from 20 °C to 50 °C, the height of the mercury in the thermometer will slowly rise as shown in Fig. 1.7.

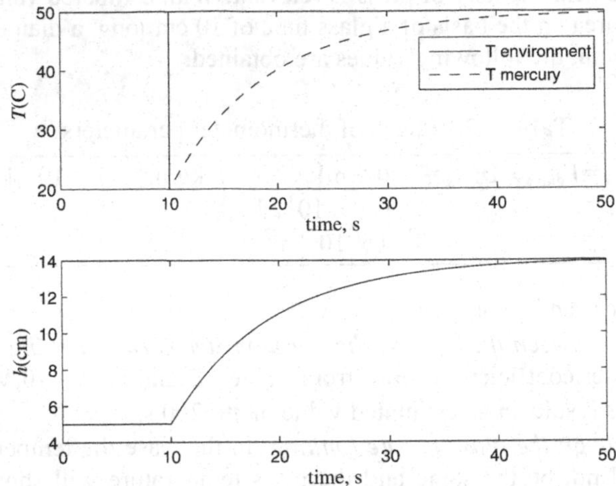

Fig. 1.7. Thermometer response to a step change in environmental temperature.

The dynamic behavior is in agreement with the interpretation from the simple behavioral model as depicted in Fig. 1.5. The thermometer behaves like a first-order system with a mercury height which varies linearly with the temperature.

The speed at which this will take place depends on the heat transfer coefficient from the environment to the glass. In air the temperature change is gradual, in boiling water the temperature rise is almost instantaneous. Calibration thermometers always use mercury as a liquid. The accuracy is approximately 0.1 °C. The thermometer should always be fully submerged, the scale of these thermometers is not linear.

1.4 Use of Diagrams

1.4.1 Diagrams

In Fig. 1.2 where the modeling steps were discussed, three deliverables from the modeling process were mentioned:

- context or environmental model
- behavioral model
- specific model of simulation

Diagrams are a good means to illustrate these deliverables: the environmental model is illustrated by a context diagram showing the connections with the environment, the behavioral model is supported by data-flow diagrams showing the functions and internal structure and also for simulations often graphical tools are used.

The utilization of graphical techniques is not new. Context diagrams in the field of process control are usually called input–output diagrams. Signal-flow diagrams, electric analogies or bond graphs have been applied in process dynamics before to clarify structures. Some simulation packages use block diagrams to input the models into the software. The method used in the sequel is "borrowed" from the field of informatics (Yourdon, 1989).

The representation of the functional structure is done by using so-called data flow diagrams (DFDs, Fig. 1.8). There is an essential difference with signal flow diagrams (van der Grinten, 1970). In the latter type of diagrams, first a list is made of all variables and subsequently the relationships are identified (Rademaker, 1971). In developing the DFDs, the reverse order is followed.

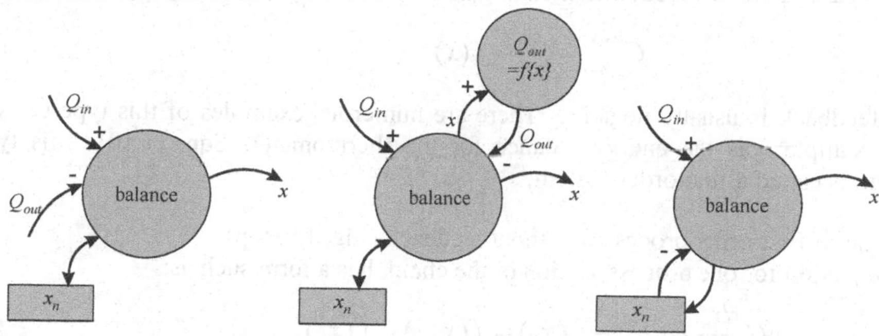

Fig. 1.8. Examples of balances; left: integrator, middle: first-order with external feedback, right: first-order with internal feedback.

A DFD consists of functions represented by circles, interconnected by arrows representing the relationships by which the variables are transported. The sign next to the arrow indicates whether the variable has a positive or negative impact on a function. DFDs can be hierarchically structured. Every function can be further detailed into sub-DFDs. At the lowest level, every function is defined by its equation. Differential equations are represented by a buffer, because they have a "status memory". The new state is determined by the present state and a new input or external influence. At start-up, this memory should have an initial value. If the state only occurs in the balance in the derivative (left-hand side of balance), then a double arrow is used (Fig. 1.8, left) and if a state also appears as a state feedback (right-hand side of balance), then two single arrows are used (Fig. 1.8, right). By this method distinction can be made between balances with and without feedback .

The graphical representation ensures that the structure of the model is shown and clarified. It also ensures that missing relationships are shown and which relationships have to be worked out in more detail. The behavioral model has the same inputs and outputs as the environmental model. Throughout the book these diagrams will be used.

1.4.2 Interpretation of Behavioral Diagrams

During the model design phase, but also during the model analysis phase, data-flow-diagrams are useful. The diagrams consist of linked balances and additional equations. The behavior can be derived from the way the balances are serially interconnected in chains and linked in parallel into loops. The chains will correspond to convective and conductive behavior. Every loop can be associated with a closed loop gain. They generate the so-called poles of the system. In addition, parallel influences can act on a chain or loop. They generate the so-called zeros of the system. Convective, conductive, closed-loop and parallel process behavior will be discussed in a later chapter.

Systems can be arranged, depending on their DFD-structure into several categories:

- balance without feedback (Fig. 1.8, left)

The differential equation has the following form, such as:

$$C\frac{dx}{dt} = Q_{in} - Q_{out} \tag{1.14}$$

The process concerns a pure integrator. An example of this system is a tank or surge drum.

- Balance with feedback (Fig. 1.8, middle and 1.8, right)
In this case, there is a feedback, meaning that the overall differential equation becomes:

$$C\frac{dx}{dt} = Q_{in} - f(x) \tag{1.15}$$

The feedback is usually negative. There are numerous examples of this type of system; one example was the energy balance for the thermometer, Eqn. (1.10). This type of system is called a first-order system.

- A chain of first-order processes without feedback (Fig. 1.9, top).
The equation for one process section of the chain, has a form such as:

$$C\frac{dx_n}{dt} = Q_{n-1} - f(x_n) = f(x_{n-1}) - f(x_n) \tag{1.16}$$

Convective processes, such as a flow through a pipe or the liquid flow over the trays of a distillation column can be described in this way.

- A chain of interactive first-order processes (Fig. 1.9, bottom).
The equation for one process section of the chain, has the form:

$$C\frac{dx_n}{dt} = Q_{n-1} - Q_n = f(x_{n-1}, x_n) - f(x_n, x_{n+1}) \tag{1.17}$$

Conductive processes, such as heat conduction or component diffusion or the vapor flow through column trays can be described in this way.

- Loops of interactive balances (Fig. 1.10, left).
Loops are formed by mutual interaction of balances. As will be shown in a subsequent chapter, the solution can have several forms. The set of equations for two interacting balances, could have the form:

$$\begin{aligned} C_1\frac{dx}{dt} &= Q_{in,1} - f_x(x) - g_x(y) \\ C_2\frac{dy}{dt} &= Q_{in,2} + f_y(x) - g_y(y) \end{aligned} \tag{1.18}$$

Examples are interacting mass and energy balance (evaporator) or interacting component and energy balance (reactor).

- Parallel influencing of chain or loop (Fig. 1.10, right).
Inputs can act parallel to different balances. When the influences are opposite, it may lead to an inverse system response. The set of equations for two interacting balances with a parallel influence from an input variable $Q_{in,1}$, could have a form such as:

$$\begin{aligned} C_1\frac{dx}{dt} &= Q_{in,1} - Q_{in,2} - g_x(y) \\ C_2\frac{dy}{dt} &= -Q_{in,1} + f_y(x) - g_y(y) \end{aligned} \tag{1.19}$$

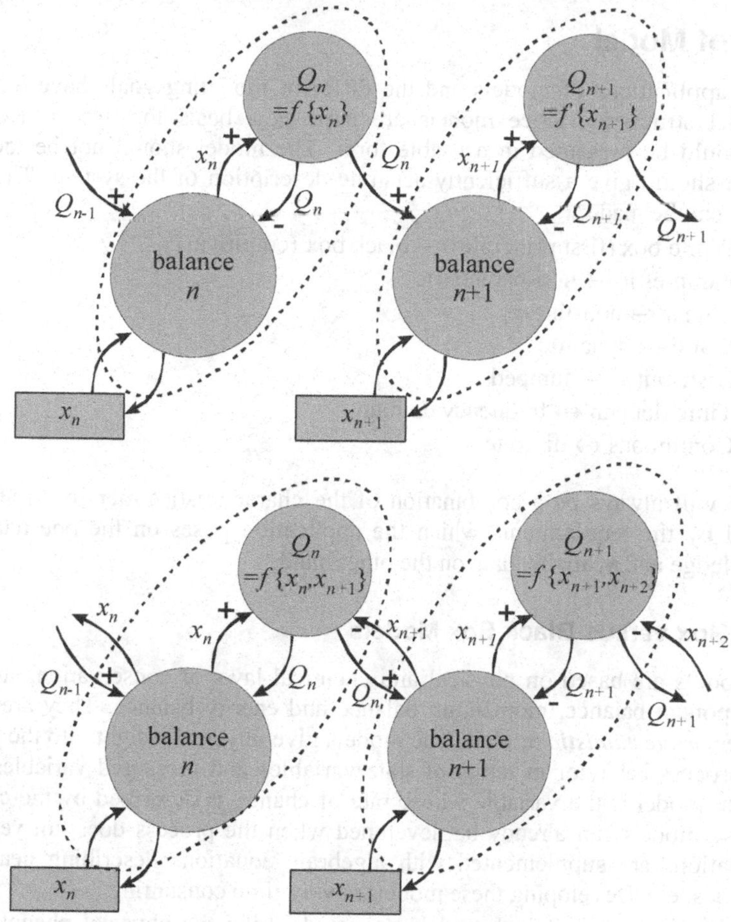

Fig. 1.9. Chain of process sections with internal feedback per section; top: only feed-forward linking, bottom: also feedback linking.

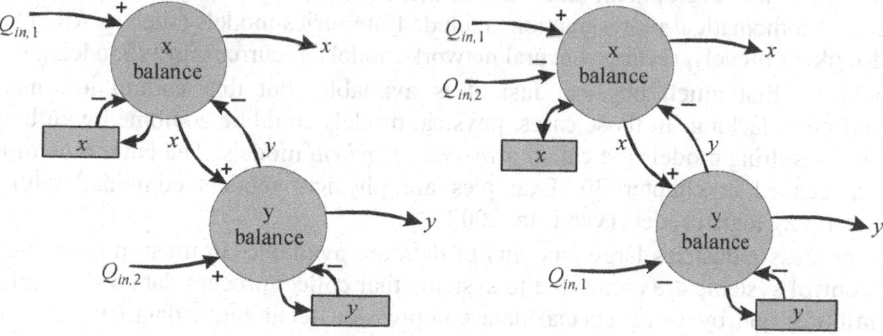

Fig. 1.10. Examples of interactive balances: left, two interactive first-order balances; right, a parallel influence of $Q_{in,1}$.

1.5 Types of Model

The different applications of models and the different modeling goals have lead to many different model structures. Since models are used as a basis for further decisions, the knowledge should be presented in a usable form. The model should not be too complex, nevertheless it should give a sufficiently accurate description of the system. The following classifications can be made:

i. White box (first principles)↔ black box (empirical)
ii. Parametric ↔ non-parametric
iii. Linear ↔ non-linear
iv. Static ↔ dynamic
v. Distributed ↔ lumped
vi. Time domain ↔ frequency domain
vii. Continuous ↔ discrete

The model will always be a combination of the characteristics mentioned above which will be posed by the requirements which the application poses on the one hand, and the process knowledge and available data on the other hand.

1.5.1 White Box versus Black Box Models

White box models are based on physical and chemical laws of conservation, such as mass balance, component balance, momentum balance and energy balance. They are also called *first principles or mechanistic* models. The models give physical insight into the process and explain the process behavior in terms of state variables and measured variables. The state variable of the model is the variable whose rate of change is described by the conservation balance. These models can already be developed when the process does not yet exist. The dynamic equations are supplemented with algebraic equations describing heat and mass transfer, kinetics, etc. Developing these models is very time consuming.

Black box models or empirical models do not describe the physical phenomena of the process, they are based on input/output data and only describe the relationship between the measured input and output data of the process. These models are useful when limited time is available for model development and/or when there is insufficient physical understanding of the process. Mathematical representations include time series models (such as ARMA, ARX, Box and Jenkins models, recurrent neural network models, recurrent fuzzy models).

It could be that much physical insight is available, but that certain information or understanding is lacking. In those cases, physical models could be combined with black box models; the resulting models are called *gray box or hybrid* models. The latter type of models will be discussed in chapter 30. Examples are physical models combined with neural network or fuzzy logic models (van Lith, 2002).

In the process industries large amounts of data are available. In most modern industries, process control systems are connected to systems that collect process data on a regular basis (e.g. 1 minute) and by using special data compression techniques, data can be stored for years. Generally, it is difficult to discover dependencies amongst the data, even for an experienced engineer.

Especially in an operating environment where quality has become more important than quantity, there is a strong desire to develop input–output models that can be used in advanced control applications, in order to develop control strategies for quality improvement. These models are usually discrete linear transfer function (difference equation) type models, which provide a representation of the dynamic behaviour of the process at discrete sampling intervals. They are in general much simpler to develop than theoretical models, both their structure and parameters can empirically be identified from plant data.

1.5.2 Parametric versus Non-parametric Models

For the large group of dynamic lumped models, which are used for optimization and control, also a classification between parametric and non-parametric models can be made. This classification resembles the classification in white-box and black-box models. The latter distinction is based on the difference in knowledge content, whereas the former distinction refers to the form of the set of equations.

Parametric models are more or less white box or first principle models. They consist of a set of equations that express a set of quantities as explicit functions of several independent variables, known as "parameters". Parametric models need exact information about the inner structure and have a limited number of parameters. For instance, for the description of the dynamics, the order of the system should be known. Therefore, for these models, process knowledge is required. Examples are state space models and (pulse) transfer functions. Non-parametric models have many parameters and need little information about the inner structure. For instance, for the dynamics, only the relevant time horizon should be known. By their structure, they are predictive by nature. These models are black box and can be constructed simply from experimental data. Examples are step and pulse response functions.

Structural considerations concern the order (parametric models) or the time-horizon (non-parametric models) of the model. Usually, the model is identified by minimizing the error between data and model. This can lead to over-modeling by including noise or eventualities. The common strategy is to use separate data sets for model identification and for testing. Statistical tests can be applied to test the parameter significance.

1.5.3 Linear versus Non-linear Models

Linear models have a property that is called superposition. If the model is of the form $y=f(u)$, then the property of superposition states that:

$$\alpha f(u_1) + \beta f(u_2) = f(\alpha u_1 + \beta u_2) \tag{1.20}$$

where α and β are constants. Linear dynamic models, such as time series models, are often used as an approximation of the true relationship between the process input and process output. Linear models often provide an accurate description of reality provided the operating range is limited. In that case, a first-order Taylor series approximation is appropriate, where the first-order derivative is used to describe the behavior around the operating point. Linear models are much easier to handle mathematically and easier to interpret as the relationship between input and output is explicit.

It is generally not easy to solve non-linear models analytically, but sophisticated numerical methods embedded in commercial software packages are available to deal with several classes of non-linear models.

Sometimes, linear techniques can be used to describe non-linear process behavior. An example is a fuzzy model, discussed in chapter 28, which is a combination of local linear models in distinct operating areas. Developing a non-linear model requires much insight and understanding of the developer as to what mechanism underlies the observed data. Application of empirical techniques for modeling non-linear process behavior has therefore become very popular, such as the application of neural networks, described in chapter 27.

1.5.4 Static versus Dynamic Models

In a static model, the most recent output or dependent variables depend on the most recent values of the input or independent variables. In dynamic models, the state variables describe the change in dependent variable as a function of independent variables and time, the response is called the system transient. In some cases, dynamic modeling may be required, for example when predicting future values of variables. Another example is the modeling of a batch reactor. In this type of reactor, the conditions change with time, hence a static model would not be very useful. When a continuous process is operated at constant conditions, however, the description of the independent variables could very well be static, in which case the time derivative does not appear in the model.

In chemical engineering, static models are, among others, used in optimization and process design, whereas dynamic models find their application in process control and prediction of the values of future process variables.

1.5.5 Distributed versus Lumped Parameter Models

In some cases, the independent and dependent process variables can vary along a spatial coordinate. A well-known example is a tubular reactor, where temperatures, concentrations and other process variables vary with the axial coordinate of the reactor. In the absence of micro-mixing, the system variables could even vary with the radial coordinate of the reactor. Spatial variations of process variables lead to complex models and consequently the solution of the model is difficult. Models that account for spatial variations are called *distributed parameter* models. A widely used method that approximates the behavior of distributed parameter systems is the division of the spatial coordinate into small sections, within each section the system properties are assumed to be constant. Each section can then be considered as an ideally mixed section and the entire process is approximated by a series of ideally mixed sub-systems. Such a system approximation is called a *lumped parameter* system. In the case of a chain process, such as a distillation tray column, usually every tray is considered to be lumped, i.e. process conditions on each tray are averaged and assumed constant.

The advantage of a lumped parameter model over a distributed parameter model, is that the lumped parameter model is much easier to solve. The modeler should ensure, however, that the lumped parameter representation is an adequate approximation of the true process behaviour. It may be difficult to determine a-priori whether a lumped parameter description is valid or not. There are, however, some criteria that can help the modeler, such as the Péclet number in systems with mass transport (Westerterp *et al.*, 1984).

1.5.6 Time Domain versus Frequency Domain

The goal of basic process control is to eliminate or to compensate disturbances. The influence of these disturbances is determined by the amplitude and the frequency of the changes. Therefore, control engineers like to describe systems in the frequency or Laplace domain. The input–output relation between an input δu and the output δy can be written as a transfer function. This is a function in the frequency ω:

$$\frac{\delta y(\omega)}{\delta u(\omega)} = \frac{Q(\omega)}{P(\omega)} \tag{1.21}$$

$\delta u(\omega)$ and $\delta y(\omega)$ describe the amplitude of the signals as a function of the frequency. The frequency domain offers the possibility to compose in an easy way the overall input–output relation of a complex system consisting of several interconnected subsystems, from the separate transfer functions.

1.6 Continuous versus Discrete Models

In a continuous model, a variable has a value at any given instant of time. In a discrete model, a variable has only values at discrete instances in time, for example, every second. Variables can also have continuous or discrete values with respect to a spatial coordinate, this type of behavior will not be considered here.

Discrete time models describe the state of the system at given time intervals and are therefore useful for efficient computation. When using computers for calculation of the model output, continuous models need to be discretized, since numerical solution methods require discrete models.

Discretization of model equations refers to the approximation of (usually) the first and higher order derivatives in the model. There are several difference approximations that can be made, a forward difference approximation, a backward and a central difference approximation. A backward approximation can be written as:

$$\frac{dy}{dt} = \frac{y_k - y_{k-1}}{\Delta t} \tag{1.22}$$

where y is the variable to be discretized, t is the continuous time variable and k is the discrete time step. The approximation is valid as long as the time step is sufficiently small. When the time-shift with Δt is described by the use of the operator z, then:

$$y_k - y_{k-1} = (1 - z^{-1})y_k \tag{1.23}$$

The input–output relationship between an input u_k and the output y_k can be written as a pulse transfer function. This is a function in the shift-operator z:

$$\frac{y_k(z)}{u_k(z)} = \frac{Q(z)}{P(z)} \tag{1.24}$$

For prediction of a process variable, any type of model can be used. It is good practice to start with a linear model. Only if the prediction is poor and adding more model complexity does not improve the prediction, would one consider developing a non-linear model.
For process control purposes, a linear model is preferred, although incorporating non-linear models in process control applications is often no major problem.

Files referred to in this chapter:
F0106.m: used to calculate relationships in Fig. 1.6.

References

Eykhoff, P. (1974) *System Identification*, John Wiley and Sons.
van der Grinten, P.M.E.M. (1970) Process control loops, Prisma Technica.
Hangos, K. and Cameron, I. (2003) *Process modeling and model analysis*, Academic Press.
van Lith, P.F. (2002) *Hybrid Fuzzy First Principles Modeling*, University of Twente.
Rademaker, O. (1971) *Teaching Modeling Principles in Automatic Control*. IFAC Workshop on higher education in automatic control. Dresden.
Westerterp, K.R., van Swaaij, W.P.M. and Beenackers, A.A.C.M. (1994) *Chemical Reactor Design and Operation*, John Wiley.
Yourdon, E. (1989) *Modern structured analysis*, Prentice-Hall Int.

2 Process Modeling Fundamentals

As indicated in the previous chapter, models are used for a variety of applications, such as study of the dynamic behavior, process design, model-based control, optimization, controllability study, operator training and prediction. These models are usually based on physical fundamentals, conservation balances and additional equations. In this chapter the conservation balances for mass, momentum, energy and components are introduced.

2.1 System States

2.1.1 Conservation Laws

To describe a process system we need a set of variables that characterize the system and a set of relationships that describe how these variables interact and change with time. The variables that characterize a state, such as density, concentration, temperature, pressure and flow rate, are called state variables. They can be derived from the conservation balances for mass, component, energy and momentum.

The principle of conservation is based on the fundamental physical law that mass, energy and momentum can neither be formed from nothing nor disappear into nothing. This law is applicable to every defined system, open or closed. In process engineering, a system is usually a defined volume, process unit, or plant. The system extent may be restricted to a phase or even a bubble or particle. The considered volume is not necessarily constant. For open systems, the mass, energy and momentum flows passing through the system boundary should also be taken into account. The equations that relate the state variables to other state variables and to the various independent variables of the considered system are called the state equations.

Consider the open system shown in Fig. 2.1. In an open system the change of the mass is based on the difference of the mass between inflow and outflow. The mass in a closed system is always constant.

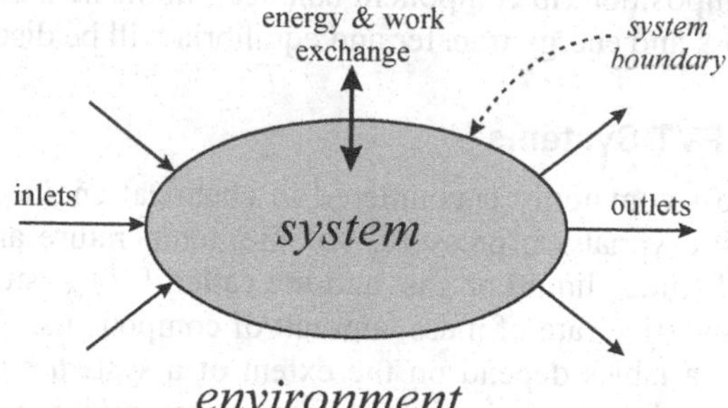

Fig. 2.1. Interaction between system and its environment.

$$\begin{bmatrix} \text{time rate of change} \\ \text{of mass} \\ \text{within system} \end{bmatrix} = \begin{bmatrix} \text{net mass flow} \\ \text{through boundary} \end{bmatrix} \tag{2.1}$$

The component balance is a special case of the mass balance. The sum of the component balances of all components should agree with the mass balance. However, there can be a net production or consumption of a component within a system.

$$\begin{bmatrix} \text{rate of change} \\ \text{of component mass} \\ \text{within system} \end{bmatrix} = \begin{bmatrix} \text{net flow of} \\ \text{component mass} \\ \text{through boundary} \end{bmatrix} + \begin{bmatrix} \text{net production of} \\ \text{component mass through} \\ \text{reaction within system} \end{bmatrix} \tag{2.2}$$

The energy balance is the result of the first law of thermodynamics. For an open system all kinds of environmental influences should be taken into account.

$$\begin{bmatrix} \text{rate of change} \\ \text{of internal +} \\ \text{kinetic + potential} \\ \text{energy within system} \end{bmatrix} = \begin{bmatrix} \text{net flow of internal +} \\ \text{kinetic + potential} \\ \text{energy through boundary} \end{bmatrix} + \begin{bmatrix} \text{heat flow by} \\ \text{conduction + radiation} \\ \text{through boundary} \end{bmatrix}$$
$$+ \begin{bmatrix} \text{(electrical) heat} \\ \text{from sources} \\ \text{within system} \end{bmatrix} + \begin{bmatrix} \text{work done} \\ \text{by system on} \\ \text{environment} \end{bmatrix} \tag{2.3}$$

Newton's second law states that a body will accelerate proportional to the force performed on the body in the direction of the force. The momentum balance is the general case of this law allowing mass variation with time.

$$\begin{bmatrix} \text{net change of} \\ \text{momentum} \\ \text{in time} \end{bmatrix} = \begin{bmatrix} \text{net force} \\ \text{working on} \\ \text{boundary} \end{bmatrix} \tag{2.4}$$

It can be thought of as a dynamic force balance. In process engineering systems, usually only hydraulic forces are considered. They are the result of gas, liquid or particle movements. In this book only the principles are mentioned for flow systems.

The state equations describe the behavior of the process with time. The accumulation is described by the terms on the left-hand side of the differential equation. The right-hand side contains of the production and transport terms. The transport terms are caused by differences compared to an equilibrium situation in pressure (momentum balance), temperature (in energy balance) and composition (in component balance). In the next sections of this chapter, the relationships for mass and energy transfer and equilibria will be discussed.

2.1.2 Equilibrium in PVT Systems

The type of system most commonly encountered in chemical engineering applications has the primary characteristic variables: pressure, volume, temperature and composition. Such systems are made up of fluids, liquid or gas, and are called PVT systems. The conservation laws concern the accumulation rate of mass, amount of components, energy and momentum of such a system. The variables depend on the extent of a system and are therefore called extensive variables. Extensive properties are additive. When multiple systems are combined to a new system, the new value of the variable will be the sum of the initial ones. In contrast, temperature, pressure, specific volume and composition are conditions imposed upon or exhibited by the system. These are intensive variables. When systems are combined to a new system, the new value of the variable will be the equilibrium value of the initial ones.

The relationship between in the intensive and the extensive variable is the mass. Every mass specific property can be considered as an intensive variable. The conservation balances describe the accumulation rate based on mass, components, energy and momentum exchanges with the outside world in terms of intensive variables. In a PVT system these intensive variables are not independent, but are interrelated by thermodynamic relationship. Any homogeneous PVT mixture at equilibrium requires uniformity of constant temperature and pressure throughout the system and uniformity of constant composition throughout each phase. Each phase can be described by the Gibbs-Duhem relationship:

$$-\hat{S}dT + \hat{V}dP - \sum x_i d\mu_i = 0 \tag{2.5}$$

in which:
$\hat{S}$ specific entropy
T absolute temperature
$\hat{V}$ specific volume
P absolute pressure
μ chemical potential
x molar fraction
From this equation the equilibrium relationships between the pressure, temperature and composition can be derived.

For example, in the case of the evaporation of one component, the relationship between pressure and temperature can be obtained. Assume that an infinitesimal amount of energy is added to the system, causing a very small change dT and dP in both phases, then:

$$-S_V dT + V_V dP = -S_L dT + V_L dP \tag{2.6}$$

The indices V and L refer to the liquid and vapor phase respectively. The temperature and pressure for both phases are assumed to be equal. This relationship can be rewritten as:

$$\frac{dP}{dT} = \frac{S_V - S_L}{V_V - V_L} \tag{2.7}$$

Since $\Delta S = \Delta H_{vap}/T$ and the molar liquid volume can be ignored compared with the molar vapor volume ($V_V >> V_L$):

$$\frac{dP}{dT} = \frac{S_V - S_L}{V_V - V_L} = \frac{\Delta H_{vap}}{T \cdot V_V} \tag{2.8}$$

This is the relationship that Clausius-Clapeyron found experimentally. By use of the ideal gas law for the vapor phase $V_V = RT/P$, it follows that:

$$\frac{dP}{dT} = \frac{\Delta H_{vap} P}{RT^2} \tag{2.9}$$

By integration, the relationship between the pressure and temperature of a boiling liquid can be derived:

$$\ln P = -\frac{A}{T} + B \tag{2.10}$$

2.1.3 Degrees of Freedom

The state of a PVT-system is fixed when the temperature, pressure and composition of all phases are fixed. The temperature and the pressure are assumed uniform throughout the system. However, for equilibrium states these state variables are not independent. The number of independent states is given by the phase rule and is called the number of degrees of freedom and is determined by the difference between the number of thermodynamic states and the number of relationships between these states.

The phase-rule variables of a system are the temperature, pressure and n–1 mole fractions in each phase. Therefore, the number of states is: $2+(n-1)\pi$.

The phase-rule equations are:

- all chemical potentials for each phase μ_1 to μ_π are equal: this results in $n(\pi-1)$ equations,
- for all independent chemical reactions $\sum_i \mu_i dx_i = 0$, resulting in r equations.

Therefore the number of degrees of freedom becomes:

$$\begin{aligned} F &= states - equations \\ &= \left(2+(n-1)\,\pi\right)-\left((\pi-1)\,n-r\right) \end{aligned} \tag{2.11}$$

or

$$F = 2-\pi+n-r \tag{2.12}$$

in which:

F number of degrees of freedom
π number of phases
n number of components
r number of independent chemical reactions

2.1.4 Example of States: A Distillation Column

In this example a tray of a distillation column to separate two components is considered. It is assumed that the vapor phase is in equilibrium with the liquid phase and that the mass of the vapor can be neglected. Therefore, the temperature and the pressure are uniform and the vapor mass balance can be omitted. The seven variables at a tray i are:

P_i pressure
T_i temperature
M_i liquid mass
x_i fraction of the most volatile component in the liquid phase
y_i fraction of the most volatile component in the vapor phase
L_i liquid flow
V_i vapor flow

For the thermodynamics we have to consider four variables ($2+(n-1)\pi= 4$): P, T, x and y. The number of thermodynamic relationship is two ($n(\pi-1) = 2$). In that case, according to the phase law, there are two degrees of freedom. It is advisable to choose the pressure P and the liquid fraction x as the independent variables and the temperature T and vapor fraction y as the dependent variables of the thermodynamic relationship. The independent variables will be determined by the balances.

The pressure P and the vapor flow V are the result of the energy balance and the pressure drop relationship. The liquid fraction x follows from the liquid component balance. Finally, the tray mass M and the liquid flow L are determined by the mass balance and the tray hydraulics. This relationship does not determine any intensive variable. The relationships between the variables are shown in Fig. 2.2.

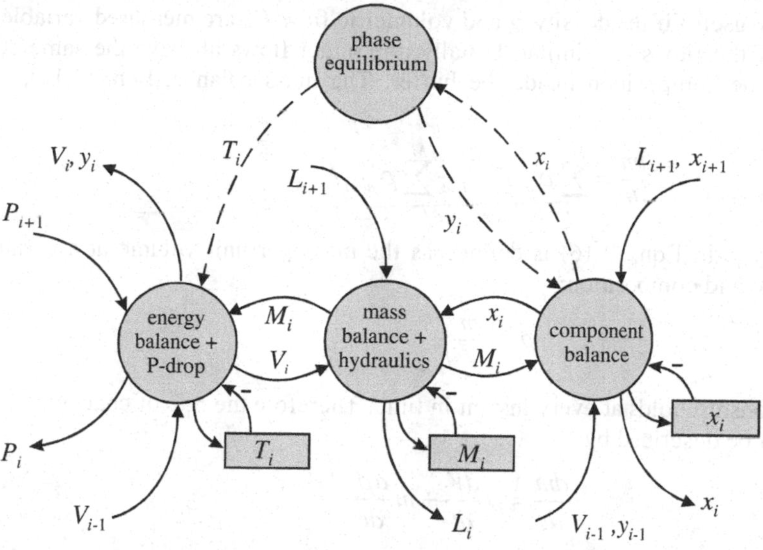

Fig. 2.2. Tray model with energy, mass and component balance.

2.2 Mass Relationship for Liquid and Gas

2.2.1 Mass Balance

For a liquid or gas buffer, shown in Fig. 2.3, the mass balance holds. This equation relates the rate of change in mass m to the difference between inlet mass flow and outlet mass flow:

$$\frac{dm}{dt} = F_{m,in} - F_{m,out} \tag{2.13}$$

In case there are N inlet flows and M outlet flows, this equation may be written as:

$$\frac{dm}{dt} = \sum_{i=1}^{N} F_{m,in,i} - \sum_{j=1}^{M} F_{m,out,j} \tag{2.14}$$

Often the volumetric flow is used in stead of the mass flow:

$$F_{m,i} = \rho_i F_{v,i} \tag{2.15}$$

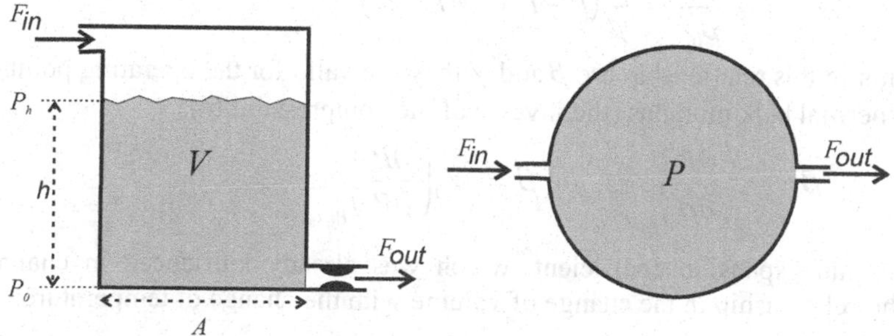

Fig. 2.3. Liquid buffer with variable volume and gas buffer with fixed volume.

This is only useful if the density ρ_i and volumetric flow $F_{v,i}$ are measured variables and if the densities of the flows are similar. Usually, the outlet flows all have the same composition, similar to the composition inside the buffer. The mass balance, Eqn. (2.14), can then be written as:

$$\frac{dm}{dt}=\sum_{i=1}^{N}\rho_{in,i}F_{v,in,i}-\sum_{j=1}^{M}\rho_{out,j}F_{v,out,j} \tag{2.16}$$

The density ρ in Eqn. (2.16) is defined as the mass per unit volume at a certain pressure, temperature and composition:

$$\rho=\left(\frac{m}{V}\right)_{P_0,T_0,x_0} \tag{2.17}$$

This relationship holds at every instant in time. Therefore the rate of change in the mass of a system can be described by:

$$\frac{dm}{dt}=\rho\frac{dV}{dt}+m\frac{d\rho}{dt} \tag{2.18}$$

Combined with Eqn. (2.14) this results in:

$$\rho\frac{dV}{dt}+V\frac{d\rho}{dt}=\sum_{i=1}^{N}F_{m,in,i}-\sum_{j=1}^{M}F_{m,out,j} \tag{2.19}$$

The density depends on the state variables: the pressure P, temperature T and composition x. In the operating point P_0, T_0, x_0 it holds that:

$$\rho_0=f(P_0,T_0,x_0) \tag{2.20}$$

For changes in the state in the operating point, the specific mass will change. For example, if the pressure rises, the density will increase. If the change in independent variable is small (in this case the pressure), the change in the dependent variable will depend linearly on the change in independent variable.

If the pressure or temperature changes, the change in relative density for a pure component can be calculated from the following equation:

$$\frac{\rho-\rho_0}{\rho_0}=\frac{1}{\rho_0}\left(\frac{\partial\rho}{\partial P}\right)_{P_0,T_0,x_0}(P-P_0)+\frac{1}{\rho_0}\left(\frac{\partial\rho}{\partial T}\right)_{P_0,T_0,x_0}(T-T_0) \tag{2.21}$$

The partial derivatives are constant in the operating point:

$$\frac{\rho-\rho_0}{\rho_0}=\frac{1}{\beta}(P-P_0)-\gamma(T-T_0) \tag{2.22}$$

The constants in this relationship are β and γ, they are valid for the operating point P_0, T_0, x_0. β is the isothermal bulk modulus (the inverse of the compressibility):

$$\beta=\rho_0\left(\frac{\partial P}{\partial\rho}\right)_{P_0,T_0,x_0} \quad or\ \beta=-V_0\left(\frac{\partial P}{\partial V}\right)_{P_0,T_0,x_0} \tag{2.23}$$

γ is the thermal expansion coefficient, which was already introduced in chapter one, it describes the relationship in the change of volume with the change of temperature:

$$\gamma=\frac{1}{V_0}\left(\frac{\partial V}{\partial T}\right)_{P_0,T_0,x_0} \quad or\ \gamma=-\frac{1}{\rho_0}\left(\frac{\partial\rho}{\partial T}\right)_{P_0,T_0,x_0} \tag{2.24}$$

Hence for changes in the state variables P and T at constant composition, the changes in density can be described by:

$$\frac{\partial \rho}{\partial t}=\frac{\partial \rho}{\partial P}\frac{dP}{dt}+\frac{\partial \rho}{\partial T}\frac{dT}{dt}=\rho\left(\frac{1}{\beta}\frac{dP}{dt}-\gamma\frac{dT}{dt}\right) \tag{2.25}$$

Liquid Accumulation

The starting point for the description of the accumulation in a liquid vessel, as shown in Fig. 2.3, is Eqn. (2.14). If there is only one inlet flow and one outlet flow, this equation can be written as:

$$\rho\frac{dV}{dt}+V\frac{d\rho}{dt}=F_{m,in}-F_{m,out} \tag{2.26}$$

In this equation the density ρ is a function of pressure and temperature, as described in Eqn. (2.21).

The parameter β, defined in Eqn. (2.23), can be used if the pressure change is gradual, such that the heat produced during this change can also be removed, ensuring that the temperature remains constant. Usually this is hard to realize, and a correction has to be applied. If the pressure change is fast, such that all heat remains in the system, the adiabatic bulk modulus should be used:

$$\beta_{ab}=\frac{c_p}{c_v}\beta \tag{2.27}$$

in which c_p is the specific heat at constant pressure and c_v the specific heat at constant volume. This modulus also accounts for the temperature effect. For liquids the ratio c_p / c_v is slightly larger than 1. The expansion coefficient γ for most liquids is of the order of 10^{-3} K^{-1}. Exceptions are water vapor, ethylene oxide and mercury.

If the temperature and pressure effects can be neglected, Eqn. (2.26) can be reduced to:

$$\rho\frac{dV}{dt}=F_{m,in}-F_{m,out} \tag{2.28}$$

If the inlet and outlet density are the same, the equation becomes:

$$\frac{dV}{dt}=F_{v,in}-F_{v,out} \tag{2.29}$$

Gas accumulation

For an ideal gas volume it holds that:

$$\rho=\frac{nM}{V}=\frac{PM}{RT} \tag{2.30}$$

in which:
n number of moles
M molecular weight, kg/mole
V volume, m^3
P absolute pressure, N/m^2
R gas constant, N.m/mole.K
T absolute temperature, K

Most gases behave according to this equation over a large range of temperatures and pressures. The exception is helium. A gas that is subject to a polytropic process can be described with the relationship:

$$P = C\rho^n \tag{2.31}$$

in which:

$n = 1$	for an isothermal process
n = ratio of specific heats	for an adiabatic process
$n = 0$	for an isobaric process
n = infinity	for an iso-volumetric process

The value of β, defined in Eqn. (2.23), can be found from this relationship.

The bulk modulus of a gas is related to the gas pressure as given in Eqn. (2.23), combination with Eqn. (2.31) gives:

$$\beta = \rho_0 \left(nC\frac{\rho^n}{\rho} \right)_{P_0,T_0,x_0} = nP_0 \tag{2.32}$$

One should keep in mind that for phase equilibria, such as the equilibrium between liquid and vapor, there are additional relationships between pressure and temperature. The number of degrees of freedom for the states decreases, owing to the additional relationships between temperature and pressure (phase rule of Gibbs).

For an accumulation gas tank with fixed volume, shown in Fig. 2.3, with one inlet flow and one outlet flow, Eqn. (2.26) holds. As the gas buffer has a fixed volume, $dV/dt = 0$, hence:

$$V\frac{d\rho}{dt} = F_{m,in} - F_{m,out} \tag{2.33}$$

From the ideal gas law at constant temperature, Eqn. (2.30), it follows that:

$$\frac{d\rho}{dt} = \frac{M}{RT}\frac{dP}{dt} \tag{2.34}$$

Substitution of this equation into Eqn. (2.33) results in:

$$\frac{MV}{RT}\frac{dP}{dt} = F_{m,in} - F_{m,out} \tag{2.35}$$

In this equation, the ratio MV/RT represents the mass capacity $Cap_{m,gas}$ of the gas at constant temperature in a fixed volume, hence:

$$Cap_{m,gas}\frac{dP}{dt} = F_{m,in} - F_{m,out} \tag{2.36}$$

A similar equation can also be derived for the volumetric flow. At varying pressure but constant temperature and composition, it follows from Eqn. (2.25) that:

$$\frac{d\rho}{dt} = \frac{d\rho}{dP}\frac{dP}{dt} = \frac{\rho}{\beta}\frac{dP}{dt} \tag{2.37}$$

Substitution into Eqn. (2.33) results in:

$$\frac{\rho V}{\beta}\frac{dP}{dt} = \rho_{in}F_{v,in} - \rho F_{v,out} \tag{2.38}$$

If the densities are the same, the simplified equation becomes:

$$\frac{V}{\beta}\frac{dP}{dt} = F_{v,in} - F_{v,out} \tag{2.39}$$

In this equation, V/β represents the compression capacity of the gas, $Cap_{v,gas}$, at a certain temperature and fixed volume. This equation is only valid in a certain operating range.

The formulation of this type of capacity is analogous to the formulation of the capacity for mechanical systems (mass) and electrical systems (condenser). This equation can be used for mass flows, volumetric flows and molar flows. The units of the capacity have to be adjusted accordingly.

2.2.2 Fluid Momentum Balance

Flow induction is often called inertia, since this phenomena is a result of liquid inertia. For the reference volume in Fig. 2.4, the momentum balance holds:

$$\sum F_{external} = \frac{dmv}{dt} = m\frac{dv}{dt} + v\frac{dm}{dt} \tag{2.40}$$

The external forces are:

$$\sum F_{external} = P_1A_1 - P_2A_2 \tag{2.41}$$

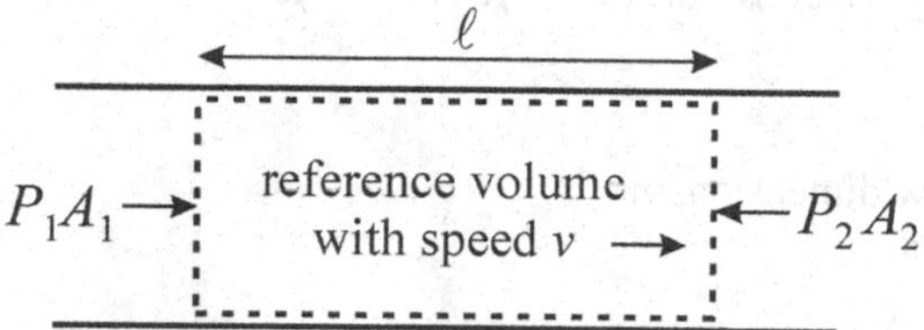

Fig. 2.4. Pipe with reference volume.

In case $A_1=A_2=A$ this equation can be simplified to:

$$\sum F_{external} = A(P_1 - P_2) = A\Delta P \tag{2.42}$$

Since the mass of the reference volume is constant, combination of Eqns. (2.40) and (2.42) results in:

$$A\Delta P = m\frac{dv}{dt} \tag{2.43}$$

Substitution of the equations for the mass $m = \rho lA$ and the velocity $v = F_v/A$ results in:

$$m\frac{dv}{dt} = \rho lA\frac{dF_v/A}{dt} \tag{2.44}$$

Thus for a pipe with length l:

$$\Delta P = \frac{\rho l}{A}\frac{dF_v}{dt} \tag{2.45}$$

This is an equation of the induction type (accumulation of the flow variable), similar to electrical systems. The induction constant is $\rho l/A$. Notice that the flow F_v will not be immediately present upon application of a pressure drop across the induction, rather, it is being built up with a rate proportional to the pressure difference.

2.2.2 Properties of Liquid and Gas Mass Transfer

Characterization of mass transport

Liquids and gases (generally fluids) are not capable of passing on static pressures. If a fluid is subject to shear stress as a result of flow, the shear stress will lead to a continuous deformation. The resistance against this deformation is called viscosity. For gases and so-called Newtonian fluids at constant pressure and temperature, the viscosity is independent of the shear stress:

$$\tau = \mu \frac{dv}{dy} \tag{2.46}$$

in which:

τ shear stress, N/m^2
μ dynamic viscosity, kg/m.s
dv/dy velocity gradient, s^{-1}

Viscosity can apparently be viewed as conduction of impulse, analogous to thermal conduction or material diffusion (as will be discussed in the sequel). The flow pattern inside a body or along a body with diameter d (for example a tube) depends on the flow velocity v and can be characterized by the Reynolds number:

$$N_{Re} = \frac{inertia\ forces}{viscous\ forces} = \frac{\rho d^2 v^2}{\mu dv} = \frac{\rho vd}{\mu} \tag{2.47}$$

in which:

ρ density, kg/m^3
d characteristic flow dimension, m
v flow velocity, m/s

This dimensionless number indicates the ratio of inertia forces from the flow ($\sim \rho d^2 v^2$) and the viscous forces ($\sim \mu\, d\, v$). For low Reynolds numbers (for a pipe $N_{Re} < 1400$), the viscous forces dominate and the flow is laminar. In laminar flow, the flow lines remain separated for the entire length of the pipe. A flow line is a line that indicates the direction of the flow in that particular point. The velocity at the wall is virtually equal to zero, and increases with the distance from the wall. In this regime, the flow velocity varies linearly with the pressure drop, (compare to the law of Ohm for electrical systems, in which the current varies linearly with the voltage difference). If the Reynolds number is high, (for a pipe $N_{Re} > 3000$), the flow is turbulent. Small disturbances cause a turbulence (eddies) that can result in instabilities. In this case the fluid flow is proportional to the square root of the pressure difference. In process plants, fluid flow is usually turbulent flow.

Resistance to flow

A flow resistance dissipates energy and can manifest itself in different forms. If the flow is laminar, a linear relationship exists between pressure drop ΔP and flow F:

$$\Delta P = RF \tag{2.48}$$

For the general case, the resistance R can be given by:

$$R = \frac{32\mu l}{A d_h^{\,2}},\ d_h = \frac{4 * area}{circumference} \tag{2.49}$$

in which d_h is the hydraulic diameter and l the length of the pipe. For a pipe with diameter d it holds that $A = \pi d^2/4$ and $d_h = d$. In this case the resistance is:

$$R = \frac{128\mu l}{\pi d^4} \tag{2.50}$$

Figure 2.5 shows a pipe with a flow restriction in the form of an orifice. The flow through the orifice is turbulent. The velocity will increase from point A to point B.

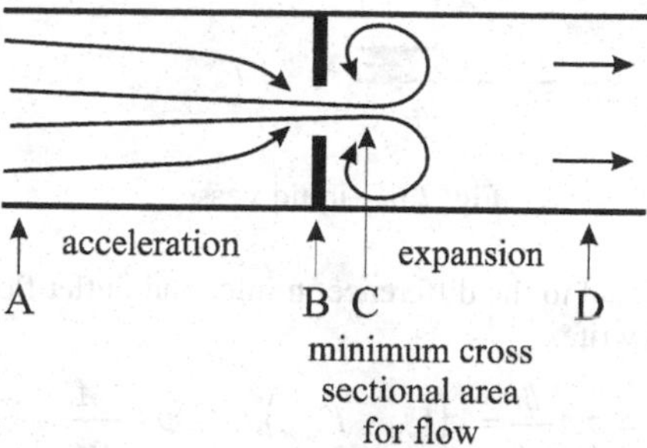

Fig. 2.5. Pipe line with flow restriction (orifice).

At point C the turbulence will increase and hence there will be dissipation of energy. At point C the additional turbulence has disappeared. The pressure drop between point A and B and between points C and D will be small. The pressure drop across the opening will therefore be P_A-P_D. It can be further assumed that $v_C >> v_A = v_D$, and it holds that $A_C = C_{cor} A_B$ and $v_B A_B = v_C A_C$. According to Bernoulli, we may write:

$$P_A - P_D = \frac{\rho}{2} v_c^{\ 2} \tag{2.51}$$

For the flow velocity through the opening one may write:

$$v_B = \frac{F}{A_{res} C_{cor}} \tag{2.52}$$

A_{res} is the area of the opening of the restriction, and C_{cor} is a correction factor between 0 and 1, depending on the type of opening. After substitution we may write:

$$\Delta P = \frac{\rho}{2}\left(\frac{F}{A_{res} C_{cor}}\right)^2 \quad or \quad F = A_{res} C_{cor} \sqrt{\frac{\rho}{2} \Delta P} \tag{2.53}$$

As can be seen from Eqn. (2.53), the flow can be determined by measuring the pressure drop and taking the square root.

2.2.3 Example of a Flow System: A Vessel

In this example the dynamics of a tank filled with liquid is considered, as shown in Fig. 2.6. The input of the system is a flow which can vary. We are interested how the outlet flow responds to changes in the inlet flow. Let us assume that the outlet flow is not subject to under or over pressure. The balance for the height of the liquid follows directly from the mass balance, Eqn. (2.26) for constant density:

$$\rho \frac{dV}{dt} = \rho A \frac{dh}{dt} = \rho\left(F_{v,in} - F_{v,out}\right) \tag{2.54}$$

The symbols are shown in Fig. 2.6.

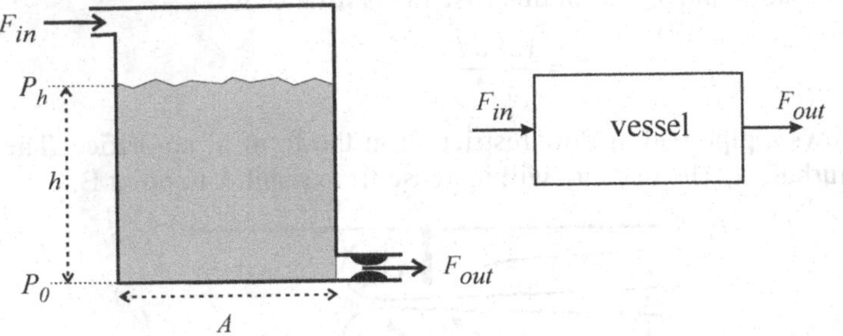

Fig. 2.6. Liquid vessel.

The accumulation in mass is equal to the difference in inlet and outlet flow. If the vessel is considered as a hydraulic system, we may write:

$$Cap\frac{d(P_h - P_0)}{dt} = \rho A\frac{dh}{dt} = \rho(F_{v,in} - F_{v,out}), \quad Cap = \frac{A}{\rho g} \tag{2.55}$$

The outlet flow can be given by:

$$F_{v,out} = \beta h^x \tag{2.56}$$

In this equation x is a coefficient which should be determined experimentally. It will depend on the flow situation and generally has a value between 0 and 1. After substitution the following balance results:

$$\frac{dh}{dt} = \frac{1}{A}(F_{v,in} - \beta h^x) \tag{2.57}$$

As mentioned in the previous chapter, there are two methods to determine the unknown parameters x and β: through curve fitting or through specific experiments. In both cases the system can be simplified by setting $F_{v,in} = 0$. This corresponds to draining the vessel for which:

$$\frac{dh}{dt} + \frac{\beta}{A}h^x = 0 \tag{2.58}$$

Through simulation two cases are investigated: laminar and turbulent outflow. In the case of laminar outflow the exponent x is approximately equal to one, in the case of turbulent outflow it is approximately equal to two. The values of β are chosen such that the flow at the start of both processes is the same. The results are shown in Fig. 2.7.

Initially the decrease in level is the same, but the tank with turbulent outflow empties faster. The outflow in case of turbulent flow decreases linearly with time. Especially the time to empty the tanks is different. In case of turbulent flow this time is two minutes, in case of laminar outflow this time is larger than 5 minutes.

The general solution of the equation is not simple. In case $x = 1$, the balance becomes:

$$\frac{dh}{dt} + \frac{\beta}{A}h = 0 \tag{2.59}$$

with initial conditions: $h(t = 0) = h_0$ and $F_{v,out}(t = 0) = F_0$. The solution of this equation is:

$$h(t) = h_0 exp\left(-\frac{\beta}{A}t\right) \tag{2.60}$$

It can be concluded that the vessel is theoretically empty when t approaches infinity. The term A/β is the time constant of the system and has the dimension of time.

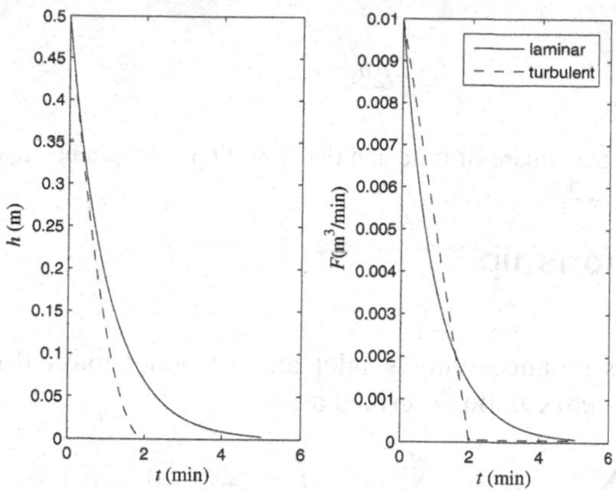

Fig. 2.7. Simulation results of emptying liquid tank with laminar and turbulent outflow.

For values $0 < x < 1$, the solution is not an exponential function. Substitution of

$$h = z^{1/(1-x)} \tag{2.61}$$

results in:

$$\frac{dz}{dt} = -\frac{\beta}{A}(1-x) \tag{2.62}$$

The solution of this equation is:

$$z(t) = -\frac{\beta}{A}(1-x)t + C \tag{2.63}$$

With the initial conditions $h(t = 0) = h_0$, $z(t = 0) = h_0^{1-x}$ and $F_{v,out}(t = 0) = F_0$ this becomes:

$$z(t) = -\frac{\beta}{A}(1-x)t + h_0^{1-x} \tag{2.64}$$

Back transformation from z to h results in:

$$h(t) = \left(-\frac{\beta}{A}(1-x)t + h_0^{1-x}\right)^{\frac{1}{1-x}} \quad with\ \beta = F_0 h_0^{-x} \tag{2.65}$$

The time to empty the vessel is the time to reach $h(t) = 0$:

$$\frac{\beta}{A}(1-x)t_{empty} = h_0^{1-x} \tag{2.66}$$

After substitution of β this time becomes:

$$t_{empty} = \frac{A h_0^{1-x}}{\beta}\frac{1}{1-x} = \frac{A h_0}{F_0}\frac{1}{1-x} \tag{2.67}$$

If the values of h_0, F_0 and t_{empty} are measured, the values of x and β can be reconstructed. In Fig. 2.7 the value of x was assumed to be equal to 0.5. This gives for $h(t)$:

$$h(t) = \left(-\frac{\beta}{2A}t + h_0^{1/2}\right)^2 \tag{2.68}$$

and for t_{empty}:

$$t_{empty} = 2\frac{Ah_0}{F_0} \tag{2.69}$$

The level $h(t)$ depends on the square of time and the flow $F(t) \approx h^{1/2}$, hence it varies linearly with time. This can also be seen in Fig. 2.7.

2.3 Energy Relationship

2.3.1 Energy Balance

Analogous to the mass balance with N inlet and M outlet mass flows (Eqn. (2.14)), the energy balance for a system can be described as:

$$\frac{dE}{dt} = \sum_{i=1}^{N} F_{m,in,i}\hat{E}_{in,i} - \sum_{j=1}^{M} F_{m,out,j}\hat{E}_{out,j} + Q + W \tag{2.70}$$

E is the total energy, which is equal to the sum of internal (intrinsic) energy U, kinetic energy K_E and potential energy P_E. $\hat{E}$ is the total energy per unit mass.[1]

$$E = U + K_E + P_E \tag{2.71}$$

The terms on the right-hand side of the energy balance, Eqn. (2.70), refer to entering convective energy flows, the leaving convective energy flows, the net heat flux Q that enters the system, with:

$$Q = Q_{conduction} + Q_{radiation} + Q_{electric} \tag{2.72}$$

and the net amount of work W that acts upon the system with:

$$W = W_S + W_E \tag{2.73}$$

in which:
W_S applied mechanical work, J/s
W_E expansion energy, J/s

If the pressure is constant, we may write for the expansion energy:

$$W_E = -P\frac{dV}{dt} \tag{2.74}$$

In most thermal applications, the energy balance can be further simplified:

- $K_E \approx 0$
  because the flow velocities are often small, the contribution of the kinetic energy can be ignored.
- $P_E \approx 0$
  because differences in height are often small, the contribution of the potential energy can be ignored.

The result of these simplifications is a balance for the internal energy. From thermodynamics it is known that the enthalpy H is the sum of the internal energy and the product PV. Using this thermodynamic relationship:

$$\frac{dU}{dt} = \frac{dm\hat{H}}{dt} - \frac{dPV}{dt} = \sum_{i=1}^{N} F_{m,in,i}\hat{H}_{in,i} - \sum_{j=1}^{M} F_{m,out,j}\hat{H}_{out,j} + Q + W \tag{2.75}$$

[1] Variables with a hat denote specific variables, i.e. per unit mass.

For many liquids, because pressure differences are often small, it holds that $d(P/\rho)/dt \approx 0$.

The result of these simplifications is an enthalpy balance. This balance does not account for mechanical changes but is valid for most thermal systems:

$$\frac{dm\hat{H}}{dt} = \sum_{i=1}^{N} F_{m,in,i}\hat{H}_{in,i} - \sum_{j=1}^{M} F_{m,out,j}\hat{H}_{out,j} + Q + W \tag{2.76}$$

For a stationary situation ($dH/dt = 0$) in which case the meaning of the enthalpy balance can be interpreted as: the energy of the outlet flows is equal to the energy of the inlet flows plus the supplied energy through heat or work.

The enthalpy H is a function of temperature, pressure, phase, and composition. Usually the pressure dependency can be neglected.

Temperature dependency

The specific enthalpy $\hat{H}_i$ of a substance i depends on the temperature T with the specific heat capacity c_P:

$$\left(\frac{\partial \hat{H}_i(T)}{\partial T}\right)_P = c_{P,i}(T) \tag{2.77}$$

The absolute specific enthalpy at a certain temperature is related to a reference temperature T_{ref} according:

$$\hat{H}_i(T) = \hat{H}_i(T_{ref}) + \int_{T_{ref}}^{T} c_{P,i}(T)dT \tag{2.78}$$

However, not the absolute enthalpy, but only the contribution of the enthalpy flux is of interest in the energy balance.:

$$\Delta\hat{H}_i = \int_{T_1}^{T_2} c_{P,i}(T)dT \tag{2.79}$$

Example

If a liquid mass flow F_m of a component i with a constant specific heat $c_{P,i}$ is heated up from a initial T_{init} to an operating temperature T, then the enthalpy flux can be written as:

$$Q = F_{m,i}\Delta\hat{H}_i = F_{m,i}c_{P,i}(T - T_{init}) \tag{2.80}$$

Phase dependency

If in the temperature trajectory a transfer of phase is included, for instance, from liquid to vapor at boiling temperature T_{bp} with a heat of vaporization $\hat{H}_{i,vap}$, then the absolute specific enthalpy becomes:

$$\hat{H}_i(T) = \hat{H}_i(T_{ref}) + \int_{T_{ref}}^{T_{bp}} c_{P,i}(T)dT + \hat{H}_{vap,i} + \int_{T_{bp}}^{T} c_{P,i}(T)dT \tag{2.81}$$

Only the contribution of the enthalpy change is of interest in the energy balance.

Example

If a liquid mass flow F_m of a component i with a constant specific heat $c_{P,i}$ is heated up from T_{init} to the boiling point T_{bp} and evaporated, then the enthalpy flux can be written as:

$$Q = F_{m,i}\Delta\hat{H}_i = F_{m,i}\left(c_{P,i}(T_{bp} - T_{init}) + \hat{H}_{vap,i}\right) \tag{2.82}$$

Composition dependency

The specific enthalpy $\hat{H}$ depends not only on the temperature but also on the composition. When the composition changes as a result of reaction, then this dependency should also be taken into account:

$$\hat{H} = \hat{H}(T, n_a, n_b, ...) \tag{2.83}$$

where n_a and n_b are the number of moles of components A and B. More specifically, n_a and n_b can be replaced by n_i to indicate an arbitrary component i. By dividing the number of moles n by the volume V, a composition dependent enthalpy can be defined:

$$\frac{\partial m\hat{H}(T, C_i)}{\partial n_i} = \frac{\partial \rho \hat{H}(T, C_i)}{\partial C_i} = \tilde{H}_i(T) \quad [\mathrm{Jmol^{-1}}] \tag{2.84}$$

where i = component a … n. $\tilde{H}_i(T)$ is the specific partial molar enthalpy of a component i at a certain temperature, which equals the heat of formation of that specific component. The heat of formation $\tilde{H}_i(T)$ is usually defined at a reference temperature of 25 C. For any arbitrary temperature, without phase transitions, we may write:

$$\tilde{H}_i(T) = \tilde{H}_{formation,i}(T_{ref}) + \int_{T_{ref}}^{T} c_{Pmol,i}(T)dT \tag{2.85}$$

in which c_{Pmol} is the heat capacity per mole. The relationship between $\hat{H}$ and $\tilde{H}$ becomes:

$$\rho(T, C)\hat{H}(C, T) = \sum_{i=a}^{n} C_i \tilde{H}_i(T) \quad [\mathrm{Jkg^{-1}}] \tag{2.86}$$

Combination of Eqn. (2.86), which describes the relationship of $\hat{H}$ with the formation enthalpy at a temperature T_{ref}, with Eqn. (2.81), which describes its temperature dependency, leads to a general expression for $\hat{H}$:

$$\hat{H}(C, T) = \frac{1}{\rho(C, T)} \sum_{i=a}^{n} C_i \tilde{H}_i(T_{ref}) + c_P(T - T_{ref}) \tag{2.87}$$

Only the contribution of the enthalpy change is of interest in the energy balance.

Example

Assume that the components A and B are transferred to C and D according to the stochiometric equation:

$$\nu_a A + \nu_b B \rightarrow \nu_c C + \nu_d D \tag{2.88}$$

in which ν_i are the stoichiometric coefficients. The concentration change of all components during the reaction can be written as:

$$\frac{dC_i}{dt} = \nu_i r \tag{2.89}$$

in which r is the reaction rate. Then for the enthalpy flux due to the reaction, we may write:

$$Q_{reaction} = \sum_{i=a}^{d} \tilde{H}_i \frac{dn_i}{dt} = V \sum_{i=a}^{d} \tilde{H}_i \frac{dC_i}{dt} = V r \sum_{i=a}^{d} \nu_i \tilde{H}_i \tag{2.90}$$

The change $\Delta H_{reaction}$ is defined as:

$$\Delta H_{reaction}(T) = \sum_{i=a}^{d} \nu_i \tilde{H}_i(T) \tag{2.91}$$

Combining Eqns. (2.90) and (2.91) gives the heat flux due to the reaction:

$$Q_{reaction} = r(T)V\Delta H_{reaction} \tag{2.92}$$

In the reactor example, section 2.4.4, the use of the temperature dependency Eqn. (2.77) and the composition dependency Eqn. (2.87) are used to solve the enthalpy balance.

2.3.2 Thermal Transfer Properties

Heat transfer takes place through three principal mechanisms: convection, conduction and radiation.

Convective heat transfer

At an interface between gas and liquid or solid or between liquid and solid, convective heat transfer can take place when those media have a temperature difference. It can be in the form of free convection, such as in the case of a central heating radiator, or it can be forced convection, for example, an air flow from a blower. Owing to the resistance to heat transfer, a heat gradient will occur. In the model that describes this interface transfer, it is assumed, as shown in Fig. 2.8, that the temperature gradient restricts itself to a boundary layer. The bulk outside this boundary layer has a uniform temperature as a result of convection or mixing. Fig. 2.8 shows two possibilities.

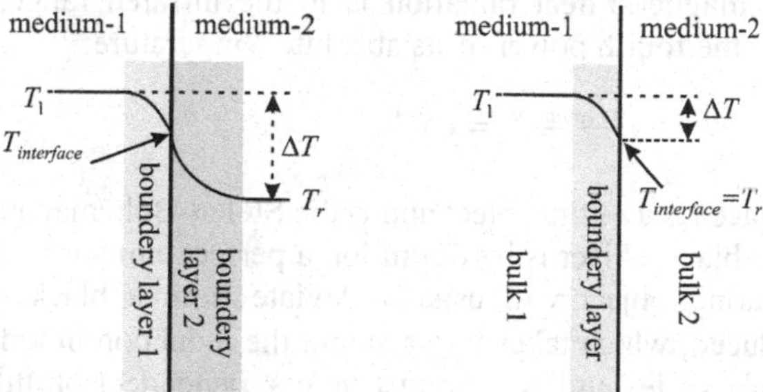

Fig. 2.8. Model for convective heat transport (stationary situation). Left: both media show a gradient. Right: the gradient is restricted to one medium.

The left situation will occur at gas-liquid or liquid-liquid interfaces. Both media show a gradient. When in a gas or liquid temperature differences exist, natural convection flow will raise and eliminate these differences. The right situation will occur at the interface between gas or liquid and a conductive solid.

The heat flow $Q_{convection}$ per unit area A at a temperature difference ΔT on the boundary layer can be given by:

$$\frac{Q_{convection}}{A} = \alpha(T_l - T_r) \tag{2.93}$$

The thermal resistance for heat transfer is $1/\alpha$.

Thermal conduction

Conduction takes place within stagnant gas or liquid layers (layers, which are sufficiently thin that no convection as a result of temperature gradients can occur), solids or on the boundary between solids. It is assumed that the molecules stay in their position and the heat is passed on. The heat flow per unit area as a result of conduction is determined by the Fourier relationship:

$$\frac{Q_{conduction}}{A} = -\lambda \frac{dT}{dx} \tag{2.94}$$

The model describes the heat transport Q in terms of the thermal conductivity λ and the temperature gradient dT/dx (Fig. 2.9). If at the same heat flow, the thermal conductivity of a material is larger, then the temperature gradient, i.e. the temperature differences, will be smaller. The equation only holds when a steady state is established and the gradient is constant.

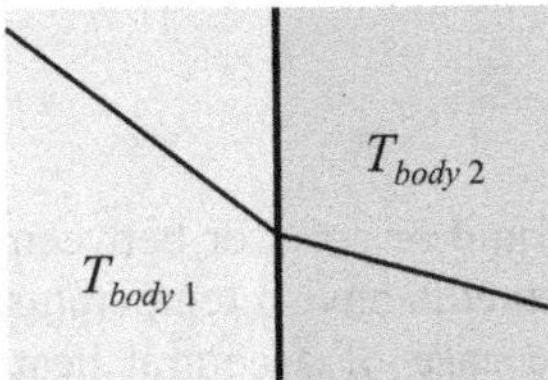

Fig. 2.9. Model for conductive heat transport (stationary situation).

Thermal radiation

Heat transport can also take place from one object to another object through radiation. The wavelength of electromagnetic heat radiation is in the infrared range. An object radiates energy proportional to the fourth power of its absolute temperature:

$$\frac{Q_{radiation}}{A} = \sigma T^4 \tag{2.95}$$

in which A is the surface area of the object and σ the Stefan-Boltzmann constant (56.7×10^{-9} W/m^{2}K^4). A so-called black object is by definition a perfect emitter or absorber of radiation. Since a normal radiating object will usually deviate from a black object, the emission coefficient ε is introduced, which takes into account the reduction in radiation. The emission coefficient of materials varies and is, amongst others, dependent on the surface roughness. Radiation is not only different from conduction and convection in a mathematical sense, but especially in its increased sensitivity for temperature. It is the dominating factor for of heat transfer in furnaces and cryogenic vacuum processes. In case of free convection at room temperature, radiation can amount to as much as 50% of the total heat transfer, depending on the emission coefficients of the objects:

$$\frac{Q_{radiation}}{A} = \varepsilon \sigma T^4 \tag{2.96}$$

If radiation falls on an object, the absorbed fraction is called absorption. If the temperature of the object is different from its environment, the emission and absorption will be different. In the general case of radiation exchange between an object i with temperature T_1 and its black environment with temperature T_2, the net exchange can be described by:

$$Q_{radiation} = \sigma A \left(\varepsilon_i T_1^4 - \alpha_i T_2^4 \right) \tag{2.97}$$

in which

$$\varepsilon_i = \int_0^1 \varepsilon_\lambda df_{\lambda T_1} \ and \ \alpha_i = \int_0^1 df_{\lambda T_2} \tag{2.98}$$

The values of ε_i and α_i are the area under a curve that describes ε and α as a function of f, which depends on λT. If ε does not depend on the wavelength, the area is called grey, in which case $\varepsilon_i = \alpha_i = \varepsilon_\lambda$. The total emission and absorption show a decrease for an increase in temperature. If the temperature T_1 differs not much from T_2, α_i can be expressed as $\varepsilon_i\,(T_2/T_1)^m$ where m is an experimental parameter. It can be shown that the relationship in Eqn. (2.97) can then be approximated by:

$$Q_{radiation} = \sigma A \varepsilon_{av}\left(1+\frac{m}{4}\right)\left(T_1^4 - T_2^4\right) \tag{2.99}$$

in which ε_{av} is determined by the average of T_1 and T_2. For metals, the fitting parameter m is approximately equal to 0.5, for non-metals the value is small and negative.

If two objects radiate towards each other, the net heat exchange depends on the areas that view each other. In this case the view factor $F_{1,2}$ is introduced. $F_{1,2}$ is the fraction of radiation that is emitted by the surface A_1 and received by the surface A_2. Since the net exchange between the two surfaces is the same, it holds that $A_1F_{1,2} = A_2F_{2,1}$. The final equation becomes then:

$$Q_{radiation} = \sigma A_1 F_{1,2} \varepsilon_{av}\left(1+\frac{m}{4}\right)\left(T_1^4 - T_2^4\right) \tag{2.100}$$

If the temperature difference between T_1 and T_2 is small, this relationship is often written as a linear relationship:

$$Q_{radiation} = \sigma A_1 F_{1,2} \alpha_{radiation}\left(T_1 - T_2\right) \tag{2.101}$$

This equation resembles the equations for heat exchange through convection and conduction. The non-linearities are accounted for by introduction of the coefficient $\alpha_{radiation}$.

Combination of thermal convection, conduction and capacity

When heat is transported by a temperature difference through a solid object, for example a wall as indicated in Fig. 2.10, or when an object is heated up, the temperature of the object is not uniformly distributed, but has a temperature gradient.

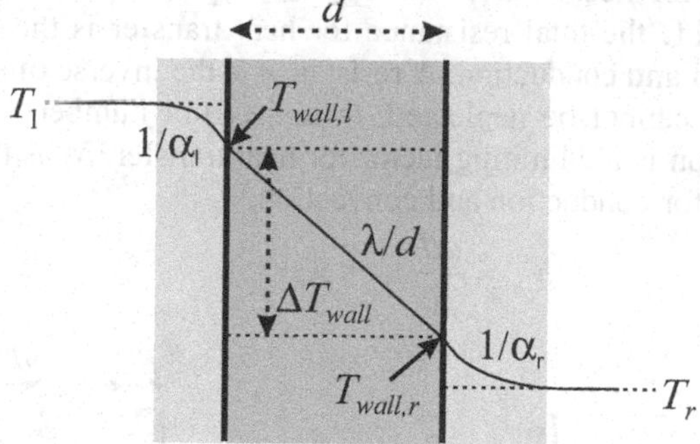

Fig. 2.10. Model for combined convective and conductive heat transport through a wall (stationary situation).

Also in the boundary layers around the object, temperature gradients will occur owing to the heat transfer resistance.

Three possibilities will be discussed:

- The heat capacity of the object can be neglected. An example is the heating of a reactor in which the heating of the wall can be neglected compared to the heating of the reactor contents.

- The heat capacity can not be neglected and convection is the limiting factor in the heat transfer.
- The heat capacity can not be neglected and conduction is the limiting factor in the heat transfer.

If the heat capacity can be neglected, then in a stationary situation, the process can be described by:

$$\frac{Q}{A} = \alpha_l\left(T_l - T_{wall,l}\right) = \alpha_r\left(T_{wall,r} - T_r\right) = \frac{\lambda}{d}\Delta T_{wall} \tag{2.102}$$

which states that the heat entering the wall is equal to the heat conducted through the wall and leaving the wall. When a overall heat transfer coefficient is defined according to:

$$\frac{Q}{A} = U(T_l - T_r) \tag{2.103}$$

then from the equation for the total temperature difference:

$$\left(T_l - T_r\right) = \left(T_l - T_{wall,l}\right) - \left(T_{wall,l} - T_{wall,r}\right) - \left(T_{wall,r} - T_r\right) \tag{2.104}$$

it follows that:

$$\frac{1}{U} = \frac{1}{\alpha_l} + \frac{d}{\lambda} + \frac{1}{\alpha_r} \tag{2.105}$$

in which:

U overall heat transfer coefficient, J/m^2.s
α_l heat transfer coefficient from T_l to the plate, J/m^2.s
α_r heat transfer coefficient from the plate to T_r, J/m^2.s
d plate thickness, m
λ heat conduction coefficient through the plate, J/m.s

The thermal resistance for heat transfer through a flat plate is therefore d/λ. For a plate or wall, shown in Fig. 2.11, the total resistance for heat transfer is the sum of the individual resistance of convection and conduction. A resistance is the inverse of a transfer coefficient.

If the heat capacity cannot be neglected, then the Biot number N_B determines whether convection or conduction is the limiting factor for heat transfer. N_B is the dimensionless ratio between the resistance for conduction and convection:

$$N_B = \frac{\alpha d}{\lambda} \tag{2.106}$$

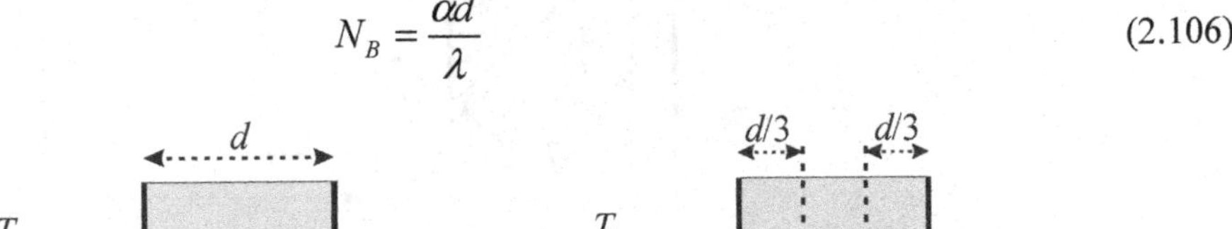

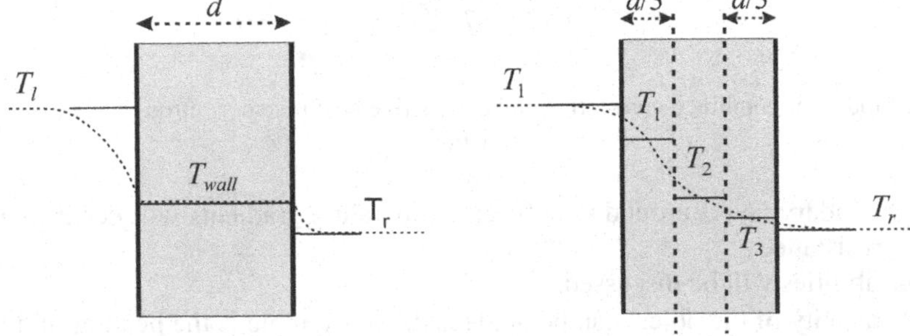

Fig. 2.11. Two approximations for the conduction through a plate. Left: one section with a uniform temperature distribution. Right: three sections shaping a temperature gradient.

If the Biot number is small ($N_B < 0.1$), the resistance of the wall is small and the temperature will be almost uniform. This is shown in Fig. 2.11 left.

The resistances concentrate at the wall surface, at that position the temperature gradients are the largest. The resistance of conduction has been ignored. The heat capacity of the wall, C_{wall}, determines the heat accumulation in the wall. This system representation is called a lumped system representation.

The heat balance for the wall becomes for this case:

$$\frac{C_{wall}}{A}\frac{dT_{wall}}{dt} = \alpha_l(T_l - T_{wall}) - \alpha_r(T_{wall} - T_r) \tag{2.107}$$

If the Biot number is large ($N_B > 0.1$), the resistance of the wall is substantial and consequently a temperature gradient will be present. In this case one could divide the wall in several equivalent sections, each representing a certain capacity. This representation is called a distributed system representation. This is shown in Fig. 2.11 right, for a wall divided in three sections. In one section the temperature is then assumed to be uniform. A model for a wall with heat transfer area A, thickness d and specific heat capacity $\hat{C}_{wall}$ ($JK^{-1}m^{-3}$), consisting of three sections, can be described by three balance equations for T_1, T_2 and T_3. The temperatures are supposed to be positioned in the middle of the sections (Fig. 2.12).

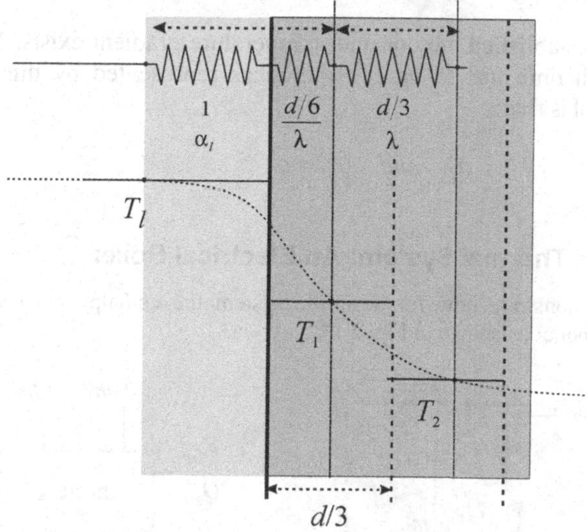

Fig. 2.12. Heat transfer resistances.

The three balances become:

$$\begin{aligned}
\hat{C}_{wall}\frac{d}{3}\frac{dT_1}{dt} &= U_l(T_l - T_1) - \frac{\lambda}{d/3}(T_1 - T_2) \ , \quad U_l = 1\bigg/\left(\frac{1}{\alpha_l} + \frac{d/6}{\lambda}\right) \\
\hat{C}_{wall}\frac{d}{3}\frac{dT_2}{dt} &= \frac{\lambda}{d/3}(T_1 - T_2) - \frac{\lambda}{d/3}(T_2 - T_3) \\
\hat{C}_{wall}\frac{d}{3}\frac{dT_3}{dt} &= \frac{\lambda}{d/3}(T_2 - T_3) - U_r(T_3 - T_r) \ , \quad U_r = 1\bigg/\left(\frac{1}{\alpha_r} + \frac{d/6}{\lambda}\right)
\end{aligned} \tag{2.108}$$

$1/U_l$ and $1/U_r$ are the heat resistances from the environment to the middle of the boundary sections. When no gradient is present in the boundary layer ($\alpha = \infty$) then the resistance is only determined by the conduction through the wall.

Multiple sections are necessary to describe the change of gradient in the wall. When a temperature rise occurs at T_l, the temperature will increase section after section. For a wall divided into n sections, the temperature of an arbitrary section i (not at the boundary) is given by:

$$\hat{C}_{wall}\frac{d}{n}\frac{dT_i}{dt} = \frac{\lambda}{d/n}(T_{i-1} - T_i) - \frac{\lambda}{d/n}(T_i - T_{i+1}) = \frac{\lambda}{d/n}(T_{i-1} - 2T_i + T_{i+1}) \quad (2.109)$$

Since i indicates the section location, this can be rewritten as:

$$\hat{C}_{wall}\frac{dT(x,t)}{dt} = \lambda\frac{(T_{i-1} - 2T + T_{i+1})}{\Delta x^2} \quad (2.110)$$

When $\Delta x \to \infty$:

$$\hat{C}_{wall}\frac{dT(x,t)}{dt} = \lambda\frac{d^2T(x,t)}{dx^2} \quad (2.111)$$

If no steady state is established, no constant temperature gradient exists. Then the change of the temperature with time has to be considered as represented by this equation. For the boundary layers it holds that:

$$U_l = \alpha_l \quad and \quad U_r = \alpha_r \quad (2.112)$$

2.3.3 Example of a Thermal System: An Electrical Boiler

This example will demonstrate how for a simple system the enthalpy balance can be derived. It concerns an electrical boiler as shown in Fig. 2.13.

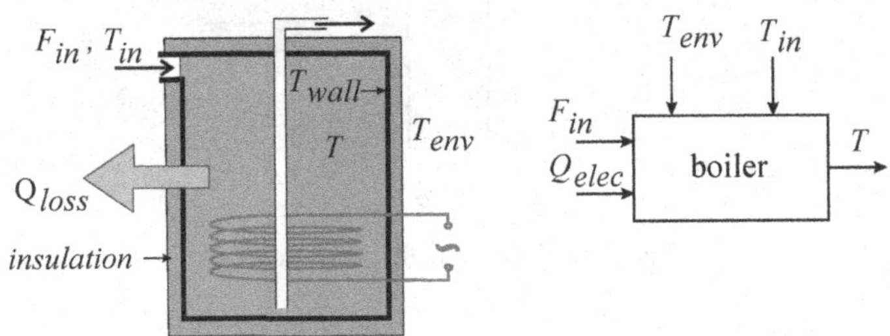

Fig. 2.13. Electrical boiler with heat loss to the environment.

The boiler is completely filled with water and has one ingoing flow and one outgoing flow. The wall of the boiler is made of metal. The boiler loses heat to the environment. To limit the loss, it is insulated at the outside. The mass balance of the water is:

$$\frac{dm}{dt} = V\frac{d\rho}{dt} = F_{m,in} - F_{m,out} \quad (2.113)$$

The enthalpy balance, Eqn. (2.76) can be written as:

$$\frac{dH}{dt} = F_{m,in}\hat{H}_{in} - F_{m,out}\hat{H}_{out} + Q_{electric} - Q_{loss,wall} \tag{2.114}$$

with

$$\frac{dH}{dt} = \frac{dm\hat{H}}{dt} = V\frac{d\rho\,\hat{H}}{dt} \tag{2.115}$$

owing to the fact that V is constant. The density ρ and enthalpy $\hat{H}$ are both temperature dependent. In the case of water, the density at 15 °C is 999 kg/m^3, at 80 °C it is 971 kg/m^3, a difference of 2.8%. For the specific enthalpy (enthalpy per unit mass), only the sensible heat has to be taken into account. The enthalpy is always determined with respect to a reference temperature. It is then the amount of heat above that temperature:

$$\hat{H}_i(t) = c_{P,i}\left(T - T_{ref}\right) \tag{2.116}$$

In the region from T_{ref} to T, no phase changes may take place. Also the specific heat is temperature dependent: for water at 15 C it is 4.1885 J/K and at 80 °C it is 4.1991 J/K, a difference of 0.3%. Depending on the required model accuracy, variations in the density ρ and specific heat c_P may be ignored or should be taken into account. If changes are ignored, the mass balance becomes:

$$V\frac{d\rho}{dt} = F_{m,in} - F_{m,out} = 0 \quad \Rightarrow F_{m,in} = F_{m,out} \tag{2.117}$$

The enthalpy balance can then be simplified to:

$$V\rho\, c_P\frac{dT}{dt} = F_{m,in}c_P\left(T_{in} - T\right) + Q_{electric} - Q_{loss} \tag{2.118}$$

If it is assumed that the wall is made of metal, heat conduction will be good and can be described by:

$$C_{wall}\frac{dT_{wall}}{dt} = Q_{loss,wall} - Q_{loss,env} \tag{2.119}$$

in which C_{wall} is the heat capacity of the wall and:

$$\begin{aligned} Q_{loss,wall} &= \alpha_{water}\left(T - T_{wall}\right) \\ Q_{loss,env} &= \alpha_{insulation}\left(T_{wall} - T_{env}\right) \end{aligned} \tag{2.120}$$

$Q_{loss,wall}$ is the heat loss from the water to the wall, $Q_{loss,env}$ is the heat loss from the wall to the environment. Usually $\alpha_{water} >> \alpha_{insulation}$, since heat transport through the water is much better than through the insulation. In that case it is reasonable to assume that $T_{wall} \approx T$, in which case Eqns. (2.118) and (2.119) can be combined:

$$\left(V\rho\, c_P + C_{wall}\right)\frac{dT}{dt} = F_{m,in}c_P\left(T_{in} - T\right) + Q_{electric} - \alpha_{insulation}\left(T - T_{env}\right) \tag{2.121}$$

The model, originally consisting of two differential equations, is now reduced to one differential equation.

Recall the assumptions that were made to derive this model:

- the temperature in the liquid is the same everywhere
- the physical properties of the water are constant
- the volume V of the tank is constant
- the heat transfer to the wall and heat conduction through the wall are much better than the heat transfer from the wall to the environment.

In a stationary situation, Eqn. (2.121) can be reduced to:

$$Q_{electric} = F_{m,in} c_P (T_{in} - T) + \alpha_{insulation} (T - T_{env}) \tag{2.122}$$

The electrical heating warms up the liquid to a temperature T and compensates for heat losses.

2.4 Composition Relationship

2.4.1 Component Balance

According to Eqn. (2.2), the component balance of component k for a considered process system or phase can be described as a concentration balance with concentration C_k [mole.m^{-3}] and volumetric flows F_v:

$$\frac{dVC_k}{dt} = \sum_{i=1}^{N} F_{v,in,i} C_{in,k} - \sum_{j=1}^{M} F_{v,out,j} C_k + J_{transfer} + J_{reaction} \tag{2.123}$$

where $i = 1$, N is the number of inlet flows and $j = 1$, M the number of outlet flows. As a partial mass balance with mass fraction x_k [kg.kg^{-1}] and mass flows F_m this equation can be written as:

$$\frac{dx_k m}{dt} = \sum_{i=1}^{N} x_{in,k} F_{m,in,i} - \sum_{j=1}^{M} x_k F_{m,out,j} + J_{transfer} + J_{reaction} \tag{2.124}$$

or as a partial molar balance with molar fraction x_k [mole.mole^{-1}] and molar flows F_{mol}:

$$\frac{dx_k M}{dt} = \sum_{i=1}^{N} x_{in,k} F_{mol,in,i} - \sum_{j=1}^{M} x_k F_{mol,out,j} + J_{transfer} + J_{reaction} \tag{2.125}$$

In chemical engineering the concentration balance is often used for chemical transformation processes, because the reaction rate equation is described as a function of the concentration (mole per volume). Separation processes such as distillation and extraction are usually based on equilibria of different liquid or vapor phases which are expressed as a function of the ratio of partial molar composition (mole per mol). In that case partial molar balances are more linked up with this description.

The number of required balances should be one less than the number of components. The sum of all components should agree with the mass balance.

2.4.2 Component Equilibria

Liquid–liquid or liquid–gas equilibria

The stationary distribution of a component j between two phases can be described by a distribution coefficient K_j which is a function of the temperature. x_j is the molar or weight fraction in one of the phases and y_j the fraction in the other.

$$y_j = K_j(T)\, x_j \quad \text{with} \quad K_j(T) = k_{0,j} e^{-\Delta E_j / RT} \tag{2.126}$$

The line which describes the relationship between y and x is called the equilibrium curve. This relationship is often linear over a certain range with a slope m_j.

$$m_j = \left(\frac{\partial y_j}{\partial x_j}\right)_T \tag{2.127}$$

Assume an amount w_a of substance A contains a solved component j with the molar fraction $x_{j,initial}$. This amount is mixed with an amount w_b of a solvent B. When the distribution coefficient for component j between B and A is K_i , then j will be extracted from the initial fraction $x_{j,init}$ to result in a much lower final fraction $x_{j,final}$, according to:

$$x_{j,final} = \frac{w_a}{w_a + K_j w_b} x_{j,initial} \tag{2.128}$$

Vapor–liquid equilibria at boiling point

If the vapor behaves as an ideal gas (limited pressure) and the vapor phase is in complete equilibrium with the liquid phase (equal temperature and pressure), then ideal gas-liquid behavior can be assumed and the equations of Raoult, Dalton and Antoine are applicable.

For a non-ideal mixture the adapted Raoult equation can be used to calculate the partial pressure P_j of component j from its liquid fraction x_j and the pure component vapor pressure P_j^0, which only depends on the temperature. According to Dalton, the vapor fraction at equilibrium y_j is the ratio between the partial pressure and the total pressure.

$$P_j = x_j \gamma_j P_j^0 \quad \text{and} \quad y_j = {}^{P_j}\!/_{P} \quad \text{for} \quad j = 1 \ldots N \tag{2.129}$$

in which γ_j is the activity coefficient. The sum of the partial vapor pressures should equal the total pressure:

$$P = \sum_{j=1}^{N} P_j \tag{2.130}$$

The Antoine equation gives the vapor pressure P_j^0 of the pure component j.

$$P_j^0 = exp\left(A_j - \frac{B_j}{C_j + T}\right) \quad \text{for} \quad j = 1 \ldots N \tag{2.131}$$

When the pressure P and the liquid compositions x_1 to x_N or vapor compositions y_1 to y_N are known, from the preceding gas-liquid equilibrium equations, the compositions of the other phase and de temperature T can be iteratively solved. Eqn. (2.130) can be used as the iteration criterion that should be fulfilled. The ratio between the fraction in the gas and liquid phase can be expressed in a distribution coefficient:

$$K_j(P) \equiv \frac{y_j}{x_j} = \gamma_j \frac{P_j^0}{P} \tag{2.132}$$

For a binary mixture, it is sometimes permissible to assume constant relative volatility. The relative volatility α of component x_1 to x_2 is defined as:

$$\alpha(P) = \frac{K_1(P)}{K_2(P)} = \frac{y_1 x_2}{x_1 y_2} \tag{2.133}$$

Because for a binary mixture $x_1 + x_2 = 1$ and $y_1 + y_2 = 1$, this can be rewritten as:

$$y_1 = \frac{\alpha(P)x_1}{1+(\alpha(P)-1)x_1} \tag{2.134}$$

The slope of the equilibrium curve at a certain liquid composition x_1 can be written as:

$$m(x_1) = \frac{\partial y_1}{\partial x_1}(P) = \frac{\alpha}{(1+(\alpha-1)x_1)^2} \tag{2.135}$$

2.4.3 Component Transfer Properties

Interface component transfer

When components are transferred across a liquid-liquid or a liquid-gas interface A, in each phase, owing to the resistance to component transfer, a concentration gradient will occur. In the model that describes this interfacial transfer, it is assumed, as shown in Fig. 2.14, that the concentration gradient restricts itself to a boundary layer.

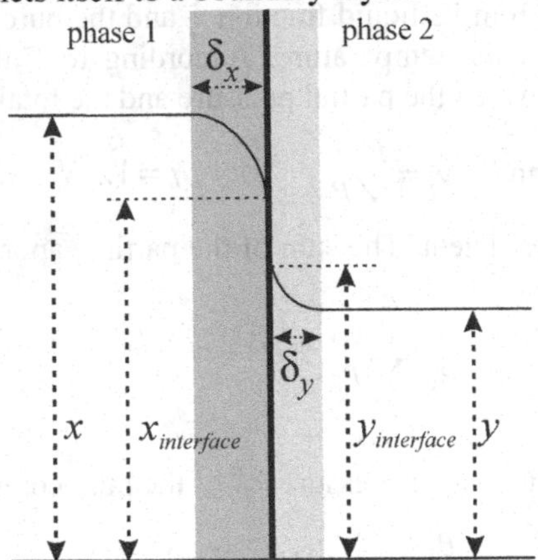

Fig. 2.14. Model for convective component transport (stationary situation).

The bulk outside this boundary layer has an uniform composition by mixing or convection. The concentrations of the two phases at the interface are usually unequal and are assumed to be at equilibrium.

The flow J_{phase} of component j per unit area A at a composition difference on the boundary layer can be given by:

$$\frac{J_{phase}}{A} = k_x(x_j - x_{j,interface}) = k_y(y_{j,interface} - y_j) \tag{2.136}$$

in which x_j and y_j are the bulk concentrations of phase-1 and phase-2, $x_{j,interface}$ and $y_{j,interface}$ are the concentrations at the interface.

The component transfer resistance in phase-1 is $1/k_x$ and in phase-2 it is $1/k_y$. If the equilibrium equation $m_j = \partial y_j / \partial x_j$ holds in the range considered, there is no need to solve the interfacial compositions, since in that case the rate of transfer is proportional to $(K_j x_j - y_j)$:

$$\frac{J_{phase}}{A} = K_y\left(K_j x_j - y_j\right) = K_y\left(y_j^{eq} - y_j\right) \tag{2.137}$$

in which K_j is the overall component transfer coefficient for phase-2. From:

$$\frac{J_{phase}}{A} = k_y\left(y_{j,interface} - y_j\right) = K_y\left(y_j^{eq} - y_j\right) \tag{2.138}$$

it follows that:

$$\begin{aligned} \frac{1}{K_y} &= \frac{1}{k_y}\frac{\left(y_j^{eq} - y_j\right)}{\left(y_{j,interface} - y_j\right)} = \frac{1}{k_y} + \frac{1}{k_y}\frac{\left(y_j^{eq} - y_{j,interface}\right)}{\left(y_{j,interface} - y_j\right)} \\ &= \frac{1}{k_y} + \frac{1}{k_x}\frac{\left(y_j^{eq} - y_{j,interface}\right)}{\left(x_j - x_{j,interface}\right)} = \frac{1}{k_y} + \frac{m_j}{k_x}\frac{\left(x_j - x_{j,interface}\right)}{\left(x_j - x_{j,interface}\right)} \end{aligned} \tag{2.139}$$

in which m_j is the slope of the equilibrium line of component j. Thus:

$$\frac{1}{K_y} = \frac{1}{k_y} + \frac{m_j}{k_x} \tag{2.140}$$

In case of a liquid phase at boiling point which is in equilibrium with the vapor phase, the component transfer can be described by:

$$\frac{1}{K_G} = \frac{1}{k_G} + \frac{m_j}{k_L} \tag{2.141}$$

with K_G is the gas phase overall coefficient, k_g is the gas phase, k_l is the liquid phase and m_j is the slope of the equilibrium curve.

Component diffusion

Diffusion takes place within stagnant gas or liquid layers (layers in which no convection occurs). The component flow per unit area as a result of conduction is determined by the Fick's law:

$$\frac{J_{diffusion}}{A} = -D\frac{dC}{dx} \tag{2.142}$$

The model describes the component transport J in terms of the diffusivity D and the composition gradient dC_j/dx of component j. If no steady state is established, no constant concentration gradient exists. Then the change of the concentration with time has to be considered as represented by the equation:

$$\frac{\partial C(x,t)}{\partial t} = D\frac{\partial^2 C(x,t)}{\partial x^2} \tag{2.143}$$

Reaction kinetics

Chemical reactions produce or consume components. The transformation is expressed in moles, because of the stoichiometric relationship between produced and consumed components. The reaction rate r_j for component j is defined as the moles of component j produced or consumed per unit of time and per volume.

$$\frac{J_{reaction}}{V} = r_j \tag{2.144}$$

This reaction rate is positive when a component is produced and negative when it is consumed. When component i is consumed or produced k times faster than component j, then $r_i = k \cdot r_j$. The reaction rate depends on the temperature and component concentrations, according to:

$$r_j = k_j \cdot f\left(C_j^{\alpha}, C_k^{\beta}, \ldots\right) \quad \text{with} \quad k_j = k_{0j} e^{-E_j/(RT)} \tag{2.145}$$

The order of the reaction equals the sum of the coefficients (α, β,...). For a first-order reaction the reaction rate equation becomes:

$$r_j = k_j C_j \tag{2.146}$$

2.4.4. Example of a component system: A CISTR

In a stirred reactor as shown in Fig. 2.15 a component A is converted to component B according the stochiometric relationship:

$$A \rightarrow B \tag{2.147}$$

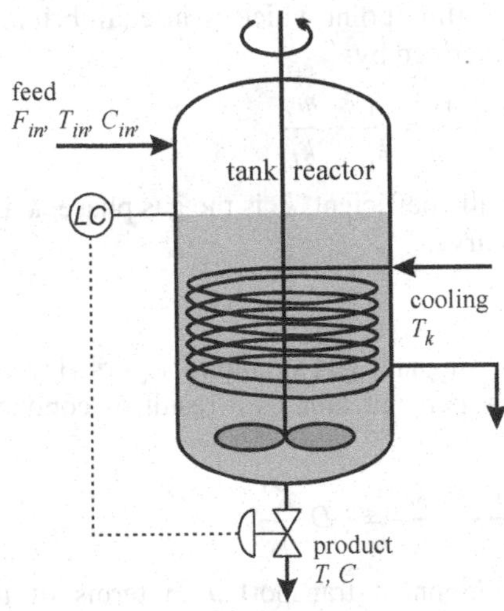

Fig. 2.15. Chemical reactor with single component conversion.

The reaction is exothermal and the components are dissolved. The following assumptions are made:

- the reactor is well mixed and can be considered as a lumped system
- in the operating range the density and the heat capacities do not vary as a function of temperature and composition

First the mass, component balances for the components A and B and the energy balance will be derived from the general balances that are defined above. Next, it will be shown that the energy balance for a reaction can be obtained more directly by extracting the reaction heat from the enthalpy as a separate term.

Mass balance

The general equation for the mass balance is Eqn. (2.14):

$$\frac{dm}{dt} = \sum_{i=1}^{N} F_{m,in,i} - \sum_{j=1}^{M} F_{m,out,j} \tag{2.148}$$

Since the reactor has one entering and one effluent flow, the balance can be simplified to:

$$\frac{dm}{dt} = \rho \frac{dV}{dt} = \rho F_{v,in} - \rho F_{v,out} \tag{2.149}$$

Component balances

The general equation for the component balance of component k is Eqn. (2.123):

$$\frac{dVC_k}{dt} = \sum_{i=1}^{N} F_{v,in,i} C_{in,k} - \sum_{j=1}^{M} F_{v,out,j} C_k + J_{transfer} + J_{reaction} \tag{2.150}$$

Since for the reaction it holds that:

$$\begin{aligned} J_{transfer} &= 0 \\ J_{reaction} &= -k_{0,AB} e^{-E_{AB}/(RT)} C_A V = -r_{AB}(T) V \end{aligned} \tag{2.151}$$

the component equation for component A can be simplified to:

$$\frac{dVC_A}{dt} = F_{v,in} C_{in,A} - F_{v,out} C_A - r_{AB}(T) V \tag{2.152}$$

Substitution of the mass balance derived previously:

$$\begin{aligned} \frac{dVC_A}{dt} &= V \frac{dC_A}{dt} + C_A \frac{dV}{dt} = V \frac{dC_A}{dt} + C_A \left(F_{v,in} - F_{v,out}\right) \\ &= F_{v,in} C_{in,A} - F_{v,out} C_A - r_{AB}(T) V \end{aligned} \tag{2.153}$$

gives after elimination of equal terms:

$$V \frac{dC_A}{dt} = F_{v,in} \left(C_{in,A} - C_A\right) - r_{AB}(T) V \tag{2.154}$$

A similar component equation can be derived for component B:

$$V \frac{dC_B}{dt} = -F_{v,in} C_B + r_{AB}(T) V \tag{2.155}$$

Energy balance

The general equation for the energy balance is Eqn. (2.76):

$$\frac{dH(C,T)}{dt} = \sum_{i=1}^{N} F_{m,in,i} \hat{H}_{in,i}(C,T) - \sum_{j=1}^{M} F_{m,out,j} \hat{H}_j(C,T) + Q + W \tag{2.156}$$

For the reactor with one entering and effluent flow and no exchange of work W, this equation becomes:

$$\frac{dH(C,T)}{dt} = F_{m,in}\hat{H}_{in}(C,T) - F_{m,out}\hat{H}(C,T) + Q \tag{2.157}$$

Rewriting of the left-hand side term, by using the temperature dependency Eqn. (2.77) and the composition dependency Eqn. (2.84) of the enthalpy H gives:

$$\begin{aligned}\frac{dH(C,T)}{dt} &= \frac{\partial H}{\partial T}\frac{dT}{dt} + \frac{\partial H}{\partial C_A}\frac{dC_A}{dt} + \frac{\partial H}{\partial C_B}\frac{dC_B}{dt} \\ &= c_P \rho V \frac{dT}{dt} + \tilde{H}_A V \frac{dC_A}{dt} + \tilde{H}_B V \frac{dC_B}{dt}\end{aligned} \tag{2.158}$$

Substitution of component balances for $\frac{dC_A}{dt}$ and $\frac{dC_B}{dt}$ gives:

$$\begin{aligned}\frac{dH(C,T)}{dt} = c_P \rho V \frac{dT}{dt} &+ \tilde{H}_A\left(F_{v,in}\left(C_{in,A} - C_A\right) - r_{AB}(T)V\right) \\ &+ \tilde{H}_B\left(-F_{v,in}C_B + r_{AB}(T)V\right)\end{aligned} \tag{2.159}$$

Rewriting of the right-hand side terms is possible by using the relationship between the enthalpy $\hat{H}$ and partial molar enthalpy $\tilde{H}_i$ of the components Eqn. (2.88):

$$\begin{aligned}F_{v,in}\rho\hat{H}_{in}(C_{in},T_{in}) &= F_{v,in}\left(C_{in,A}\tilde{H}_A(T) + c_P\rho(T_{in} - T)\right) \\ F_{m,out}\rho\hat{H}(C,T) &= F_{v,out}\left(C_A\tilde{H}_A(T) + C_B\tilde{H}_B(T)\right)\end{aligned} \tag{2.160}$$

and of the cooling term:

$$Q = -Q_{cool} = AU(T_{cool} - T) \tag{2.161}$$

Substitution of the result for the left-hand side Eqn. (2.159) and the right-hand side Eqn. (2.160) and (2.161) into the initial balance Eqn. (2.157) results in:

$$\begin{aligned}&c_P\rho V\frac{dT}{dt} + \tilde{H}_A\left(F_{v,in}\left(C_{in,A} - C_A\right) - r_{AB}(T)V\right) + \\ &\tilde{H}_B\left(-F_{v,in}C_B + r_{AB}(T)V\right) = F_{v,in}\left(C_{in,A}\tilde{H}_A(T) + c_P\rho(T_{in} - T)\right) - \\ &F_{v,out}\left(C_A\tilde{H}_A(T) + C_B\tilde{H}_B(T)\right) - Q_{cool}\end{aligned} \tag{2.162}$$

Elimination of equal terms finally results in:

$$c_P\rho V\frac{dT}{dt} + r_{AB}(T)V\left(-\tilde{H}_A + \tilde{H}_B\right) = c_P\rho F_{v,in}(T_{in} - T) - Q_{cool} \tag{2.163}$$

The difference between the partial enthalpies of the feed A and the product B is the reaction enthalpy, according to Eqn. (2.91):

$$\left(\tilde{H}_B(T) - \tilde{H}_A(T)\right) = \Delta H_{reaction\ A\to B}(T) = \Delta H_{AB}(T) \tag{2.164}$$

It can be concluded that the energy balance can be simplified to:

$$c_P \rho V \frac{dT}{dt} = c_P \rho F_{v,in} \left(T_{in} - T\right) - r_{AB} V \Delta H_{AB} - UA\left(T - T_{cool}\right) \tag{2.165}$$

The terms $c_P \rho F_{v,in}\left(T_{in} - T\right)$ and $-r_{AB} V \Delta H_{AB}$ are contributions of enthalpy changes. The first indicate the heating-up of the feed and the second the reaction heat by the conversion.

Alternative energy balance

An alternative approach is to consider the specific enthalpy $\hat{H}$ not as a function of the temperature and composition $\hat{H}(C,T)$ but only as a function of temperature $\hat{H}(T)$. Then the reaction heat has to be accounted for as a separate term. For the reactor with one entering flow and one effluent flow, the basic Eqn. (2.76) can be written as:

$$\frac{dH(C,T)}{dt} = F_{m,in} \hat{H}_{in}(C,T) - F_{m,out} \hat{H}(C,T) - Q_{cool} \tag{2.166}$$

which can be modified to:

$$\frac{dH(T)}{dt} = F_{m,in} \hat{H}_{in}(T) - F_{m,out} \hat{H}(T) + Q_{reaction} - Q_{cool} \tag{2.167}$$

Rewriting of the left-hand side term gives:

$$\begin{aligned} \frac{dH(T)}{dt} &= \rho \frac{dV\hat{H}(T)}{dt} = \rho V \frac{d\hat{H}(T)}{dt} + \rho \hat{H}(T) \frac{dV}{dt} \\ c_P \rho V \frac{dT}{dt} &+ \rho \hat{H}(T) \frac{dV}{dt} = c_P \rho V \frac{dT}{dt} + \hat{H}(T)\left(F_{m,in} - F_{m,out}\right) \end{aligned} \tag{2.168}$$

and substitution of the mass balance results in:

$$\begin{aligned} c_P \rho V \frac{dT}{dt} &+ \hat{H}(T)\left(F_{m,in} - F_{m,out}\right) = \\ & F_{m,in} \hat{H}_{in}(T) - F_{m,out} \hat{H}(T) + Q_{reaction} - Q_{cool} \end{aligned} \tag{2.169}$$

Elimination of equal terms gives:

$$\begin{aligned} c_P \rho V \frac{dT}{dt} &= \rho F_{v,in} \left(\hat{H}_{in}(T) - \hat{H}(T)\right) + Q_{reaction} - Q_{cool} \\ &= c_P \rho F_{v,in} \left(T_{in} - T\right) - r_{AB}(T) V \Delta H_{AB} - Q_{cool} \end{aligned} \tag{2.170}$$

which equals the result of the equation derived Eqn. (2.165). The first method is based on the elaboration of the enthalpy of components. However, considering that only changes in enthalpy deliver a first contribution to the energy balance, this alternative method is as effective as the first one.

Files referred to in this chapter:
F0207par.m: parameter definition file
F0207.mdl: simulation for laminar and turbulent outflow
F0207plot.m: file to plot simulation results.

3 Extended Analysis of Modeling for Process Operation

In the analysis phase, the system boundaries, the degrees of freedom for control (the inputs) and the variables that have to be controlled (the outputs) are established on the basis of the operation of the process and the goals of the operation. The result is a so-called environmental model. This model is a good starting point for the design of a behavioral model and a global control scheme. For the first application, the desired level of detail has to be established on the basis of the formulated goal. For the design of a global control scheme for the process, additional knowledge is required with respect to the static behavior and dynamic relationships between process inputs and outputs.

3.1 Environmental Model

Process behavior is the description of the state variables to variations in the control and disturbance variables. The environmental model gives an indication of which input and output variables are of interest. As is indicated in Fig. 3.1, the environmental model is a model of the process in a form that is relevant for the control application.

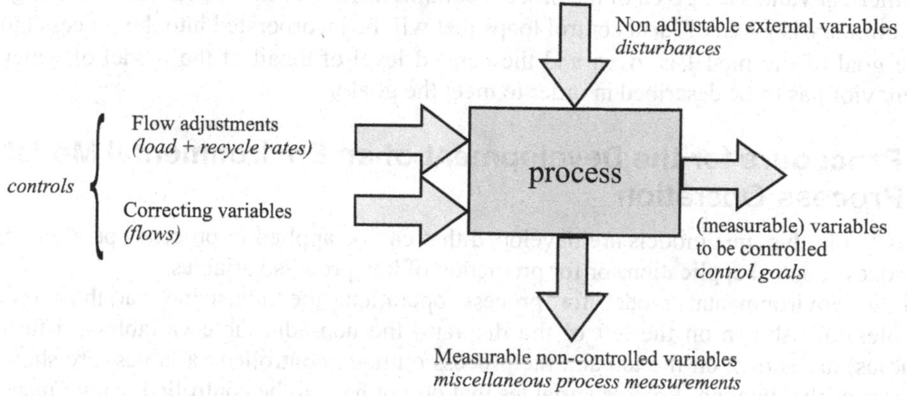

Fig. 3.1. Environmental model of the process.

The control variables are divided into flows that can or have to be directly adjusted and flows that can be used for control loops.

The process in the context of process control is shown in Fig. 3.2. The process outputs that have to be controlled are compared to their respective target or desired values, controllers adjust the selected process inputs.

The environmental model is the product of a problem analysis. It forms the basis for the development of behavioral models of the process. This holds for physical models, in which the internal dynamic mechanism is described by physical laws, as well as empirical (black box) models, in which only "overall" dynamic relationships between process inputs and outputs are formulated. The environmental model can also be used for the development of a process control scheme. By evaluating the static (power of control) and dynamic relationships (speed of control) between process inputs and outputs, the process control scheme can be selected. In general no dynamic model of the process is required yet at this stage. This is the starting point that is used in the chapters on process control.

Process Dynamics and Control: Modeling for Control and Prediction. Brian Roffel and Ben Betlem.

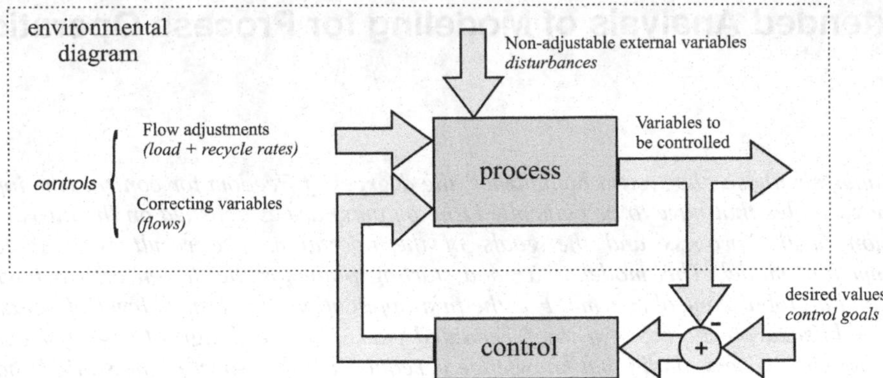

Fig. 3.2. Environmental model in the context of process control.

The environmental model consists of an environmental diagram in which incoming and outgoing information flows are indicated, supplemented with a description in which:

- a brief description is given of the goal of the process unit (no process control goals)
- a schematic diagram is shown of the process unit, in which the boundaries of the system are clearly shown
- numerical values are given of the process conditions (to be established from the design)
- a summary is given of ideal control loops that will be incorporated into the process model
- the goal of the model is given and the desired level of detail of the model of which the behavior has to be described in order to meet the goal

3.2 Procedure for the Development of an Environmental Model for Process Operation

In this book, dynamic models are developed that can be applied in process operation, either for process control applications or for prediction of key process variables.

In an environmental model for process operation, the adjustable variables (control variables) are shown on the left of the diagram, the non-adjustable variables (disturbance variables) are shown on the top and the process outputs (controlled variables) are shown on the right of the diagram. Process variables that do not have to be controlled but are measured (for external use) are output variables pointing downward. The procedure for the development of the environmental model is shown in Table 3.1.

Table 3.1. Design procedure for environmental diagram.

	Environmental model
i.	Determine the goal and mode of operation of the process unit
ii.	Determine the process boundaries and the external disturbances
iii.	Determine the goal of the model and the associated level of detail
iv.	Establish the controlled variables (conditions, mass and qualities)
v.	Select, if required, the correcting variables for the throughput and recycle loops
vi.	Determine the correcting variables
vii.	Develop the environmental model and determine the degrees of freedom

3.2.1 Goal and Mode of Operation of the Process

As a first step, the goal and mode of operation of the process has to be established. The operation of the process can usually be well described by means of the driving force and the phenomena associated with it.

Example: Evaporator

The process is an evaporator in which a liquid is evaporated; the vapor flow is discharged into a vapor supply system. The evaporator contains a heat exchanger, the heat transfer area remains under the liquid level (Fig. 3.3). Heat for the evaporation process is delivered by condensing steam in the heat exchanger. Depending on the goal of the evaporator, the inputs and outputs of the environmental model will be different, resulting in a different control scheme.
Two different goals will be considered.

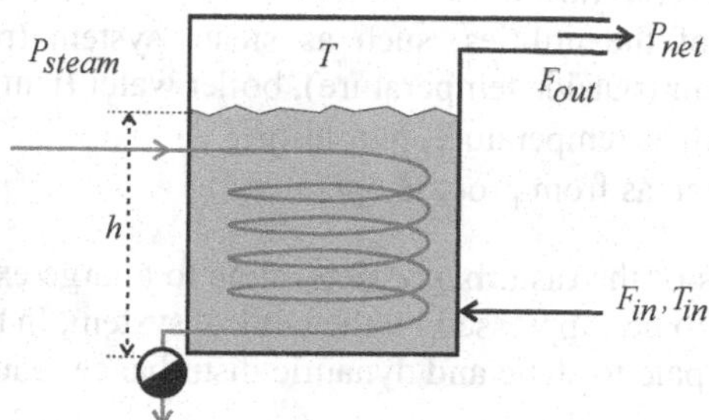

Fig. 3.3. Evaporator.

Goals:

1. An adjustable feed flow F_{in} has to be evaporated. The evaporation is determined by the desired flow. It may be assumed that the required amount of heat is not limiting.
2. A maximum flow F_{out} has to be evaporated. The evaporation is then determined by the amount of steam that is available.

The driving force of the delivered amount of heat Q in the process unit is determined by the product of the overall heat transfer coefficient U, the heat transfer area and the temperature difference between the condensing steam and the evaporating liquid:

$$Q = U \cdot A \cdot \left(T_{\text{condensing vapor}} - T_{\text{evaporating liquid}}\right) \tag{3.1}$$

The only way to change the vapor flow is therefore to change the temperature difference. Since the contribution of the sensible heat can be neglected, heat transfer is determined by the difference between the temperature of the condensing steam and the evaporation temperature of the liquid in the vessel. The evaporation can therefore be affected by increasing the pressure on the steam side or lowering the pressure on the evaporating liquid side. It should be noted that a condensation pot is essential, since it separates the vapor and condensed steam. Without it, no pressure can be built up.

No special process conditions are required as is often the case in reactors. The liquids on the condensation side and evaporation side are at boiling point: boiling pressure and temperature.

3.2.2 Process Boundaries and External Disturbances

The chosen system boundaries have to agree with the boundaries as they are shown on the process flowsheet. Investigate whether the given process boundaries are also the right ones for the process control scheme. Pay special attention to the sources of energy and material, which belong specifically to the process and which can be manipulated. They can lead to the selection of a larger area for the process boundaries. Specifically the separation between what can be influenced and cannot be influenced (external influences) determines the system boundaries.

Example: Room temperature control

For room temperature control that is provided by a central heating system, the system consists of a room and heating system from a control point of view. The temperature measurement is located in the room whereas the manipulated variable is the gas valve of the heating furnace.

A distinction can be made between internal and external disturbances. Examples of internal disturbances are fouling of pipes or catalyst poisoning by coke. These are examples of disturbances that do not come from the outside and therefore they do not appear in the environmental model. In case they are important for control, they have to be represented in the internal behavioral model.

Examples of external disturbances are:

- inlet and outlet material flows that are determined by other process units.
- composition of inlet material flows
- impact of conditions of the utilities, such as steam system (pressure), cooling water (temperature), cooling air (outside temperature), boiler water (temperature)
- weather conditions, such as temperature, humidity
- human disturbances, such as from process operators

The frequency contents of the disturbance determine to a large extent in which frequency range the disturbance has to be suppressed by the control system. In the discussion on control systems, attention will be paid to static and dynamic disturbance reduction.

Example: Evaporator

In the case of the evaporator as shown in Fig. 3.3, the most important disturbances are:

- the inlet temperature of the feed
- the pressure of the vapor supply system in which the evaporated liquid is discharged
- the steam pressure.
  In case of the first goal, where the steam flow is used to evaporate the desired feed flow, the influence of the steam pressure is limited. It depends on the desired level of detail whether changes in steam pressure will be taken into account. However, for the second goal, fluctuations in the steam pressure are relevant.

During the development of the environmental model, it is advisable to idealize the interfacing between the process and the control system as much as possible. Assume that the communication takes place via variables with a maximum possible level of abstraction. It is not necessary to take measurement and control specifications already into account. For example, for correcting variables one can take material flows or energy flows and for controlled variables one could take pressures and mass fractions. It is advisable, however, to use measurable variables, which can also be measured in reality, for example the mass instead of the level in a vessel.

Sometimes it is possible to divide the system into smaller independent subsystems. In this case it is much easier to analyze the separate subsystems.

3.2.3 Goals and Level of Detail of Modeling

The goal of a model can be very diverse. In this book it is usually a study of the response of the process to changes in throughput or changes of disturbances for a control or controllability analysis.

The level of detail is directly dependent on the modeling goal and has already an impact on the choices for the environmental model. The choices can be divided into the frequency domain or time domain one is interested in:

- Relatively fast changes (from seconds to minutes time base).
  On this time scale, pressure changes have to be taken into account in order to describe relatively fast control behavior. This calls for a high level of detail. For the environmental model this means that limited abstractions of inputs and disturbances are possible. For example, pressure variations in condensers and boiler cannot be neglected. This choice has even larger consequences for the behavioral model. Sometimes valves and pumps have to be modeled and heat capacities of walls and distillation trays have to be taken into account.
- Relatively slow changes (from minutes to hours time scale).
  Quality changes in processes with interactions between mass and energy flows fall into this category. To describe the behavior of quality changes, it is sufficient to take mass and energy flows as inputs into account.
- Relatively slow and smooth changes (from hours to days).
  When process changes are slow, such as is the case with catalyst decay or fouling of polymer reactors, all other process phenomena can be considered constant.

If process dynamics with a different time horizon have to be modeled at the same time, so-called stiff systems emerge. In order to compute a solution, the smallest time constant has to be taken into account, whereas for the calculation of the new stationary values, the entire horizon has to be taken into account. The way to get around this problem is to divide the system into hierarchical subsystems.

Example: Evaporator

In case of the evaporator shown in Fig. 3.3, the modeling goal depends on the goal of the operation.

1. Adjustable vapor throughput.
   The modeling goal is to investigate vapor flow behavior as a function of throughput variations. It is therefore not required to take fast pressure changes into account. The time scale of relevant changes is in the order of the residence time.
2. Maximum vapor flow.
   The modeling goal is to investigate variations in the produced vapor flow as a function of variations in the steam pressure. The level of detail depends entirely on the speed of pressure variations. If pressure variations are fast, a detailed model should be developed. However, if the variations of the pressure arc limited, the level of detail is similar to the previous case.

3.2.4 Checklist for Controlled Variables

The first step is to determine which variables have some form of self-regulation. A variable is self-regulating when an accumulation or process condition returns to a new steady state value without an external intervention. This should be a value within the operating region, without exceeding capacity limits. Some examples of self-regulation are:

- liquid accumulations that are controlled by means of overflow weirs. Examples are liquid on the trays of a distillation column, liquid in a kettle reboiler.
- interacting gas accumulations.
  Example: coupled pressure vessels in series, where the pressure drop is small compared to the absolute pressure. It is sufficient if the pressure in one of the vessels is controlled.

- Coupled liquid accumulations.
  Example: liquid levels in process units, where the liquid is part of a closed loop system. One of the liquid levels should be left uncontrolled, since the total amount of liquid is constant.
- Components or temperature in chained processes.
  In chained process it is sufficient if the composition or temperature at both ends are controlled.
  Example: in a distillation column, it is not required to control the composition at every tray, because they follow from the component balances that describe the working lines.
- Conditions or concentrations which are dependent.
  The maximum number of conditions or concentrations that can be controlled is determined by the phase rule of Gibbs (chapter 2). This rule states that the number of degrees of freedom F in a closed system is equal to two plus the number of components n minus the number of phases π that are in equilibrium with each other: ($F = 2 + n - \pi$). The number of degrees of freedom agrees with the number of controlled conditions or components.
  Example 1: of a pure boiling liquid of which vapor and liquid are in equilibrium, the temperature or pressure can be controlled but not both. ($F = 2+1-2 = 1$).
  Example 2: in the top of a distillation column, two components are present. In this case, pressure, temperature and composition can not all be controlled but only two of the three variables.
- Through sufficient feedback, a variable stays within acceptable limits.
  Example: in an evaporator, the pressure and temperature will increase when more liquid is evaporated. Consequently, the temperature difference across the heat-exchanging surface will decrease and the pressure will stay within acceptable limits.

Variables that have to be controlled:

- *Non self-adjusting process conditions*
  This concerns conditions that should be maintained in order to ensure efficient and safe process operation. These are also the variables that are candidates for process optimization. Examples are: temperature, pressure, acidity or residence time in a reactor, pressure and temperature in a stripper, oxygen concentration in a burner or the speed of rotation of a mixer.
- *Non self-adjusting material accumulations*
  Every vessel or process unit can be represented by its material contents, consisting of one or more phases. Select variables that are representative for material contents, such as pressure for a vapor or gas, level for liquid in a vessel or the intermediate level for two separate liquid phases. For control purposes, solids consisting of powder of granular material can usually be considered to behave as a liquid.
- *Qualities of end or intermediate products* that are determined by a hierarchically higher control layer.

Some state variables such as pressures and levels fall into more than one of the above categories. They can represent the accumulation of vapor or liquid, in addition they can represent a process condition. A liquid level is a good measure for material accumulation. In case of a buffer tank it does not have to be controlled tightly, since the purpose of the tank is to smooth fluctuations in the flow. However, the level can also represent the residence time in a reactor, which is given by the accumulation (kg) divided by the throughput (kg/s). The residence time is, with pressure and temperature, an important process condition that has to be controlled.

It is important to select easily interpretable variables as controlled variables. For the environmental model, measurement problems should not be considered, however, do not select variables that cannot be measured. Also indicate which variables are ideally controlled. Control becomes then part of the equipment and falls within the framework of the process. The setpoint of the controller has become a new degree of freedom.

Only assumptions that relate to the inputs and outputs are relevant at the level of the environmental diagram. Assumptions with respect to self-regulating accumulations can be made later during the development of the behavioral model. Often this cannot be avoided, since the contribution of different terms has to be evaluated.

For the purpose of problem analysis, it can be assumed that the measurements are ideal. It is important, however, to take into consideration that not all variables can be measured online. This is often the case with quality measurements.

Example: Evaporator

In case of an evaporator, shown in Fig. 3.3, the controlled variables are:

- *non self-adjusting process conditions:*
  not applicable.
- *non self-adjusting accumulations:*
  the level at the evaporation side is undetermined and should be maintained within acceptable limits. The level at the condensate side of the evaporator is regulated by the condensate pot. In fact, this is considered as an ideal control loop.
- *qualities to be controlled:*
  Not applicable.

3.2.5 Throughput and Recycles

The process throughput is, similar to the product quality, usually determined by a higher hierarchical control layer. If the throughput (product quantity) of the part of the process for which the model is being developed, is undetermined, it has to be set by one of the process flows.

In practice, it is recommended to set the throughput with one of the most important feed flows. Often this is the flow that feeds the heart of the process (usually the reactor). This flow is then the sum of all the feed flows and recycles. The product flow can only be used if the process contains few accumulations and side flows.

If recycle flows are present in the process or process unit, it is recommended to control these flows directly with the help of a flow controller or flow ratio controller. An example is the reflux flow in distillation or the solvent that is re-circulated in case of absorption or extraction. Usually the recycle flow determines the separation efficiency or conversion of the reaction. If the quality can be measured, the corresponding flow can be used for quality control.

Example: Evaporator

For the evaporator, shown in Fig. 3.3, the throughput has to be set, depending on the goal of the operation. In case of the evaporator with adjustable throughput, it is recommended to use the vapor flow or liquid supply to achieve this. In case of the evaporator with maximum energy consumption, the energy supply is limiting. The energy supply has therefore to be set at its maximum, usually the valve should be wide open.

3.2.6 Checklist for Correcting Variables

The adjustable process inputs are usually flows. They can be adjusted by valves or pumps. In case of a valve the flow is adjusted by means of the valve position, which can be considered as an adjustable resistance. Flow is a result of the pressure difference, which is affected by the flow itself (Fig. 3.4a). In case of a pump the situation is slightly more complicated since the pump generates a flow as well as a pressure difference depending on the speed of rotation (Fig. 3.4b).

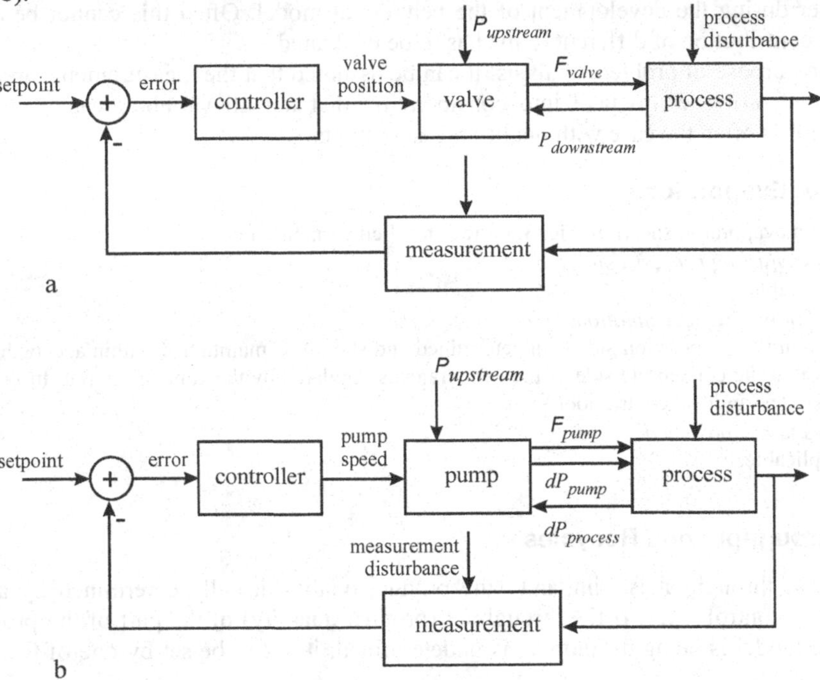

Fig. 3.4. Simple feedback control with a. valve and b. pump.

For the problem analysis, it is not required to include the measurement and control or correcting devices (Fig. 3.5). In case of the control devices, it is usually assumed that the flow follows directly from the controller. This is in fact an abstract representation of reality, in which it is assumed that the valve or pump includes an internal, ideal flow controller.

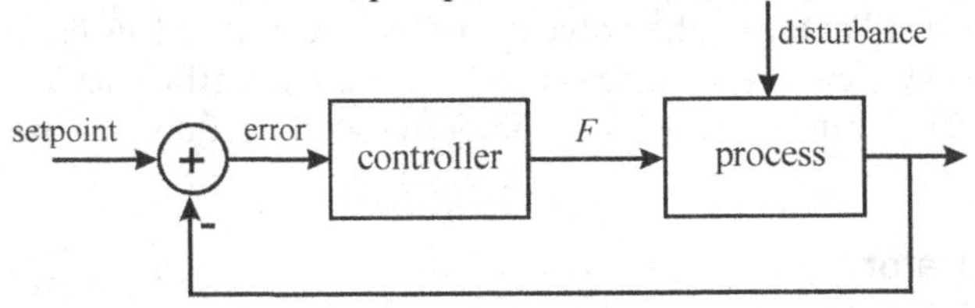

Fig. 3.5. Abstract representation of a control loop for the environmental model.

This assumption is justified, since in case of large flow variations as a result of pressure variations, the valve or pump is equipped with a flow controller, which serves as a secondary controller for another controller. In addition, the valve or pump will be equipped with an internal device for control of the valve position (the valve positioner) or the speed of rotation.

A far-reaching form of abstraction is to assume that an energy flow can be adjusted instead of the flow of a heating medium (for example steam) or the cooling medium (for example water). It is then assumed that the energy flow can also be adjusted in practice. This level of abstraction depends very much on the level of detail that is desired for the model. If not much detail is required, a large amount of abstraction is allowed for the process inputs. The output of a controller is often a pneumatic signal or a digital signal that still has to be scaled, or a pulse signal adjusting the speed of rotation of a pump. The description of the interface between process and environment should be abstract with a minimum number of connections. Generally, flows can be assumed as process inputs. It is then assumed that every control loop possesses an ideal flow controller as slave controller. As a result, the controls are made independent of the state of the process or the adjacent process and consequently the process units are decoupled.

If one chooses a less abstract correcting variable (for example the valve position, valve opening or the pneumatic pressure), the pressure drop across the valve has to be taken into account in the internal model. The pressure drop is determined by the mass accumulation in the process unit and the external pressure. This leads usually to a much more complicated model. A controlled variable (the accumulation) is internally coupled with the correcting flow, also additional disturbances (from adjacent process units) should be taken into account. Such a detailed model is only useful if pressure control or pressure propagation through the equipment has to be modeled. These are relatively fast phenomena.

If the control interface itself has to be studied, then it should be described by a more detailed model. What holds for a flow, also holds on a lower level of control for the valve positioner.

A simple rule-of-thumb for the selection of adjustable process variables is:

Every incoming flow that has not been determined externally, should be controlled, except when self-regulation of the adjustable mass- or energy accumulation fixes the inlet flow.

There are numerous restrictions for dealing with manipulated variables, they will be discussed in chapter 33.

3.2.7 Environmental Diagram and Degrees of Freedom

All information has now been collected in order to create the environmental diagram. In principle this diagram should be balanced. This means that the number of variables that can be adjusted has to agree with the number of variables that can be controlled.

Example: Evaporator

The adjustable variables can be separated into two categories, namely the flow that is used for control of the throughput and flows that are used for control loops. In case of the evaporator, shown in Fig. 3.3, three flows can be adjusted: the steam flow, the feed flow and the vapor flow. The vapor flow should not be used, since a valve in the vapor line causes an undesirable pressure drop. A valve is only useful if the vapor should be delivered at a desired pressure. The condensate flow does not have to be adjusted either, the condensate pot contains a level controller.

The remaining adjustable variables that are useful, are different for each operational goal:

1. Adjustable vapor throughput
   The steam flow and the feed flow can be used.
2. Maximum vapor flow
   No flow is required to adjust the throughput. The energy supply determines the throughput, since the maximum amount of energy that is available should be used. Only the feed flow remains.

Since items ii) to vii) have been addressed (Table 3.1), the final environmental diagram of the uncontrolled process can be drawn (Fig. 3.6a).

For the first goal in which an adjustable flow is produced, the diagram of the controlled process is shown in Fig. 3.6b, for the second goal in which the maximum amount of energy is used, the control diagram is shown in Fig. 3.7b. In the last process, the steam flow pipeline is fully open, such that the pressure (and temperature) on the steam side are maximal.

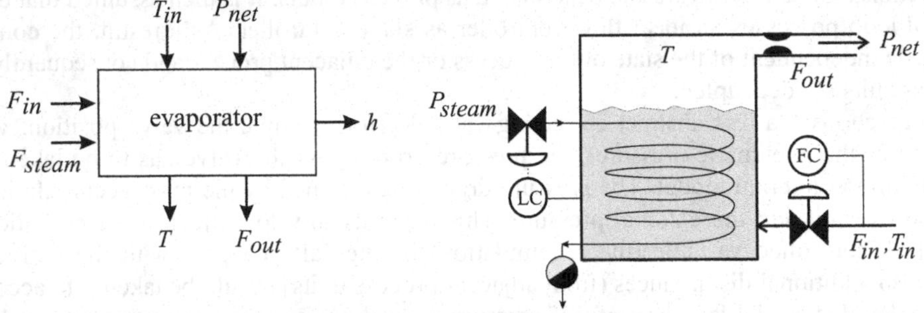

Fig. 3.6. Evaporator with adjustable throughput.

From the previous figures it is clear that an evaporator with fixed evaporating surface has advantages when the second goal has to be realized in which the throughput should be maximized. The flow can easily be adjusted to the energy supply. When the steam pressure increases, the transferred amount of heat automatically increases.

The first goal, however, is more difficult to realize. If the throughput is increased, the required energy is not automatically adjusted. The pressure in the vessel will increase in order to produce a higher vapor flow, since a higher pressure drop between the vessel and the vapor net is required. The steam pressure will therefore have to increase even more, in order to ensure that the required amount of heat is delivered. In case of this evaporator it is interesting to investigate how the vapor flow reacts to changes in the steam pressure.

As was mentioned in section 3.2.6, control loops can be made part of the system and therefore of the model. The control loops fall then within the system boundaries. Often these are control loops, of which the behavior is assumed to be ideal, which means that they are fast with respect to the other dynamic phenomena in the system.

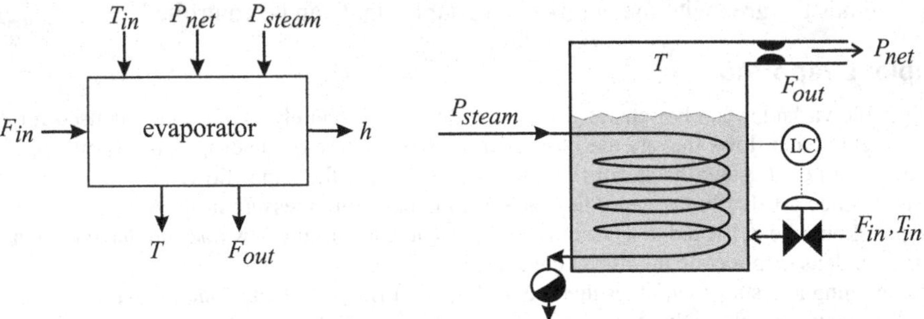

Fig. 3.7. Evaporator with maximum vapor throughput.

The setpoint of the ideal controller has then become an input variable, which is imposed from the outside. The controlled variable is eliminated this way (see Fig. 3.8).

By including control loops between the correcting variables and controlled variables, the degrees of freedom are used. The degrees of freedom are not eliminated, they are shifted to the setpoints of the control loops.

A central issue is: when is the model complete? When correcting variables have not been taken into account, the system is usually undetermined. If important state variables have been forgotten, they will also not appear in the behavioral model and in any control scheme. For example, if, for simplicity reasons, in the case of distillation the column pressure has not been taken into account as state variable, the pressure will not appear in the behavioral model and pressure control will not be part of the system.

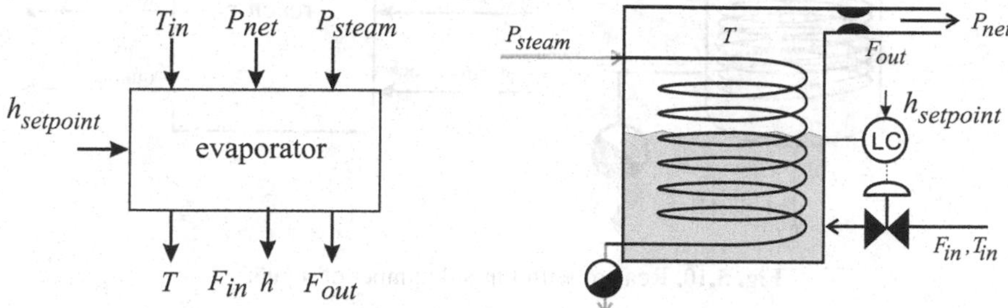

Fig. 3.8. Evaporator accepting the maximum amount of heat, including ideal level control.

Completeness is hard to achieve! By verifying checklists for the manipulated variables and controlled variables, omissions can be avoided as much as possible. When developing the behavioral model, it is still required to feedback to the environmental model.

In the developed model, it is already determined how many degrees of freedom for control are available: the number of correcting variables minus the variable used for control of the throughput. System behavior is determined only if these variables are given. The disturbance variables do not play a role, since they are already determined externally.

After all the variables have been determined (items iv t/m/ vii), one has to check whether the number of correcting variables agrees with the number of behavioral variables. They have to match more or less, since every flow can usually be associated with a (component) mass or energy accumulation. If there are more flows than variables to be controlled, this can be solved by using flow or flow ratio controllers, which adjust the excess number of flows. Figure 3.9 shows this for a polymerization reactor. When the monomer flow is used for throughput control, and the cooling oil flow for pressure control (largest impact on the pressure), then the purge flow can be fixed of controlled in ratio with the monomer flow.

In some cases, the number of controlled variables exceeds the number of manipulated variables. This situation is shown in Fig. 3.10.

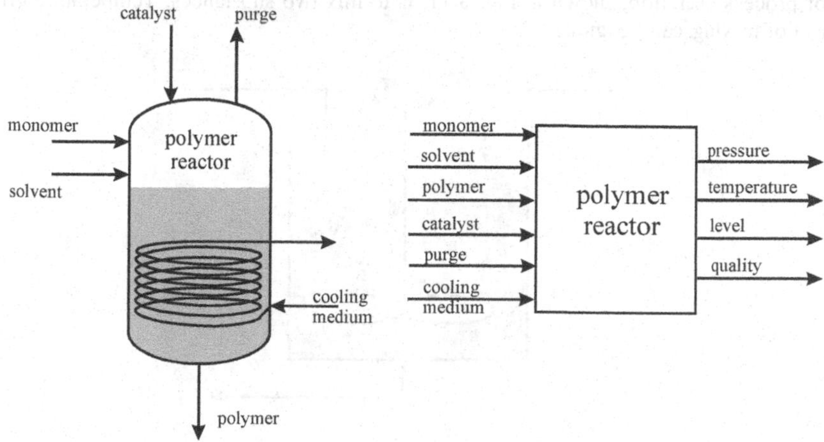

Fig. 3.9. Polymer reactor with many inputs.

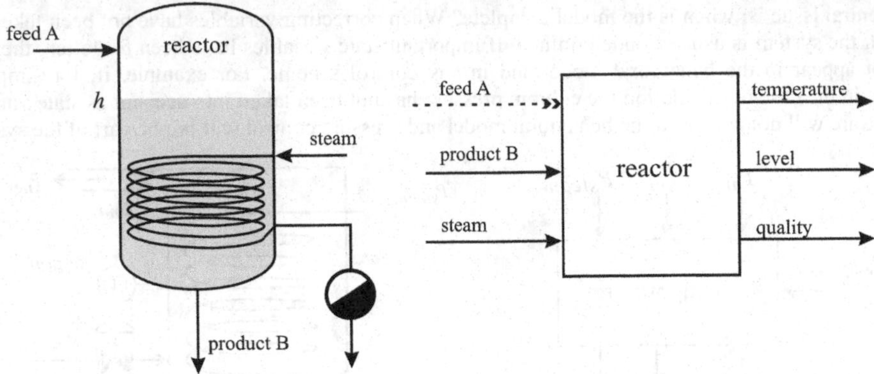

Fig. 3.10. Reactor with limited number of inputs.

The feed flow is used for adjustment of the throughput, consequently the number of controlled variables exceeds the number of manipulated variables by one. One possibility is to leave one controlled variable free, but it is better to make the level or temperature control a slave controller of a quality controller by means of a cascade control structure (see chapter 33). The choice is determined by the variable with the highest impact on quality, either the temperature or the residence time.

For the polymerization reactor shown in Fig. 3.9, quality is represented by the molecular weight distribution of the polymer that is produced. In the separation train this cannot be affected anymore. For most chemical reactors, however, maximum conversion and selectivity are in some cases more important rather than a specific final quality.

3.3 Example: Mixer

A mixer is a simple example that demonstrates the principle of interacting balances. Behavior is determined by two differential equations: one total mass balance and one component balance. If the level is constant by means of an ideal level controller, the number of balances reduces to one. This will be demonstrated in a subsequent chapter. This chapter limits itself to the environmental model.

3.3.1 Operation and Goal of the Process

The goal of process operation, shown in Fig. 3.11, is to mix two substances. Temperature effects, for example heat of mixing, can be ignored.

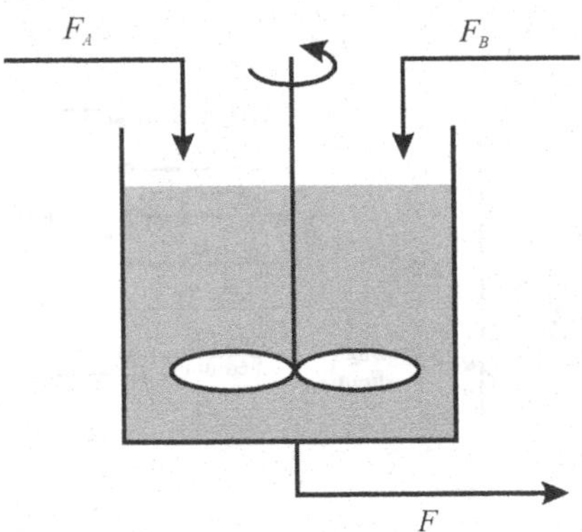

Fig. 3.11. Mixing process for liquids A and B.

3.3.2 Purpose and Level of Detail of the Model

Goal of the study is to investigate the behavior of the concentration to changes in the throughput and mixing ratio. The process is considered to be ideally mixed, i.e. it is assumed that the components can be well mixed and that the mixing time is relatively short compared to the residence time. Hence, the level of detail which will be observed is in the order of the magnitude of the residence time.

3.3.3 Environmental Diagram

The process is not subject to any disturbances, there are three adjustable flows and two measurable state variables, namely the level h and the volume fraction x of the mixture. The final composition is self-regulating, however, the level is not self-regulating, The environmental diagram is shown in Fig. 3.12.

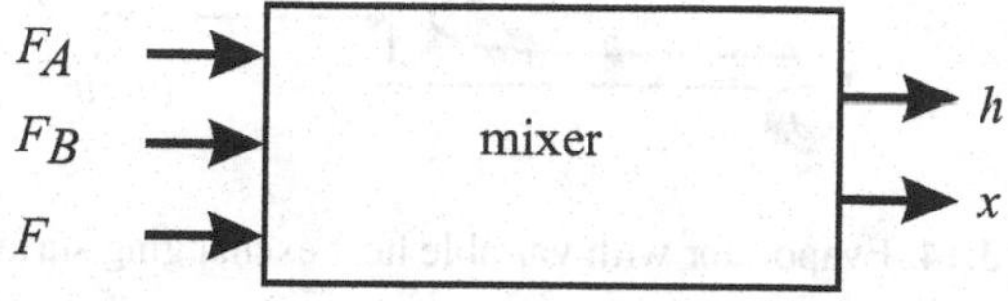

Fig. 3.12. Standaard environmental diagram of the mixer.

If a level controller and flow ratio controller are installed as shown in Fig. 3.13, and if it is assumed that both controllers work relatively fast, the environmental diagram can be simplified. The controllers have become part of the system. The setpoints have become the new degrees of freedom in stead of the flows, which are now adjusted by the controllers. The variables that are controlled may be assumed constant.

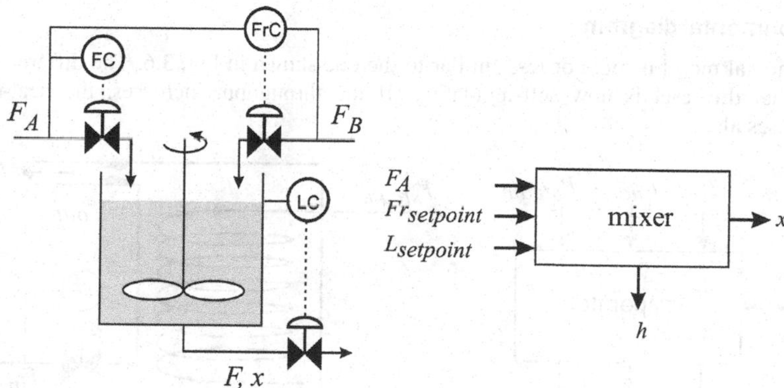

Fig. 3.13. Mixing process with ideal controllers and corresponding environmental diagram.

3.4 Example: Evaporator with Variable Heat Exchanging Surface

The evaporator shown in Fig. 3.14, is a suitable process which can be used to demonstrate the basic principles of interaction between energy and mass balances. This will be worked out further in a subsequent chapter. In this chapter the environmental diagram will be developed. This diagram is almost similar to the one used earlier in this chapter, however, the behavior is strongly different.

3.4.1 Operation and Goal of the Process

The process concerns an evaporator that evaporates a liquid and discharges the vapor to a vapor system. In the evaporator a heat exchanger is mounted from the top to the bottom, as shown in Fig. 3.14. This implies that the heat-exchanging surface will depend on the liquid level. The heat required for the evaporation is delivered by condensing steam.

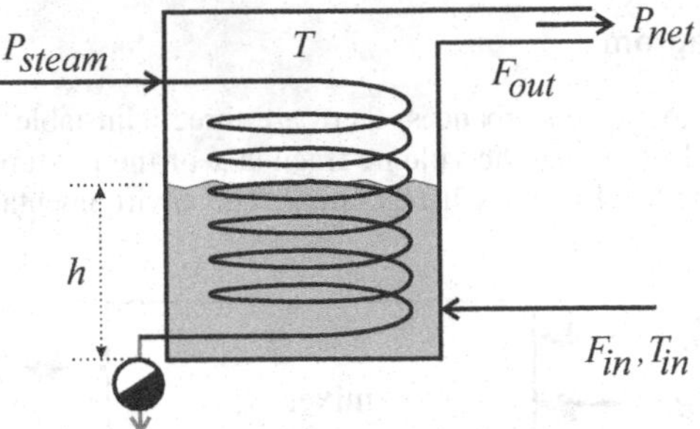

Fig. 3.14. Evaporator with variable heat exchanging surface.

3.4.2 Purpose and Level of Detail of the Model

The purpose of the model is to investigate the behavior of the liquid level and vapor flow to changes in process throughput. The behavior of the level is important in order to know whether it is sufficiently self-regulating. The behavior of the vapor flow can be compared to an evaporator with fixed heat exchanging surface.

3.4.3 Environmental diagram

The environmental model is more or less similar to the one shown in Fig. 3.6. The difference is, that it is assumed that the level is now self-regulating. If the throughput increases, the heat-exchanging surface increases also.

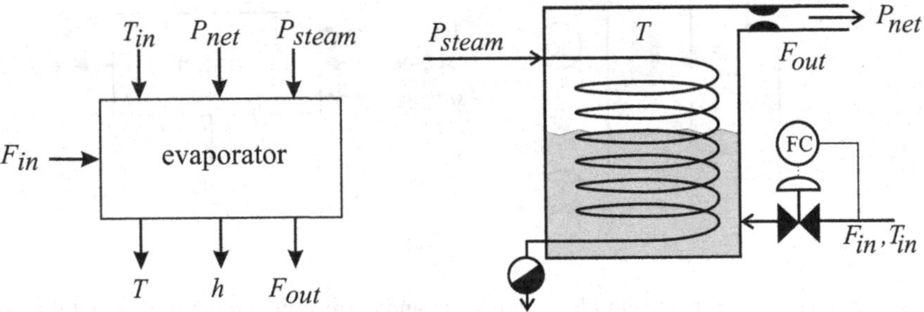

Fig. 3.15. Evaporator with ideal controller.

As mentioned earlier, the evaporator with fixed heat exchanging surface is particularly suitable to deliver a maximum vapor flow. The evaporator with variable heat exchanging surface is particularly suitable to accommodate a variable throughput, since the heat-exchanging surface adapts itself to these changes. A level controller would be incorrect to use. Only if a maximum vapor flow is required, a level controller could ensure that the maximum surface is used. In case of these evaporators it is interesting to investigate how the vapor flow reacts to changes in process throughput.

4 Design for Process Modeling and Behavioral Models

The behavioral model is the internal model of the process, which indicates how the outputs are determined from the inputs, given the assumptions that have been made. The behavioral model will be developed further by using data flow diagrams. This will clarify the model structure and ensure consistency and completeness of the model.

4.1 Behavioral Model

In the environmental model it is determined what the goals of the equipment are and how the equipment functions. This leads to the control goals that have to be realized and shows the relationships (inputs and outputs) between process and environment.

The behavioral model is based on the environmental model and consists of hierarchical data flow diagrams, supplemented by functional descriptions (equations), assumptions and optionally a list (data-dictionary) of all the variables and constants.

The procedure for the derivation of the behavioral model is shown in Table 4.1.

4.1.1 Division into Sub-models

A process unit could consist of one or more pieces of equipment. If an environmental model was developed for a combination of pieces of equipment, as a first step a division could be made into sub-models. One sub-model would then describe the behavior of one piece of equipment. The model that shows the interaction between the sub-models is the hierarchically highest data flow diagram.

Generally, this physical subdivision ensures a minimum number of connections between the sub-models, while the number of functional connections within the sub-model is maximal.

Table 4.1. Design procedure for behavioral model.

	Behavioral model
i.	make division into sub-models
ii.	create list of assumptions
iii.	derive balances for the description of the accumulations
iv.	simplify balances by eliminating products of variables in the derivative term by using the chain rule
v.	add additional equations
vi.	rearrange the system
vii.	develop the data flow diagrams by using the equations
viii.	verify whether the demands from the checklist are met
ix.	under special circumstances, sometimes alternative models can be developed

When dividing functions into hierarchical groups, one should try to minimize the coupling and maximize the functional relationship. This is a rule that should not only be adhered to at the highest level, it should also be a guide when one increases the level of detail of the model.

Process Dynamics and Control: Modeling for Control and Prediction. Brian Roffel and Ben Betlem.

Every function is described by one equation. In order to avoid redundancy, functional descriptions are only necessary at the lowest level where the function is worked out completely. If necessary, one could use a data dictionary, which shows all the variables and constants with a description, values and units, which is common for information systems.

4.1.2 Assumptions

Phenomena can often be ignored or the description can be simplified on the basis of quantitative data. Three types of simplifications can be distinguished:

- contributions of terms to equations may be so small that they can be omitted with respect to other terms.
- for static variables it may often be assumed that the variable is constant in its operating region. For example: the density is assumed to be independent of temperature or composition.
- dynamic variables can sometimes be considered to be static. This means that capacities are ignored, often with respect to other capacities. Examples are: ignoring heat capacities of the wall of a boiler and heating coil with respect to the heat capacity of the liquid, or ignoring the mass of the vapor with respect to the mass of the liquid in an evaporator. Care should be taken when ignoring capacities, since they can affect control behavior in the high-frequency range.

Simplifications are made in order to ensure that one still has insight into the model. Especially when the model has to be solved analytically, simplifications are often required. Simplifications should be judged for every situation separately. For example, in the case of the evaporator the boil-up effect has not been taken into account in order to maintain insight into the model.

For the steps iii) and v) one has to ensure that ignoring phenomena or terms and making simplifications are justified.

Example: Evaporator with constant heat exchanging surface

a. liquid and vapor are fully in equilibrium, $T = T_{liquid} = T_{vapour}$ and $P = P_{liquid} = P_{vapour}$,
b. liquid is ideally mixed,
c. mass of vapor can be neglected compared to the mass of the liquid,
d. the boil-up effect can be ignored, in other words, the vapor volume in the liquid is constant,
e. the steam transfers its heat at the condensation temperature,
f. energy transfer due to super heating can be ignored,
g. the heat of evaporation ΔH_{vap}, the specific heat c_p and the density ρ of the water are temperature independent within the operating region,
h. the heat capacity of the vessel wall can be ignored with respect to the heat capacity of the water (heat capacity = Mc_p),
i. the heat capacity of the heating coil can be ignored,
j. the heat transfer coefficient U_{hta} does not depend on the steam flow,
k. heat losses to the surroundings can be ignored.

4.1.3 Balances

The balances have been discussed in detail in chapter two. Briefly, a balance represents the change in accumulation (Eqn. 4.1). The left-hand side of this equation represents the accumulation of the state variable of a fundamental variable X. This state variable represents the process behavior. Its value is determined by the right-hand side of the equation, which consists of inlet and outlet flows and generation and consumption terms. The outlet flows and the consumption terms give a negative contribution to the accumulation. The magnitude of the accumulated variable can only be affected by the terms in the right-hand side. Usually, the flows in these terms are the manipulated variables which influence the accumulation variable, i.e. the process behavior. The balance equation in its general format is:

$$\frac{d\begin{bmatrix}\textit{accumulation of } X\\ \textit{within the system}\end{bmatrix}}{dt} = \sum\begin{bmatrix}\textit{flow of } X \textit{ into or}\\ \textit{out the system}\end{bmatrix} + \sum\begin{bmatrix}\textit{amount of } X \textit{ generated or consumed}\\ \textit{within the system per unit of time}\end{bmatrix} \tag{4.1}$$

The balances relate to fundamental variables: component mass, energy and impulse. For these variables the laws of conservation hold: in a stationary situation the sum of the right-hand side terms is zero, i.e. the laws of conservation of mass, energy and impulse.

The balances are the kernel of the dynamic model. A brief summary of the results from chapter two is:

Mass balance of a system with one input and one output.
The mass is the accumulating variable. The flows are adjustable variables or disturbances of the system:

$$\frac{dM}{dt} = F_{in} - F_{out} \tag{4.2}$$

Component balance of a reactor, in which component A is formed or reacts with rate of reaction r:

$$\frac{d(c_A V)}{dt} = \pm \sum_{inlet\,/\,outlet\ i} c_{A,i} F_i \pm V \sum_{reactions\ k} r_k \tag{4.3}$$

Energy balance of a reactor, in which kinetic and potential energy remains unchanged:

$$\frac{\partial MH}{\partial T} \cdot \frac{dT}{dt} = \pm \sum_{inlet\,/\,outlet\ i} H_i F_i \pm V \sum_{reactions\ k} r_k \cdot \Delta H_k \pm \sum_{heat\ sources\ l} Q_l \pm \sum_{shaft\ work\ m} W_m \tag{4.4}$$

Example: Evaporator with constant heat exchanging surface
The mass balance is:

$$\rho A_{vessel} \frac{dh}{dt} = F_{in} - F_{out} \tag{4.5}$$

and the energy balance is:

$$c_p \rho A_{vessel} \frac{dhT}{dt} = c_p F_{in} T_{in} - c_p F_{out} T - \Delta H_{vap} F_{out} + U_{hta} A_{hta} (T_{steam} - T) \tag{4.6}$$

4.1.4 Simplification of Balances

The derivatives in the differential equations often contain product terms of variables. These can often be reduced to the derivative of a single variable by substitution of another differential equation. The advantage is that an equation is obtained that is better interpretable. In chapter two this has been applied to the energy and component balances.

Example: Evaporator with constant heat exchanging surface

The energy balance contains a derivative term that is a product of height and temperature. This can be simplified by writing the derivative as:

$$\frac{d(hT)}{dt} = h\frac{dT}{dt} + T\frac{dh}{dt} \tag{4.7}$$

Subsequently, the mass balance can be combined with Eqn. (4.7). This yields an expression for the temperature:

$$c_p \rho A_{vessel} h \frac{dT}{dt} = c_p F_{in}(T_{in} - T) - \Delta H_{vap} F_{out} + U_{hta} A_{hta}(T_{steam} - T) \tag{4.8}$$

This equation can be interpreted much better. In a stationary situation the added heat is equal to the heat that is required to warm up and evaporate the feed.

4.1.5 Additional Equations

The additional equations are usually equations to describe: equilibria, mass and energy transfer, rate of reaction and definition of variables.

Example: Evaporator with constant heat exchanging surface

Additional equations are required to describe F_{out} as a function of the pressure difference between the pressure in the vessel and the pressure in the net:

$$F_{out} = c_v \sqrt{(P - P_{net})} \tag{4.9}$$

Owing to the vapor-equilibrium, the relationship between pressure and temperature can be given by the law of Clausius-Clapeyron:

$$P = P_{ref} exp\left(\frac{\Delta H_{vap}}{R}\left(\frac{1}{T_{ref}} - \frac{1}{T}\right)\right) \tag{4.10}$$

4.1.6 Rearranging the System

The purpose of rearranging the system is to ensure that

- every function produces one variable
- the order of the system is correct

Actions to rearrange the system are: split up, merge and rearrange functions. The order of the system has to be equal to the number of required balances. In the case of the evaporator all equations have already been rearranged. In case of a mixer (next chapter) this is still required. There is an overlap between the balance for the density and the component composition, since the density depends on the component composition.

The way in which the model is going to be used in the future, determines the necessity of this step. If the model is going to be used for an analytical solution or to get insight into the functioning of the process, the highest requirements have to be met with respect to the way in which the model is structured. If, however, the model equations serve as input for simulation studies, the requirements are strongly dependent on the software package that is used.

4.1.7 Data-flow-diagram

The design of data flow diagrams has been discussed in chapter two. They consist of functions represented by a circle, connected by arrows representing the data that is transported. The sign next to the arrow indicates whether the function has a positive or negative contribution to another function. Differential equations can be recognized by a square, indicating a status buffer since such an equation has a memory.
When arranging the system, it is good practice to ensure that cause-effect relationships go from left to right in the diagram.

The input and output arrows can be rearranged in agreement with the environmental model or with the diagram of the hierarchically higher level.

Example: Evaporator with constant heat exchanging surface

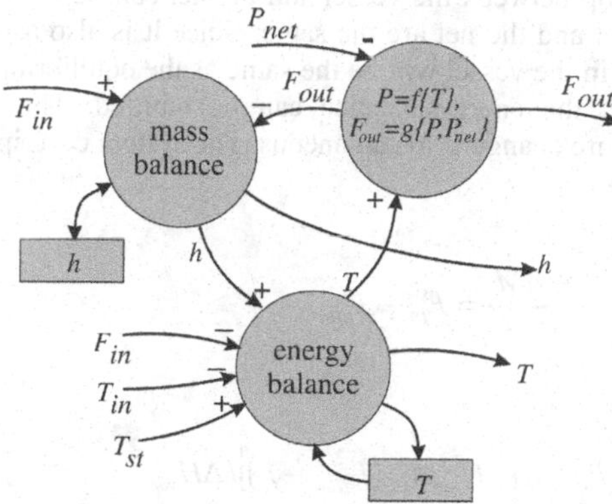

Fig. 4.1. Data-flow-diagram of an evaporator.

If one process in the diagram corresponds to one equation, which produces one variable, a completeness check of the degrees of freedom in the environmental model as well as completeness of the behavioral model is automatically achieved.
This limits the chance that dependencies are omitted when deriving transfer functions; an error that is often made.

4.1.8 Checklist

For the realized model, the following list of checks is important. All requirements should be met.

- for phase equilibria with multiple components, it is important to investigate by how many balances the system is determined,
- two functions may not produce the same output variable,
- all outputs of the model have to be described as functions of the input variables. This means that the inputs and outputs of the environmental model should be the same as the ones of the behavioral model and that all outputs have to be connected via functions with the inputs,
- the number of variables should be equal to the number of equations plus the number of model inputs (disturbances plus external controls).
  $N_{\text{variables}} = N_{\text{equations}} + N_{\text{inputs}}$.
  This requirement is always met if the previous requirement is met, i.e. if every equation produces one variable.

The number of equations should be minimally equal to the number of correcting variables. Every additional variable that is introduced in the internal model, also introduces an additional equation. If the process model contains single input-single output controllers, every correcting variable determines one controlled variable. The new variable (is the new degree of freedom) in the control algorithm is the setpoint of the controller.

4.1.9 Alternative Models

It is often interesting to change the order of a system to see how it affects the dynamics. This way, one gets more feel for what is important in the model, which equation really affects the dynamics.

Example: Evaporator with fixed heat exchanging surface
Suppose that the pressure drop between the vessel and the net can be ignored. Then one may assume that the pressure in the vessel and the net are the same. Since it is also assumed that the system is in equilibrium, the temperature in the vessel will be the same as the equilibrium temperature belonging to that pressure. Consequently, the energy balance can be omitted. The system is immediately at equilibrium and the temperature change is instantaneous. The system description is therefore reduced to the following set of equations
Mass balance:

$$\frac{dM}{dt} = F_{in} - F_{out} \tag{4.11}$$

Static energy balance:

$$F_{out} = \left[c_p F_{in}(T_{in} - T) + U_{hta} A_{hta}(T_{steam} - T)\right] / \Delta H_{vap} \tag{4.12}$$

The temperature depends directly on the pressure on the pressure in the net:

$$P_{net} = P_{ref} \exp\left(\frac{\Delta H_{vap}}{R}\left(\frac{1}{T_{ref}} - \frac{1}{T}\right)\right) \tag{4.13}$$

The pressure build-up, and consequently the temperature build-up, that is required to generate a larger vapor flow, has been omitted.
The system is shown in Fig. 4.2. From a comparison with Fig. 4.1, it is apparent that the relationships have been reversed. The energy balance no longer determines the temperature, but the vapor flow and the equilibrium relationship determines the temperature. Because the time constant for the temperature build-up is in the same order of magnitude as the residence time, it cannot be omitted.

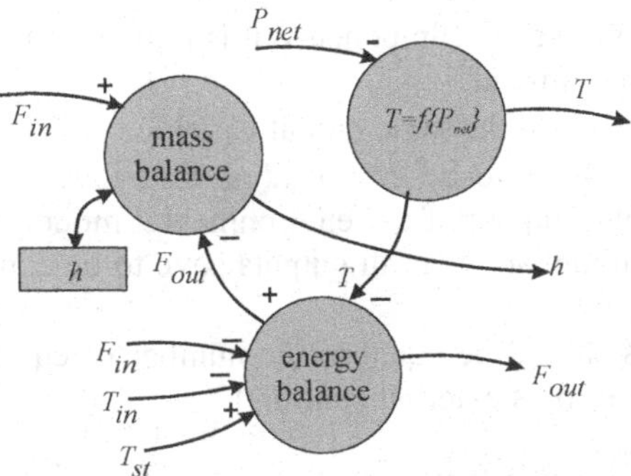

Fig. 4.2. Evaporator with static energy balance.

The evaporator with variable heat exchanging surface has the same behavioral model as shown in Fig, 4.1. The energy balance, however, has to be adapted to accommodate the changing heat transfer area:

$$c_p \rho A_{vessel} \frac{dhT}{dt} = c_p F_{in} T_{in} - c_p F_{out} T - \Delta H_{vap} F_{out} + \frac{U_{hta} A_{hta}}{L_{hta}} h(T_{steam} - T) \tag{4.14}$$

4.2 Example: A Mixer

Consider a mixer, as shown in Fig. 4.3.

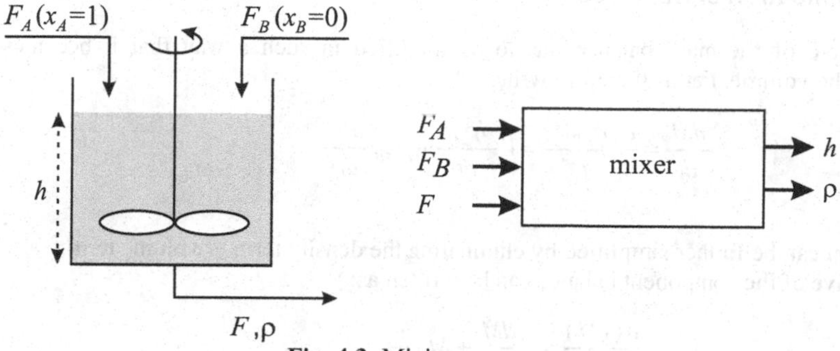

Fig. 4.3. Mixing process.

4.2.1 Division into Sub-systems

Owing to the simplicity of the process this is not required.

4.2.2 Assumptions

a. The fluid is ideally mixed, meaning that the concentration in the vessel is uniform. Only the dynamics in the order of magnitude of the residence time are considered. It is assumed that the mixing time constant is relatively small.
b. Energy effects do not play a role, neither owing to slight temperature differences of the flows, nor owing to the heat of mixing.
c. The compounds A and B have different densities. The density of the mixture depends linearly on the volume fractions of both components.

4.2.3 Balances

When a flow F_A, which contains a volume fraction x, is mixed with a flow F_B with component B, the volume fraction cannot be represented by:

$$x = \frac{F_A}{F_A + F_B} \tag{4.15}$$

This is only statically correct. The liquid, as well as the component in the liquid are accumulating variables. The system has to be described by two differential equations.
A mixer with two components is described by two differential equations. The choice of the balance equation is free, as long as the number is correct and as long as they are independent.
Mass balance:

$$\frac{dM}{dt} = \rho_A F_A + \rho_B F_B - \rho_{AB} F_{AB} \tag{4.16}$$

in which F is the volumetric flow and ρ the density.
Component balance:

$$\frac{d(yM)}{dt} = \rho_A F_A - y\rho_{AB} F_{AB} \tag{4.17}$$

in which y is the weight fraction.

4.2.4 Simplification of Balances

The derivative of the mass balance has to be modified in such a way that it becomes a balance containing the volume. Let us therefore write:

$$\frac{dM}{dt} = \frac{d(\rho_{AB}V)}{dt} = V\frac{d\rho_{AB}}{dt} + \rho_{AB}\frac{dV}{dt} \tag{4.18}$$

The equation can be further simplified by eliminating the density term or volume term.
The derivative of the component balance can be written as:

$$\frac{d(yM)}{dt} = y\frac{dM}{dt} + M\frac{dy}{dt} \tag{4.19}$$

After substitution of the mass balance, a simpler balance is obtained:

$$\rho_{AB}V\frac{dy}{dt} = (1-y))\rho_A F_A - y\rho_B F_B \tag{4.20}$$

4.2.5 Additional Equations

Apart from the two balances, there are two additional algebraic equations. The first one indicates what the relationship is between the density and the weight fraction. Since the volume fraction x is defined as:

$$\rho_{AB} = x \cdot \rho_A + (1-x) \cdot \rho_B \text{ with } \rho_{AB} = \frac{x}{y}\rho_A \text{ and}$$
$$\rho_{AB} = \frac{1-x}{1-y}\rho_B \tag{4.21}$$

the relationship for the weight fraction is:

$$\rho_{AB} = \frac{\rho_A \rho_B}{(1-y)\cdot \rho_A + y \cdot \rho_B} \tag{4.22}$$

The second algebraic equation relates the height in the vessel to the volume:

$$h = \frac{V}{A_{vessel}} \tag{4.23}$$

4.2.6 Rearranging

The equation for the density can be used to rewrite the mass balance or component balance. In this case we rewrite the mass balance as:

$$V\frac{\rho_A-\rho_B}{(1-y)\cdot\rho_A+y\cdot\rho_B}\frac{dy}{dt}+\frac{dV}{dt}=\frac{\rho_A}{\rho_{AB}}F_A+\frac{\rho_B}{\rho_{AB}}F_B-F_{AB} \tag{4.24}$$

After substitution of the component balance one gets:

$$\frac{\rho_A-\rho_B}{(1-y)\cdot\rho_A+y\cdot\rho_B}\left\{(1-x)\frac{\rho_A}{\rho_{AB}}F_A-x\frac{\rho_B}{\rho_{AB}}F_B\right\}+\frac{dV}{dt}= \\ \frac{\rho_A}{\rho_{AB}}F_A+\frac{\rho_B}{\rho_{AB}}F_B-F_{AB} \tag{4.25}$$

which results in:

$$\frac{dV}{dt}=F_A+F_B-F_{AB} \tag{4.26}$$

This is in fact a volume balance, which results in this case because it was assumed that the density depends linearly on the concentrations of the individual components.

4.2.7 Data-flow-diagram

After substitution of the volume balance in the mass balance, a model results with two equations, in which the volume is expressed as a function of the level.
The density balance becomes:

$$A_{vessel}h\frac{d\rho_{AB}}{dt}=F_A(\rho_A-\rho_{AB})+F_B(\rho_B-\rho_{AB}) \tag{4.27}$$

and the volume balance:

$$A_{vessel}\frac{dh}{dt}=F_A+F_B-F_{AB} \tag{4.28}$$

The behavioral model can now be developed:

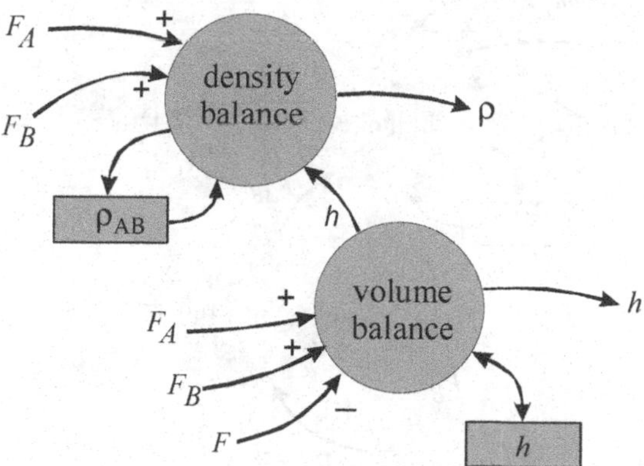

Fig. 4.4. Behavioral model of the mixer.

4.2.8 Checklist

No special items need to be addressed.

4.2.9 Alternative Model

The mixer can also be considered with one additional assumption: the level is ideally controlled, it may therefore be assumed constant

The environmental diagram for this situation is shown in Fig. 4.5.

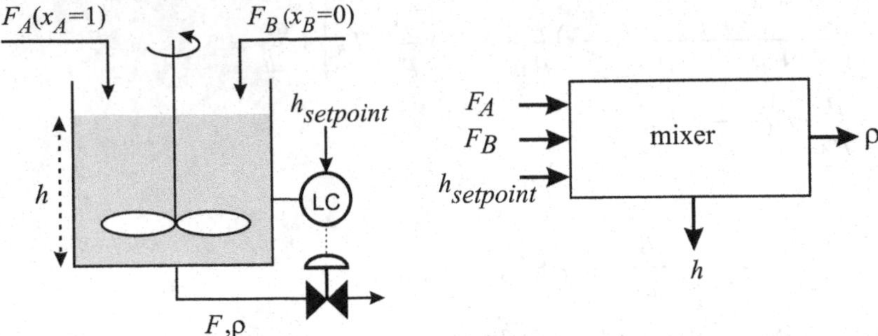

Fig. 4.5. Mixer with controlled level.

The equations can be simplified by this simplification. The volume balance can now be omitted. The order of the system is decreased by one, since addition of the controller deletes one balance. To show the impact of the controller more clearly, it could be introduced separately in the behavioral model. Because of this change, the density balance has been rewritten. The density balance becomes:

$$A_{vessel}\, h_{setpoint}\, \frac{d\rho_{AB}}{dt} = F_A\rho_A + F_B\rho_B - F_{AB}\,\rho_{AB} \tag{4.29}$$

The algebraic equation for the ideal level controller is:

$$F_{AB} = F_A + F_B \tag{4.30}$$

Figure 4.6 shows the behavioral model corresponding to these equations.

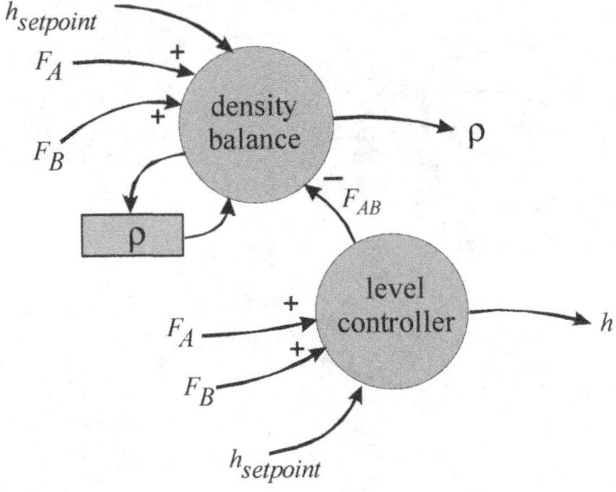

Fig. 4.6. Behavioral model of mixer with ideal level control.

5 Transformation Techniques

In this chapter, two transformation techniques will be introduced that are useful for model analysis. Continuous as well as discrete transfer functions, which are derived by using these transformations, will be discussed and conversion between these domains will be illustrated.

5.1 Introduction

Once a model has been defined, we would like to get more insight into the behavior of the process variable we are interested in when the correcting variable or the disturbance variable changes. This information could be obtained in different ways.

One could define the model and select a suitable simulation technique. Another method could be to try to find an analytical solution of the model. However, one could also use the Laplace transformation technique, which gives a good insight into the response of the process variable. This transformation technique is suitable for the continuous time domain. An advantage over simulation is that the behavior as a function of the (design) parameters can be studied.

The discrete counterpart of the Laplace transform is the z-transform. Both techniques, together with some examples will be discussed in this chapter.

5.2 Laplace Transform

Most chemical engineering processes can be modeled by nonlinear equations. However, to analyze the behavior or to control complex systems, linear methods are more suitable. A nonlinear system can be adequately approximated by a linear system in its operating point. This approximation makes the system much better to understand and process behavior can be more easily interpreted. Additionally, it enables the design of effective linear controllers.

The Laplace transformation is rather traditional and is more a tool for analysis of process behavior, whereas the z-transformation is more modern and is especially suitable for computer implementation of control systems or for discrete simulation. The coherence between the methods is indicated in Fig. 5.1.

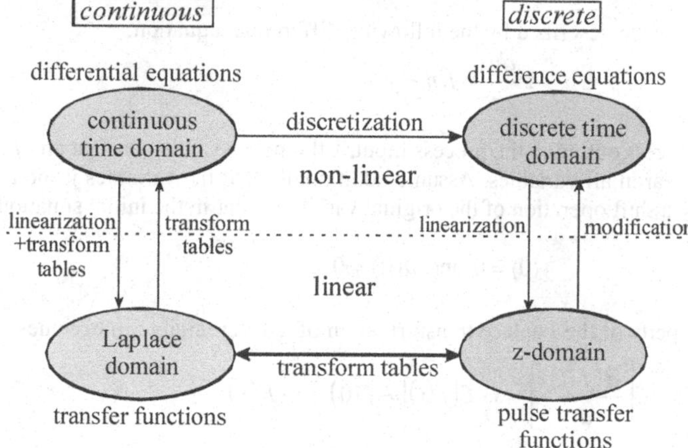

Fig. 5.1. Domain transformations.

Process Dynamics and Control: Modeling for Control and Prediction. Brian Roffel and Ben Betlem.

The Laplace transformation is defined by:

$$\mathcal{L}[f(t)] = \lim_{\substack{\varepsilon \to 0^+ \\ T \to \infty}} \int_{\varepsilon}^{T} f(t)e^{-st}dt = F(s) \tag{5.1}$$

where $f(t)$ is the continuous time domain function. From the definition, it can be noticed that Laplace transformation is a transformation from the time domain to the s-domain, where s is a variable in the complex s-plane: $s = a+jb$.

Some of the properties of Laplace transform that are useful, are the following:

- The Laplace transform is a linear operation:

$$\mathcal{L}[\tau f(t)] = \tau F(s) \tag{5.2}$$
$$\mathcal{L}[f(t)+g(t)] = F(s)+G(s)$$

- Assuming constant steady state, the transformation of a first-order derivative can be given by:

$$\mathcal{L}\left[\frac{df(t)}{dt}\right] = s\mathcal{L}f(t) - f(0) \tag{5.3}$$

in which: $f(0) = f(t=0)$

- In general, the transformation of a nth-order derivative can be given by:

$$\mathcal{L}\left[\frac{d^n(f(t))}{dt^n}\right] = s^n \cdot \mathcal{L}[f(t)] - s^{n-1}f(0) - s^{n-2}f'(0) - \cdots - sf^{(n-2)}(0) - f^{(n-1)}(0) \tag{5.4}$$

- The transformation of an integration can be given by:

$$\mathcal{L}\left[\int f(t)dt\right] = \frac{1}{s}\mathcal{L}[f(t)] = \frac{1}{s}F(s) \tag{5.5}$$

- The transformation of a time delay function can be given by:

$$\mathcal{L}[f(t-\theta)] = e^{-\theta s}F(s) \tag{5.6}$$

Example

Suppose a process can be described by the following differential equation:

$$\tau\frac{dy}{dt} = Ku - y \tag{5.7}$$

in which y is the process output, u the process input, τ the process time constant and K the process gain. This equation is linear in all variables. Assume additionally that the variables y and u by are chosen in such a way, e.g. by a shift operation of the original variables, that in the initial situation it holds that:

$$y(0) = 0 \text{ and } u(0) = 0 \tag{5.8}$$

In that case the property of the Laplace transformation of a differential term becomes:

$$\mathcal{L}\left[\frac{d(f(t))}{dt}\right] = s \cdot \mathcal{L}[f(t)] - f(0) = s \cdot F(s) \tag{5.9}$$

Assume $f(0) = 0$. This is, for example, the case when $f(t)$ is a deviation variable. In that case all function values are relative to a steady state value. Applying Eqns (5.9) and (5.2)–(5.7) results in:

$$\begin{aligned} &\mathcal{L}\left[\tau \frac{d(y(t))}{dt}\right] = \mathcal{L}[Ku(t) - y(t)] \Rightarrow \\ &\tau s \mathcal{L}(y(t)) = K\mathcal{L}[u(t)] - \mathcal{L}[y(t)] \Rightarrow \\ &\frac{y(s)}{u(s)} = \frac{K}{\tau s + 1} \end{aligned} \tag{5.10}$$

which is called the process transfer function. The Laplace transform is an elegant way of writing the process input–output relationships in the form of a transfer function. In later chapters we shall make use of some of the special properties of Laplace transforms.

5.3 Useful Properties of Laplace Transform: limit functions

Back transformation from the transfer function to a differential equation in the time-domain, can be achieved by substitution of $s{\cdot}x(s) = dx(t)/dt$. The variable s is independent of the time and indicates more or less the rate of change. This can be explained by the final-value theorem and the initial value theorem applied to a variation in the input variable of a system. The final steady state value can be found by taking $s \rightarrow 0$:

$$\lim_{t\to\infty}(y(t)) = \lim_{s\to 0} s(G_P \cdot u(s)) \tag{5.11}$$

The initial value at the time the step takes place can be found by taking $s \rightarrow \infty$:

$$\lim_{t\to 0}(y(t)) = \lim_{s\to\infty} s(G_P \cdot u(s)) \tag{5.12}$$

The initial slope of the time response curve can be determined easily, as differentiation corresponds to multiplication by s:

$$\lim_{t\to 0}\left(\frac{dy(t)}{dt}\right) = \lim_{s\to\infty} s(sG_P \cdot u(s)) \tag{5.13}$$

Example

For a step change applied to the first-order system of Eqn. (5.10), the initial and final values become:

$$\lim_{t\to\infty}(y(t)) = \lim_{s\to 0} s\left(\frac{K}{1+\tau s}\frac{A}{s}\right) = KA \tag{5.14}$$

$$\lim_{t\to 0}(y(t)) = \lim_{s\to\infty} s\left(\frac{K}{1+\tau s}\frac{A}{s}\right) = 0 \tag{5.15}$$

$$\lim_{t\to 0}\left(\frac{dy(t)}{dt}\right) = \lim_{s\to\infty} s\left(s\frac{K}{1+\tau s}\frac{A}{s}\right) = \frac{KA}{\tau} \tag{5.16}$$

This is in agreement with the response of first-order differential equation on a step-change with a zero initial value.

$$y(t) = KA(1 - \exp(-t/\tau)) \tag{5.17}$$

$$y'(t) = \frac{KA}{\tau}\exp(-t/\tau) \tag{5.18}$$

5.4 Transfer Functions

In the time domain an input–output relationship can be described by a differential equation relating the output $y(t)$ to the input $u(t)$. Also for multi-variable processes, for all possible input–output combinations such a relationship can be derived. For Laplace-transform linear relationships are required. Linearization of non-linear relationships will be discussed in the next chapter. The general linear input–output relationship in the time domain has the form:

$$\begin{aligned} &a_n \frac{d^n y}{dt}(t) + \cdots + a_1 \frac{dy}{dt}(t) + a_0 y(t) \\ &= b_m \frac{d^m u}{dt}(t-\theta) + \cdots + b_1 \frac{du}{dt}(t-\theta) + b_0 u(t-\theta) \end{aligned} \tag{5.19}$$

in which θ is the time shift between the input and output. The left-hand side of this relationship is the homogenous part of the differential equation, which determines the eigen behavior of the system. All interactive input–output relationships of the same system have an equal homogenous part. The right-hand side determines how the input influences this behavior. The solution of these equations can be found in e.g. Steward (2003).
Suppose that for y and u it holds that:

$$\begin{aligned} y(0) &= y'(0) = y^{(n-1)}(0) = 0 \\ u(0) &= u'(0) = u^{(m-1)}(0) = 0 \end{aligned} \tag{5.20}$$

then the Laplace transform of this input–output relationship becomes:

$$\begin{aligned} &a_n s^n y(s) + \cdots + a_1 s y(s) + a_0 y(s) \\ &= \left(b_m s^m u(s) + \cdots + b_1 s u(s) + b_0 u(s)\right) e^{-\theta s} \end{aligned} \tag{5.21}$$

The conditions of Eqn. (5.20) for y and u are fulfilled when y and u are deviation variables, which represent the deviation of a variable compared to a steady state value. From Eqn. (5.21) it can be concluded that the Laplace operator s can be associated with a "rate of change" operator.

A transfer function is defined as the ratio between the output and the input in the Laplace domain:

$$\frac{y(s)}{u(s)} = \frac{\left(b_m s^m + \cdots + b_1 s + b_0\right) e^{-\theta s}}{a_n s^n + \cdots + a_1 s + a_0} = G_P(s) \tag{5.22}$$

in which G_P concerns the transfer function for the process P for the particular input–output relationship of u to y. The transfer function can be written as a ratio of two polynomials in s:

$$G_P(s) = \frac{y(s)}{u(s)} = \frac{Q(s)\, e^{-\theta s}}{P(s)} \tag{5.23}$$

For physically realistic systems the power of s of the denominator P should be higher or at least equal to the power of s of the numerator Q. The power in an exponent of a delay term $e^{-\theta s}$ is always negative, since physically predictive systems cannot be realized. One of the mathematical advantages of Laplace transformation is, that it enables the input–output description of serial and parallel subsystems and description of signals easily.

Laplace transformation can be used to analyze system behavior for control purposes and to solve a system analytically for different kinds input signals.

For several types of time function, such as pulse, step, ramp, wave changes, the Laplace transformations are shown in Table 5.1. The time functions belonging to a particular Laplace transform are given in Table 5.2.

Table 5.1. Laplace transforms of various functions

Signal type	Continuous time function: $f(t)$ with $t \geq 0$	Laplace transform
Unit pulse with time A and amplitude 1/A	$\delta_A(t)$	$\frac{1}{A}\frac{1-e^{-sA}}{s}$
Unit pulse with $A \to \infty$	$\delta(t_0)$	1
Unit step	1	$\frac{1}{s}$
Ramp	t	$\frac{1}{s^2}$
Power	t^n	$\frac{n!}{s^{n+1}}$
Wave	$\sin(\omega t)$, $\cos(\omega t)$	$\frac{\omega}{s^2+\omega^2}$, $\frac{s}{s^2+\omega^2}$
Exponential	$e^{-t/\tau}$	$\frac{\tau}{\tau s+1}$
Exponential with power	$t^n e^{-t/\tau}$	$\frac{n!\cdot\tau^{n+1}}{(\tau s+1)^{n+1}}$
Exponential with wave	$e^{-t/\tau}\sin(\omega t)$	$\frac{\tau^2\omega}{(\tau s+1)^2+\tau^2\omega^2}$

Both polynomials $P(s)$ and $Q(s)$ can in principle be factored into their roots:

$$G_P(s) = \frac{Q(s)\,e^{-\theta s}}{P(s)} = \frac{(s-n_m)(s-n_{m-1})\cdots(s-n_1)\,e^{-\theta s}}{(s-p_n)(s-p_{n-1})\cdots(s-p_1)} \tag{5.24}$$

An nth order polynomial has n roots. The roots of $P(s)$ are called the poles and the roots of $Q(s)$ the zeros. The roots may have imaginary parts.

For a transfer function without time delay θ, the solution in the time domain can be found by writing $y(s)$ as function of $u(s)$. Because $u(s)$ can also be written as a ratio of polynomials in s, it follows that:

$$y(s) = G_P(s)u(s) = \frac{Q(s)}{P(s)}u(s) = \frac{Q(s)}{P(s)}\frac{r(s)}{q(s)} \tag{5.25}$$

Table 5.2. Inverse Laplace transforms.

Process type	Continuous process $G_P(s)$	Continuous time function: f(t)
	$\frac{1}{(\tau s+1)}$	$\frac{1}{\tau}e^{-t/\tau}$
	$\frac{1}{(\tau_1 s+1)(\tau_2 s+1)}$	$\frac{\tau_1 e^{-t/\tau_1} - \tau_2 e^{-t/\tau_2}}{\tau_1 - \tau_2}$
	$\frac{1}{(\tau s+1)^n}$	$\frac{1}{(n-1)!}\frac{t^{n-1}}{\tau^n}e^{-t/\tau}$
delay	$\lim_{n\to\infty}\frac{1}{(\frac{\theta}{n}s+1)^n} = e^{-\theta s}$	$H(t-\theta)$
step response second-order $\zeta>1$	$\frac{1}{s(\tau_1 s+1)(\tau_2 s+1)}$	$1-\frac{\tau_1^2 e^{-t/\tau_1} - \tau_2^2 e^{-t/\tau_2}}{\tau_1 - \tau_2}$
step response n^{th} order $\zeta=1$	$\frac{1}{s(\tau s+1)^n}$	$1-\left(1+\frac{t}{\tau}+\frac{1}{2!}\left(\frac{t}{\tau}\right)^2+\cdots+\frac{1}{(n-1)!}\left(\frac{t}{\tau}\right)^{n-1}\right)e^{-t/\tau}$
step response second-order $0\le\zeta<1$	$\frac{1}{s(\tau^2 s^2+2\zeta\tau s+1)}$	$1-\frac{e^{-\zeta t/\tau}}{\sqrt{1-\zeta^2}}\sin(\omega t+\tan^{-1}(\omega)),\ \omega=\frac{\sqrt{1-\zeta^2}}{\tau}$

Using partial-fractions expansion (also called Heaviside expansion) a sum of terms is obtained:

$$\begin{aligned} y(s) = \frac{Q(s)}{P(s)}\frac{r(s)}{q(s)} &= \frac{C_i}{(s-p_{n+k})} + \frac{C_{i-1}}{(s-p_{n+k-1})} + \cdots + \frac{C_1}{(s-p_1)} \\ &= \sum_{i=1}^{n+k}\frac{C_i}{(s-p_i)} \end{aligned} \tag{5.26}$$

$n+k$ is the order of the polynomial Pq, as P is n^{th} order and q is k^{th} order. This agrees with the super-position property of differential equations. The total solution is the sum of the individual solutions. The solution in the time domain follows from back-transformation of the individual terms:

$$y(t) = \mathcal{L}^{-1}\left(\frac{Q(s)}{P(s)}u(s)\right) = \sum_{i=1}^{n+k} C_i e^{p_i t} \tag{5.27}$$

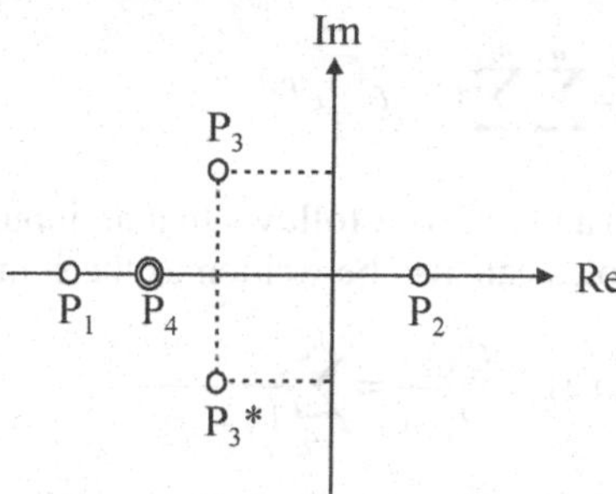

Fig. 5.2. Location of poles in the complex plane.

Depending on the sign and composition of the poles, they entail different properties of the solutions:

- poles with a negative real part (P_1 in Fig. 5.2)
  When the real part of the pole is negative than the solution $\exp(p_i t) \rightarrow 0$ for $t \rightarrow \infty$. The result is a damped signal, moving the system to a new stationary state.
- poles with a positive real part(P_2 in Fig. 5.2)
  When the real part of the pole is positive than the solution $\exp(p_i t) \rightarrow \infty$ for $t \rightarrow \infty$. The result is an always increasing signal.
- poles with an imaginary part (P_3 in Fig. 5.2)
  In this case the poles appear always in conjugate pairs consisting of a complex and its adjunct complex pole, resulting in an oscillatory behaviour,

$$\begin{aligned} y(t) &= \mathcal{L}^{-1}\left(\frac{C}{(s-p)} + \frac{C^*}{(s-p^*)}\right) \\ &= Ce^{pt} + C^* e^{p^*t} \end{aligned} \tag{5.28}$$

Assume $p = p_{\text{Re}} + jp_{\text{Im}}$, then by means of the Euler equation $e^{j\omega t} = \cos(\omega t) + j\sin(\omega t)$ it can be derived that:

$$\begin{aligned} y(t) &= Ce^{(p_{\text{Re}}+jp_{\text{Im}})t} + C^* e^{(p_{\text{Re}}-jp_{\text{Im}})t} \\ &= Ce^{p_{\text{Re}}t}\left[\cos(p_{\text{Im}}t) + j\sin(p_{\text{Im}}t)\right] + C^* e^{p_{\text{Re}}t}\left[\cos(-p_{\text{Im}}t) + j\sin\right. \\ &= \left[C + C^*\right]e^{p_{\text{Re}}t}\cos(p_{\text{Im}}t) + \left[jC - jC^*\right]e^{p_{\text{Re}}t}\sin(p_{\text{Im}}t) \\ &= C_1 e^{p_{\text{Re}}t}\cos(p_{\text{Im}}t) + C_2 e^{p_{\text{Re}}t}\sin(p_{\text{Im}}t) \end{aligned} \tag{5.29}$$

The solution consists of an oscillating part, with a wave frequency depending on the size of the imaginary part of the poles pair and a damping part that depends on the sign and the size of the real part of the poles pair.

- coinciding poles (P_4 in Fig. 5.2)
  When a pole occurs twice, then two solutions can be found:

$$\begin{aligned} y(t) &= \mathcal{L}^{-1}\left(\frac{Q(s)}{(s-p)^2}\right) = \mathcal{L}^{-1}\left(\frac{C_1}{(s-p)} + \frac{C_2}{(s-p)^2}\right) \\ &= C_1 e^{pt} + C_2 t e^{pt} \end{aligned} \tag{5.30}$$

In general, when $P(s)$ is a nth order polynomial with n poles p_k of which p_k occurs n_k times as coinciding pole, then the solution becomes:

$$y(t)=\sum_{k=1}^{n}\sum_{j=1}^{n_k}C_{k,j}t^{j-1}e^{p_k t} \tag{5.31}$$

In accordance with Eqn. (5.25) and (5.26) it follows that an input–output relationship defined by a transfer function with n poles can also be written as the sum of these poles:

$$\frac{y(s)}{u(s)}=G_P(s)=\frac{Q(s)}{P(s)}=\sum_{i=1}^{n}\frac{C_i}{(s-p_i)} \tag{5.32}$$

A minimum requirement for asymptotic stability of this relationship is that all n poles have negative real parts. In that case for a small deviation the system will return to its stationary point y=0. Because, all interactive input–output relationships of a system have an equal denominator $P(s)$, this consequently means that all input–output relationships of the system are stable. We may conclude that a system is stable if the relationship:

$$P(s)=0 \tag{5.33}$$

has negative real roots. The denominator polynomial is the Laplace transformed homogenous part of the differential Eqn. (5.19) and corresponds to the characteristic polynomial of the original differential equation and Eqn. (5.33) corresponds to the characteristic equation. The poles of Eqn. (5.32) are equal to the eigen values of the differential equation. In Chapter 7 the stability of a system in an operating point will be derived directly from the homogenous part of the differential equation.

In the description of the solution or the discussion of stability, the time delay plays no role. For a time delay θ the solution becomes:

$$y(t)=\sum_{k=1}^{n}\sum_{j=1}^{n_k}C_{k,j}t^{j-1}e^{p_k(t-\theta)} \tag{5.34}$$

Example

In case of a step change in u with magnitude A, $\delta u(s)$=A/s. Equation (5.10) becomes then:

$$y(s)=\frac{K}{1+\tau s}u(s)=\frac{K}{1+\tau s}\cdot\frac{A}{s}=AK\left(\frac{1}{s}-\frac{\tau}{1+\tau s}\right) \tag{5.35}$$

Back transformation into the time domain gives:

$$y=\left(1-e^{-t/\tau}\right)K\,A \tag{5.36}$$

The Laplace domain offers the possibility to compose the overall transfer functions of a complex system, which consists of several interconnected subsystems, in an easy way. An example is shown in Fig. 5.3.

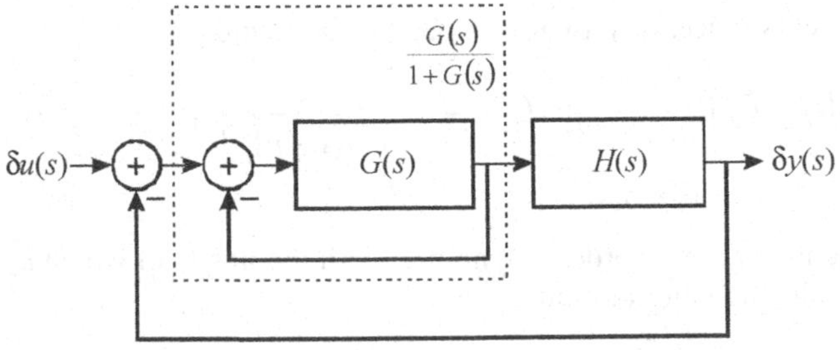

Fig. 5.3. Block diagram in Laplace domain.

The transfer function of the system shown in Fig. 5.3 can be found by exploring all possible paths.

$$y(s) = H(s) \cdot \frac{G(s)}{1+G(s)} (u(s) - y(s)) \tag{5.37}$$

This leads to the following transfer function between δy and δu:

$$\frac{y(s)}{u(s)} = \frac{H(s)G(s)}{1+G(s)+H(s)G(s)} \tag{5.38}$$

In general the transfer function of a system with feedback can be given by:

$$G_{system}(s) = \frac{direct\ transfer\ function\ from\ u\ to\ y}{1+closed\ loop\ transfer\ functions} \tag{5.39}$$

For a more elaborate treatment of Laplace transform and the initial and final value theorems, the reader is referred to the many textbooks on process control.

Another transfer function often encountered in the process industries is a so-called process dead-time:

$$y(s) = e^{-\theta s} u(s) \tag{5.40}$$

The variations in y show a delay of time θ with respect to u.

The combination of transfer functions shown in Eqn. (5.10) and (5.40) are often encountered in the process industry. Equation (5.10) could be the response of a well-mixed vessel, whereas Eqn. (5.40) is typically found in case of fluid flow through pipelines. For a pipeline feeding into a vessel the overall transfer function between process output and process input could then be written as:

$$\frac{y}{u} = \frac{K e^{-\theta s}}{1+\tau s} \tag{5.41}$$

5.5 Discrete Approximations

As mentioned before, the Laplace transform is a convenient way of representing a process model. However, in today's world of computers discrete time representations of systems is preferred. The simplest way is a first-order discretization:

$$\frac{dx}{dt} \approx \frac{x_k - x_{k-1}}{\Delta t} \tag{5.42}$$

and

$$\frac{d^2x}{dt^2} \approx \frac{x_k - 2x_{k-1} + x_{k-2}}{\Delta t^2} \tag{5.43}$$

Note: x is a deviation variable, Δt is the discretization interval.
For a first-order system of Eqn. (5.10) one obtains:

$$\frac{\tau}{\Delta t}[y_k - y_{k-1}] = K\, u_{k-1} - y_k \tag{5.44}$$

The input variable u is given a time index k–1 rather than k, since we know that it is physically impossible for u to have an impact on y at time k. Equation (5.44) can be written as:

$$y_k = K_{lag}\ y_{k-1} + K(1 - K_{lag})u_{k-1} \tag{5.45}$$

in which $K_{lag} = \tau/(\tau+\Delta t)$ and Δt = discretization interval. The new state depends for K_{LAG} on the last state and for 1-K_{LAG} on the latest input.

The expression for K_{lag} is valid as long as Δt is relatively small compared with τ. If this is not the case, a better expression for K_{lag} is:

$$K_{lag} = e^{-\Delta t/\tau} \tag{5.46}$$

Example

The difference between the determinations of K_{lag} is illustrated in Table 5.3; where Δt=5.

Table 5.3. Comparison of expressions for K_{lag}.

τ	$e^{-\Delta t/\tau}$	$\tau/(\tau+\Delta t)$
15	0.72	0.75
20	0.78	0.80
25	0.82	0.83
30	0.86	0.86

5.6 *z*-Transforms

Equations, such as Eqn. (5.45), are not very convenient for representation in a control block diagram. The *z*-transform is an elegant way of representing discrete transfer functions. The backward shift operator z^{-1} is defined as:

$$y(k-1) = z^{-1}\ y(k) \tag{5.47}$$

Introduction into Eqn. (5.45), gives:

$$y(k) = K_{LAG}\ z^{-1}\ y(k) + K\,(1 - K_{LAG})z^{-1}\ u(k) \tag{5.48}$$

or

$$\frac{y(k)}{u(k)} = \frac{K\,(1 - K_{LAG})\,z^{-1}}{1 - K_{LAG}\ z^{-1}} \tag{5.49}$$

The *z*-transform for a pure time delay θ is:

$$\frac{y(k)}{u(k)} = z^{-f-1} \tag{5.50}$$

where $f = \theta/T$, the number of whole sample periods of delay. T is the sampling time, which is generally chosen equal to the sampling time $\Delta\, t$.

To demonstrate the transformation from Laplace to *z*-notation consider the following first-order transfer function with dead time:

$$\frac{\delta\, y}{\delta\, u} = \frac{2.2\, e^{-10s}}{1 + 20s} \tag{5.51}$$

If the discretization interval is 2 minutes, $K_{LAG} = e^{-2/20} = 0.90$, the *z*-transform is:

$$\frac{y(k)}{u(k)} = \frac{2.2(1-0.90)\,z^{-6}}{1-0.90z^{-1}} = \frac{0.22\,z^{-6}}{1-0.90z^{-1}} \tag{5.52}$$

This function is called the pulse transfer function and as shown above, it can be derived directly from a discrete equation. The relationship $y(k)/u(k)$ is identical to the relationship $y(z)/u(z)$. $y(k)$ stands for the magnitude of signal y at an arbitrarily sample moment kT, whereas $y(z)$ stands for the total sampled signal y as defined.

Figure 5.1 indicates that there is another route to obtain this function, starting with the transfer function in the Laplace domain. Consider the hold-process combination shown in Fig. 5.4.

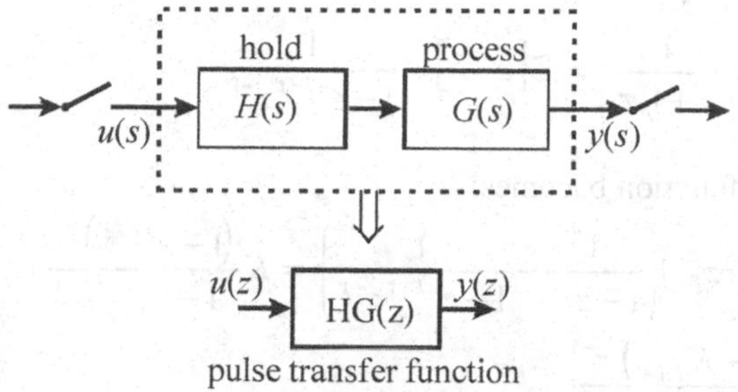

Fig. 5.4. Pulse transfer function derivation from process with hold element.

The process $G(s)$ is a continuous, but is sampled and controlled discretely as is usual for computer controlled systems. To supply the process with a continuous input signal, the discrete signal is equipped with a hold element $H(s)$. The total pulse transfer function is usually denoted as $HG(z)$ and can be derived according to Eqn. (5.53).

$$\frac{y(z)}{u(z)} = \mathcal{Z}\left[\mathcal{L}^{-1}[H(s)G(s)]\right] = HG(z) \tag{5.53}$$

Several kinds of hold elements exist. The best known and most used is the zero-order hold with the transfer function:

$$H(s) = \frac{1-e^{-Ts}}{s} \tag{5.54}$$

This function holds the signal for one control time step with length T.

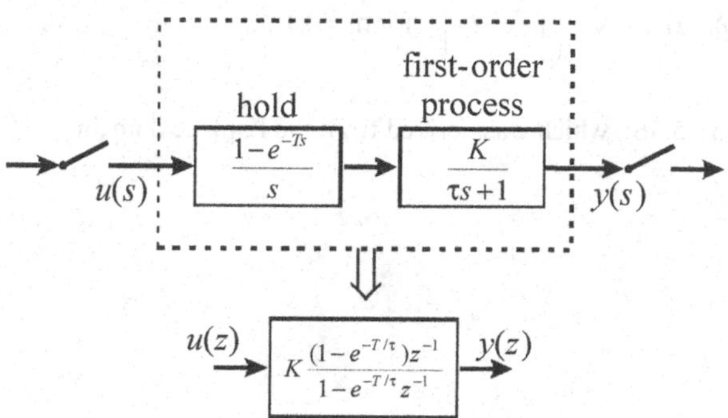

Fig. 5.5. Pulse transfer function derivation for a first-order lag system.

In case of a first-order time lag of Eqn. (5.10), the pulse transfer function can be found as follows (Fig. 5.5).

$$HG(z) = Z\left[\frac{1-e^{-Ts}}{s}\frac{K}{\tau s+1}\right] = \left(1-z^{-1}\right)Z\left[\frac{K}{s(\tau s+1)}\right]$$
$$= K\left(1-z^{-1}\right)Z\left[\frac{1}{s}-\frac{1}{s+1/\tau}\right] \tag{5.55}$$

The transformations can be derived from the standard transforms:

$$Z\left[\frac{1}{s}\right] = Z[\text{unit step}] = \frac{1}{1-z^{-1}}$$
$$Z\left[\frac{1}{s+1/\tau}\right] = Z\left[e^{-t/\tau}\right] = \frac{1}{1-e^{-T/\tau}z^{-1}} \tag{5.56}$$

and the pulse transfer function becomes:

$$HG(z) = K\left(1-z^{-1}\right)\left\{\frac{1}{1-z^{-1}}-\frac{1}{1-e^{-T/\tau}z^{-1}}\right\} = K\frac{\left(1-e^{-T/\tau}\right)z^{-1}}{1-e^{-T/\tau}z^{-1}}$$
$$= \frac{K\left(1-K_{LAG}\right)z^{-1}}{1-K_{LAG}\,z^{-1}} \tag{5.57}$$

This result agrees with Eqn. (5.49), which was derived from discretization.

For several types of time functions, such as pulse, step, ramp, wave changes, etc., the z-transformations are given.

Example

In case of a step change in y with magnitude A, $y(z)=A/(1-z^{-1})$. Applied to the first-order pulse response function:

$$\delta T(z) = K_P\frac{\left(1-e^{-T/\tau}\right)z^{-1}}{1-e^{-T/\tau}z^{-1}}\delta T_{in}(z) = K_P\frac{\left(1-e^{-T/\tau}\right)z^{-1}}{1-e^{-T/\tau}z^{-1}}\frac{A}{1-z^{-1}}$$
$$= KA\left(\frac{1}{1-z^{-1}}-\frac{1}{1-e^{-T/\tau}z^{-1}}\right) \tag{5.58}$$

The inverse z-transform yields:

$$\delta y(nT) = K\,A\left(1-e^{-\frac{nT}{\tau}}\right) \quad \text{for n} = 0,1,2,... \tag{5.59}$$

This is the same as Eqn. (5.36), which was derived from the Laplace domain.

Table 5.4. Pulse transfer functions for the responses of Fig. 5.6. ($\delta_i = e^{-T/\tau_i}$).

Continuous process $G_P(s)$	Pulse transfer functions	Relationships between discrete and continuous parameters
A1. first-order with delay $\theta = k \cdot T$ $\dfrac{K_P \cdot e^{-\theta s}}{\tau s + 1}$	$\dfrac{a_0}{1-b_1 z^{-1}} z^{-k-1}$	$a_0 = K_p(1-\delta)$ $b_1 = \delta$
A2. first-order with delay $\theta = kT - \Delta$ $\dfrac{K_P \cdot e^{-\theta s}}{\tau s + 1}$	$\dfrac{a_0 - a_1 z^{-1}}{1-b_1 z^{-1}} z^{-k-1}$	$a_0 = K_p(1-\delta^{1-\Delta})$, $a_1 = a_0 - K_p(1-b_1)$ $b_1 = \delta$
B. second-order overdamped, with delay $\dfrac{K_P \cdot e^{-\theta s}}{(\tau_1 s + 1)(\tau_2 s + 1)}$	$\dfrac{a_0 - a_1 z^{-1}}{1-b_1 z^{-1} - b_2 z^{-2}} z^{-k-1}$	$a_0 = K_p\left(\dfrac{\tau_1(1-\delta_1) - \tau_2(1-\delta_2)}{\tau_1 - \tau_2}\right)$ $a_1 = a_0 - K_p(1-b_1-b_2)$ $b_1 = \delta_1 + \delta_2$, $b_2 = -\delta_1\delta_2$
C. second-order underdamped with delay $\dfrac{K_P \cdot e^{-\theta s}}{\tau^2 s^2 + 2\xi\tau s + 1}$	$\dfrac{a_0 - a_1 z^{-1}}{1-b_1 z^{-1} - b_2 z^{-2}} z^{-k-1}$	$a_0 = K_p\left(1 - \dfrac{\delta}{\eta}\sin\left(\dfrac{T}{\tau}\eta + \cos^{-1}(\xi)\right)\right)$ $a_1 = a_0 - K_p(1-b_1-b_2)$ $b_1 = 2\delta^{\xi}\cos\left(\dfrac{T}{\tau}\eta\right)$, $b_2 = -\delta^{2\xi}$ $\eta = \sin(\cos^{-1}(\xi))$
D. second-order with nonminimum phase with delay $\dfrac{K_P(\tau_0 s + 1) \cdot e^{-\theta s}}{(\tau_1 s + 1)(\tau_2 s + 1)}$	$\dfrac{a_0 - a_1 z^{-1}}{1-b_1 z^{-1} - b_2 z^{-2}} z^{-k-1}$	$a_0 = K_p\left(\dfrac{\tau_1(1-\delta_1) - \tau_2(1-\delta_2) + \tau_0(\delta_1 - \delta_2)}{\tau_1 - \tau_2}\right)$ $a_1 = a_0 - K_p(1-b_1-b_2)$ $b_1 = \delta_1 + \delta_2$, $b_2 = -\delta_1\delta_2$
E. lead-lag with delay $\dfrac{K_P(\tau_0 s + 1) \cdot e^{-\theta s}}{(\tau_1 s + 1)}$	$\dfrac{a_0 - a_1 z^{-1}}{1-b_1 z^{-1}} z^{-k-1}$	$a_0 = K_p\left(\dfrac{\tau_0}{\tau_1}\delta\right)$, $a_1 = a_0 - K_p(1-b_1)$ $b_1 = \delta$

Example

In case of a step change in y with magnitude A, $y(z) = A/(1-z^{-1})$. Applied to the first-order pulse response function:

$$\delta T(z) = K_P \frac{\left(1-e^{-T/\tau}\right) z^{-1}}{1-e^{-T/\tau} z^{-1}} \delta T_{in}(z) = K_P \frac{\left(1-e^{-T/\tau}\right) z^{-1}}{1-e^{-T/\tau} z^{-1}} \frac{A}{1-z^{-1}}$$
$$= KA\left(\frac{1}{1-z^{-1}} - \frac{1}{1-e^{-T/\tau} z^{-1}}\right) \tag{5.60}$$

The inverse z-transform yields:

$$\delta y(nT) = K\,A\left(1-e^{-\frac{nT}{\tau}}\right) \quad \text{for } n = 0,1,2,... \tag{5.61}$$

This is the same as Eqn. (5.36), which was derived from the Laplace domain.

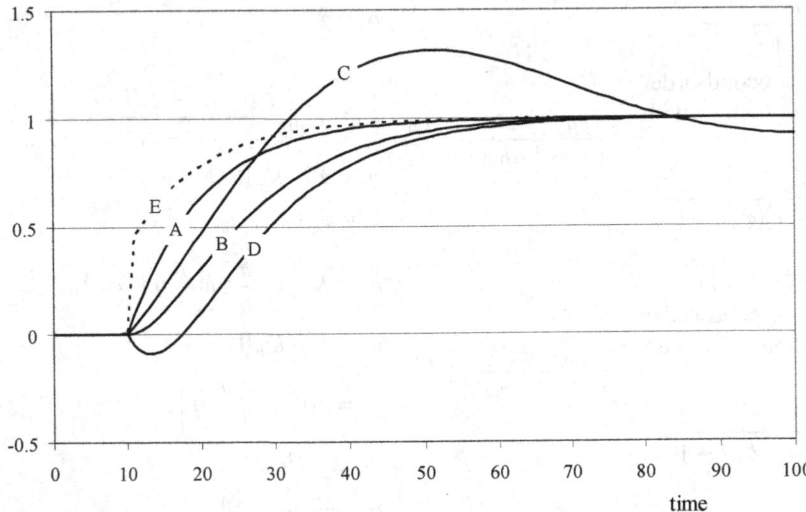

Fig. 5.6. Step responses of systems mentioned in Table 5.4.

Table 5.4 lists several pulse transfer functions of other type of processes. Other interesting properties of the impulse transfer functions are:

- For a series of continuous processes, the pulse transfer function can be characterized by:

$$G_1G_2...G_N(z) = \mathcal{Z}[G_1(s)\cdot G_2(s)\cdot ...G_N(s)] \neq G_1(z)\cdot G_2(z)\cdot ...G_N(z) \tag{5.62}$$

thus:

$$HG(z) \neq H(z)\cdot G(z) \tag{5.63}$$

- The initial and final value theorem for the first-order system become:

$$\lim_{t\to 0}(\delta y(t)) = \lim_{z\to\infty}\left(K\frac{\left(1-e^{-T/\tau}\right)z^{-1}}{1-e^{-T/\tau}z^{-1}}\frac{A}{1-z^{-1}}\right) = 0 \tag{5.64}$$

and

$$\lim_{t\to\infty}(y(t)) = \lim_{z\to 1}(1-z^{-1})\left(K\frac{\left(1-e^{-T/\tau}\right)z^{-1}}{1-e^{-T/\tau}z^{-1}}\frac{A}{1-z^{-1}}\right) = K\,A \tag{5.65}$$

respectively.

- The initial slope can be found indirectly, by first deriving the pulse response function of the differentiated first-order function. In the Laplace domain, differentiation corresponds to multiplying by s.

$$\begin{aligned}\lim_{t\to 0}\left(\frac{\delta y(t)}{dt}\right) &= \lim_{z\to\infty}\left(Z\left[\frac{1-e^{-Ts}}{s}s\frac{K}{\tau s+1}\right]\cdot\frac{A}{1-z^{-1}}\right)\\ &= \lim_{z\to\infty}\left((1-z^{-1})K\frac{1/\tau}{1-e^{-T/\tau}z^{-1}}\frac{A}{1-z^{-1}}\right) = \frac{K\,A}{\tau}\end{aligned} \tag{5.66}$$

which agrees with the result from the Laplace domain.

Example

An input–output relationship for a process is given by:

$$11\frac{dy}{dt} = 2u - y \tag{5.67}$$

What are the equations for the Laplace and z-transform?

Applying the equations derived in this chapter, we get for the Laplace transform:

$$\frac{\delta y}{\delta u} = \frac{2}{11s+1} \tag{5.68}$$

where d indicates the deviation from steady state.

For the z-transform, if we choose a discretization interval of two time steps:

$$\frac{y_k}{u_k} = \frac{2(1-0.834)z^{-1}}{1-0.834\,z^{-1}} = \frac{0.332\,z^{-1}}{1-0.834\,z^{-1}} \tag{5.69}$$

References

Steward, J. (2003) *Calculus*, Thomson, Brooks and Cole, Belmont, CA, USA.

6 Linearization of Model Equations

In this chapter another useful tool will be described for model analysis: linearization of model equations. When combined with Laplace or z-transform, the process model can be decomposed into linear transfer functions and the dynamic response of the variable of interest can easily be obtained and compared with the solution from a simulation study. The advantage of working with linear transfer functions is that process variable interaction can be visualized and better understood than in the case of a simulation study.

6.1 Introduction

Many process models are non-linear. When the model consists of only one differential equation, the solution may not be too complicated. However, when multiple differential equations are involved, it may be difficult to get an insight into the dynamic process behavior. Linearization of model equations can be a helpful tool for the following reasons:

- linear systems can be analyzed by using available powerful mathematical tools
- system behavior can be analyzed because relationships between process inputs and process outputs are expressed in terms of process variables and/or operating conditions
- it is relatively easy to investigate the stability of a linear system

Laplace transformation is another useful tool. When used in combination with linearization it can help us to write the model equations in transfer function format. Thus, to apply Laplace transform to non-linear equations, they should first be linearized.

6.2 Non-linear Process Models

Three process models will be considered in order to demonstrate the procedure, followed in linearization, a level process with free outflow, an evaporator with variable level and a chemical reactor with first-order reaction.

6.2.1 Level Process with Free Outflow

Let us first consider a simple level process with outflow through a fixed restriction as shown in Fig. 6.1. It may be assumed that the cross sectional area of the tank is constant and given by A, in addition the density of the liquid is constant and given by ρ. No heat effects are involved, i.e. the temperature in the tank is constant. Let us assume that the flow of liquid out of the tank is a turbulent flow and can be described by:

$$F_{out} = \beta\sqrt{h} \tag{6.1}$$

where F is the volumetric flow. The mass balance for the tank can now be given by:

$$A\frac{dh}{dt} = F_{in} - \beta\sqrt{h} \tag{6.2}$$

Process Dynamics and Control: Modeling for Control and Prediction. Brian Roffel and Ben Betlem.

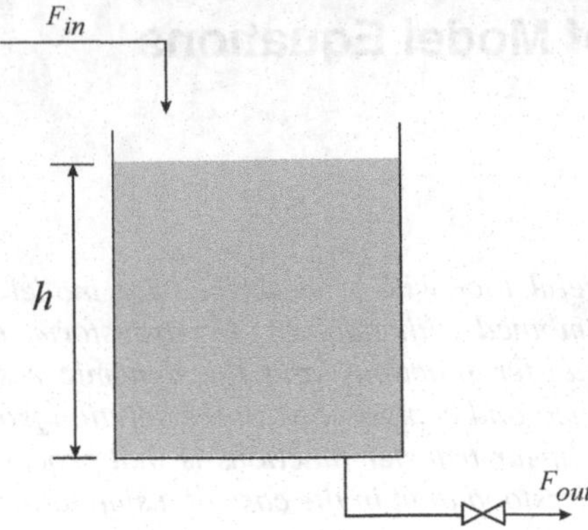

Fig. 6.1. Variation of the level in a tank.

We are interested in describing the variation of the liquid level as a function of a variation in the inlet flow F_{in}.

6.2.2. Evaporator with variable level

The evaporator was already introduced in previous chapters and will also be taken to demonstrate the linearization procedure. The process is shown in Fig. 6.2. The assumptions and an elaborate treatment of the model development procedure are given in chapter 15. Here we are only interested in the (non-linear) equations describing the process for the purpose of linearization.

Since the level may vary, a mass balance, under the assumption of constant physical properties, may be written as:

$$\rho A \frac{dh}{dt} = F_{in} - F_{out} \tag{6.3}$$

where ρ is the liquid density, A the cross sectional area of the tank, h the liquid level and F the mass flow. The subscripts 'in' and 'out' refer to the inlet and outlet conditions respectively.

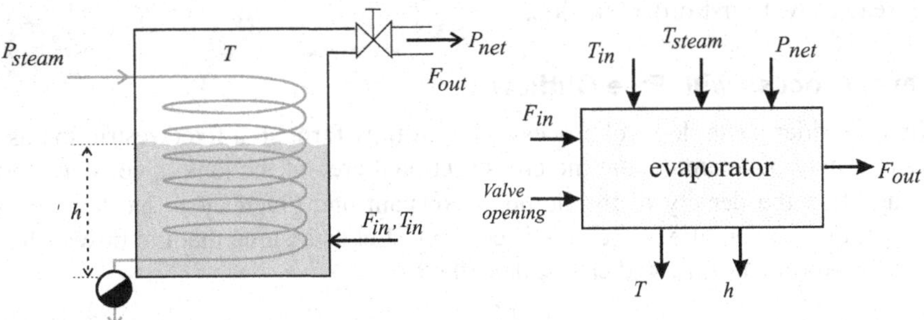

Fig. 6.2. Evaporator with variable heat transfer surface.

The energy balance can be written as:

$$\rho c_p A h \frac{dT}{dt} = c_p F_{in}(T_{in} - T) - F_{out}\Delta H + UA\frac{h}{h_{max}}(T_{steam} - T) \tag{6.4}$$

where T is the liquid temperature in the tank, c_p is the specific heat of the liquid, h_{max} is the maximum height of the heat transfer area (top of the heat exchanger), T_{steam} is the steam temperature, UA is the product of heat transfer coefficient and heat transfer area and ΔH is the heat of vaporization.

Additional equations are required to complete the model description. The first one is the relationship between the outlet flow and the pressure, taking assumption four into account:

$$F_{out} = c_V\sqrt{P - P_{net}} \tag{6.5}$$

Since there is only one component present in the tank (boiling pure liquid), there is also a relationship between the pressure in the tank and the temperature of the vapor (and liquid, which is the same). This relationship can be well described by the law of Clausius-Clapeyron:

$$P = P_{ref}exp\left(\frac{\Delta H}{R}\left(\frac{1}{T_{ref}} - \frac{1}{T}\right)\right) \tag{6.6}$$

The model now consists of Eqns. (6.3)–(6.6).

6.2.3 Chemical reactor with first-order reaction

The third process that will be considered is a chemical reactor in which a first-order reaction A → B takes place. The reactor is shown in Fig. 6.3. We are interested in the conversion in the reactor, i.e. in the outlet concentration of component A. So as not to complicate the model too much, some assumptions are made:

- the reactor is ideally mixed,
- heat losses to the environment can be ignored,
- physical properties may be assumed constant,
- the first-order reaction can be described by an Arrhenius-type equation,
- the mass of the heating coil and steam inside the coil may be ignored compared with the mass of the liquid in the reactor.

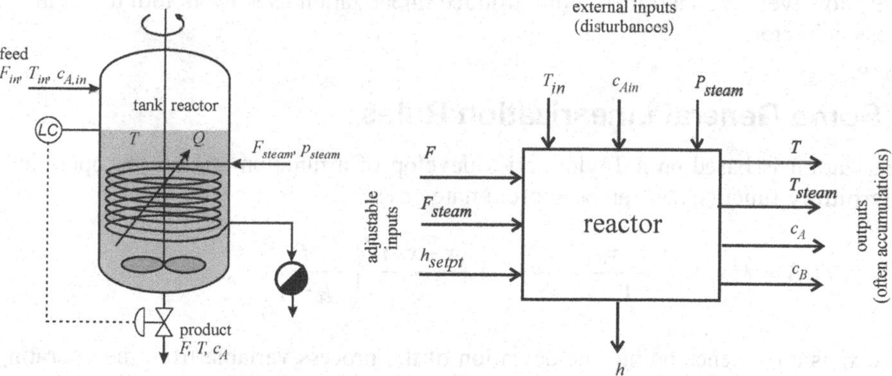

Fig. 6.3. Chemical reactor with first-order reaction.

If the level is variable, the model will become somewhat complicated since we have to deal with a total mass balance, a component balance and an energy balance. For the purpose of demonstrating linearization, it will suffice to assume the level to be constant. This eliminates the total mass balance, since the inlet flow will be equal to the outlet flow, i.e. $F_i = F$.

Models for chemical reactors will be discussed in detail in chapter 8, here we will give the model equations without further derivation or comment.

The component balance for the reactor can be given as:

$$V\frac{dc_A}{dt} = F(c_{Ain} - c_A) - Vke^{-E/RT}c_A \tag{6.7}$$

and the energy balance by:

$$\rho V c_p \frac{dT}{dt} = F\rho c_p (T_{in} - T) + Vke^{-E/RT}c_A \Delta H + Q \tag{6.8}$$

in which the variables with their respective steady-state values are:

V	reactor volume	$= 5\ m^3$
c_A	outlet concentration of component A	$= 200.13\ kg/m^3$
c_{Ain}	inlet concentration of component A	$= 800\ kg/m^3$
F	total volumetric flow	$= 0.005\ m^3/s$
k	pre-exponential constant	$= 18.75\ s^{-1}$
E	activation energy for the reaction	$= 30$ kJ/mol
T	reactor temperature	$= 413$ K
T_{in}	temperature of inlet flow	$= 353$ K
ρ	density	$= 800\ kg/m^3$
c_p	specific heat	$= 1.0$ kJ/kg.K
ΔH	heat of reaction (exothermic)	$= 5.3$ kJ/kg
Q	heat supplied to the reactor	$= 224.1$ kJ/sec
R	gas constant	$= 0.00831$ kJ/mol.K

As can be seen, both model equations are non-linear. The question is now, how can we get an impression of the behavior of the outlet concentration c_A when the reactor throughput changes.

One way is to simulate both equations and to give a change in the flow *F*. However, this does not provide us with much insight.

Another way is try to find an analytical solution for both equations. This is not a trivial matter, however. We shall therefore introduce linearization as a useful tool to get insight into process behavior.

6.3 Some General Linearization Rules

Linearization is based on a Taylor series develop of a function around the operating point. Any arbitrary function *f(x)* can be approximated by:

$$f(x) = f(x_0) + \frac{(x-x_0)}{1!}\left(\frac{df}{dx}\right)_{x_0} + \frac{(x-x_0)^2}{2!}\left(\frac{d^2 f}{dx^2}\right)_{x_0} + \ldots.. \tag{6.9}$$

where x_0 is a reference point. The deviation of the process variable from the operating point can be given by $\delta x = x - x_0$, in which δx is called a deviation variable. By assuming that second and higher order terms can be neglected, we may approximate Eqn. (6.9) by (see also Fig. 6.4):

$$f(x) = f(x_0) + \left(\frac{df}{dx}\right)_{x_0} \delta x \tag{6.10}$$

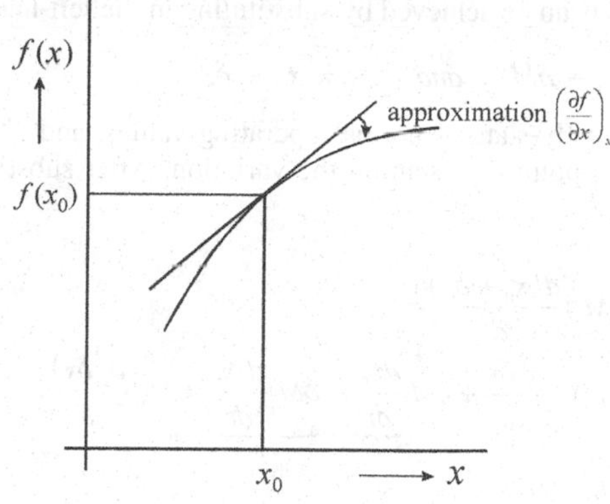

Fig. 6.4. Approximation of a function in a reference point.

The original function is now approximated by its tangent to the curve in the reference point. The result of Eqn. (6.10) can now be used in the general model differential equation:

$$\frac{dx}{dt} = f(x) = f(x_0) + \left(\frac{df}{dx}\right)_{x_0} \delta x \tag{6.11}$$

In the operating (reference) point x_0, $\delta x = 0$, hence we may write:

$$\frac{dx_0}{dt} = f(x_0) \tag{6.12}$$

Let us now introduce a new variable $x = x_0 + \delta x$, Eqn. (6.11) may then be written as:

$$\frac{d(x_0 + \delta x)}{dt} = \frac{dx_0}{dt} + \frac{d(\delta x)}{dt} = f(x_0) + \left(\frac{df}{dx}\right)_{x_0} \delta x \tag{6.13}$$

which can be written, using Eqn. (6.12), as:

$$\frac{d(\delta x)}{dt} = \left(\frac{df}{dx}\right)_{x_0} \delta x \tag{6.14}$$

This is the general equation for the linearized process variable x, written in deviation variable format δx. The deviation variable δx is zero in the steady-state operating point.

It often happens that the process model is somewhat more complicated, i.e. the derivative dx/dt is not only a function of x but also, for example, of y. In that case, the linearized model equation (6.14) can be rewritten as:

$$\frac{d(\delta x)}{dt} = \left(\frac{\partial f(x,y)}{\partial x}\right)_{x_0,y_0} \delta x + \left(\frac{\partial f(x,y)}{\partial y}\right)_{x_0,y_0} \delta y \tag{6.15}$$

Example

Suppose a simple mixing process can be described by the following equation:

$$M\frac{dx}{dt} = F(x_{in} - x) \tag{6.16}$$

in which M = mass in the tank, F = mass flow to the tank, x = concentration and all variables can vary.

Linearization of Eqn. (6.16) can be achieved by substituting in the left-hand side of the balance:

$$M = M_0 + \delta M \quad and \quad x = x_0 + \delta x \tag{6.17}$$

in which M_0 and x_0 are the steady-state or normal operating values, and δM and δx are the deviation variables around the operating point representing the variation. After substitution, we get for the left-hand side of the equation:

$$\begin{aligned} M\frac{dx}{dt} &= (M_0 + \delta M)\frac{d(x_0 + \delta x)}{dt} = \\ &= M_0\underbrace{\frac{dx_0}{dt}}_{=0} + M_0\frac{d\delta x}{dt} + \delta M\underbrace{\frac{dx_0}{dt}}_{=0} + \underbrace{\delta M\frac{d\delta x}{dt}}_{\approx 0} \approx M_0\frac{d(\delta x)}{dt} \end{aligned} \tag{6.18}$$

The right-hand side of Eqn. (6.16) can be written as:

$$\begin{aligned} &(F_0 + \delta F)[(x_{in0} + \delta x_{in}) - (x_0 + \delta x)] = \\ &F_0 x_{in0} + F_0\delta x_{in} - F_0 x_0 - F_0\delta x + x_{in0}\delta F + \delta F\,\delta x_{in} - x_0\delta F - \delta F\,\delta x = \\ &\underbrace{F_0(x_{in0} - x_0)}_{=0} + F_0(\delta x_{in} - \delta x) + (x_{in0} - x_0)\delta F + \underbrace{\theta(\delta^2)}_{\approx 0} = \\ &F_0(\delta x_{in} - \delta x) + (x_{in0} - x_0)\delta F \end{aligned} \tag{6.19}$$

Combination of Eqns. (6.18) and (6.19) results in:

$$M_0\frac{d(\delta x)}{dt} = F_0(\delta x_{in} - \delta x) + (x_{in0} - x_0)\delta F \tag{6.20}$$

When Eqn. (6.16) is written as:

$$\frac{dx}{dt} = \frac{F}{M}(x_{in} - x) \tag{6.21}$$

linearization becomes:

$$\frac{d(\delta x)}{dt} = \left(\frac{F_0}{M_0}\right)(\delta x_{in} - \delta x) + \frac{(x_{in} - x)_0}{M_0}\delta F - \underbrace{\frac{(x_{in} - x)_0 F_0}{M_0^2}\delta M}_{=0} \tag{6.22}$$

The third term of the right-hand side of Eqn. (6.22) equals zero, since in steady state $F_0(x_{in0} - x_0) = 0$, hence the result of Eqns. (6.20) and (6.22) is the same.

6.4 Linearization of Model of the Level Process

Equation (6.2) can now be linerarized, using Eqn. (6.14). The result is:

$$A\frac{d(\delta h)}{dt} = \delta F_{in} - \frac{1}{2}\beta h_0^{-1/2}\delta h \tag{6.23}$$

In the steady-state situation, $dh_0/dt = 0$, hence we may write Eqn. (6.2) as:

$$F_{in,0} = \beta\sqrt{h_0} \tag{6.24}$$

Combination of Eqn. (6.23) and (6.24) and introduction of the Laplace transform results in:

$$\frac{\delta h}{\delta F_{in}} = \frac{K}{\tau s + 1} \tag{6.25}$$

in which

$$\tau = 2Ah_0 / F_{in,0}\,, K = 2h_0 / F_{in,0} \tag{6.26}$$

Equation (6.25) is a first-order transfer function with process gain K and time constant τ. The time constant has the dimension of time (seconds or minutes), the process gain has the dimension of s/m^2.

As can be seen from Eqn. (6.26), the time constant is equal to two times the residence time in the steady-state situation.

Example

Given are the following steady-state data:

Table 6.1. Tank variables.

parameter	value
A	0.07 m^2
F_{in}	1.5×10^{-3} m^3/s
h	0.5 m

From steady-state data it can be calculated that the constant β in Eqn. (6.1) is equal to 0.002121 m$^{2.5}$/s. The residence time in steady-state is $V/F_{in,0} = Ah_0/F_{in,0} = 0.07 \times 0.5/1.5 \times 10^{-3} = 23.2$ s, hence the time constant of the response is $2 \times 23.3 = 46.6$ s and the process gain $K = 666.67$ s/m^2.

The non-linear process model and the linearized process model are simulated in Matlab Simulink and stored in file F0605.mdl. The simulation setup will be explained in chapter 8. Here we are only interested in comparing the response of the linearized and non-linear models.

The response for a 5% step change in inlet flow is given in Fig. 6.5 for the original non-linear model of Eqn. (6.1) and the linearized model of Eqn. (6.25).

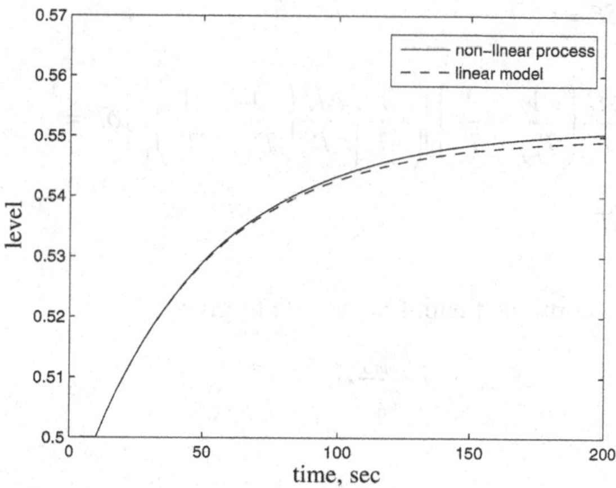

Fig. 6.5. Response of level process using original and linearized model.

6.5. Linearization of the Evaporator Model

Linearization of Eqn. (6.3) results in:

$$\rho A s \delta h = \delta F_{in} - \delta F_{out} \tag{6.27}$$

in which δ is a variation around the steady-state value.

Since we are only interested in changes of F_{out} as a result of changes to F_{in}, changes in T_{steam} and T_{in} will not be considered. Linearization of the energy balance, Eqn. (6.4) results then in:

$$\left(\rho c_p A h_0 s + c_p F_{in0} + \frac{UAh_0}{h_{max}}\right)\delta T = -c_p(T_0 - T_{in0})\delta F_{in} - \Delta H \delta F_{out} + \frac{UA}{h_{max}}(T_{steam0} - T_0)\delta h \tag{6.28}$$

where the index '0' refers to the steady-state operating point.

Equation (6.4) can be written as:

$$F_{out} = f(P - P_{net}) \tag{6.29}$$

Linearization of Eqn. (6.4) gives therefore:

$$\begin{aligned}\delta F_{out} &= \frac{\partial f}{\partial (P - P_{net})_{P_0, P_{net0}}}(\delta P - \delta P_{net}) \\ &= \frac{\frac{1}{2}c_v}{\sqrt{P_0 - P_{net0}}} = \frac{1}{2}\frac{F_{out0}}{(P_0 - P_{net0})}(\delta P - \delta P_{net})\end{aligned} \tag{6.30}$$

Equation (6.5) is written as:

$$P = f(T) \tag{6.31}$$

Linearization gives therefore:

$$\begin{aligned}&\delta P = \left(\frac{\partial f}{\partial T}\right)_{T_0}\delta T = \\ &P_{ref}exp\left(\frac{\Delta H}{R}\left(\frac{1}{T_{ref}} - \frac{1}{T_0}\right)\right)\left(\frac{\partial}{\partial T}\left\{\frac{\Delta H}{R}\left(\frac{1}{T_{ref}} - \frac{1}{T_0}\right)\right\}\right)\delta T = \\ &P_0 \frac{\Delta H}{RT_0^2}\delta T\end{aligned} \tag{6.32}$$

Equation (6.30) can be combined with Eqn. (6.31) to give:

$$\delta F_{out} = \beta \frac{F_{out0}}{T_0}\delta T \tag{6.33}$$

in which

$$\beta = \frac{1}{2}\frac{P_0}{(P_0 - P_{net0})}\frac{\Delta H}{RT_0} \tag{6.34}$$

which is a dimensionless constant.

6.6 Normalization of the Transfer Function

In the operating point we may write that $F_{in0} = F_{out0} = F_0$. The following time constant is introduced:

$$\tau_1 = \frac{\rho A h_0}{F_0} \tag{6.35}$$

which enables us to write the mass balance as:

$$\tau_1 s \frac{\delta h}{h_0} = \frac{\delta F_{in}}{F_0} - \frac{\delta F_{out}}{F_0} \tag{6.36}$$

The time constant τ_1 is the residence time in the tank in the steady-state situation.

After introduction of the following time constants,

$$\tau_2 = \frac{\rho c_p A h_0}{c_p F_0 + \beta \Delta H \dfrac{F_0}{T_0} + \dfrac{UAh_0}{h_{max}}} \tag{6.37}$$

and

$$\tau_3 = \frac{\rho c_p A h_0 T_0}{\dfrac{UAh_0}{h_{max}} \left(T_{steam0} - T_0\right)} \tag{6.38}$$

the energy balance can be written as:

$$(\tau_2 s + 1)\delta T = -\frac{\tau_2}{\tau_1}(T_0 - T_{in0})\frac{\delta F_{in}}{F_0} + \frac{\tau_2}{\tau_3} T_0 \frac{\delta h}{h_0} \tag{6.39}$$

The time constant τ_2 is the total heating time and the time constant τ_3 is the heating time of the vessel compared with the heating time of the coil.
Substitution of the mass balance and the equation for the outgoing flow finally results in:

$$\left(\tau_1 s(\tau_2 s + 1) + \frac{\tau_2}{\tau_3}\beta\right)\frac{\delta T}{T_0} = \left(-\tau_2 \frac{(T_0 - T_{in0})}{T_0} s + \frac{\tau_2}{\tau_3}\right)\frac{\delta F_{in}}{F_0} \tag{6.40}$$

The transfer function from inlet flow δF_{in} to temperature δT becomes then:

$$\frac{\delta F_{out}}{\delta F_{in}} = \frac{-\tau_3 s(T_0 - T_{in0})/T_0 + 1}{\dfrac{\tau_1\tau_3}{\beta} s^2 + \dfrac{\tau_1\tau_3}{\tau_2\beta} s + 1} \tag{6.41}$$

The dynamics of this transfer function are derived and discussed in detail in chapter 15. As can be seen, the dynamic and static relationship between δF_{in} and δF_{out} is expressed in terms of process variables, which gives us insight in this relationship.

6.7. Linearization of the Chemical Reactor Model

Linearization of the reactor model of Eqns. (6.7) and (6.8) is somewhat more complicated. Equation (6.7) can be written as:

$$V\frac{dc_A}{dt} = Fc_{Ain} - Fc_A - Vke^{-E/RT}c_A \\ = f_1(F,c_{Ain}) - f_2(F,c_A) - f_3(c_A,T) \tag{6.42}$$

Using Eqn. (6.11), Eqn. (6.42) can now be linearized:

$$V\frac{d(\delta c_A)}{dt} = \left(\frac{\partial f_1}{\partial F}\right)_0 \delta F + \left(\frac{\partial f_1}{\partial c_{Ain}}\right)_0 \delta c_{Ain} - \left(\frac{\partial f_2}{\partial F}\right)_0 \delta F - \\ \left(\frac{\partial f_2}{\partial c_A}\right)_0 \delta c_A - \left(\frac{\partial f_3}{\partial c_A}\right)_0 \delta c_A - \left(\frac{\partial f_3}{\partial T}\right)_0 \delta T \tag{6.43}$$

which can be rewritten as:

$$V\frac{d(\delta c_A)}{dt} = c_{Ain0}\delta F + F_0 \delta c_{Ain} - c_{A0}\delta F - F_0 \delta c_A - Vke^{-E/RT_0}\delta c_A \\ - Vke^{-E/RT_0} c_{A0}\left(\frac{E}{RT_0{}^2}\right)\delta T \tag{6.44}$$

Rearranging terms and introducing the Laplace operator results in:

$$\delta c_A = \frac{K_1 \delta F}{\tau_c s + 1} + \frac{K_2 \delta c_{Ain}}{\tau_c s + 1} - \frac{K_3 \delta T}{\tau_c s + 1} \tag{6.45}$$

with

$$\tau_c = \frac{V}{F_0 + Vke^{-E/RT_0}},\ K_1 = \frac{c_{Ain0} - c_{A0}}{F_0 + Vke^{-E/RT_0}} \\ K_2 = \frac{F_0}{F_0 + Vke^{-E/RT_0}},\ K_3 = \frac{Vke^{-E/RT_0} c_{A0}(E/RT_0{}^2)}{F_0 + Vke^{-E/RT_0}} \tag{6.46}$$

After substitution of the steady-state values into the time constant and process gains of Eqn. (6.46) we get:

$$\tau_c = 250\ \text{s},\ K_1 = 3\times10^4,\ K_2 = 0.25,\ K_3 = 3.174 \tag{6.47}$$

The second equation (energy balance) of the reactor model can be rewritten as:

$$\rho V c_p \frac{dT}{dt} = F\rho c_p (T_{in} - T) - Vke^{-E/RT} c_A \Delta H + Q \\ = F\rho c_p T_{in} - F\rho c_p T - Vke^{-E/RT} c_A \Delta H + Q \\ = f_1(F,T_{in}) - f_2(F,T) - f_3(c_A,T) + f_4(Q) \tag{6.48}$$

Using Eqn. (6.11), Eqn. (6.48) can be written as:

$$\rho V c_p \frac{d(\delta T)}{dt} = \left(\frac{\partial f_1}{\partial F}\right)_0 \delta F + \left(\frac{\partial f_1}{\partial T_{in}}\right)_0 \delta T_{in} - \left(\frac{\partial f_2}{\partial F}\right)_0 \delta F - \\ \left(\frac{\partial f_2}{\partial T}\right)_0 \delta T - \left(\frac{\partial f_3}{\partial c_A}\right)_0 \delta c_A - \left(\frac{\partial f_3}{\partial T}\right)_0 \delta T + \delta Q \tag{6.49}$$

which can be rewritten as:

$$V\rho c_p \frac{d(\delta T)}{dt} = \rho c_p (T_{in} - T)\delta F + F\rho c_p \delta T_{in} - F\rho c_p \delta T + Vke^{-E/RT} \Delta H \delta c_A + Vke^{-E/RT} c_A \Delta H \left(\frac{E}{RT_0^{\ 2}} \right) \delta T + \delta Q \tag{6.50}$$

Rearranging terms and introducing the Laplace operator results in:

$$\delta T = \frac{K_4 \delta F}{\tau_T s + 1} + \frac{K_5 \delta T_{in}}{\tau_T s + 1} + \frac{K_6 \delta c_A}{\tau_T s + 1} \tag{6.51}$$

with

$$\begin{aligned}
\tau_T &= \frac{V\rho c_p}{F_0 \rho c_p - Vke^{-E/RT_0} c_{A0} \Delta H (E/RT_0^{\ 2})} \\
K_4 &= \frac{\rho c_p (T_{in0} - T_0)}{F_0 \rho c_p - Vke^{-E/RT_0} c_{A0} \Delta H (E/RT_0^{\ 2})} \\
K_5 &= \frac{F_0 \rho c_p}{F_0 \rho c_p - Vke^{-E/RT_0} c_{A0} \Delta H (E/RT_0^{\ 2})} \\
K_6 &= \frac{Vke^{-E/RT_0} \Delta H}{F_0 \rho c_p - Vke^{-E/RT_0} c_{A0} \Delta H (E/RT_0^{\ 2})}
\end{aligned} \tag{6.52}$$

Substitution of the steady-state values in the time constant and process gains of Eqn. (6.52) results in:

$$\tau_T = 1091.8\ s,\ K_4 = -1.31 \times 10^4,\ K_5 = 1.09,\ K_6 = 0.022 \tag{6.53}$$

The response of the change in reactor outlet concentration c_A to a change in reactor throughput F can now be obtained by combining Eqns. (6.45) and (6.51) while setting changes in c_{Ain} and T_{in} to zero:

$$\delta c_A = \frac{K_1}{\tau_c s + 1} \delta F - \frac{K_3}{\tau_c s + 1} \left[\frac{K_4}{\tau_T s + 1} \delta F + \frac{K_6}{\tau_T s + 1} \delta c_A \right] \tag{6.54}$$

This equation can be rearranged to:

$$\frac{\delta c_A}{\delta F} = \frac{(K_1 - K_3 K_4)}{(1 + K_3 K_6)} \frac{\dfrac{K_1 \tau_T}{(K_1 - K_3 K_4)} s + 1}{\dfrac{\tau_c \tau_T}{(1 + K_3 K_6)} s^2 + \dfrac{(\tau_c + \tau_T)}{(1 + K_3 K_6)} s + 1} \tag{6.55}$$

which is a pseudo-first-order transfer function.
After substituting values for the time constants and gains, Eqn. (6.55) can be written as:

$$\frac{\delta c_A}{\delta F} = 6.69 \times 10^4 \frac{457.5 s + 1}{2.55 \times 10^5 s^2 + 1255.5 s + 1} \tag{6.56}$$

The non-linear and linear process models are simulated in Matlab Simulink and stored in file F0606.mdl. The design of the simulation will be explained in chapter 8.

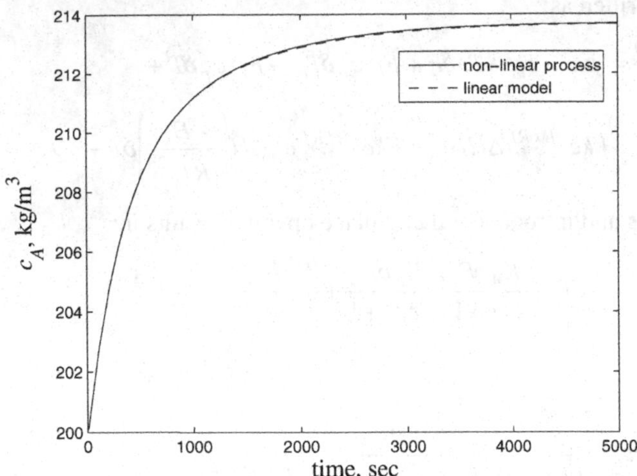

Fig. 6.6. Response of reactor concentration using original and linearized model.

The response, shown in Fig. 6.6 is obtained for a step change of 5% in the reactor throughput. As can be seen, the reactor outlet concentration increases by about 9%, there is not much difference between the responses of the two models. It can be concluded that linearization is a powerful tool for the analysis of the response of non-linear systems and is helpful in understanding the dynamics of the process.

Files referred to in this chapter:
F0605par.m: parameter definition file
F0605.mdl: model to generate Fig. 6.5
F0605plot.m: plot file for Fig. 6.5
F0606.mdl: model to generate Fig. 6.6
F0606par: parameters for model F0606
F0606plot.m: plot file for Fig. 6.6.

7 Operating points

7.1 Introduction

The operating point of a system is an equilibrium point in which the energy (and/or mass and/or momentum) supplied to a system per unit time is equal to the dissipated energy (and/or mass and/or momentum) per unit time. The location of the operating point is determined by the process operating conditions on the basis of system characteristics. The way in which the stability can be determined is illustrated by an exothermal reaction. It will also be illustrated how the transition between operating points takes place.

7.2 Stationary System and Operating Point

The operating region of a system is an area in which the state of the system can be varied during the operation. This area is determined during the design of the system. If all system inputs are constant (disturbances and controls of the system), every stable system will eventually move towards a stationary condition (steady state). By changing the controls of the system, the location of the stationary point can be influenced. In many cases, the system never reaches a steady state as a result of the presence of disturbances.

The point where the system should be maintained or operated is called the operating point. Characteristics of the operating point are:

- it is an equilibrium point;
- not every system can be kept in a steady state condition, the operating point is then determined by the time average values of the system inputs. The system remains then as close as possible to the operating point. Linearization of the system in that point gives the smallest possible error, in another operating point the error will be larger.
- the combination of variables that determine the operating point can be used for linearization and as initial condition for system integration.

A system is in equilibrium if all balances are in equilibrium, i.e. if the variables that describe the system, such as temperature, pressure and flow, do not change. This means that the sum of the flows of mass, energy and momentum that are produced by the system or are supplied to the system is equal to the sum of the flows that are dissipated by the system or discharged from the system. If these flows are constant, the system remains in a stationary condition.

The flows that are produced (the rate of production) or dissipated (rate of consumption) are usually dependent on the state of the system:

$$\begin{aligned} &\sum \mathit{Flows\ in/out}\ (\mathit{system\ \ state},\, \mathit{state\ \ environment}) + \\ &\sum \mathit{Flows\ \ produced/dissipated}\ (\mathit{system\ \ state}) = 0 \end{aligned} \qquad (7.1)$$

For a closed system, an operating point can be graphically depicted by plotting the rate of production and the rate of consumption against the state of the system. The lines that describe the flows are called characteristics. This is shown in Fig. 7.1: the operating point is the intersection of both lines. It holds that the produced mass, energy or momentum is equal to the consumed mass, energy or momentum.

Process Dynamics and Control: Modeling for Control and Prediction. Brian Roffel and Ben Betlem.

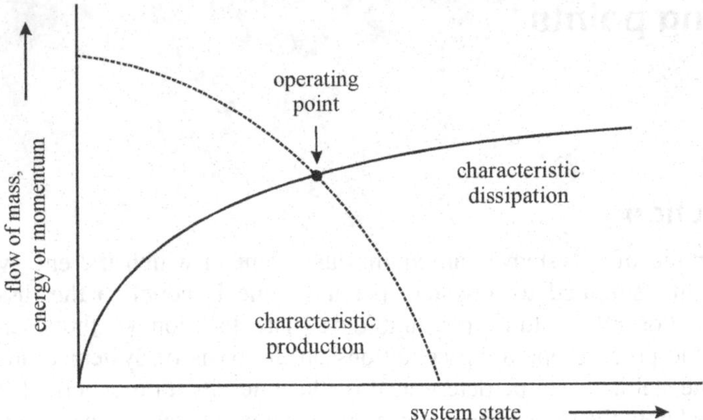

Fig. 7.1. Operating point.

The characteristic lines are important for design purposes. A design is always based on stationary relationships. It is difficult to establish these relationships for equipment based on laws of physics, since it is generally affected by other processes and equipment. What is the liquid height a pump can deliver at a certain flow rate, or what is the torque of a gas engine at a certain speed of rotation? This depends on many factors which are not all known. However, for most systems the stationary relationship can be measured without too much difficulty. For pumps and engines these characteristics are usually specified by the pump or engine manufacturer.

7.3 Flow Systems

Axial and centrifugal pumps deliver a flow F and a height (or pressure) ΔP. The flow and the height depend on the type of pump. The combination of flow and height is called pump characteristic (see Fig. 7.2). It can be mathematically represented by:

$$\Delta P^n + F^m = constant \tag{7.2}$$

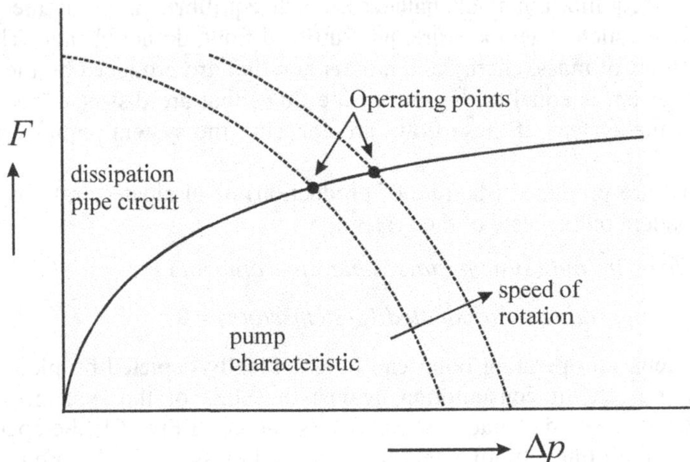

Fig. 7.2. Operating points of a pump and pipe line.

The equation holds for a constant speed of rotation. When the height increases, the flow decreases and vice versa. The constant in the equation depends on the speed of rotation. The higher the speed of rotation, the higher the flow and the height. The powers n and m range from 1 to 3. Piston pumps deliver a flow that is almost linear with the speed of rotation.

The pipe line and equipment dissipate the flow energy. In the process industry, flows are usually turbulent; in this case it holds that:

$$F = C\sqrt{\Delta P} \tag{7.3}$$

The dissipation of the flow energy can be described by this relationship and is shown in Fig. 7.2. The intersection between pump and pipeline characteristic is the operating point.

7.4 Chemical System

Figure 7.3 shows a continuous, ideally mixed tank reactor. In the reactor, an exothermal reaction takes in which a reactant with concentration C_{in} is converted.

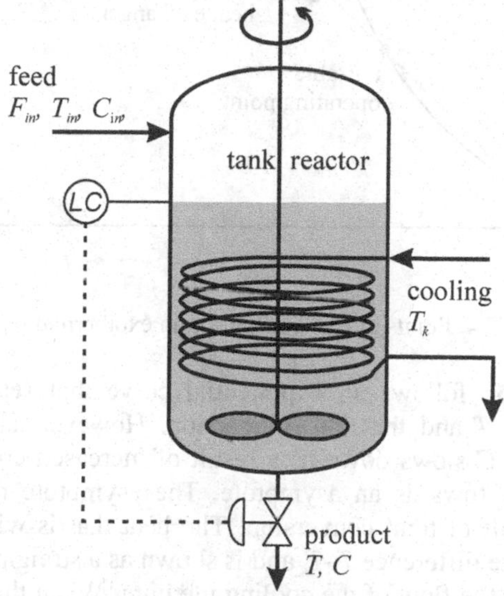

Fig. 7.3. Exothermal, ideally mixed tank reactor.

The generated heat is withdrawn by a cooling coil. It is assumed that the temperature increase of the cooling medium is small compared with the temperature difference between the reactor contents and the cooling T_c, as a result of which the temperature difference across the cooling coil can be ignored. The level in the reactor is ideally controlled, which means that the inlet flow F_{in} is equal to the outlet flow.

The heat that is generated as a result of the reaction, $Q_{reaction}$, is:

$$Q_{reaction} = k_0 e^{-E/RT} C\Delta H_r \tag{7.4}$$

in which:

k_0 pre-exponential kinetic constant
E activation energy
R gas constant
T reaction temperature
C concentration of the reactant
ΔH_r heat of reaction

The heat that is withdrawn by the cooling coil, Q_{he}, can be given by:

$$Q_{he} = U_{he} A_{he} \left(T_c - T\right) \tag{7.5}$$

with:

U_{he} heat transfer coefficient of the heat exchanger per unit area
A_{he} surface area of the heat exchanger
T_c average temperature of the cooling medium, assuming that $T–T_c >> (T_{c,out}–T_{c,in})$, otherwise one should use the logarithmic temperature difference.

The heat that is generated and the heat that is dissipated are shown in Fig. 7.4.

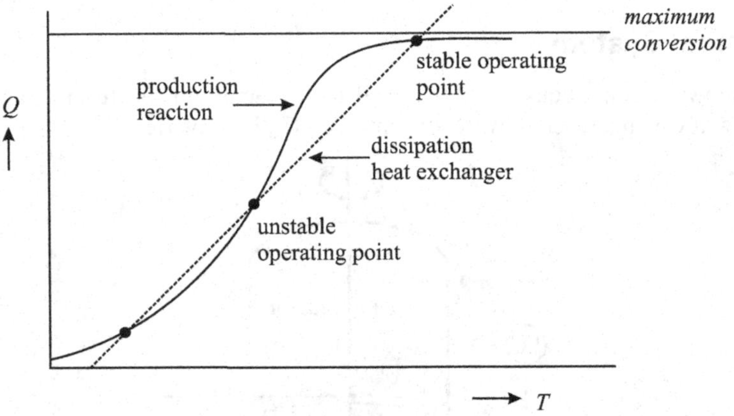

Fig. 7.4. Points of equilibrium for an exothermal reaction.

The heat of production follows an exponential curve that represents the relationship between the temperature T and the rate of reaction. However, at high temperatures, the increase in concentration C slows down as a result of increased conversion and the rate of reaction increases slowly towards an asymptote. The asymptote represents the maximum heat production as a result of total conversion. The heat that is withdrawn is more or less linear with the temperature difference $T–T_c$ and is shown as a straight curve. The slope of this line can be influenced by the flow of the cooling medium. When the cooling flow increases, the heat transfer coefficient increases. Both effects lead to increased cooling. The intersections between the two lines are the points of equilibrium. Often there are three operating points, as the steady state operating point, usually the middle point is chosen. At a low temperature the conversion is too small, high temperatures are undesirable, because the efficiency of the catalyst decreases or catalyst decay is too pronounced.

It is interesting to investigate whether these points of equilibrium are asymptotically stable or unstable. To do this, the linearization theory from the previous chapter will be used.

It is assumed that the volume V, the density ρ and specific heat capacity c_p are constant (independent of the time, temperature and composition). As will be explained in more detail in chapter 12, the dynamic equations for this system are the component balance:

$$V\frac{dC}{dt} = F_{in}\left(C_{in} - C\right) - rCV \tag{7.6}$$

with F_{in} = reactor feed flow and:

$$r = k_0 e^{-E/RT} \tag{7.7}$$

The energy balance is:

$$c_p \rho V \frac{dT}{dt} = c_p \rho F_{in} (T_{in} - T) + rCV\Delta H_r - U_{he} A_{he} (T - T_c) \tag{7.8}$$

For a strong exothermal reaction, it may be assumed that the heat required to warm up the reactants is much smaller than the heat of reaction, i.e.:

$$c_p \rho F_{in} (T_{in} - T) << rCV\Delta H_r \approx U_{he} A_{he} (T - T_k) \tag{7.9}$$

The energy balance can then be simplified to:

$$c_p \rho V \frac{dT}{dt} = rCV\Delta H_r - U_{he} A_{he} (T - T_c) \tag{7.10}$$

7.5 Stability in the Operating Point

The component balance and energy balance are highly interactive. When the temperature increases, the rate of reaction increases according to an Arrhenius equation. However, the increased conversion due to the temperature rise will slow down the rate of reaction. The stability should therefore be considered for an equilibrium point, taking into account the interaction between the temperature T and conversion C. The other input variables are assumed to be constant: F_{in}, C_{in}, T_{in} and T_c.

In Chapter 5 is was stated that a linear system based on deviation variables, is asymptotically stable if the roots of the characteristic equation of an input-output relationship have negative real parts. The roots of the characteristic equation correspond to the poles of the transfer function or the eigen values of the homogenous part of the differential equation. Therefore, it is necessary to derive the characteristic linear input-output relationship for the process in an operating point. Because only the homogenous part of the differential equation is of interest, the inputs of the differential equations can be ignored. Next, from the roots of the characteristic equation, conditions for stability can be formulated. The following three steps will be performed:

Step 1: linearization of the state equations
The state equation for the concentration C can be given by Eqns. (7.6) and (7.7).
When F_{in}, C_{in} and V are constant, linearization gives:

$$V \frac{d\delta C}{dt} = \left(\frac{\partial (F_{in}(C_{in} - C) - rCV)}{\partial C} \right)_0 \delta C + \left(\frac{\partial (F_{in}(C_{in} - C) - rCV)}{\partial T} \right)_0 \delta T \tag{7.11}$$

This can also be written as:

$$V \frac{d\delta C}{dt} = -(F_{in} + r_0 V)\,\delta C - \left(\frac{\partial r}{\partial T} \right)_0 C_0 V \delta T \tag{7.12}$$

with

$$\left(\frac{\partial r}{\partial T} \right)_0 = \left(\frac{\partial \left(k_0 e^{-E/RT} \right)}{\partial T} \right)_0 = k_0 e^{-E/RT_0} \frac{E}{RT_0^2} = r_0' \tag{7.13}$$

Substitution gives for the linearized equation for C:

$$V\frac{d\delta C}{dt} = -(F_{in} + r_0V)\,\delta C - r_0'C_0V\delta T \tag{7.14}$$

The state equation for the temperature is given by Eqn. (7.10). When T_c and V are constant, linearization gives:

$$c_p\rho V\frac{d\delta T}{dt} = \left(\frac{\partial(rCV\Delta H_r - U_{he}A_{he}(T-T_k))}{\partial C}\right)_0 \delta C + \left(\frac{\partial(rCV\Delta H_r - U_{he}A_{he}(T-T_c))}{\partial T}\right)_0 \delta T \tag{7.15}$$

The linearized equation for T becomes now:

$$c_p\rho V\frac{d\delta T}{dt} = r_0V\Delta H_r\delta C + (r_0'C_0V\Delta H_r - U_{he}A_{he})\,\delta T \tag{7.16}$$

Step 2: Substitution in order to obtain the characteristic equation.
The substitution to arrive at the characteristic system equation can be done as follows. Suppose the system can be described by the following state equations:

$$\begin{aligned}\dot{C} &= aC + bT\\ \dot{T} &= cC + dT\end{aligned} \tag{7.17}$$

Differentiation of the state T gives:

$$\ddot{T} = c\dot{C} + d\dot{T} \tag{7.18}$$

Substitution of the state equation for C results in:

$$\ddot{T} = c\dot{C} + d\dot{T} = c(aC + bT) + d\dot{T} = acC + bcT + d\dot{T} \tag{7.19}$$

The term cC is known from the state equation for T. This will be used a second time, without differentiation:

$$\ddot{T} = acC + bcT + d\dot{T} = a(\dot{T} - dT) + bcT + d\dot{T} \tag{7.20}$$

thus

$$\ddot{T} = (a+d)\dot{T} + (bc - ad)T \tag{7.21}$$

which can be written as:

$$\ddot{T} - (a+d)\dot{T} - (bc - ad)T = 0 \tag{7.22}$$

and the characteristic equation becomes:

$$s^2 - (a+d)s - (bc - ad) = 0 \tag{7.23}$$

This relation holds for C and T. The linearized state equation for C was given in Eqn. (7.14), therefore the coefficients a and b are:

$$\begin{aligned}a &= -(F_{in} + r_0V)/V\\ b &= -r_0'C_0\end{aligned} \tag{7.24}$$

The linearized equation for T is given by Eqn. (7.16), therefore the coefficients c and d can be written as:

$$c = r_0 \Delta H_r / c_p \rho$$
$$d = \left(r_0' C_0 V \Delta H_r - U_{he} A_{he}\right) / c_p \rho V \tag{7.25}$$

Equation (7.21) can now be constructed, but this is not required.

Step 3: Formulation of the conditions for stability
A system is stable if the solutions of the characteristic equation, Eqn. (7.23), contain a negative real part. For a second-order system, one required and sufficient condition for stability is that all terms in the characteristic equation have the same sign. The poles of the second-order normalized (making the last term equal to +1) characteristic equation:

$$As^2 + Bs + 1 = 0 \tag{7.26}$$

are:

$$p_{1,2} = \frac{-B \pm \sqrt{B^2 - 4A}}{2A} \tag{7.27}$$

The poles have negative real parts when A and B are positive. Thus, we can formulate two conditions for stability from the characteristic Eqn. (7.23).

Condition 1:

$$bc - ad < 0 \tag{7.28}$$

Substitution of the expressions for *c* and *d* results in:

$$-\frac{r_0' C_0 r_0 \Delta H_r}{c_p \rho} + \frac{\left(F_{in} + r_0 V\right)\left(r_0' C_0 V \Delta H_r - U_{he} A_{he}\right)}{c_p \rho V^2} < 0 \tag{7.29}$$

or

$$-r_0' C_0 V r_0 V \Delta H_r + \left(F_{in} + r_0 V\right)\left(r_0' C_0 V \Delta H_r - U_{he} A_{he}\right) < 0 \tag{7.30}$$

This can be further simplified to:

$$F_{in} r_0' C_0 V \Delta H_r - \left(F_{in} + r_0 V\right) U_{he} A_{he} < 0 \tag{7.31}$$

and finally results in:

$$\frac{F_{in}}{F_{in} + r_0 V} r_0' C_0 V \Delta H_r - U_{he} A_{he} < 0 \tag{7.32}$$

This stationary condition is shown in Fig. 7.4. The condition for stability is that the slope of the *S*-curve, describing the heat of reaction, should be smaller than the slope of the line, describing the heat withdrawal by the cooling coil. The term $F_{in}/\left(F_{in} + r_0 V\right)$ can be simplified by using the static version of the component balance (7.6):

$$r_0 C_0 V = F_{in}\left(C_{in} - C_0\right) \tag{7.33}$$

Combination of the two expressions results in:

$$\frac{F_{in}}{F_{in} + r_0 V} = \frac{F_{in}}{F_{in} + F_{in}\left(C_{in} - C_0\right)/C_0} = \frac{C_0}{C_{in}} \tag{7.34}$$

This term indicates the relative conversion. In case of limited conversion, $C_0/C_{in} \approx 1$, which implies that the curve shows an exponential behavior:

$$r_0' C_0 V \Delta H_r - U_{he} A_{he} < 0 \tag{7.35}$$

In case of an arbitrary temperature rise, the increase in the change of the heat of reaction with temperature $r_0' C_0 V \Delta H_r$ has to be smaller than the increase in the change of the heat removal with temperature $U_{he}A_{he}$. This result is understandable. If in the unstable operating point in Fig. 7.4 the temperature increases, the heat production will increase more than the heat removal. This will result in an additional temperature increase.

In case of a temperature increase in a stable operating point, the heat removal will be larger than the heat production, therefore the reactor will cool down again to its stable operating point.

If $C_0/C_{in} << 1$, the increase in the heat production is limited owing to the fact that the conversion increase does not depend exponentially on the temperature. Consequently, process operation is always stable.

Condition 2:
This is the condition for the dynamic part of the equation.

$$a + d < 0 \tag{7.36}$$

Substitution of the expressions for a and d gives:

$$-(F_{in} + r_0 V) + \frac{r_0' C_0 V \Delta H_r}{c_p \rho} - \frac{U_{he} A_{he}}{c_p \rho} < 0 \tag{7.37}$$

which can be written as:

$$-c_p \rho (F_{in} + r_0 V) + r_0' C_0 V \Delta H_r - U_{he} A_{he} < 0 \tag{7.38}$$

If $F_{in} >> r_0 V$ (no reaction limitation), condition two is less stringent than condition one, since $c_p \rho (F_{in} + r_0 V)$ is positive. This term represents the heat capacity of the reactor, capable of absorbing variations in the heat production. Generally, the condition for stability is met when condition one is met.

7.6 Operating Point Transition

The previous two sections dealt with how to determine the operating point and the stability of the operating point. This section will deal with the transition from one operating point to another operating point. This will be illustrated for the chemical reactor. The component balance for the reactant can be described by Eqns. (7.6) and (7.7) and the energy balance by Eqn. (7.10). The following data will be used:

Table 7.1. Reactor variables.

parameter	value	parameter	value
V	$1\ m^3$	k_0	$32\ s^{-1}$
F_{in}	$1 \times 10^{-3}\ m^3 \cdot s^{-1}$	R	$8.31\ J \cdot K^{-1}$
C_{in}	1 kg/kg	E	42 kJ
ρ	$1 \times 10^3\ kg \cdot m^{-3}$	T_c	300 K
c_p	$1\ kJ \cdot kg^{-1} K^{-1}$	$U_{he}A_{he}$	$5\ kJ \cdot K^{-1}$
ΔH_r	$2 \times 10^3\ kJ \cdot kg^{-1}$		

The operating point is moved to a 10% higher throughput. Figure 7.5 shows the characteristics for the heat production and energy dissipation by the heat exchanger.

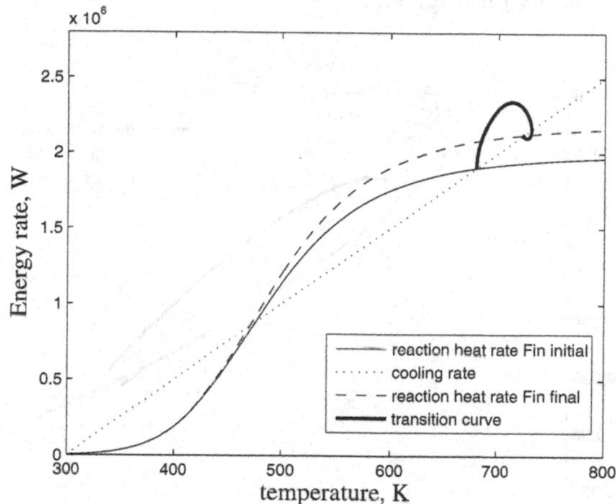

Fig. 7.5. Transition as a result of change in feed flow in exothermal reactor.

Both points are stable operating points. The bold line shows the transition that follow from the step response after the increase in production. The step response of the concentration and the temperature are shown in Fig. 7.6.

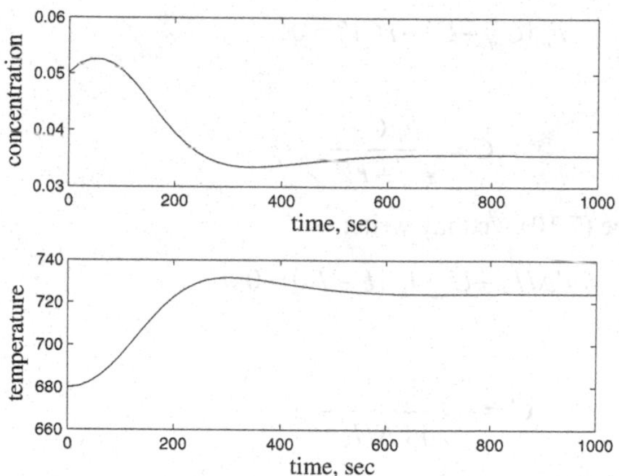

Fig. 7.6. Step response of C and T to feed change.

When the feed rate is increased, first the feed fraction (reactant) will increase. This increased fraction leads to a higher rate of reaction, and consequently the production will increase as well as the heat production. Consequently, the temperature will rise, as a result of which the rate of reaction will increase, leading to a lower feed fraction. Eventually, a stationary point is found, in which the conversion has increased. As can be seen, there is an inverse response of the composition. First, it increases; subsequently it decreases to a value lower than the initial value. This type of behavior is often found in chemical reactors.

For the design and determination of stability, a figure with the static characteristics is very useful. In order to show the transition from one operating point to another operating point, a picture of the behaviour of the combined state equations in a phase plane is very useful. The transition for this case is shown in Fig. 7.7.

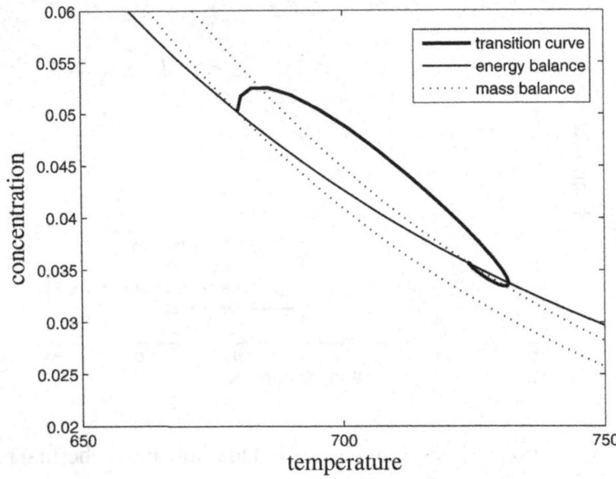

Fig.7.7. Phase plane of the transition during a feed change.

The working lines are the solutions of the steady state equations for the reactor. For the mass balance we may write:

$$F_{in}(C_{in} - C) - rCV = 0 \tag{7.39}$$

thus

$$C = \frac{F_{in}C_{in}}{F_{in} + rV} \tag{7.40}$$

For the energy balance (7.10) we may write:

$$rCV\Delta H_r - U_{he}A_{he}(T - T_c) = 0 \tag{7.41}$$

or

$$C = \frac{U_{he}A_{he}(T - T_c)}{rV\Delta H_r} \tag{7.42}$$

Figure 7.7 shows the interaction between the state variables C and T clearly. The concentration has an inverse response and concentration as well as temperature show some overshoot.

Files referred to in this chapter:
F0705.m: generation of Fig. 7.5
F0706par.m: definition of reactor parameters
F0706.mdl: generation of data for Fig. 7.6
F0706plot.m: generation of Fig. 7.6
F0707.m: generation of Fig. 7.7

8 Process Simulation

Process simulation in Matlab Simulink is explained to enable the user to develop linear and non-linear models and compare them. First, the use of Simulink is introduced by designing a model for a tank with free outflow of liquid. Next, the simulation of a chemical reactor dscribed by a component balance and energy balance is developed and explained.

8.1 Using Matlab Simulink

Simulink is a program that runs as part of the MATLAB environment (Mathworks, 2000). Simulink and MATLAB provide, among others, a tool for dynamic system simulation. Simulink provides a graphical user interface that can be used in building block diagrams, performing simulations as well as analyzing results. It is assumed that you have MATLAB Rxx[1] installed on your computer in the following directory: C:\MATLABRxx. If your software is installed in another directory you have to change all path references accordingly. This chapter was written as a tutorial. For more information on MATLAB and/or Simulink you have to consult the user manual.

To start MATLAB, double click the corresponding icon, after a few moments the MATLAB Command Window will appear as shown in Fig. 8.1. In the command mode you can type a MATLAB command to which we will come back later. Now type Simulink. What appears is the Simulink library browser, as shown in Fig. 8.2. You now havc various options to select building blocks for your simulation. The blocks that are used mostly are in the following groups:

- *continuous*: this group contains, amongst others, an integrator, differentiator, first-order transfer function.
- *math operations*: in this category you will find blocks for addition, multiplication, and mathematical functions, such as sin, exp, log.
- *sinks*: this group contains the endpoints for process variables, such as, for example, the scope.
- *sources*: in this category you will find input signals. The most important ones are the step, constant and sine wave.

8.2 Simulation of the Level Process

In a simulation, process parameters will be assigned numeric values. Rather than using these in the simulation, it is preferred to use variable names in the simulation and assign values to them in a separate file. To do so, select "File, New, m-file" in the Matlab command window. This opens the Matlab editor which enables you to assign values to parameters as shown in Fig.8.3.

---

[1] The examples were obtained with release xx=14SP1

Process Dynamics and Control: Modeling for Control and Prediction. Brian Roffel and Ben Betlem.

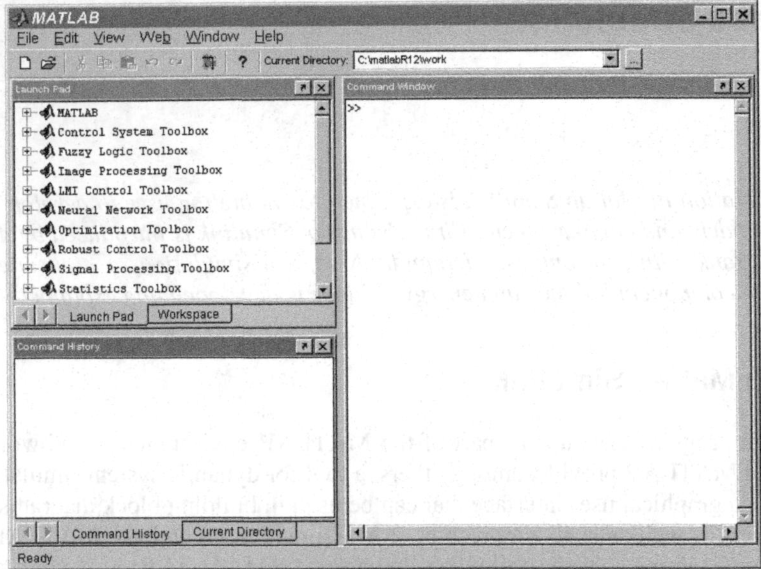

Fig. 8.1. MATLAB Command Window.

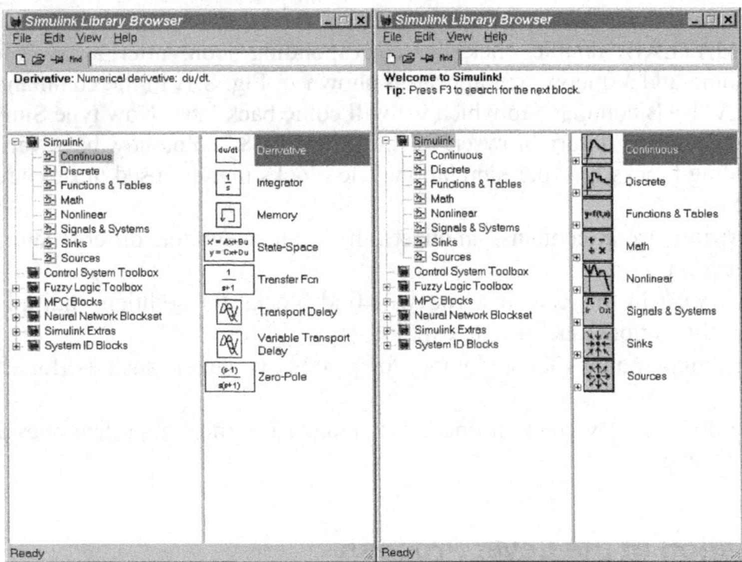

Fig. 8.2. Simulink library browser.

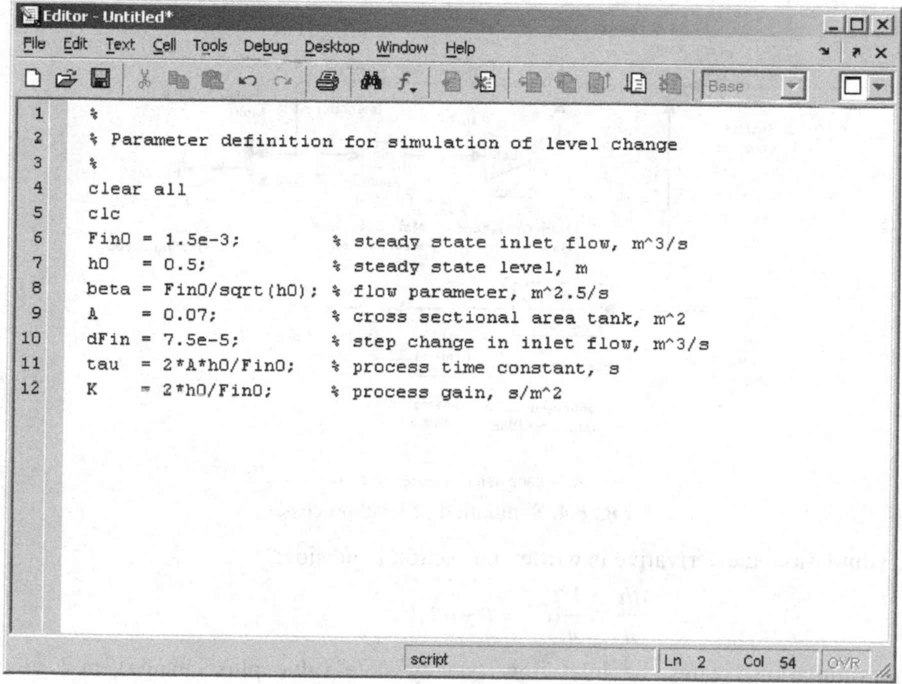


```
%
% Parameter definition for simulation of level change
%
clear all
clc
Fin0 = 1.5e-3;           % steady state inlet flow, m^3/s
h0   = 0.5;              % steady state level, m
beta = Fin0/sqrt(h0);    % flow parameter, m^2.5/s
A    = 0.07;             % cross sectional area tank, m^2
dFin = 7.5e-5;           % step change in inlet flow, m^3/s
tau  = 2*A*h0/Fin0;      % process time constant, s
K    = 2*h0/Fin0;        % process gain, s/m^2
```

Fig. 8.3. Definition of simulation parameters in an m-file.

Next you have to indicate that you are going to build a new model. This is possible by clicking the left-most option on the toolbar of the Simulink Library Browser menu (Create a new model). Let us assume you would like to build the following model:

$$A\frac{dh}{dt} = F_{in} - \beta\sqrt{h} \tag{8.1}$$

with steady state values: $F_{in} = 1.5 \times 10^{-3}$ m^3/s, $A = 0.07$ m^2 and $h_0 = 0.5$ m, as defined in the parameter list, in this case it could be named F0803par.m.

As shown in chapter six, the linearized model relating changes in the inlet flow to changes in the level can be given by:

$$\frac{\delta h}{\delta F_{in}} = \frac{666.67}{46.6s+1} \tag{8.2}$$

From the library browser, shown in Fig. 8.2, we can drag various blocks to the empty model form. For example, the block “integrator” can be positioned on the empty model form, by positioning the mouse on the block, pressing (and holding down) the left mouse button and dragging the block to the form. The Simulink “directory” contains various types of blocks, such as, for example, Continuous, Discrete, Math type blocks and so forth. The model equations (8.1) and (8.2) are shown in Fig. 8.4. For every type of block it is shown where the block was obtained from, for example, the block Gain1 was selected from Simulink\Math Operations\Gain.

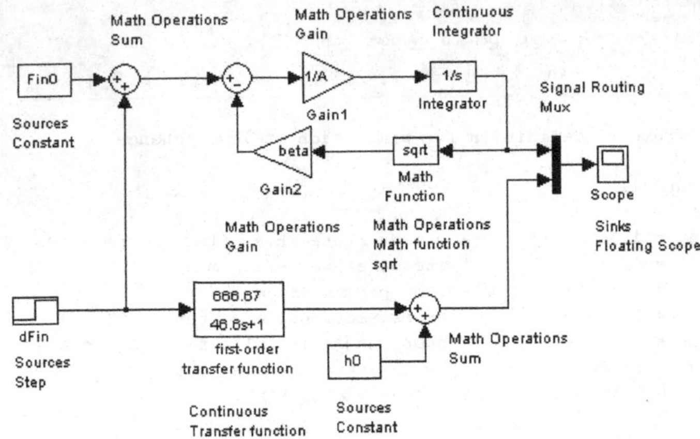

Fig. 8.4. Simulation of level process.

In the simulation, the derivative is written on the left hand side:

$$\frac{dh}{dt} = \frac{1}{A}\left(F_{in} - \beta\sqrt{h}\right) \tag{8.3}$$

The simulation creates the inlet flow (steady state value plus change), and subtracts $\beta\sqrt{h}$, then multiplies it by $1/A$ (Gain1), which then results in the derivative of the level. In order to obtain the value of the level, a simple integration is required, given by the 1/s block. The output is $h(t)$, integration starts at an initial value h_0.

The first-order transfer function, which represents the linearized model, has the change in inlet flow as input, the output is then the change in outlet flow, adding the steady state value of the level results in the actual level. This is then compared with the output from the non-linear process.

There are several ways to monitor the results of the simulation. The easiest method is to route the output signal (level) to a floating scope. If one would like a better graph for reports or other documents, one could route the level also to a block from the Simulink Sinks library and choose To File, in addition to the Scope. Double clicking the File block would show Fig. 8.5.

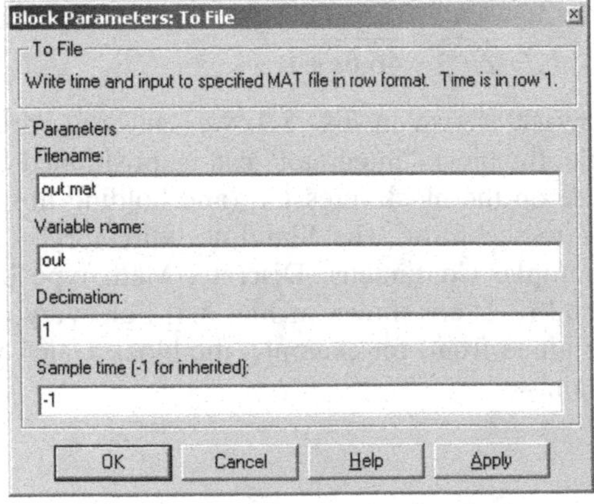

Fig. 8.5. File block data.

One could choose as file name, for example out.mat and as variable name out. The level would then be stored as follows: row one in out.mat would contain values of the time, row two would contain the level from the non-linear process simulation and row three the values of the level from the linearized process model. If one would like to plot the results, one could type:

```
load out;
out = out';
time = out(:,1);
y1 = out(:,2);
y2 = out(:,3);
plot(time, y1,'b',time,y2,'r')
xlabel('time, sec','Fontsize', 15)
ylabel('level', 'Fontsize',15)
grid
```

Save the picture by selecting File Export in the picture window. One could save the picture as an enhanced metafile or jpg file that can easily be inserted in, for example, a Word document.

Another elegant way of saving the simulation to the workspace and then plotting the results proceeds as follows. Double click on the scope and select the Parameters option (second from the left). Select Data History and select the Save data to workspace option. Enter an appropriate name and select Array as format. Subsequently, select Apply and OK. After you run your simulation again, a variable with the name *y* will exist. The file y.mat contains three column, the first one is the time, the second one variable y1, the second one variable y2. You can now use the plot command to visualize the process variables, for example `plot(y(:,1),y(:,2),'r',y(:,1),y(:,3),'b')`.

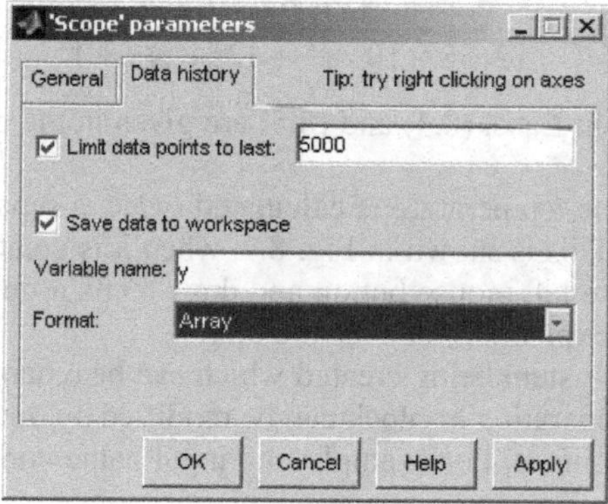

Fig. 8.6. Scope parameters entry.

Hints

Any novice to Simulink will make errors initially, as a result of which the simulation will not run. Therefore the following suggestions may be useful:

- integrators should have an initial condition.
- if one is interested in the dynamic behavior of a process, one should start at the steady state value of the simulation. If one is not sure of all initial conditions, then select at least a long enough simulation time, such that the process reaches steady state. Once the steady state is reached, an input disturbance can be given.
- at the end of a simulation, process signals are usually constant. This holds for most unit operations in the chemical industry, processes in which level changes occur are an excep-

tion. If no constant values are reached, it could be that: (i) the simulation time is too short, or (ii) the process could be unstable.

- if the simulation is unstable, a plus or minus sign could be incorrect. If all signs are correct, choose another solver or another simulation tolerance (simulation → simulation parameters …).
- if the output of the simulation is not a smooth curve, the tolerance should be lowered. (sometimes choosing another solver will help).
- ensure that the model is divided into smaller sub-models and that sub-models are clear. If possible each sub-model should represent a (dynamic) equation or function.
- define a parameter list as an m-file. Run this file before running the simulation. This will enable you to use parameters in the simulation diagram rather then numeric values.
- give blocks, sub-systems, signals and inputs and outputs meaningful names.

8.3 Simulation of the Chemical Reactor

The model for the chemical reactor was given in Eqns. (6.3) and (6.4) and will be repeated here:

$$V\frac{dc_A}{dt} = F(c_{Ain} - c_A) - Vke^{-E/RT}c_A \tag{8.4}$$

and

$$\rho V c_p \frac{dT}{dt} = F\rho c_p (T_{in} - T) + Vke^{-E/RT}c_A \Delta H + Q \tag{8.5}$$

The linearized model, relating changes in the reactor flow to changes in the concentration of component A, could be given by:

$$\frac{\delta c_A}{\delta F} = 6.69 \times 10^4 \frac{457.6s + 1}{2.55 \times 10^5 s^2 + 1254.4s + 1} \tag{8.6}$$

The steady-state values in Eqns. (8.4) and (8.5) are given in the previous chapter, they are defined in F0809par.m.

Since the exponent of the temperature is calculated twice, a separate block is created that performs this calculation. This is shown in Fig. 8.7, when it is created, position the cursor in the top left corner, press the left mouse button and draw a box around the blocks while leaving the input T and output exp(-E/RT) outside the box.

This will result in a subsystem being created which can be renamed, for example to exponent. The text under each operation or block can be modified by double clicking the text box, selecting the text and retyping it. For example, the input name constant can easily be modified to T.

Now the mass balance can easily be created: it is shown in Fig. 8.8.

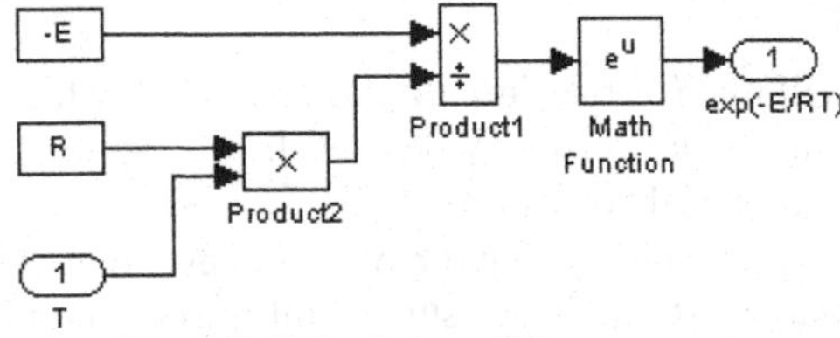

Fig. 8.7. Calculation of the exponent.

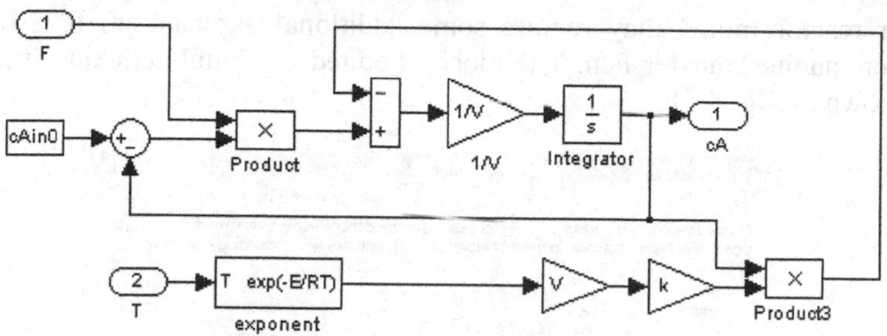

Fig. 8.8. Reactor component balance.

In this case the same procedure is followed and the mass balance is made a subsystem with inputs F and T and output c_A.

The integrator produces the value of c_A, its initial value is 200.13. The model structure closely follows Eqn. 8.4 which is rewritten in such a way that the derivative dc_A/dt becomes the left-hand side of the equation.

In a similar way, the energy balance is created as a subsystem; it is shown in Fig. 8.9.

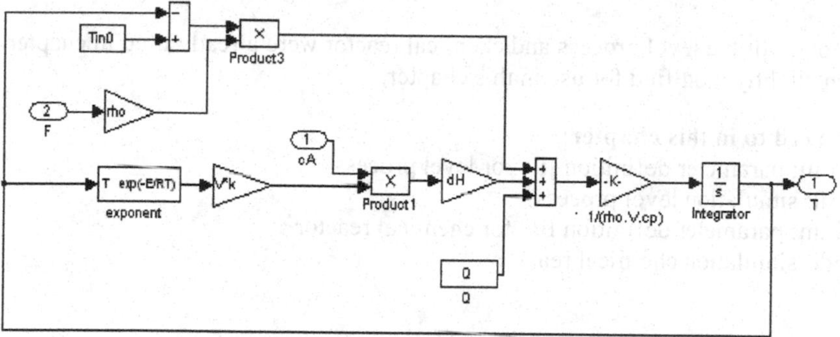

Fig. 8.9. Reactor energy balance.

The initial value of the reactor temperature is 413 K. The non-linear and linear reactor model are shown in Fig. 8.10.

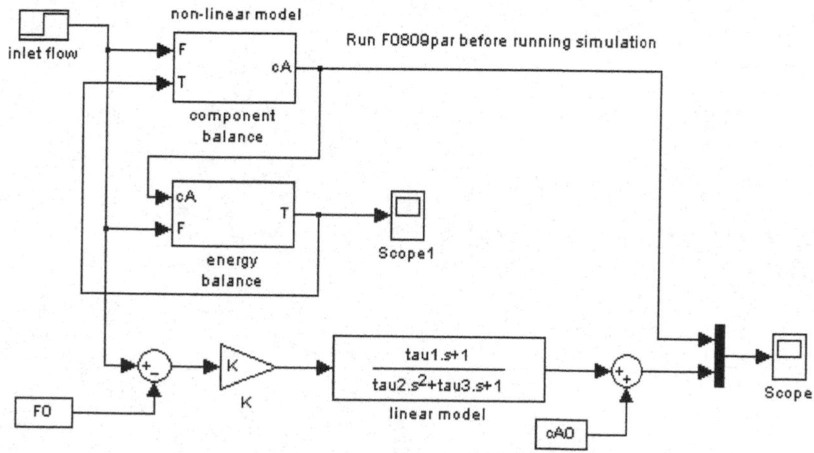

Fig. 8.10. Non-linear and linear reactor models.

The linear reactor model may require some additional explanation. It is chosen from "Simulink\Continuous\Transfer Fcn. The block is edited by double clicking it and entering the data as shown in Fig. 8.11.

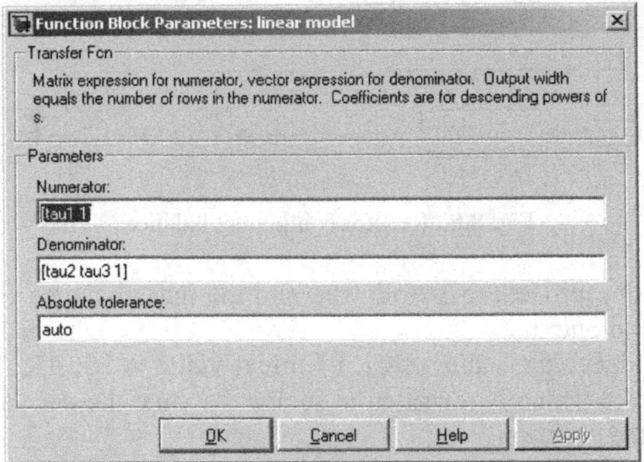

Fig. 8.11. Data input for transfer function block.

The model for the level process and chemical reactor were already used in chapter 6. They have been slightly modified for use in this chapter.

Files referred to in this chapter:
F0804par.m: parameter definition file for level process
F0804.mdl: simulation level process
F0809par.m: parameter definition file for chemical reactor
F0809.mdl: simulation chemical reactor

References

Matlab: http://www.mathworks.com/products/tech_computing/literature.shtml, (2000)

9 Frequency Response Analysis

So far, we have often characterized process behaviour by showing the step response of the process. There is, however, another way of showing the dynamics of a process: by displaying how the process would behave to sinusoidal perturbations. In that case, the ultimate output response of the process will also be a sinusoidal signal, lagging behind the original input signal and with a different amplitude. The dynamics of the process can be fully characterized by plotting the phase shift and amplitude ratio as a function of the frequency of the input signal. This method is often used to design feedback controllers. Here, we will focus on using the method for process dynamics characterization.

9.1 Introduction

Consider a first-order process with transfer function:

$$\frac{\delta y}{\delta u} = \frac{K_p}{\tau s + 1} \tag{9.1}$$

Let us assume that this process is subjected to a sinusoidal input change:

$$\delta u(t) = A \sin \omega t \tag{9.2}$$

where A is the amplitude of the sine wave and ω the frequency.
Using Laplace transform tables, Eqn. (9.2) can be transformed to the Laplace domain:

$$u(s) = \frac{A\omega}{s^2 + \omega^2} \tag{9.3}$$

hence, the Laplace transform of the output becomes:

$$y(s) = \frac{K_p}{\tau s + 1} \frac{A\omega}{s^2 + \omega^2} \tag{9.4}$$

Note that $y(s)$ is defined as $\mathcal{L}[\delta y(t)]$, the Laplace transform of $\delta y(t)$.
Equation (9.4) can be expanded into fractions and therefore be written as:

$$y(s) = \frac{B_1}{s + 1/\tau} + \frac{B_2}{s + j\omega} + \frac{B_3}{s - j\omega} \tag{9.5}$$

The constants B can now be calculated and using tables of inverse Laplace transform, the solution for $t \to \infty$ can be written as:

$$y(t \to \infty) = \frac{AK_p}{\omega^2\tau^2 + 1} \sin \omega t - \frac{AK_p \omega\tau}{\omega^2\tau^2 + 1} \cos \omega t \tag{9.6}$$

This expression can be written as:

$$a_1 \sin \omega t + a_2 \cos \omega t = a_3 \sin(\omega t + \phi) \tag{9.7}$$

in which

$$a_3 = \sqrt{a_1^{\ 2} + a_2^{\ 2}}, \quad \phi = \tan^{-1}\left(\frac{a_2}{a_1}\right) \tag{9.8}$$

resulting in:

$$\begin{aligned} y(t \to \infty) &= \frac{AK_p}{\sqrt{\omega^2\tau^2 + 1}} \sin(\omega t + \phi) \\ \phi &= \tan^{-1}(-\omega\tau) \end{aligned} \tag{9.9}$$

From Eqn. (9.9) it can be seen that the response to a sinusoidal input signal is a sinusoidal signal with the same frequency but a different angle. The output signal lags behind the input signal by an angle ϕ, which depends on the frequency ω.

The ratio between the amplitude of the input sine wave and the output sine wave, called amplitude ratio, can be written as:

$$AR = \frac{K_p}{\sqrt{\omega^2\tau^2 + 1}} \tag{9.10}$$

As can be seen from Eqn. (9.10), it also depends on the frequency ω. It can easily be shown, that these results can also be obtained by substituting $s = j\omega$ in the transfer function.

Substitution of $s = j\omega$ in the transfer function $G(s)$ results in a transfer function in the frequency domain $G(j\omega)$. Then the magnitude is equal to the amplitude ratio AR:

$$AR = |G(j\omega)| \tag{9.11}$$

and the phase angle or argument is equal to:

$$\phi = arg[G(j\omega)] \tag{9.12}$$

$G(j\omega)$ is a complex number, it can therefore be represented by a real and imaginary part:

$$G(j\omega) = Re[G(j\omega)] + j\, Im[G(j\omega)] \tag{9.13}$$

for which:

$$\begin{aligned} AR &= |G(j\omega)| = \sqrt{(Re[G(j\omega)])^2 + (Im[G(j\omega)])^2} \\ \phi &= \arg[G(j\omega)] = \tan^{-1}\frac{Im[G(j\omega)]}{Re[G(j\omega)]} \end{aligned} \tag{9.14}$$

When substituting $s = j\omega$ into Eqn. (9.1), the result is:

$$\begin{aligned} G(j\omega) &= \frac{K_p}{(j\omega\tau + 1)} \frac{(-j\omega\tau + 1)}{(-j\omega\tau + 1)} \\ &= \frac{K_p}{\omega^2\tau^2 + 1} - j\frac{K_p\omega\tau}{\omega^2\tau^2 + 1} \end{aligned} \tag{9.15}$$

Applying Eqn. (9.14) gives:

$$\begin{aligned} magnitude &= AR = \frac{K_p}{\sqrt{\omega^2\tau^2 + 1}} \\ phase\ angle\ \phi &= tan^{-1}(-\omega\tau) \end{aligned} \tag{9.16}$$

which is the same result as Eqns. (9.9) and (9.10). This method is generally applicable to linear systems.

A transfer function is often a combination of sub-transfer functions, consisting of numerator and denominator terms:

$$G(j\omega) = \frac{G_1(j\omega).G_2(j\omega)......G_n(j\omega)}{G_{n+1}(j\omega).G_{n+2}(j\omega).......G_m(j\omega)} \tag{9.17}$$

It can easily be shown (Marlin, 1995; Stephanopoulos, 1984) that the amplitude ratio and phase angle for a product of multiple transfer functions can be computed from:

$$\begin{aligned} AR &= |G_1(j\omega).G_2(j\omega).......| = |G_1(j\omega)|.|G_2(j\omega)|....... \\ \phi &= arg[G_1(j\omega).G_2(j\omega)......] = arg[G_1(j\omega)] + arg[G_2(j\omega)] + ... \end{aligned} \tag{9.18}$$

For inverse transfer functions the amplitude ratio and phase angle can be derived from the property:

$$\begin{aligned} AR &= \left|\frac{1}{G(j\omega)}\right| = \frac{1}{|G(j\omega)|} \\ \phi &= arg\frac{1}{G(j\omega)} = -arg[G(j\omega)] \end{aligned} \tag{9.19}$$

By using the multiplication and division properties, the amplitude ratio and phase angle of the above mentioned general transfer function, consisting of multiple terms, becomes:

$$\begin{aligned} AR &= |G(j\omega)| = \frac{\prod_{i=1}^{n}|G_i(j\omega)|}{\prod_{j=n+1}^{m}|G_j(j\omega)|} \\ \phi &= arg[G(j\omega)] = \sum_{i=1}^{n} arg[G_i(j\omega)] - \sum_{j=n+1}^{m} arg[G_j(j\omega)] \end{aligned} \tag{9.20}$$

with i = numerator terms and j = denominator terms. The overall amplitude ratio and phase angle can therefore easily be derived from the individual amplitude ratio and phase angle.

9.2 Bode Diagrams

The graphs in which the amplitude ratio and phase shift are plotted as a function of the frequency ω, are called Bode diagrams. In the Bode plot, $log(AR)$ and ϕ are shown as a function of ω. In case of the above mentioned first-order process, this means:

$$\begin{aligned} AR &= \frac{K_p}{\sqrt{\omega^2\tau^2+1}} \quad or \\ log(AR) &= log(K_p) - \frac{1}{2}log(\omega^2\tau^2+1) \end{aligned} \tag{9.21}$$

In case, $\omega >> 1/\tau$:

$$log(AR) = log(K_p) - log(\omega\tau) \tag{9.22}$$

thus $log(AR)$ becomes a linear function of $log(\omega\tau)$ with a slope of –1. In case of $\omega << 1/\tau$:

$$log(AR) = log(K_p) + \frac{1}{2} log(1) = log(K_p) \tag{9.23}$$

thus the gain of the process is independent of the frequency ω. In case $\omega = 1/\tau$:

$$log(AR) = log(K_p) + \frac{1}{2} log(2) \tag{9.24}$$

The phase shift of the first-order process was shown in Eqn. (9.16b).
In case, $\omega >> 1/\tau$:

$$\phi = -tan^{-1}\infty = -90^o \tag{9.25}$$

In case $\omega << 1/\tau$:

$$\phi = -tan^{-1} 0 = 0^o \tag{9.26}$$

and in case $\omega = 1/\tau$:

$$\phi = -tan^{-1} 1 = -45^o \tag{9.27}$$

In this section we shall look at Bode diagrams for a number of different processes.

9.2.1 Integrating Process

A simple example of an integrating process is a tank with constant outflow and varying inlet flow. The mass balance can be written as:

$$A\frac{dh}{dt} = F_{in} - F_{out} \tag{9.28}$$

where it is assumed that the fluid density is constant, A is the cross sectional tank area, h the liquid height and F the volumetic flow. If F_{out} is constant, the linearized model that relates changes in the inlet flow to changes in the level can be given as:

$$\frac{\delta h}{\delta F_{in}} = \frac{1}{As} = \frac{K_p}{s} \tag{9.29}$$

The frequency response can be determined by substituting $s = j\omega$ in Eqn. (9.29). The amplitude ratio and phase shift are:

$$\begin{aligned} AR &= |G(j\omega)| = \frac{|G_1|}{|G_2|} = \frac{K_p}{\omega} \\ \phi &= tan^{-1} 0 - tan^{-1}\infty = -90^o \end{aligned} \tag{9.30}$$

This result can easily be verified by using MATLAB, for the case K_p=1:

```
K = 1.0;
tau = 1.0;
H = tf(K,[tau 0]);
w = logspace(-1,1);
[mag,phase]=bode(H,w);
for i = 1:100
 AR(i) = mag(1,1,i);
 phi(i) = phase(1,1,i);
end
subplot(2,1,1),loglog(w,AR),title('AR');
subplot(2,1,2),semilogx(w,phi),title('phi');
xlabel('frequency,rad/s');
```

which results in Fig. 9.1.

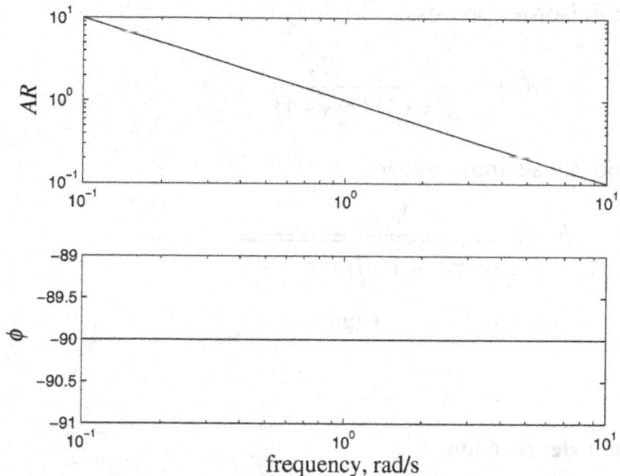

Fig. 9.1. Bode diagram of an integrating process.

9.2.2 First-order Process

The frequency information or Bode diagram for a first-order process, such as given in Eqn. (9.1) can easily be calculated using Eqn. (9.10) for the amplitude ratio and Eqn. (9.9b) for the phase angle ϕ. Assume $K_p = 1$ and $\tau = 10$, then the result as shown in Fig. 9.2 is obtained, using the process definition in MATLAB: H = tf([0 K_p],[τ 1]).

Several interesting things can be observed. For low frequencies, the process gain approaches one, the value of K_p. For relatively high frequencies, $j\omega >> 1/\tau$, the process transfer function approaches $1/j\omega$, which is an integrating process, as shown in Eqn. (9.29). In that case the phase shift ϕ approaches –90°.

The low-frequency asymptote of the amplitude ratio curve intersects the high frequency asymptote at $\omega = 0.1$ rad/s, at this point $\omega = 1/\tau$, where τ is the process time constant. This frequency is often called the corner frequency.

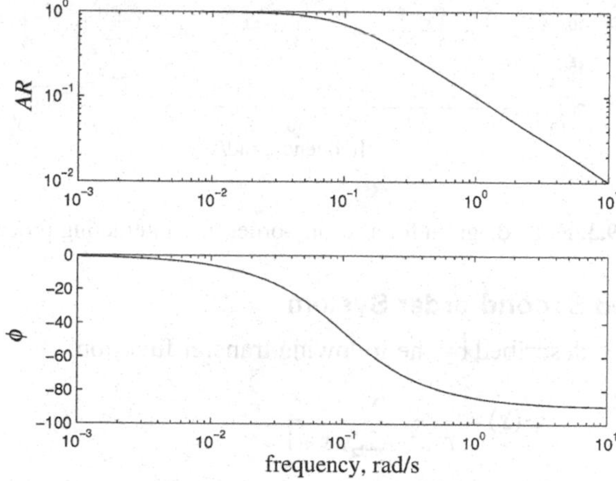

Fig. 9.2. Bode diagram of a first-order process.

9.2.3 Second-order Non-interacting System

For the following second-order process:

$$G(s) = \frac{K_1}{(\tau_1 s + 1)} \frac{K_2}{(\tau_2 s + 1)} \tag{9.31}$$

the amplitude ratio and phase angle become:

$$AR = \frac{K_1 K_2}{\sqrt{\omega^2 \tau_1{}^2 + 1}\ \sqrt{\omega^2 \tau_2{}^2 + 1}}$$
$$\phi = \tan^{-1} - \omega\tau_1 + \tan^{-1} - \omega\tau_2 \tag{9.32}$$

Example

Assume the following transfer function:

$$G(s) = \frac{2}{(10s + 1)(3s + 1)} \tag{9.33}$$

The amplitude ratio and phase angle can easily be computed using Eqn. (9.32). Matlab can also be used to construct the Bode diagram, the result is shown in Fig. 9.3.

It can be seen, that the phase angle approaches 2 × (–90) = –180°. Similarly, the maximum phase angle for a third order process will be 3 × (–90) = –270°.

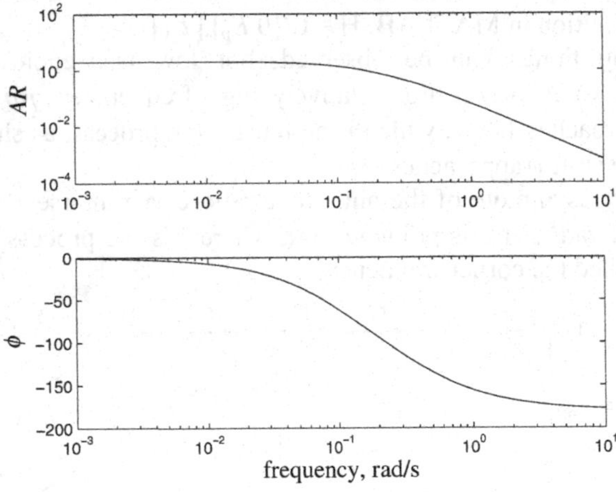

Fig. 9.3. Bode diagram for a second-order non-interacting process.

9.2.4 Underdamped Second-order System

Some processes can be described by the following transfer function:

$$G(s) = \frac{K_p}{\tau^2 s^2 + 2\xi\tau s + 1} \tag{9.34}$$

Substituting $s = j\omega$ into Eqn. (9.34) and rearranging results in:

$$G(j\omega) = \frac{G_1}{G_2} = \frac{K_p}{(1-\omega^2\tau^2) + 2j\xi\omega\tau} \tag{9.35}$$

Applying Eqn. (9.14) gives the following result for the amplitude ratio and phase shift:

$$AR = \frac{K_p}{\sqrt{(1-\omega^2\tau^2)^2 + (2\xi\omega\tau)^2}}$$
$$\phi = \tan^{-1}\left(-\frac{2\xi\omega\tau}{1-\omega^2\tau^2}\right) \tag{9.36}$$

The amplitude ratio and phase shift are shown in Fig. 9.4 for values of $\zeta = [0.1, 0.2, 1.0]$ and $\tau = 10$ s. For high frequencies the slope in the AR graph is –2, since the process is second order. The maximum phase shift is 2 × (–90) = –180°.

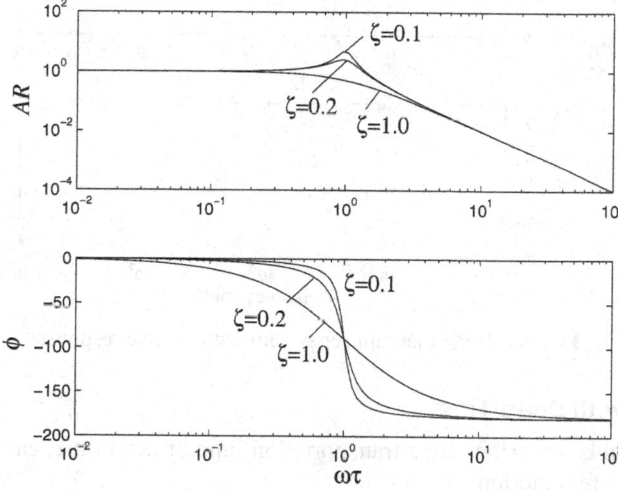

Fig. 9.4. Bode diagram for second-order under-damped process.

For values of $\zeta > 1.0$, the response is the same as that shown in section 9.2.3, since Eqn. (9.34) can be written as Eqn (9.31).

9.2.5 Inverse Response System

Consider the following transfer function:

$$G(s) = \frac{K_p(1-\tau_1 s)}{(1+\tau_2 s)} \tag{9.37}$$

Upon a step change in the input, this transfer function will show an inverse response. The amplitude ratio and phase angle can easily be derived:

$$AR = |G(j\omega)| = K_p \frac{\sqrt{\omega^2\tau_1{}^2 + 1}}{\sqrt{\omega^2\tau_2{}^2 + 1}}$$
$$\phi = tan^{-1}(-\omega\tau_1) - tan^{-1}(\omega\tau_2) \tag{9.38}$$

The Bode diagram is shown in Fig. 9.5 for K_p = 1.0, both time constants have a value of 5.0. In this case:

$$AR = |G(j\omega)| = K_p \frac{\sqrt{\omega^2\tau^2+1}}{\sqrt{\omega^2\tau^2+1}} = K_p$$
$$\phi = tan^{-1}(-\omega\tau) - tan^{-1}(-\omega\tau) = -2tan^{-1}(-\omega\tau) \tag{9.39}$$

The amplitude ratio has therefore a value of 1.0, the process gain, the maximum phase shift is –180°.

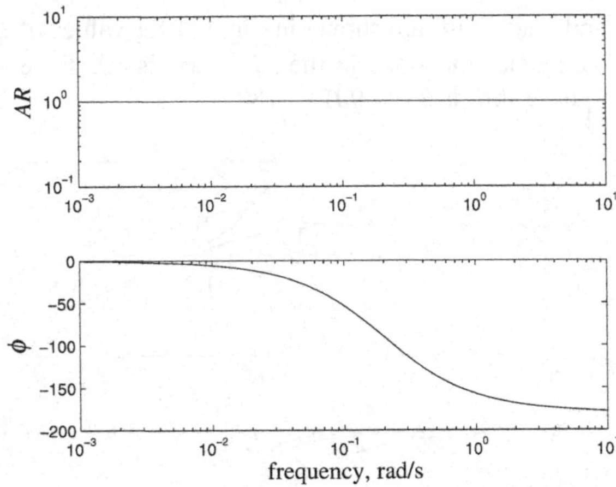

Fig. 9.5. Bode diagram for system with inverse response.

9.2.6 Process with Dead Time

A process that can be described by a transportation time or dead time, can be represented by the following transfer function:

$$G(s) = e^{-\theta s} \tag{9.40}$$

with θ the process dead time.

Substituting $s = j\omega$ into Eqn. (9.40) yields for the amplitude ratio and phase angle or phase shift:

$$AR = |\,G(j\omega)\,| = \left|e^{-j\omega\theta}\right| = \sqrt{sin^2\omega\theta + cos^2\omega\theta} = 1$$
$$\phi = arg[G(j\omega)] = tan^{-1}\frac{sin(-\omega\theta)}{cos(-\omega\theta)} = tan^{-1}tan(-\omega\theta)) = -\omega\theta \tag{9.41}$$

The Bode diagram for this process with θ= 10 s is shown in Fig. 9.6.

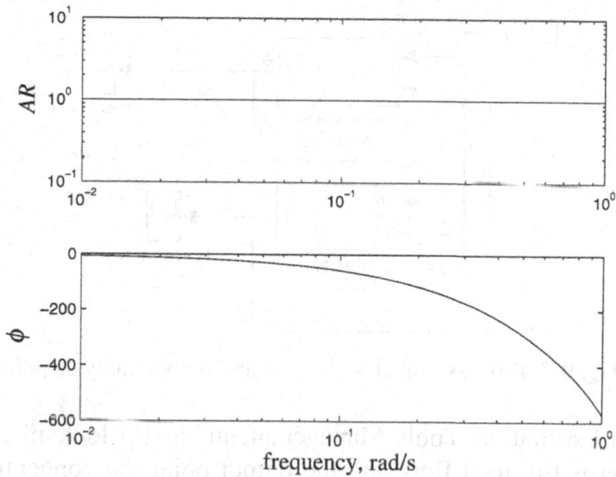

Fig. 9.6. Bode diagram for dead time process.

As can be seen, the phase shift has an exponential shape in a graph with a semi-logarithmic scale.

The process with an inverse response as well as the dead time process have an amplitude ratio which is frequency independent, they only introduce a phase shift.

The Bode plots are useful, since they can be used in controller tuning, which will be discussed in chapter 32. They also are useful if we know the frequency of the disturbance that the process is subjected to. It can easily be seen that high frequency disturbances have less impact on the process than low frequency disturbances. Controller tuning also depends on the disturbance frequency as will be shown later.

9.3 Bode Diagram of Simulink Models

The nonlinear model of the chemical reactor was already discussed in the previous chapters. How would the Bode diagram of this model look like? To answer this question, Let us open this model which is stored in file F0907A.mdl. On the menu bar, select "Tools", "Control Design", "Linear Analysis". This will open the Control and Estimation Tools Manager. Go back to the F0907A model window and position the cursor after the feed F and click the right mouse button. A menu will pop up from which you should select "Linearization points", "Input points". Next, position the cursor at the output c_A of the mass balance, click the right mouse button again and select "Linearization points", "Output points". Select the Control and Estimation Tools Manager menu and select "Plot linear analysis in a Bode response plot". The model should now show the input and output points as shown in Fig. 9.7.

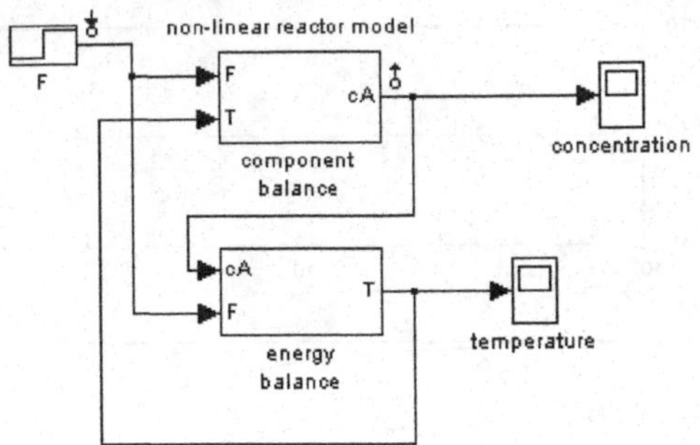

Fig. 9.7. Process model with input and output analysis points.

The Control and Estimation Tools Manager menu should look like Fig. 9.8. As can be seen, the input point is the feed flow and the output point the concentration from the mass balance. On this display, now select "Linearize Model", this will produce the Bode diagram for the linearized reactor model as shown in Fig. 9.9.

It can be seen that the phase shift approaches -90 degrees for high frequencies, i.e. this indicates first-order behavior. The process gain that was found for low frequencies was equal to 6.69×10^4 which corresponds to $20 \log_{10}(6.69 \times 10^4) = 96.5$ db, which is also the value in shown in Fig. 9.9.

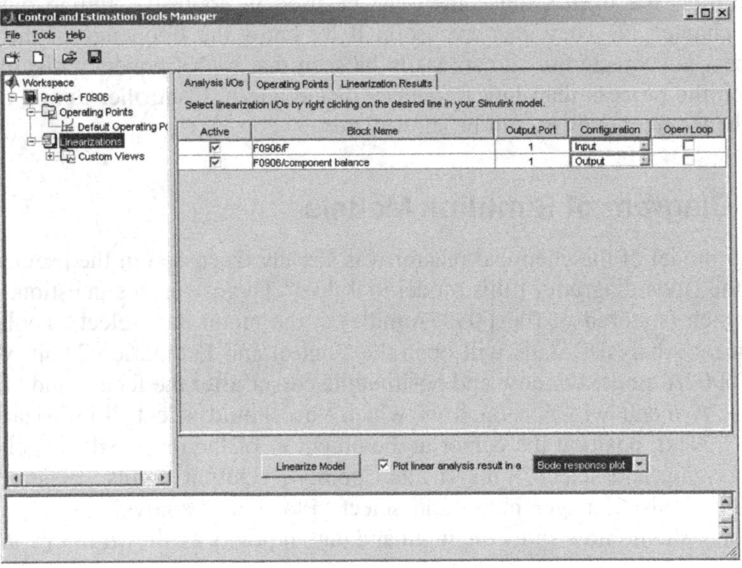

Fig. 9.8. Control and Estimation Tools Manager.

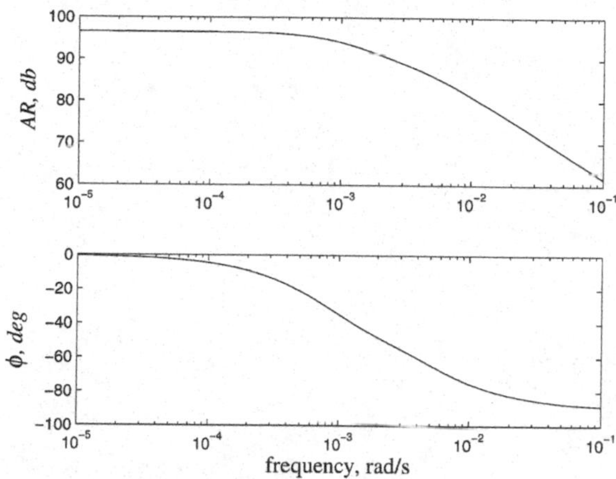

Fig. 9.9. Bode diagram for the linearized reactor model from feed to outlet concentration.

Files referred to in this chapter:

F0901: generation of Bode diagram of Fig. 9.1
F0902: generation of Bode diagram of Fig. 9.2
F0903: generation of Bode diagram of Fig. 9.3
F0904: generation of Bode diagram of Fig. 9.4
F0905: generation of Bode diagram of Fig. 9.5
F0906: generation of Bode diagram of Fig. 9.6
F0907A.mdl: reactor model for Bode analysis
F0907B.mdl: reactor model with input and output points for Bode analysis

References

Marlin, T.E. (1995) *Process Control, Designing Processes and Control Systems for Dynamic Performance*, McGraw-Hill.

Stephanopoulos, G. (1984) *Chemical Process Control, an Introduction to Theory and Practice*, Prentice Hall.

10 General Process Behavior

In the previous chapter the behavior of elementary transfer functions has been discussed in the frequency domain. Systems that are more complex are often composed of a series and/or parallel connection of these elementary transfer functions. In chapter 1 the use of diagrams was introduced to show the coherence between systems. In this chapter the behavior of the different processes will be explained in the time domain based on these diagrams, covering the entire range from elementary first-order lumped systems to complex distributed systems. In the following chapters 11–16 the behavior of different process units will be described based on this general process behavior.

10.1 Introduction

The dynamic behavior of systems can be divided into several categories:

1. Lumped processes

   These are processes or sections of processes, which can be assumed to be ideally mixed and consequently all state variables are only dependent on time and are independent of a certain location. Examples of these processes are ideally mixed reactors, evaporators and one-stage separators. These processes can be described by a combination of elementary balances:

   - accumulation process

     These systems can be described by one elementary balance which refers to the accumulation of mass, energy or components.
   - processes with interactive balances

     Many processes in the chemical industry can be characterized by the interaction between two or more balances, such as the interaction between a mass and energy balance (evaporator) or a component balance and an energy balance (reactor). The dynamic behavior can be explained by the denominator of transfer function. The balances can exhibit an interaction to the extent that the dynamic response starts to show oscillations.
   - processes with parallel balances

     In this case input variables can influence output variables along more than one path, which results in parallel balances. For example, a feed flow often has an impact on the component balance (one or more reactants), the energy balance (cooling or heating) as well as the mass balance. The balances can show opposite behavior which would result in an inverse response.

2. Distributed and chained processes

   In these processes the state variables are not only a function of time but also of the location. Distributed processes possess variables that are continuously distributed with respect to the location. Examples are heat exchangers, plug flow reactors and packed separation columns. These processes can be divided into hypothetical sections in series.

   Chained processes on the other hand possess a fixed division in sections. The division is dictated by the construction of the equipment. The best-known example is a tray column. Also combinations of serially connected unit operations can be seen as a chained process. Examples are multi-effect evaporators, connected extractor-separator combinations or a chain of reactors.

Process Dynamics and Control: Modeling for Control and Prediction. Brian Roffel and Ben Betlem.

The description of the behavior of chained processes and distributed processes is similar. Usually, every section is considered to be ideally mixed. In that case these processes can be described by series of elementary balances. The way in which the balances are coupled, determines the dynamic behavior. Depending on the way in which the balances are coupled, a distinction can be made into two types of distributed processes:

- propagation without feedback
  Temperature and composition gradients propagate through a system with the velocity of the conveying liquid or gas flow. An example is heat convection. Heated liquid or gas will propagate with the flow. The general equation for such systems is:

$$\frac{\partial x(z,t)}{\partial t} = v\frac{\partial x(z,t)}{\partial z} \tag{10.1}$$

  in which x is the temperature, concentration or mass. The change of the variable x at time t and location z depends on the velocity v.

- propagation with feedback
  Differences in temperature, composition and mass over a distance give rise to a gradient. By changing the boundary conditions, the shape of the gradient will vary. An example is heat conduction. The heat gradient in a wall will change when the temperature at one of the sides has been altered. The general equation for such systems is:

$$C\frac{\partial x(z,t)}{\partial t} = \lambda\frac{\partial^2 x(z,t)}{\partial z^2} \tag{10.2}$$

  in which x is the temperature, mass or concentration. The change of the variable x at time t and location z depends on the capacity C and the velocity of transfer λ.

10.2 Accumulation Processes

Accumulation without State Feedback

The general equation for an accumulator without state feedback is:

$$C\frac{dx}{dt} = \sum_{i=1}^{n} u_i \tag{10.3}$$

x represents the state, u the manipulated input variable and C the capacity of the system. If the sum of the inputs equals zero, the state of the system does not change.

Example: Tank

Consider a tank with an inlet flow F_{in} and outlet flow F_{out}, as shown in Fig. 10.1.

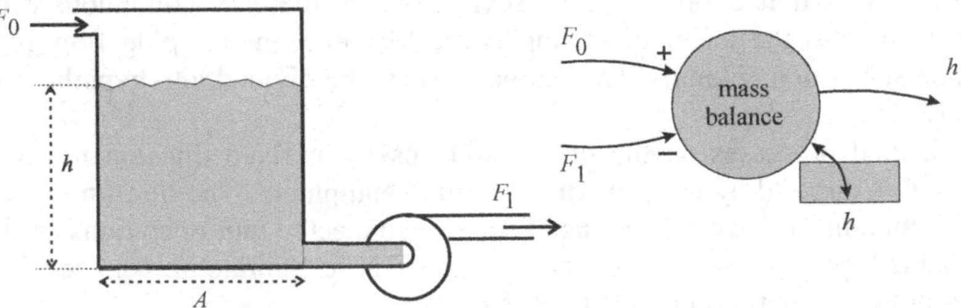

Fig. 10.1. Tank with constant outlet flow.

The outlet flow is on flow control, hence it will not change in time. Assuming constant physical properties, a mass balance can be written as:

$$\rho_f A \frac{dh}{dt} = F_0 - F_1 \tag{10.4}$$

with ρ_f = fluid density (kg/m^3), A = cross-sectional tank area (m^2), h = liquid level in the tank (m) and F = fluid flow (kg/s).

Even though this equation is linear, it can still be written in terms of deviation variables and Laplace transformed, and since $\delta F_{out} = 0$, it can be written as:

$$\frac{\delta h}{\delta F_0} = \frac{1}{\rho_f A s} = \frac{1}{\tau s} \tag{10.5}$$

in which δ= variation around the steady state.

The transfer function in Eqn. (10.5) is called an "integrator", the response for a unit step change in the inlet flow is shown in Fig. 10.2, for τ= 10.0.

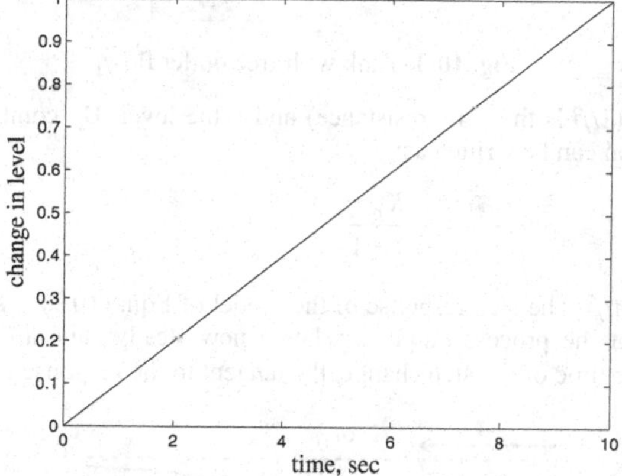

Fig. 10.2. Level response to a unit step in the input.

The main characteristic of this process is that it is not stable. Upon a step change in the process input, the output does not reach a new steady state.

Accumulation with State Feedback

The general equation for an accumulator with state feedback is:

$$\tau \frac{dx}{dt} + x = \sum_{i=1}^{n} K_i u_i \tag{10.6}$$

x is the state of the system, u the manipulated input variable and τ the time constant of the system. The time constant is determined by the ratio of the capacity (for example contents or heat capacity of the vessel) and flow or the product of capacity and flow resistance. These processes are called first-order processes. A change in the manipulated input variable u results in a driving force. The system moves to a new steady state for which it holds that:

$$x = \sum_{i=1}^{n} K_i u_i \tag{10.7}$$

All kinds of accumulation of mass, components or energy can be described by this formulation.

Example: Tank

Suppose, the outflow (Fig. 10.3) depends on the level (laminar outflow), according to:

$$F_1 = \beta h \tag{10.8}$$

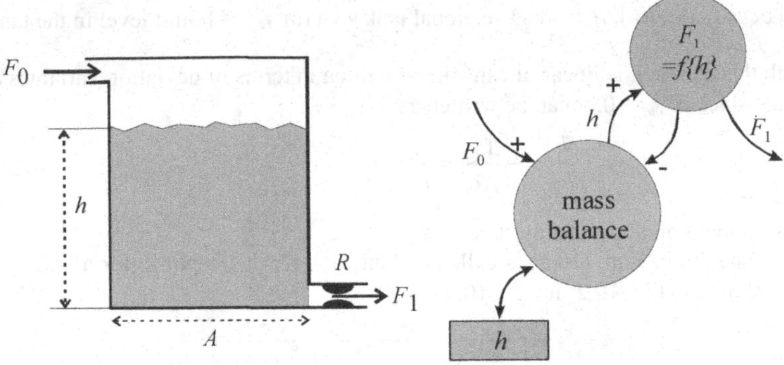

Fig. 10.3. Tank with free outlet flow.

in which β is a constant ($1/\beta$ is the flow resistance) and h the level. By combining Eqns. (10.4) and (10.8), the transfer function can be written as:

$$\frac{\delta h}{\delta F_0} = \frac{K_p}{\tau\, s + 1} \tag{10.9}$$

with $K_p = 1/\beta$ and $\tau = \rho_f A/\beta$. The step response of the model of Eqn. (10.9) is shown in Fig. 10.4. The main characteristic is that the process output reaches a new steady state after a step change in the process input. At $t = 0$, the time of the step change, the tangent to the response curve is K/τ.

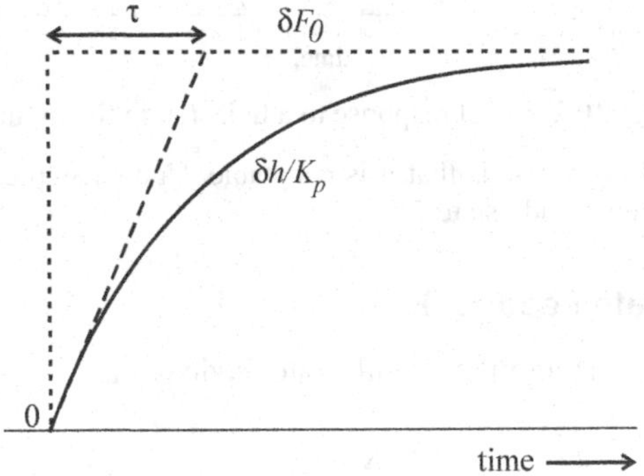

Fig. 10.4. Response of first-order process to unit step change in input.

10.3 Lumped Process with Non-interacting Balances

The general equation for a system with two non-interacting balances is:

$$\begin{aligned} \tau_1 \frac{dx}{dt} + x &= \sum_{i=1}^{n} K_i u_i \\ \tau_2 \frac{dy}{dt} + y &= \sum_{j=1}^{m} K_j u_j \end{aligned} \tag{10.10}$$

x and y represent the states, u the manipulated input variable and τ_1 and τ_2 the time constants of the system. Figure 10.5 shows the data flow diagram of this system. Sometimes, a weak connection between the balances may be present, based on the mass contents. This weak interconnection, in the data flow diagram indicated with a dashed arrow, is no real state coupling.

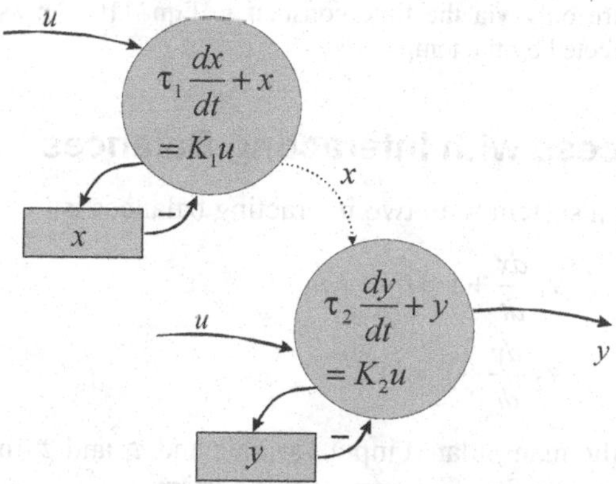

Fig. 10.5. Data flow diagram of system with two non-interacting balances.

Example: Boiler

Figure 10.6 shows a liquid heater with an electrical heating element. The outflow depends on the liquid level in the vessel as described in the previous example.

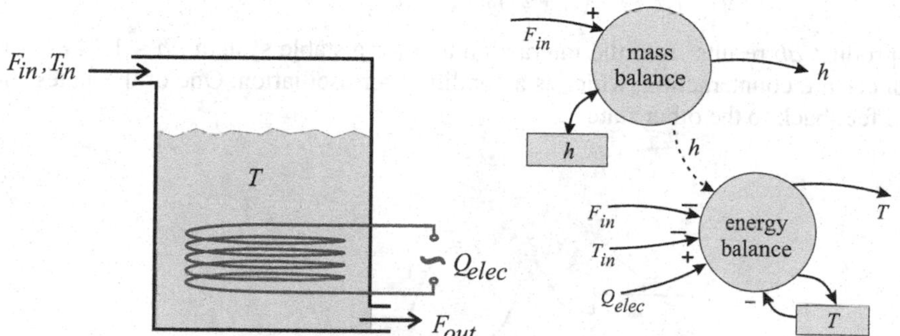

Fig. 10.6. Schematics and data flow diagram of a heater.

The heater can be described by a mass and an energy balance:

$$\begin{aligned} A\frac{dh}{dt} &= F_{in} - F_{out} = F_{in} - \beta h \\ c_p \rho A \frac{dhT}{dt} &= c_p \rho F_{in} T_{in} - c_p \rho F_{out} T + Q_{elec} \end{aligned} \tag{10.11}$$

From a substitution of the mass balance into the energy balance, it follows that:

$$c_p \rho A h \frac{dT}{dt} = c_p \rho F_{in} (T_{in} - T) + Q_{elec} \tag{10.12}$$

Consequently, the transfer function from the temperature to input flow variations becomes:

$$\frac{\delta T}{\delta F_{in}} = \frac{(T_{in} - T)_0 / F_{in,0}}{\frac{Ah_0}{F_{in,0}} s + 1} \tag{10.13}$$

As the balances are not coupled, the results for h and T are two separated first-order equations. The level affects the temperature only via the time constant in Eqn. (10.13); as can be seen from Eqn. (10.11a), the level is not affected by the temperature.

10.4 Lumped Process with Interacting Balances

The general equation for a system with two interacting balances is:

$$\begin{aligned} \tau_1 \frac{dx}{dt} + x &= ay + Ku \\ \tau_2 \frac{dy}{dt} + y &= bx \end{aligned} \tag{10.14}$$

x represents the state, u the manipulated input variable and τ_1 and τ_2 the time constants of the system. Fig. 10.7 shows a data flow diagram of this system.

The connection between the two balances is mutual. Two interactive first-order systems result in a second-order system.

The transfer function between the input u and the output y can easily be derived by Laplace transformation and substitution of one of both balances in the other.

$$\frac{y}{u} = \frac{bK}{\tau_1 \tau_2 s^2 + (\tau_1 + \tau_2) s + (1 - ab)} \tag{10.15}$$

The product ab results from the interaction and for a stable system $ab < 1$. If $ab < 0$ then the balances are counteracting, which is a condition for oscillation. One of the states shows a negative feedback to the other state.

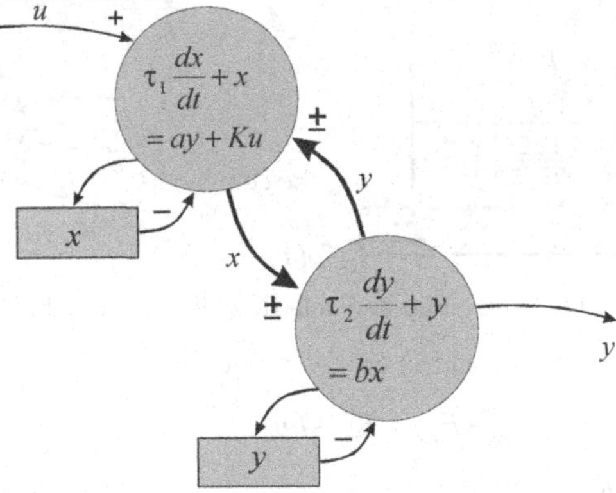

Fig. 10.7. Data flow diagram of a system with two interacting balances.

The term bK indicates the stationary influence, or gain, of the input u to the output y. The denominator may have real or complex poles. This will be investigated in more detail. The general equation for a second-order system is:

$$\frac{y}{u} = \frac{K_p}{\tau^2 s^2 + 2\varsigma\tau\, s + 1} \tag{10.16}$$

in which, K_p is the steady state gain, τ is the natural period and ς the damping factor. Expanded in its poles this equation becomes:

$$\frac{y}{u} = \frac{K_p}{\left(\tau s + \varsigma + \sqrt{\varsigma^2 - 1}\right)\left(\tau s + \varsigma - \sqrt{\varsigma^2 - 1}\right)} \tag{10.17}$$

Depending on the value of ζ, several cases can be distinguished (see also Fig.10.8):

- if $\zeta < 0$, then the system is unstable.
- if $\zeta = 0$, the poles are $p_1 = j/\tau$ and $p_2 = -j/\tau$. In that case, the system is asymptotically unstable and shows an undamped oscillation according to its eigen frequency $1/\tau$. This occurs when both state-feedback terms in the balance equations are lacking:

$$\begin{aligned} \tau_1 \frac{dx}{dt} &= ay + Ku \\ \tau_2 \frac{dy}{dt} &= bx \end{aligned} \tag{10.18}$$

- if $0 < \zeta < 1$, then the poles are complex and the system will show oscillatory behavior and is called to be under-damped. The system may have an amplitude gain of more than one when it is stimulated at its eigen frequency.
- if $\zeta = 1$, the system has two identical poles and is called to be critically damped.
- if $\zeta > 1$, the system has two real poles and is called to be over-damped.

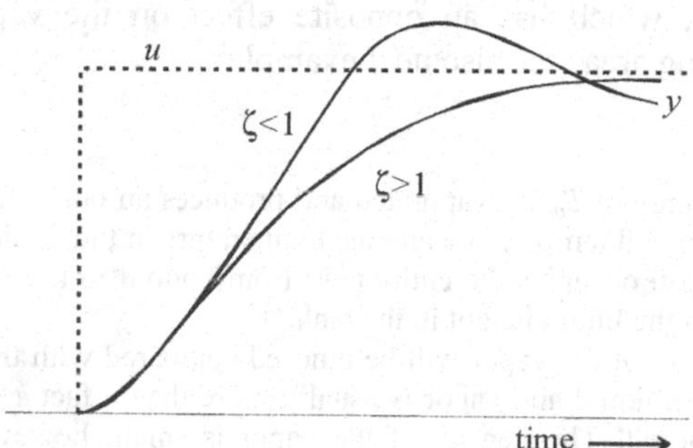

Fig. 10.8. Response to step change in inlet flow for a second-order process.

Most uncontrolled interactive systems are over-damped. Only a few processes show under-damped behavior. When the dynamic behavior of an uncontrolled system becomes oscillatory, the balances are counteracting. An example is an exothermal reaction. Also evaporators may show this behavior under certain conditions. In case of controlled systems, under-damped behavior occurs more frequently, caused by state-feedback which is introduced by the controller.

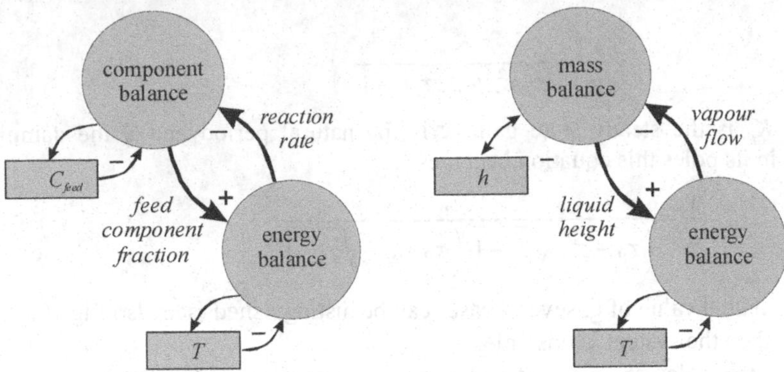

Fig. 10.9. Interacting balances, left: evaporator, right: reactor.

- A reactor has an interacting energy and component balance (Fig. 10.9 left). The component balance is a measure of the conversion that can potentially be achieved, whereas the energy balance determines the rate of conversion. In case of an exothermal reaction, a higher temperature increases the rate of reaction which subsequently decreases the feed component concentration, which has an opposite effect on the rate of reaction.
- An evaporator with a heat exchange area which varies according to the liquid height, has interacting mass and energy balances (Fig. 10.9 right). The liquid mass balance is a measure of the vapor that can potentially be generated, whereas the energy balance determines the vapor rate. A higher temperature, which is directly coupled to the pressure, will increase the vapor rate which subsequently decreases the liquid mass, which has an opposite effect on the vapor rate owing to the decreasing heating area (see also next example).

Example: Evaporator

A liquid flow F_{in} with temperature T_{in} is evaporated and produces an outlet flow F_{out}. The temperature difference between the steam coil temperature and the temperature in the liquid tank is the driving force in this process. The steam coil occupies the entire tank from bottom to top. This implies that the heat transfer area is dependent on the liquid height in the tank.

The mass and heat capacity of the vapor will be ignored compared with that of the liquid, since the difference in density between liquid and vapor is usually more than a factor of 50. However, the flow of the vapor cannot be ignored! The density of the vapor is small, however, it does have a larger volume than the liquid. Some other simplifications are that the heat capacity of the heating coil is small compared with the heat capacity of the liquid and that the physical properties are constant. Heat loss to the environment will also be ignored.

The interactions between the relationships for the evaporator are fairly complex, they are shown in Fig. 10.10. There are two dynamic relationships: the mass balance and the energy balance; however, the algebraic relationships belong to the energy balance and form one entity with it.

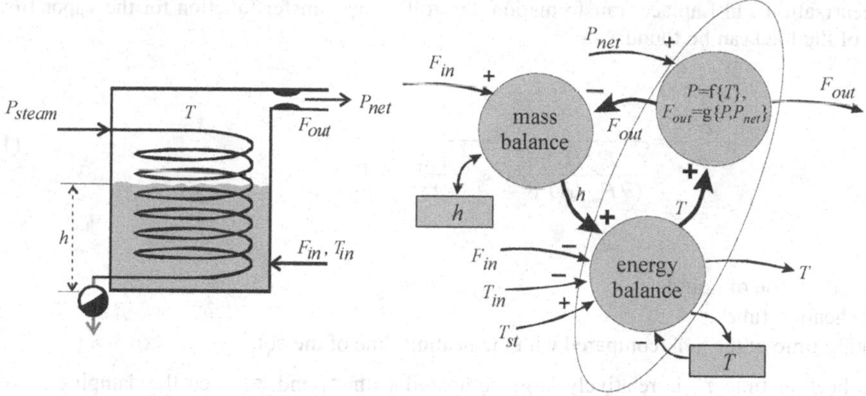

Fig. 10.10. Data flow diagram of the interaction between the relationships for an evaporator.

The evaporator can be described by the following equations (for more detail the reader is referred to chapter 15). The mass balance can be given by:

$$\rho A_c \frac{dh}{dt} = F_{in} - F_{out} \tag{10.19}$$

in which ρ = liquid density, A_c = cross-sectional area of the tank, F = mass flow and h = liquid height in the tank. The mass balance has no state-feedback of the liquid height and integrates the difference between input and output flow.

The energy balance is:

$$c_p \rho A_c h \frac{dT}{dt} = c_p F_{in}(T_{in} - T) - F_{in}\Delta H + UA_{coil}\frac{h}{h_{max}}(T_{steam} - T) \tag{10.20}$$

in which c_p = specific heat of the liquid, T the temperature of the liquid in the tank, ΔH the heat of vaporization, UA_{coil} the product of heat transfer coefficient and heat exchanging area, h_{max} the total height of the coil that can be exposed to the liquid.

The flow of vapor from the tank can be given by:

$$F_{out} = \beta\sqrt{P - P_{net}} \tag{10.21}$$

in which P_{net} = steam pressure in the central vapor system, and β a proportionality constant.

The equilibrium relationship between temperature and pressure for a liquid at boiling point is (Clapeyron):

$$P = P_{ref}\,exp\left(\frac{\Delta H}{R}\left(\frac{1}{T_{ref}} - \frac{1}{T}\right)\right) \tag{10.22}$$

If the algebraic relationships between temperature and pressure (Eqn. (10.22)) and pressure and vapor flow (Eqn. (10.21)) are used, the temperature gradient can be replaced by the gradient of F_{out}: dF_{out}/dt.

$$\frac{dT}{dt} = \frac{\partial T}{\partial P}\frac{\partial P}{\partial F_{out}}\frac{dF_{out}}{dt} \tag{10.23}$$

Hence the energy balance causes changes in the outlet flow. There are two balances which counteract each other.

After linearization and Laplace transformation, the following transfer function for the vapor flow as a function of the load can be found:

$$\frac{\delta F_{out}}{\delta F_{in}} = \frac{-\tau_3 \frac{T_{tank} - T_{in}}{T_{tank}} s + 1}{\frac{F_{out}/T}{(\partial F_{out}/\partial T)}\left(\tau_3\tau_1 s^2 + \frac{\tau_3\tau_1}{\tau_2} s\right) + 1} \tag{10.24}$$

in which:

τ_1 residence time of liquid
τ_2 total heating time
τ_3 heating time of the tank compared with the heating time of the coil

If the heating time τ_2 is relatively large compared with τ_1 and τ_3, then the damping is low and oscillatory behaviour may occur. One should realize, however, that the time constants will show a dependency on each other and can therefore not assume arbitrary values.

10.5 Processes with Parallel Balances

The general equation for a system with two parallel balances is:

$$\begin{aligned} \tau_1 \frac{dx_1}{dt} + x_1 &= K_1 u \\ \tau_2 \frac{dx_2}{dt} + x_2 &= -K_2 u \\ y &= x_1 + x_2 \end{aligned} \tag{10.25}$$

x_1 en x_2 represent the states, y is the output, u the manipulated input variable and τ_1 and τ_2 the time constants of the system. Fig. 10.11 shows a data flow diagram of this system. Equation (10.25) shows an interesting case when the equations for two accumulation processes have opposite signs for the inputs. One state variable reacts positively on an input change, the other state variable reacts negatively on an input change. The total output is the sum of the output of the two sub-systems.

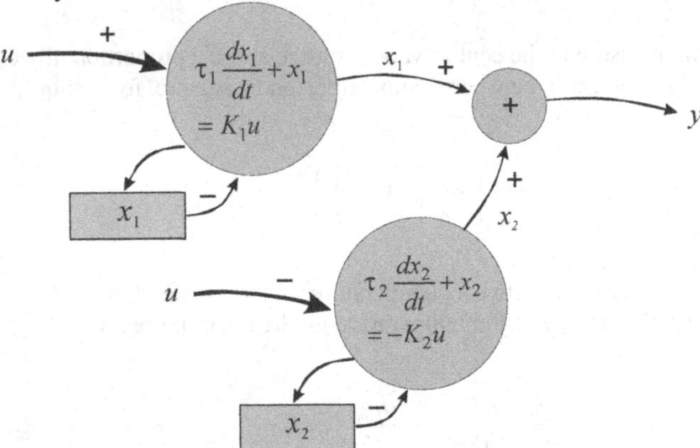

Fig. 10.11. Data flow diagram of system comprised of two parallel first-order sub-systems.

Combination of the equations in Eqn. (10.25) results in:

$$\tau_1\tau_2\frac{d^2y}{dt^2}+(\tau_1+\tau_2)\frac{dy}{dt}+y=(K_1\tau_2-K_2\tau_1)\frac{du}{dt}+(K_1-K_2)u \qquad (10.26)$$

The transfer function between the input u and the output y can easily be derived by Laplace transformation:

$$\frac{y}{u}=\frac{(K_1\tau_2-K_2\tau_1)s+(K_1-K_2)}{\tau_1\tau_2s^2+(\tau_1+\tau_2)s+1} \qquad (10.27)$$

An inverse response will result, if the two terms for the input u have an opposite sign, for example, if:

$$K_1\tau_2-K_2\tau_1<0, \quad while\ K_1-K_2>0 \qquad (10.28)$$

When a step change in the input occurs, the term du/dt dominates, since the change in input is large. If $(K_1\tau_2-K_2\tau_1)<0$, the response of y will be negative initially. In a stationary situation, this term equals zero, all other derivatives are also zero. For the stationary situation the relationship between u and y becomes then:

$$y=(K_1-K_2)u \qquad (10.29)$$

Fig. 10.12 shows the dynamic response of the system.

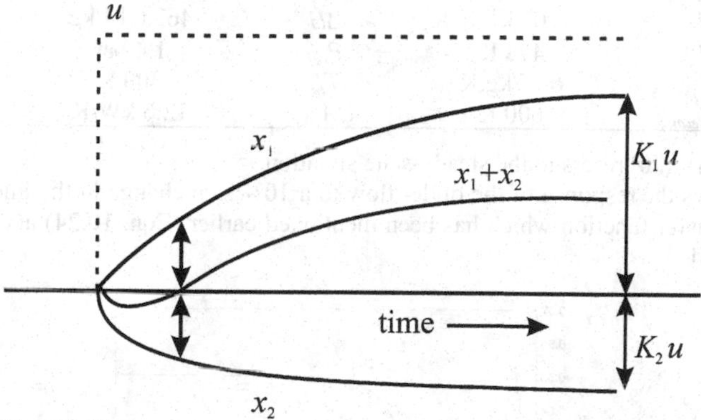

Fig. 10.12. Inverse response in system with two parallel balances.

The accumulation of x_1 is initially slow (large τ_1), but reaches a large value (large K_1), whereas the accumulation of x_2 is relatively fast (small τ_2) but limited in size (small K_2). It is obvious that for other parameter combinations, other responses will be obtained.

Example: Evaporator

An interesting phenomenon is the inverse response which can occur in an evaporator, shown in Fig. 10.13 and already described before.

Depending on how much the feed is sub-cooled, the temperature in the evaporator may drop. If more cold liquid enters the evaporator, less energy is available for evaporation and initially the outflow of vapor will decrease.

However, if the feed flow increases, the level in the evaporator starts to rise.

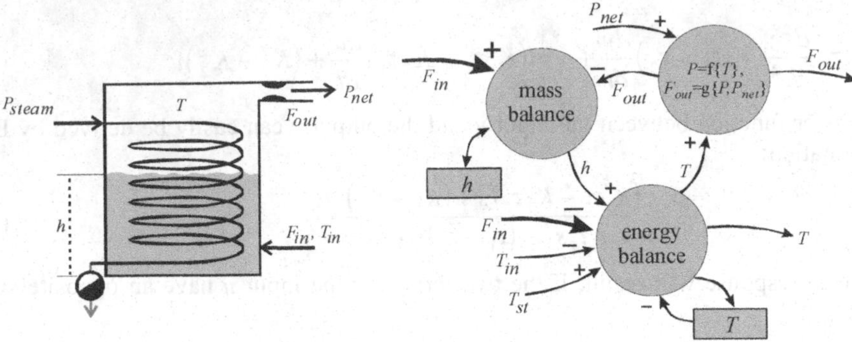

Fig. 10.13. Data flow diagram of parallel influence of F_{in} on the relationships for an evaporator.

Consequently, the heat transfer is increased which causes the vapor flow to increase and the level to decrease. For an industrial evaporator, the following data was assumed:

Table 10.1. Evaporator variables.

parameter	value	parameter	value
F_{in}	0.5 kg/s	A_c	0.1963 m^2
h_0	0.5 m	h_{max}	1.0 m
ρ	800 kg/m^3	β	1.118 kg/s.bar$^{-1/2}$
R	0.1702 kJ/kg.K	ΔH	463.0 kJ/kg
T_0	478 K	P_{ref}	1.4 bar
c_p	6 kJ/kg.K	T_{in}	300 K
T_{steam}	600 K	UA_{coil}	12.6 kW/K

in which the subscript '0' refers to the steady-state situation.

Fig. 10.14 shows the response of the outlet flow to a 10% step change in the inlet flow. The results agree with the transfer function which has been mentioned earlier (Eqn. 10.24) and describes F_{out} as a function of the load.

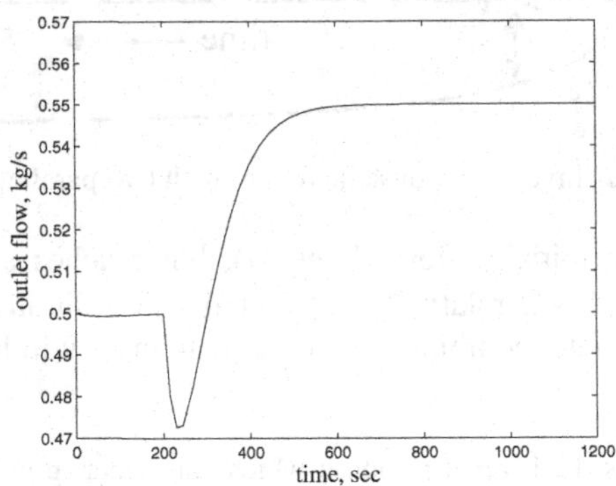

Fig. 10.14. Response of outlet flow to a 10% step change in inlet flow.

The numerator has a positive and a negative term. After a step change in the load, initially s is large, resulting in a negative response which varies linearly with the relative temperature difference and the heating time constant. Finally the vapor flow will become equal to the inlet flow.

Lead-lag

When the process is composed of two parallel transfer functions, as shown in Fig. 10.15, the response of the process output is a so-called lead-lag response as shown in Fig. 10.16.

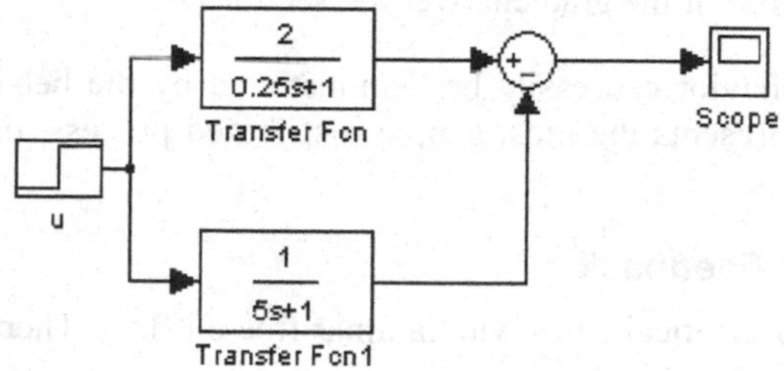

Fig. 10.15. Modified system with two parallel balances.

The response is initially determined by the part of the process with the small time constant, hence the output increases rapidly. Subsequently, the response is affected by the part of the process with the larger time constant, hence the process output will start to fall, until it reaches steady state. This type of response is often seen in different processes.

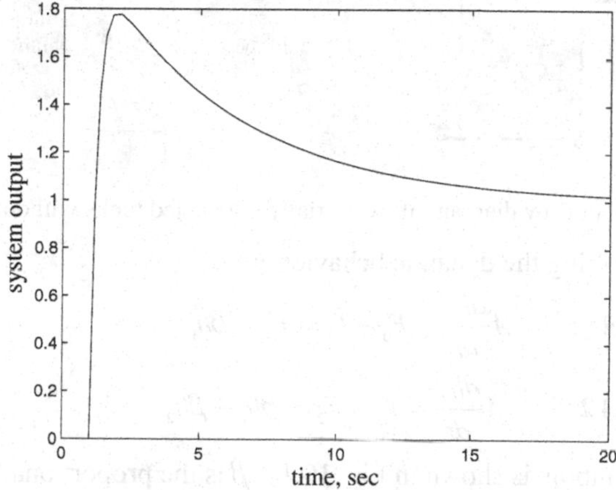

Fig. 10.16. Lead-lag response of system with two parallel balances.

Some processes exist, in which the smallest time constant can be ignored. If, for example, the time constant $\tau_l = 0.25$ s in Fig. 10.15 can be ignored, the total transfer function from u to y becomes:

$$\frac{y}{u} = 2 - \frac{1}{5s+1} = \frac{10s+1}{5s+1} = \frac{\tau_1 s+1}{\tau_2 s+1} \tag{10.30}$$

This is the expression for a first-order lead-lag function. For a unit step change in the input, the output shows a peak value equal to τ_1/τ_2 and a final value of 1.0.

10.6 Distributed Processes

Similar equipment in series (for example extraction units in series) or chains of similar sections (for example trays in a distillation columns) or equipment in which variables are a function of time and location can be described dynamically by a section model in order to characterize the distributed character of the equipment. Typical dynamic behavior of a distributed system is:

- the propagation of a gradient through the sections with the flow
- the change of the profile of the gradient over the sections

The difference in behavior can easily be demonstrated by the behavior of two identical tanks in series. This represents the most simple distributed process, only consisting of two sections.

Propagation without Feedback

Figure 10.17 shows two identical tanks with laminar free outflow. There is no feedback from the second tank to the first one.

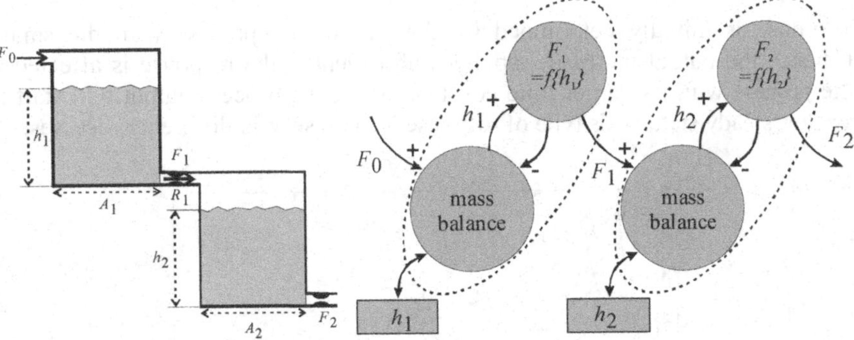

Fig. 10.17. Data flow diagram of two serially connected tanks without feedback.

The equations describing the dynamic behavior are:

$$\begin{aligned} \text{vessel 1:} \quad & A\frac{dh_1}{dt} = F_0 - F_1 = F_0 - \beta h_1 \\ \text{vessel 2:} \quad & A\frac{dh_2}{dt} = F_1 - F_2 = \beta h_1 - \beta h_2 \end{aligned} \tag{10.31}$$

The meaning of the symbols is shown in Fig. 10.17. β is the proportionality constant between the outflow from the tank and the height of the liquid in the tank. It corresponds to the inverse resistance for flow. The equations can be rewritten as:

$$\begin{aligned} \text{vessel 1:} \quad & \tau\frac{dh_1}{dt} + h_1 = \frac{F_0}{\beta} \\ \text{vessel 2:} \quad & \tau\frac{dh_2}{dt} + h_2 = h_1 \end{aligned} \tag{10.32}$$

in which $\tau = A/\beta$ = time constant of the tank. After linearization, substitution and Laplace transformation, the transfer function between the inlet flow F_0 and the height h_2 can be written as:

$$\frac{\delta h_2}{\delta F_0} = \frac{1/\beta}{(\tau s+1)^2} = \frac{1/\beta}{\tau^2 s^2 + 2\tau s + 1} \tag{10.33}$$

The damping of the system can be determined from the general equation for second-order systems, Eqn. (10.16):

$$\varsigma = \frac{1}{2}\frac{2\tau}{\sqrt{\tau^2}} = 1 \tag{10.34}$$

As can be seen, the response of two first-order systems in series without feedback is a critically damped response. This result could be expected, a disturbance at the entrance propagates through the system without damping and without amplification.

Propagation with Feedback

Figure 10.18 shows two similar tanks with laminar free outflow. They are positioned in such a way that there is feedback from the second tank to the first one. The level in the second tank has an impact on the level in the first tank. The dynamic equations can be written as:

$$\begin{aligned} &\text{vessel 1:} \quad A\frac{dh_1}{dt} = F_0 - F_1 = F_0 - \beta(h_1 - h_2) \\ &\text{vessel 2:} \quad A\frac{dh_2}{dt} = F_1 - F_2 = \beta(h_1 - h_2) - \beta h_2 = \beta h_1 - 2\beta h_2 \end{aligned} \tag{10.35}$$

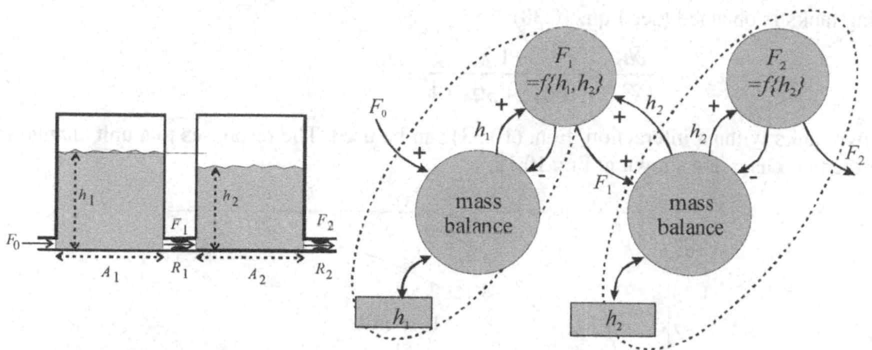

Fig. 10.18. Data flow diagram of two serially connected tanks with feedback.

The symbols are clarified in Fig. 10.18. As before, β is the proportionality constant between liquid height in the tank and the outflow. The equations can be written in the standard form:

$$\begin{aligned} &\text{vessel 1:} \quad \tau_1\frac{dh_1}{dt} + h_1 = \frac{F_0}{\beta} + h_2 \\ &\text{vessel 2:} \quad \tau_2\frac{dh_2}{dt} + h_2 = \tfrac{1}{2}h_1 \end{aligned} \tag{10.36}$$

in which τ_1 en τ_2 are the time constants with $\tau_1 = 2\tau_2 = \tau = A/\beta$. As can be seen, the time constant for the second tank is smaller. After linearization, substitution and Laplace transformation, the transfer function between the height h_2 and the inlet flow F_0 becomes:

$$\frac{\delta h_2}{\delta F_0} = \frac{1/(2\beta)}{\tau_1\tau_2 s^2 + (\tau_1 + \tau_2)s + \left(1 - \frac{1}{2}\right)} \tag{10.37}$$

To be able to make a comparison between the two systems with tanks in series, τ_1 and τ_2 are expressed in τ:

$$\frac{\delta h_2}{\delta F_0} = \frac{1/\beta}{\tau^2 s^2 + 3\tau s + 1} \tag{10.38}$$

The damping of this sytem can be determined by using equation (10.16) for second-order systems:

$$\varsigma = \frac{1}{2}\frac{3\tau}{\sqrt{\tau^2}} = \frac{3}{2} \tag{10.39}$$

The damping of the height in the second tank for the process with feedback is 3/2 times the damping of the process without feedback. As can be seen from the data flow diagram, the feedback is positive. The dynamic behavior of the process with feedback is slower than without feedback since the system is stationary when all the individual components are stationary.

Example

Using the values, $\beta = 1$ and $\tau = 10$ ($\tau_1 = 10$ en $\tau_2 = 5$) the following transfer function for the two interacting tanks is obtained (see Eqn. 10.38):

$$\frac{\delta h_2}{\delta F_{in}} = \frac{1}{100s^2 + 30s + 1} \tag{10.40}$$

For the two tanks without interaction, Eqn. (10.33) can be used. The responses to a unit change in inlet flow for the two cases, are shown in Fig. 10.19.

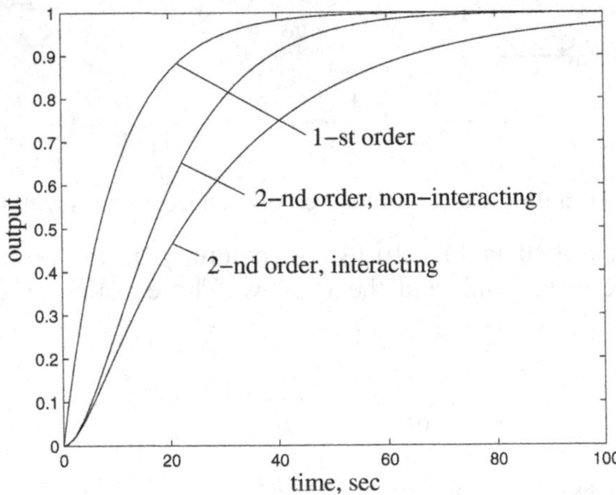

Fig. 10.19. Step responses of level to change in inlet flow for different processes.

It can be seen that, as a result of interaction, the response becomes slower.

10.7 Processes with Propagation without Feedback

The general equation for propagation without feedback for a section i of the process, can be written as:

$$\tau\frac{dx_i}{dt} = x_{i-1} - x_i \tag{10.41}$$

x_i is the state variable of section i. Figure 10.20 shows the data flow diagram for such a system.
An example for two sections is given in the previous paragraph. For n sequential sections the transfer function becomes:

$$\frac{\delta x_n}{\delta x_1} = \frac{1}{(\tau s + 1)^n} \tag{10.42}$$

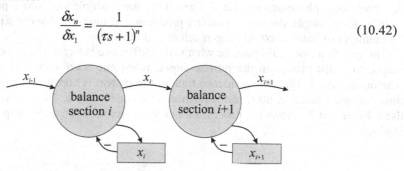

Fig. 10.20. Data flow diagram of propagation without feedback.

Example: Distillation column

In a tray distillation column, the liquid flow can be described according the data flow diagram of Fig. 10.21. It is assumed that no condensation of vapor takes place. The change in the liquid flow δL over the weir height of the tray is proportional to the change in liquid height or liquid mass on the tray δM.

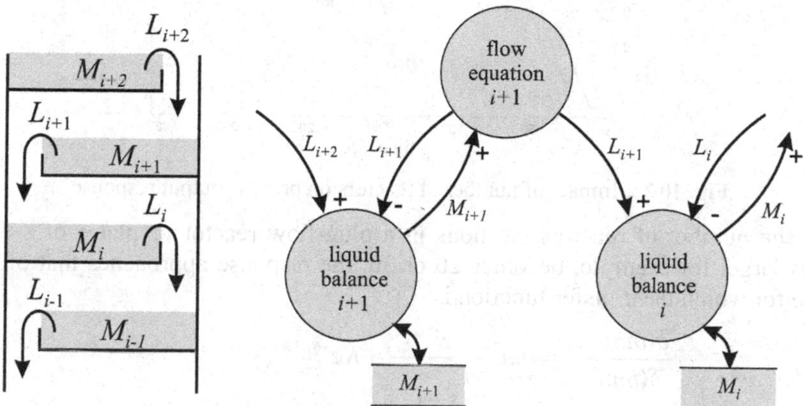

Fig. 10.21. Liquid flow through a distillation column.

The equations for the dynamic behavior become therefore:

$$\begin{aligned} \frac{dM_i}{dt} &= L_{i+1} - L_i \\ \delta L_i &= \beta\, \delta M_i \end{aligned} \tag{10.43}$$

Both equations combined results in:

$$\frac{1}{\beta}\frac{d\delta M_i}{dt} = \delta M_{i+1} - \delta M_i \tag{10.44}$$

The transfer function for the liquid flow from top to bottom becomes:

$$\frac{\delta L_n}{\delta L_0} = \frac{1}{(\tau s + 1)^n} \tag{10.45}$$

An interesting phenomenon exists, when there are multiple first-order processes in series. This is, for example, the case in rubber production in multiple, ideally stirred tank reactors. The number of reactors could be as much as eight to ten.

The question one could pose is: what is the difference between the response of the process output to a disturbance in the process input, when one reactor is used and when multiple reactors are used. The total residence time in the system is constant, i.e. when the residence time for one reactor is 60 s, the residence time per reactor is 20 s when three reactors are used. Figure 10.22 shows the process output of n reactors when a unit step change in the inlet is given.

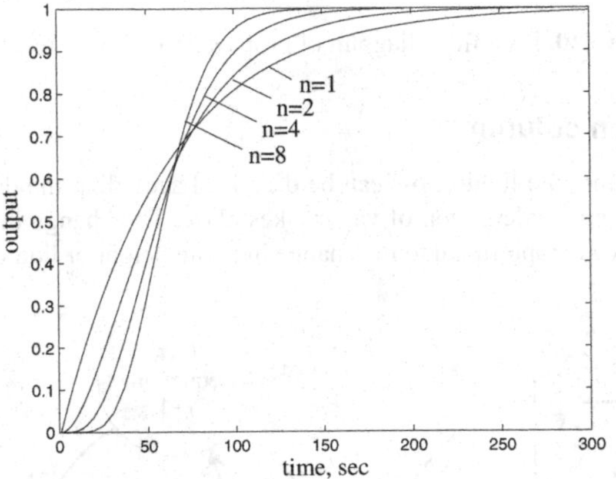

Fig. 10.22. Impact of number of reactors on process output response.

When the number of reactors, sections in a plug-flow reactor or plates of a distillation column is large, for example, becomes 20 or 30, the response approaches that of a process dead time for which the transfer function is:

$$\frac{\delta(output)}{\delta(input)} = \lim_{n \to \infty} \frac{K}{\left(\frac{\theta}{n} s + 1\right)^n} \approx K e^{-\theta s} \tag{10.46}$$

in which θ is the process dead time, which is approximately equal to n times the residence time per section. The propagation of a gradient in distributed or chained process can be described by series of first-order systems. When the number of systems or sections becomes large the behavior has a plug-flow character.

When a transfer function consists of many small time constants, these time constants can be merged in a similar manner:

$$\frac{\delta(output)}{\delta(input)} = \frac{K}{\prod_{i=1}^{n} (\tau_i s + 1)} \approx K e^{-\sum_{i=1}^{n} \tau_i s} \tag{10.47}$$

In many processes, there is often a major time constant and numerous small time constants. Consequently, the transfer function can be described by a combination of Eqns. (10.9) and (10.46):

$$\frac{\delta(output)}{\delta(input)} = \frac{Ke^{-\theta s}}{\tau s + 1} \tag{10.48}$$

10.8 Processes with Propagation with Feedback

The general equation for propagation with feedback for a section *i* can be written as:

$$\tau \frac{dx_i}{dt} = x_{i-1} - 2x_i + x_{i+1} \tag{10.49}$$

x_i is the state of section *i*. Fig. 10.23 shows the data flow diagram for such a system

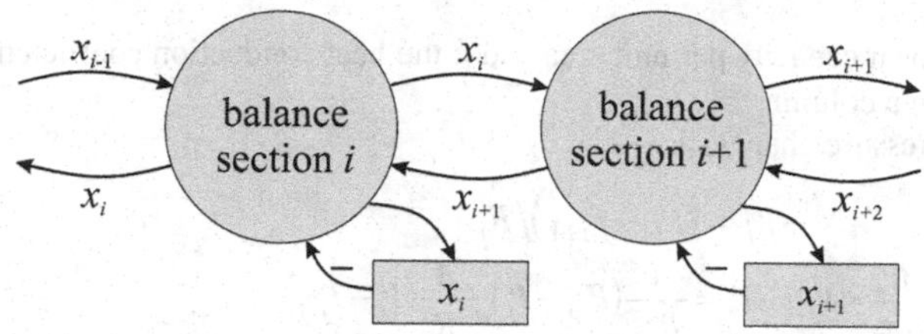

Fig. 10.23. Vapor flow in a distillation column.

Examples

There are numerous examples, in the area of liquid, vapor and heat transport.

- Mutually coupled tanks
  The dynamic behavior can be described by:

$$\begin{aligned} A\frac{dh_i}{dt} &= F_{i-1} - F_i = \beta(h_{i-1} - h_i) - \beta(h_i - h_{i+1}) \\ &= \beta(h_{i-1} - 2h_i + h_{i+1}) \end{aligned} \tag{10.50}$$

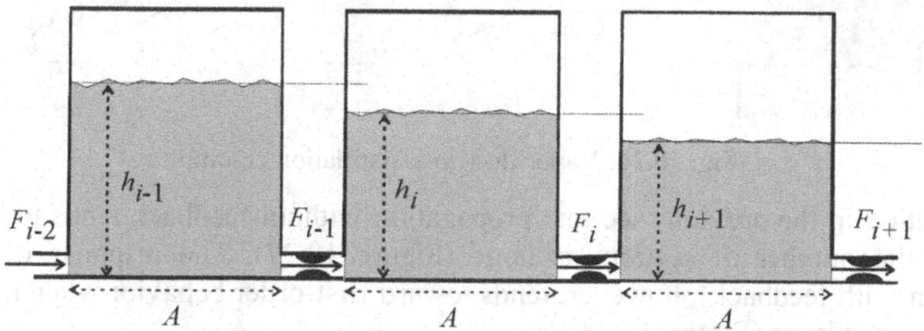

Fig. 10.24. Vapor flow in a distillation column.

- Heat transfer through a wall:

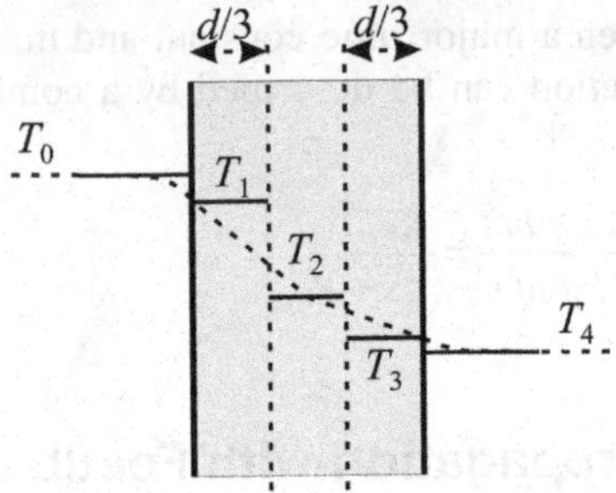

Fig. 10.25. Gradient in a wall, divided into three sections.

The change in temperature in each section can be described by:

$$C_p \frac{dT_i}{dt} = Q_{i-1} - Q_i = \lambda(T_{i-1} - T_i) - \lambda(T_i - T_{i+1}) \\ = \lambda(T_{i-1} - 2T_i + T_{i+1}) \tag{10.51}$$

in which C_p is the heat capacity per unit area and λ the heat conduction coefficient.

- Vapor flow through a column
  The equation for pressure changes is:

$$V_i = (P_i - P_{i+1})/R_i \\ C_i \frac{dP_i}{dt} = V_{i-1} - V_i = \frac{1}{R_{i-1}}(P_{i-1} - P_i) - \frac{1}{R_i}(P_i - P_{i+1}) \tag{10.52}$$

in which R_i = flow resistance, C_i = vapor capacity = $V(\partial\rho/\partial P)$, with ρ = vapor density.

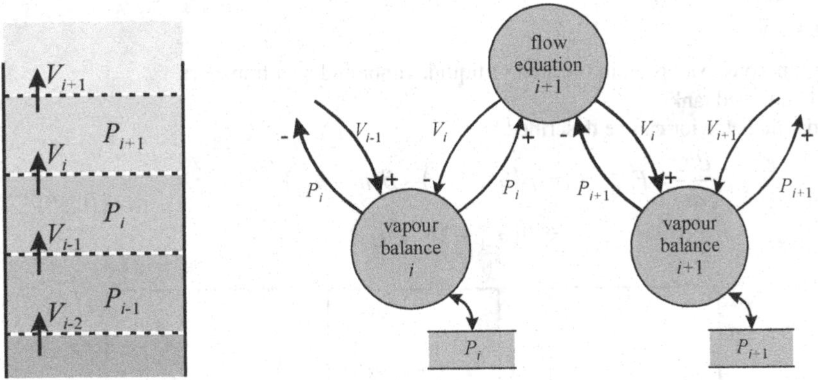

Fig. 10.26. Vapor flow in a distillation column.

As described in the previous section, propagation without feedback tends to become plug flow when the number of sections is large (Figure 10.27). The response of the process (propagation with feedback), however, tends toward first-order behavior when the number of sections becomes large (Figure 10.28).

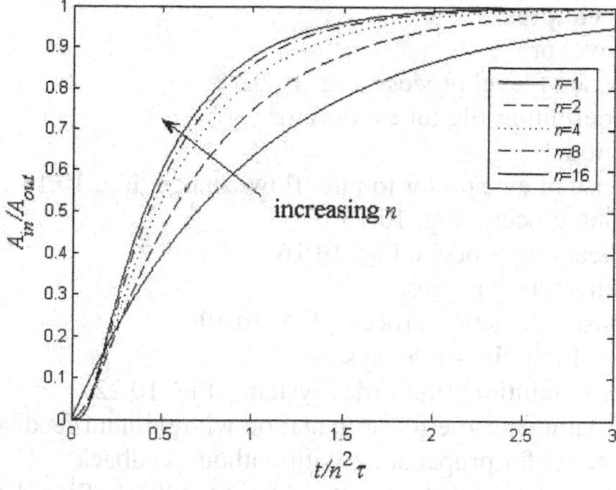

Fig. 10.27. Propagation with feedback as a function of the division in sections. The figure is normalized.

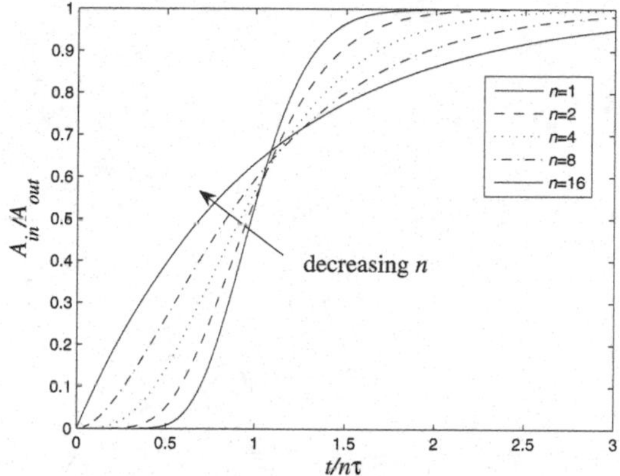

Fig. 10.28. Propagation without feedback as a function of the division in sections. The figure is normalized.

This is in accordance with section 10.6, where propagation without feedback and propagation with feedback were compared. The response of a flow is undamped, whereas the response of a gradient is strongly damped. All sections are interconnected, and the system is at equilibrium when all sections are at equilibrium. The process response becomes first order:

$$\frac{\delta(output)}{\delta(input)} = \frac{K}{(\tau s+1)} \text{ with } \tau = \tau_P(1+2+...n) \approx \tau_P \frac{n^2}{2} \tag{10.53}$$

where τ_P is the time constant of one section. For the response of the vapor flow in the distillation column $\tau_P = R_i C_i$, where R_i is the resistance for flow changes and C_i the vapor capacity. The time-constant τ increases with the square of the number of sections.

Files referred to in this chapter:
F1002.mdl: simulation level process
F1002plot.m: step response of level process, Fig. 10.02
F1014par.m: parameter definition file for evaporator
F1014.mdl: evaporator model
F1014plot.m: step response of evaporator to inlet flow change, Fig. 10.14
F1015.mdl: model lead-lag process, Fig. 10.15
F1016.m: step response lead-lag process, Fig. 10.16
F1019.mdl: simulation interacting process
F1019plot.m: step response interacting process, Fig. 10.19
F1022.mdl: simulation multiple first-order systems
F1022plot.m: step response multiple first-order systems, Fig. 10.22
F1027par.m: simulation parameter file for propagation with/without feedback
F1027.mdl: simulation model for propagation with/without feedback
F1027plot.m: step response of propagation with/without feedback, Figs. 10.27 and 10.28

11 Analysis of a Mixing Process

This chapter will analyze the mixing process in more detail. The process was already introduced in chapter 4, but some special properties of the Laplace transform and some special cases of the mixing process will be reviewed. In subsequent chapters other types of processes will be analyzed for their dynamic behavior. The purpose of the Laplace transform is to analyze how the process output of interest changes if the process input is changed. This will result in knowledge about the behavioral properties of the system, such as order, stability, integrating or non-minimum phase response behavior.

11.1 The Process

The process was already introduced in chapter 4, it is shown again in Fig. 11.1.

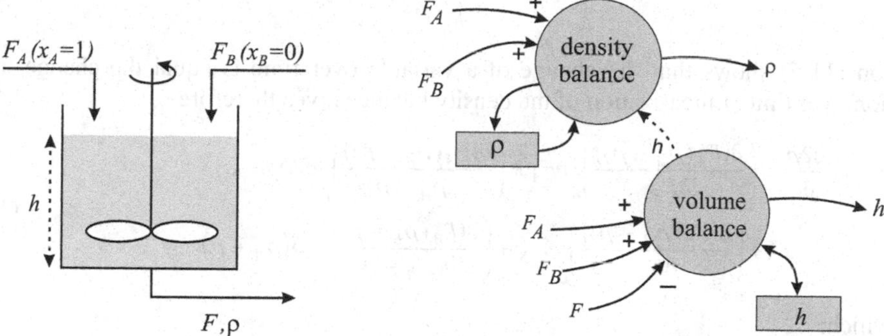

Fig. 11.1. Mixing process with behavioral model.

The mixer with two components, can be described by two basic balances, the mass balance and the component balance, in which the density is a function of the composition. Because the density ρ and the height h are defined as measurable output variables, the system is transformed into a combination of the volume and density balance. The volume balance (compare with Eqn. (4.26)) can be written as:

$$A\frac{dh}{dt} = F_A + F_B - F \tag{11.1}$$

and the density balance which can be obtained by subtracting ρ times the volume balance (Eqn. (11.1)) from the mass balance (Eqn. (4.16)), resulting in:

$$Ah\frac{d\rho}{dt} = F_A(\rho_A - \rho) + F_B(\rho_B - \rho) \tag{11.2}$$

We are interested in developing an understanding on how the density ρ and the level h vary as a function of changes in the manipulated variable F_A.

11.1.1 Linearization and Laplace Transformation

Linearization of the density balance can be achieved by substituting in the left hand-side of the balance:

Process Dynamics and Control: Modeling for Control and Prediction. Brian Roffel and Ben Betlem.

$$h = h_0 + \delta h \quad and \quad \rho = \rho_0 + \delta\rho \tag{11.3}$$

in which h_0 and ρ_0 are the steady-state or normal operating values, and δh and $\delta\rho$ are the deviation variables around the operating point representing the variation. After substitution, we get for the left-hand side of the balance:

$$\begin{aligned} Ah\frac{d\rho}{dt} &= A(h_0 + \delta h)\frac{d(\rho_0 + \delta\rho)}{dt} = \\ &= Ah_0\underbrace{\frac{d\rho_0}{\mathrm{dt}}}_{=0} + Ah_0\frac{d\delta\rho}{dt} + A\delta h\underbrace{\frac{d\rho_0}{\mathrm{dt}}}_{=0} + \underbrace{A\delta h\frac{d\delta\rho}{dt}}_{\approx 0} \approx Ah_0\frac{d\delta\rho}{dt} \end{aligned} \tag{11.4}$$

The same result can also be obtained in a somewhat different way:

$$Ah\frac{d\rho}{dt}\xrightarrow{\text{linearization}} Ah\frac{d\rho}{dt} \approx \left(\frac{\partial\left(Ah\frac{d\rho}{dt}\right)}{\partial h}\right)_0 \delta h + Ah_0\frac{d\delta\rho}{dt} \approx Ah_0\frac{d\delta\rho}{dt} \tag{11.5}$$

Equation (11.5) shows that the change of a variable over time is equal the change in the deviation over time. Linearization of the density balance gives therefore:

$$\begin{aligned} Ah_0\frac{\mathrm{d}\delta\rho}{\mathrm{d}t} = &\left(\frac{\partial(F_A(\rho_A-\rho))}{\partial F_A}\right)_0 \delta F_A + \left(\frac{\partial(F_A(\rho_A-\rho))}{\partial(\rho_A-\rho)}\right)_0 \delta(\rho_A-\rho) \\ &+ \left(\frac{\partial(F_B(\rho_B-\rho))}{\partial F_B}\right)_0 \delta F_B + \left(\frac{\partial(F_B(\rho_B-\rho))}{\partial(\rho_B-\rho)}\right)_0 \delta(\rho_B-\rho) \end{aligned} \tag{11.6}$$

from which:

$$\begin{aligned} Ah_0\frac{d\delta\rho}{dt} = &(\rho_A-\rho)_0\,\delta F_A + F_{A0}\left(\underbrace{\delta\rho_A}_{=0} - \delta\rho\right) + (\rho_B-\rho)_0\,\delta F_B \\ &+ F_{B0}\left(\underbrace{\delta\rho_B}_{=0} - \delta\rho\right) \end{aligned} \tag{11.7}$$

Introducing Laplace transformation and rearranging of terms gives:

$$(Ah_0 s + F_{A0} + F_{B0})\delta\rho = (\rho_A - \rho)_0\,\delta F_A + (\rho_B - \rho)_0\,\delta F_B \tag{11.8}$$

This equation can be simplified by introducing the static equation for the flows:

$$F_{A0} + F_{B0} = F_0 \tag{11.9}$$

which results in:

$$(Ah_0 s + F_0)\delta\rho = (\rho_A - \rho)_0\,\delta F_A + (\rho_B - \rho)_0\,\delta F_B \tag{11.10}$$

The volume balance can be dealt with in a similar manner:

$$A\frac{d\delta h}{dt} = \delta F_A + \delta F_B - \delta F \tag{11.11}$$

Laplace transformation results in:

$$As\delta h = \delta F_A + \delta F_B - \delta F \tag{11.12}$$

From the balances, Eqns. (11.10) and (11.12), it can be seen that the density is not a function of the height (the height only appears in the time constant of Eqn. (11.10)), similarly, the height is not a function of the density.

11.1.2 Determination of the Normalized Transfer Function

The outputs h and ρ can now be determined as a function of the inputs. Since the right-hand side of Eqns. (11.10) and (11.12) only contain inputs, no further rearranging is necessary. The density balance, Eqn. (11.10) can be written as:

$$\delta\rho = \frac{(\rho_A - \rho)_0}{(Ah_0 s + F_0)}\delta F_A + \frac{(\rho_B - \rho)_0}{(Ah_0 s + F_0)}\delta F_B \tag{11.13}$$

and after normalization of the deviation variable we may write:

$$\frac{\delta\rho}{\rho_0} = \left(\frac{\rho_A - \rho}{\rho}\right)_0 \frac{F_0}{(Ah_0 s + F_0)}\frac{\delta F_A}{F_0} + \left(\frac{\rho_B - \rho}{\rho}\right)_0 \frac{F_0}{(Ah_0 s + F_0)}\frac{\delta F_B}{F_0} \tag{11.14}$$

The term Ah_0/F_0 is equal to the residence time τ of the vessel. After substitution of the residence time τ, the transfer function can be written as:

$$\frac{\delta\rho}{\rho_0} = \frac{1}{(\tau s + 1)}\left(\left(\frac{\rho_A - \rho}{\rho}\right)_0 \frac{\delta F_A}{F_0} + \left(\frac{\rho_B - \rho}{\rho}\right)_0 \frac{\delta F_B}{F_0}\right) \tag{11.15}$$

The transfer function represents the relationship between one single input and one single output. For example, the transfer function between $\delta\rho$ and δF_A indicates how and how much $\delta\rho$ changes as a result of changes in δF_A.
From Eqn. (11.15) it may be derived that:

$$\frac{\delta\rho}{\delta F_A} = \frac{1}{F_0}\frac{(\rho_A - \rho)_0}{(\tau s + 1)} \tag{11.16}$$

The normalized volume balance can be written as:

$$\frac{\delta h}{h_0} = \frac{1}{\tau s}\left(\frac{\delta F_A}{F_0} + \frac{\delta F_B}{F_0} - \frac{\delta F}{F_0}\right) \tag{11.17}$$

from which:

$$\frac{\delta h}{\delta F_A} = \frac{1}{\tau s}\frac{h_0}{F_0} \tag{11.18}$$

11.1.3 Response Behavior

If the inlet flow F_A changes according to a step change, then the density approaches a new steady state value (Eqn. 11.16). The level, however, continues to change (Eqn. 11.18). The vessel will empty itself or overflow. The initial change and the final steady state value can easily be obtained with the limit rules.

For a step change in F_A of magnitude ΔF_A, we may write in the Laplace domain:

$$\delta F_A = \frac{\Delta F_A}{s} \tag{11.19}$$

A change in the density at the beginning of the response, i.e. when t→0, as a result of the step change in F_A, becomes ($\delta F_B = 0$):

$$\begin{aligned} \lim_{t\to 0}\left(\frac{d\delta\rho}{dt}\right) &= \lim_{s\to\infty}\left[s^2\left[\frac{1}{(\tau s+1)}\frac{(\rho_A-\rho)_0}{F_0}\right]\frac{\Delta F_A}{s}\right] \\ &= \frac{\Delta F_A}{F_0}\frac{(\rho_A-\rho)_0}{\tau} \end{aligned} \tag{11.20}$$

The initial slope depends on the relative amplitude of the change and the time constant. The final steady state value, when $t\to\infty$, becomes:

$$\begin{aligned} \Delta\rho = \lim_{t\to\infty}(\delta\rho) &= \lim_{s\to 0}\left[s\left[\frac{1}{(\tau s+1)}\frac{(\rho_A-\rho)_0}{F_0}\right]\cdot\frac{\Delta F_A}{s}\right] \\ &= (\rho_A-\rho)_0\frac{\Delta F_A}{F_0} \end{aligned} \tag{11.21}$$

The change in level, when $t\to 0$, as a result of a step change with magnitude ΔF_A, becomes:

$$\lim_{t\to 0}\left(\frac{d\delta h}{dt}\right) = \lim_{s\to\infty}\left[s^2\left[\frac{1}{\tau s}\frac{h_0}{F_0}\right]\cdot\frac{\Delta F_A}{s}\right] = \lim_{s\to\infty}\left[\frac{\Delta F_A}{F_0}\frac{h_0}{\tau}\right] = \frac{\Delta F_A}{F_0}\frac{h_0}{\tau} \tag{11.22}$$

Taking the limit to s is not required since the function is independent of s. This means that the change in time is time independent and is therefore constant. The final level, for $t\to\infty$, becomes:

$$\Delta h = \lim_{t\to\infty}(\delta h) = \lim_{s\to 0}\left[s\left[\frac{1}{\tau s}\frac{h_0}{F_0}\right]\cdot\frac{\Delta F_A}{s}\right] = \lim_{s\to 0}\left[\frac{\Delta F_A}{F_0}\frac{h_0}{\tau s}\right] = \pm\infty \tag{11.23}$$

The level will show a continuous change in positive or negative direction, depending on the sign of ΔF_A.

The transfer function for the density balance shows a first-order response. It does not depend on the outlet flow. If the ratio between both inlet flows remains unchanged, the density will remain unchanged, regardless of how much the outlet flow changes. The level integrates the inlet flows as well as the outlet flow. This is not a stable system.

11.2 Mixer with Self-adjusting Height

If the level is self-adjusting, then an additional equation is required that describes how the outlet flow adjusts to changes in the level. This situation is shown in Fig. 11.2.
Suppose the outlet flow can be described by:

$$F = c\sqrt{\rho h} \tag{11.24}$$

Let us now investigate how the level h changes as a result of changes in the inlet flow F_A. The behavior of the density is the same as in the previous case, since the density is not a function of the outlet flow.

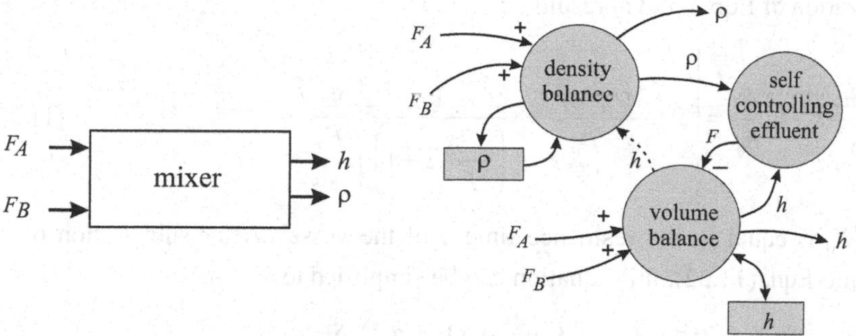

Fig. 11.2. Environmental and behavioral model for mixer with self-adjusting level.

11.2.1 Linearization and Laplace Transformation

Linearization of Eqn. (11.24) results in:

$$\delta F = \left(\frac{\partial\left(c\sqrt{\rho h}\right)}{\partial \rho}\right)_0 \delta\rho + \left(\frac{\partial\left(c\sqrt{\rho h}\right)}{\partial h}\right)_0 \delta h = \frac{c\sqrt{h_0}}{2\sqrt{\rho_0}}\delta\rho + \frac{c\sqrt{\rho_0}}{2\sqrt{h_0}}\delta h \tag{11.25}$$

This equation can be simplified by using Eqn. (11.24) in the steady state situation:

$$F_0 = c\sqrt{\rho_0 h_0} \tag{11.26}$$

which results in:

$$\delta F = \frac{c\sqrt{h_0}}{2\sqrt{\rho_0}}\delta\rho + \frac{c\sqrt{\rho_0}}{2\sqrt{h_0}}\delta h = \frac{F_0}{2}\left(\frac{\delta\rho}{\rho_0} + \frac{\delta h}{h_0}\right) \tag{11.27}$$

Substitution into the volume balance gives:

$$A\frac{d\delta h}{dt} = \delta F_A + \delta F_B - \delta F = \delta F_A + \delta F_B - \frac{F_0}{2}\left(\frac{\delta\rho}{\rho_0} + \frac{\delta h}{h_0}\right) \tag{11.28}$$

Laplace-transformation and rearranging of terms gives:

$$\left(As + \frac{F_0}{2h_0}\right)\delta h = \delta F_A + \delta F_B - \frac{F_0}{2\rho_0}\delta\rho \tag{11.29}$$

Substitution of the relationship for the density, Eqn. (11.14), results in:

$$\left(As + \frac{F_0}{2h_0}\right)\delta h = \delta F_A + \delta F_B - \frac{F_0}{2\rho_0}\left(\frac{(\rho_A - \rho)_0}{(Ah_0 s + F_0)}\delta F_A + \frac{(\rho_B - \rho)_0}{(Ah_0 s + F_0)}\delta F_B\right) \tag{11.30}$$

Rearranging of terms gives:

$$\left(As + \frac{F_0}{2h_0}\right)\delta h = \left(1 - \frac{F_0}{2\rho_0}\cdot\frac{(\rho_A - \rho)_0}{(Ah_0 s + F_0)}\right)\delta F_A + \left(1 - \frac{F_0}{2\rho_0}\cdot\frac{(\rho_B - \rho)_0}{(Ah_0 s + F_0)}\right)\delta F_B \tag{11.31}$$

11.2.2 Determination of the Normalized Transfer Function

Since we are only interested in the transfer function for changes in F_A, it is not necessary to consider changes in the other independent variable F_B.
Further normalization of Eqn. (11.31) results in:

$$\left(\frac{2Ah_0}{F_0}s+1\right)\frac{\delta h}{h_0}=\left(2-\frac{(\rho_A-\rho)_0}{\rho_0}\cdot\frac{1}{\left(\frac{Ah_0}{F_0}s+1\right)}\right)\frac{\delta F_A}{F_0} \tag{11.32}$$

The term Ah_0/F_0 is equal to the residence time τ of the vessel. After substitution of the residence time into Eqn. (11.32), this equation can be simplified to:

$$(2\tau s+1)\frac{\delta h}{h_0}=\left(\frac{2(\tau s+1)-(\rho_A-\rho)_0/\rho_0}{(\tau s+1)}\right)\frac{\delta F_A}{F_0} \tag{11.33}$$

The transfer function between the level as a function of changes in the flow F_A can now be given by:

$$\frac{\delta h}{\delta F_A}=\frac{2h_0}{F_0}\left(\frac{\tau s+1-(\rho_A-\rho)_0/\rho_0}{(\tau s+1)(2\tau s+1)}\right) \tag{11.34}$$

11.2.3 Response Behavior

It is interesting to investigate how the level reacts initially and in the long run to a step change in F_A. Let us assume the magnitude of the step change is ΔF_A.

$$\delta F_A=\frac{\Delta F_A}{s} \tag{11.35}$$

The initial behavior is then:

$$\begin{aligned}\lim_{t\to 0}\left(\frac{d\delta h}{dt}\right)&=\lim_{s\to\infty}\left[s^2\left[\frac{2h_0}{F_0}\left(\frac{\tau s+1-(\rho_A-\rho)_0/2\rho_0}{(\tau s+1)(2\tau s+1)}\right)\right]\cdot\frac{\Delta F_A}{s}\right]\\&=\lim_{s\to\infty}\left[\frac{2h_0}{F_0}\left(\frac{s^2\cdot\tau s}{\tau s\cdot 2\tau s\cdot s}\right)\right]=\frac{\Delta F_A}{F_0}\frac{h_0}{\tau}\end{aligned} \tag{11.36}$$

It can be seen that the initial behavior of the mixer with free outflow is equal to the initial behavior of the mixer with forced outflow.

However, for the stationary behavior it can be derived that:

$$\begin{aligned}\lim_{t\to\infty}(\delta h)&=\lim_{s\to 0}\left[s\left[\frac{2h_0}{F_0}\left(\frac{\tau s+1-(\rho_A-\rho)_0/2\rho_0}{(\tau s+1)(2\tau s+1)}\right)\right]\cdot\frac{\Delta F_A}{s}\right]\\&=\lim_{s\to 0}\left[\frac{2h_0}{F_0}\left(\frac{1-(\rho_A-\rho)_0/2\rho_0}{1}\right)\Delta F_A\right]=\frac{2h_0}{F_0}(1-(\rho_A-\rho)_0/2\rho_0)\Delta F_A\end{aligned} \tag{11.37}$$

After some time, a new equilibrium is achieved. If $\rho_A = \rho$, or, if the outlet flow is independent of the density, Eqn. (11.37) can be written as:

$$\lim_{t\to\infty}(\delta h) = 2\frac{\Delta F_A}{F_0}h_0 \tag{11.38}$$

If $\rho_A > 3\rho$, then the final level will be lower than the initial value of the level.

11.2.4 General Behavior

The system, described in Eqn. (11.34), is a stable system. The denominator is a series connection of two stable first-order systems. The transfer function for the density in this case is the same as in the case in which the outlet flow was not self-adjusting. The behavior of the liquid level is, however, different. The initial behavior is the same in case of disturbances, but the level is self-adjusting. The transfer function can also be written differently, by separation of terms. The process can be divided into two parallel first-order processes. The first term describes the mixing (compare with Eqn. (11.16)), with time constant τ, the second term describes the outflow of the liquid with time constant 2τ (compare with Eqn. (6.14) and (6.15)), Eqn. (11.34) can then be written as:

$$\frac{\delta h}{\delta F_A} = \frac{2h_0}{F_0}\left(\frac{(\rho_A-\rho)_0/2\rho_0}{\tau s+1} - \frac{(\rho_A-2\rho)_0/\rho_0}{2\tau s+1}\right) \tag{11.39}$$

The response in the time domain for a step of magnitude ΔF_A becomes:

$$\Delta h = \frac{2h_0}{F_0}\left(\frac{(\rho_A-\rho)_0}{2\rho_0}\left(1-e^{-t/\tau}\right) - \frac{(\rho_A-2\rho)_0}{\rho_0}\left(1-e^{-t/2\tau}\right)\right)\Delta F_A \tag{11.40}$$

Figure 11.3 shows the response of the level for a step change in F_A of magnitude ΔF_A for different values of ρ_A.

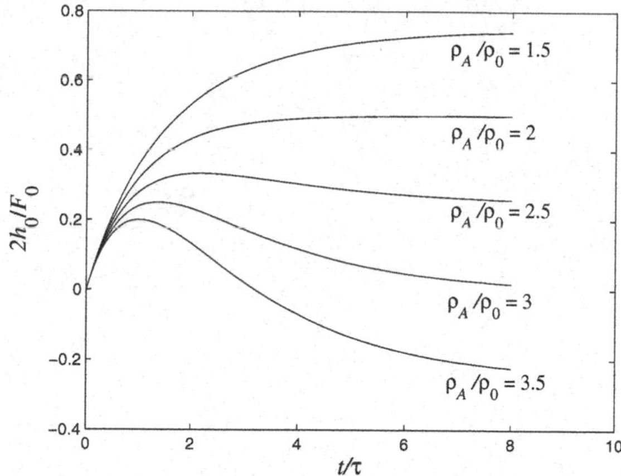

Fig. 11.3. Step response of the height in the mixer for different values of the density.

The response is the sum of two parallel processes with different time constant and different gain. The first term in Eqn. (11.39) is a fast first-order response; the second term is a slower first-order response in the opposite direction. The values used for ρ_A compared with ρ are physically not realistic; however, it shows mathematically what the response would do. If $\rho_A = 2\rho$, the transfer function is exactly a first-order process, since the second tern is equal to zero, if $\rho_A > 3\rho$, an inverse response is obtained. In that case, the slowest transfer function has a contribution in the steady state situation, which is opposite the contribution of the fastest transfer function.

12 Dynamics of Chemical Stirred Tank Reactors

In this chapter, the dynamics of ideally stirred tank reactors will be analyzed. First, the assumptions, required to limit model complexity, will be discussed. Next, various types of reaction will be considered, such as simple first-order reactions, equilibrium reactions, parallel reactions, etc. Subsequently, the analysis will be expanded to include non-isothermal reactors. Numerical examples of chemical reactors are given and the non-linear model descriptions are compared with the linearized model descriptions.

12.1 Introduction

Chemical reactors are usually the most important part of a chemical plant. They form the heart of the process where raw materials are converted into products. Modeling chemical reactors, in particular the kinetics, is generally not simple, however: it depends very much on the goals we would like to achieve. In many cases, not all raw material is converted and it may be important to monitor the concentration of unreacted components as to get an idea of the conversion that has been achieved. The reactor outlet concentration of unreacted component is often the most important variable in which we are interested.

To simplify the description of the reactor, the following assumptions have been made:

- it is assumed that the reactor is completely filled, i.e. the level is assumed to be constant
- the reactor is ideally mixed, i.e. there are no concentration gradients and the reactor concentration is the same as the outlet concentration
- the density is the same throughout the process and independent of the concentration of components and temperature

12.2 Isothermal First-order Reaction

In this section, reaction at isothermal conditions will be considered, i.e. the reaction proceeds at constant temperature.

Let the reactor throughput be F (m^3/s) and the inlet concentration be c_{Ain} (kg/m^3). The reactor volume is V (m^3). The reactor outlet concentrations are c_A (kg/m^3) and c_B (kg/m^3). The reactor is shown in Fig. 12.1.

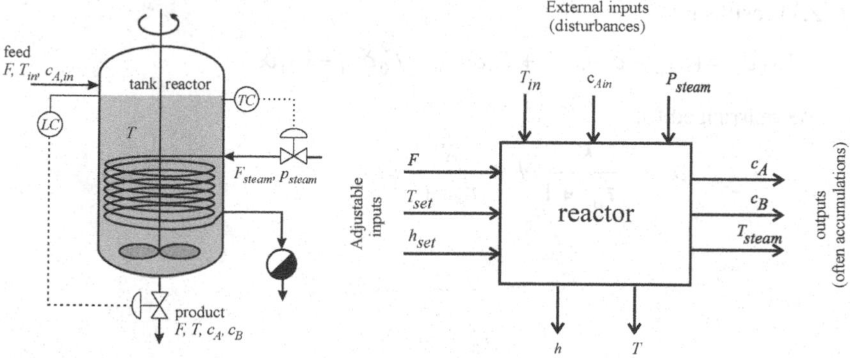

Fig. 12.1. Isothermal chemical reactor.

Process Dynamics and Control: Modeling for Control and Prediction. Brian Roffel and Ben Betlem.

Two additional assumptions are made:

- the reaction is of the following type, A → B, with a rate constant k_1
- the reaction is of first-order in component A, the reaction rate can therefore be described by $r = k_1 c_A$

In most cases it will be desired to control the conversion of component A, which is directly related to the yield of component B, hence the response of c_A to changes in c_{Ain} and F are important. The behavioral model of the reactor is shown in Fig. 12.2.

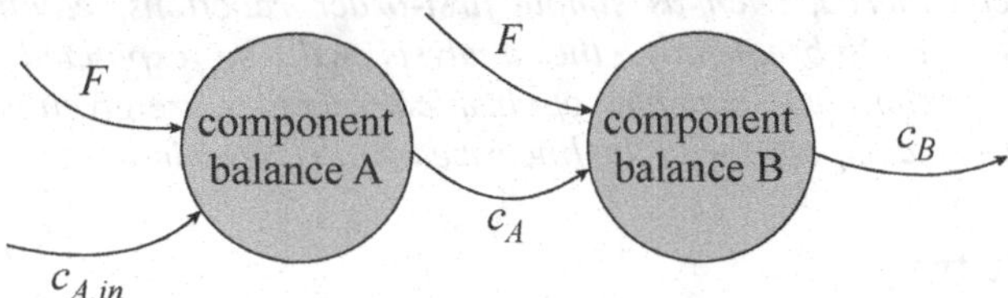

Fig. 12.2. Behavioral model of isothermal reactor.

Since there is no energy balance, nor an overall mass balance, the only dynamic balance that remains is a component balance for component A and/or B. To describe the outlet concentration of component A, the dynamic component balance of A is of interest. It can be written as:

$$V\frac{dc_A}{dt} = F(c_{Ain} - c_A) - rV = F(c_{Ain} - c_A) - Vk_1 c_A \quad (12.1)$$

The derivative term on the left-hand side of the equal sign represents the accumulation of mass of component A in kg/s; the first term on the right-hand side represents the difference between the inlet flow and outlet flow of component A; and the second right-hand side term represents the disappearance of component A through reaction.

Similarly, the component balance for species B, assuming that no B enters the reactor, can be written as:

$$V\frac{dc_B}{dt} = -Fc_B + rV = -Fc_B + Vk_1 c_A \quad (12.2)$$

12.2.1 Linearization of Model Equations

Equations (12.1) and (12.2) are simple equations and can easily be linearized. Linearization of Eqn. (12.1) results in:

$$Vs\delta c_A = (c_{Ain} - c_A)_0 \delta F + F_0 \delta c_{Ain} - F_0 \delta c_A - Vk_1 \delta c_A \quad (12.3)$$

which can be rearranged to:

$$\delta c_A = \frac{K_1}{\tau_A s + 1}\delta F + \frac{K_2}{\tau_A s + 1}\delta c_{Ain} \quad (12.4)$$

with

$$\begin{aligned}
\tau_A &= \frac{V}{F_0 + Vk_1} = \frac{\tau_R}{1 + k_1\tau_R} \\
\tau_R &= V / F_0, \text{the residence time} \\
K_1 &= \frac{(c_{Ain} - c_A)_0}{F_0 + Vk_1} = \frac{(c_{Ain} - c_A)_0 / F_0}{1 + k_1\tau_R} \\
K_2 &= \frac{F_0}{F_0 + Vk_1} = \frac{1}{1 + k_1\tau_R}
\end{aligned} \tag{12.5}$$

Linearization of Eqn. (12.2) results in:

$$Vs\delta c_B = -F_0\delta c_B - c_{B0}\delta F + Vk_1\delta c_A \tag{12.6}$$

which can be written as:

$$\delta c_B = -\frac{K_3}{\tau_R s + 1}\delta F + \frac{K_4}{\tau_R s + 1}\delta c_A \tag{12.7}$$

with

$$\begin{aligned}
K_3 &= c_{B0} / F_0 \\
K_4 &= Vk_1 / F_0 = k_1\tau_R
\end{aligned} \tag{12.8}$$

12.2.2 Model Analysis

From Eqns. (12.4) and (12.7) the transfer functions (and relationships) of interest can be obtained. From Eqn. (12.4) it can be seen that the relationship between δc_A and δF is a first-order transfer function with gain K_1 and time constant τ_A. Similarly, the relationship between δc_A and δc_{Ain} is a first-order transfer function with gain K_2 and time constant τ_A. For fast reactions (large value of k_1), τ_A approaches the value $1/k_1$, which approaches zero: in other words the response will be instantaneous. In addition, the gains K_1 and K_2 become small. It means that all raw material is almost instantaneously converted and the inlet flow and inlet concentration do not affect the outlet concentration much.

The response between δc_B and δc_{Ain} is interesting: it can be written as:

$$\frac{\delta c_B}{\delta c_{Ain}} = \frac{\delta c_B}{\delta c_A}\frac{\delta c_A}{\delta c_{Ain}} = \frac{K_2 K_4}{(\tau_R s + 1)(\tau_A s + 1)} = \frac{\dfrac{k_1\tau_R}{1 + k_1\tau_R}}{(\tau_R s + 1)(\tau_A s + 1)} \tag{12.9}$$

which is a second-order transfer function (two first-order transfer functions in series) with a time constant equal to the residence time (τ_R) and one due to reaction (τ_A). Thus the response is dependent on the residence time and the characteristics of the reaction. The reason that the two first-order transfer functions are in series is due to the fact that no feedback exists from c_B to c_A.

Example

Suppose $k_1 = 0.1\ s^{-1}$ and $\tau_R = 20$ s. It can then be calculated that $\tau_A = 6.67$ sec, $K_2 = 0.333$ and $K_4 = 2$. The model relating δc_B to δc_{Ain} is shown in Fig. 12.3.

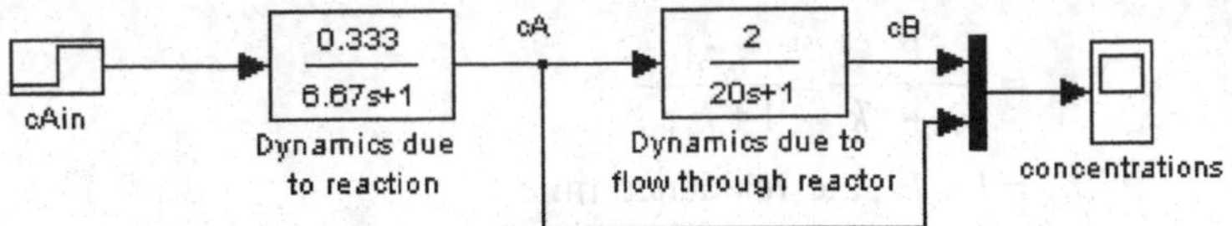

Fig. 12.3. Isothermal reactor transfer functions for changes in c_{Ain}.

The responses of c_A and c_B to changes in c_{Ain} are shown in Fig. 12.4. Clearly the first-order response of c_A and second-order response of c_B can be seen.

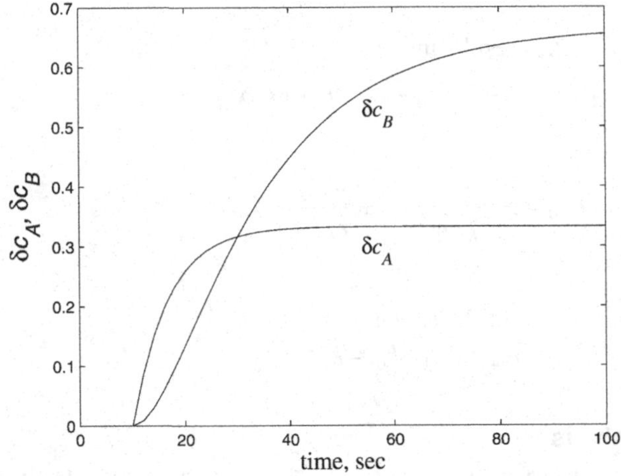

Fig. 12.4. Reactor concentration response to unit step change in c_{Ain}.

12.3 Equilibrium Reactions

In this section, the same reactor will be considered. However, it is assumed that the reaction from A to B is an equilibrium reaction, A ↔ B, with rate constant k_1 for the forward reaction and rate constant k_2 for the backward reaction. The rate of reaction can now be given by:

$$r = k_1 c_A - k_2 c_B \tag{12.10}$$

Equation (12.1) now has to be modified to

$$V\frac{dc_A}{dt} = F(c_{Ain} - c_A) - rV = F(c_{Ain} - c_A) - Vk_1 c_A + Vk_2 c_B \tag{12.11}$$

and Eqn. (12.2) can be modified accordingly:

$$V\frac{dc_B}{dt} = -Fc_B + rV = -Fc_B + Vk_1 c_A - Vk_2 c_B \tag{12.12}$$

The behavioral model will also change and is shown in Fig.12.5. It can be seen that there is interaction between the component balances.

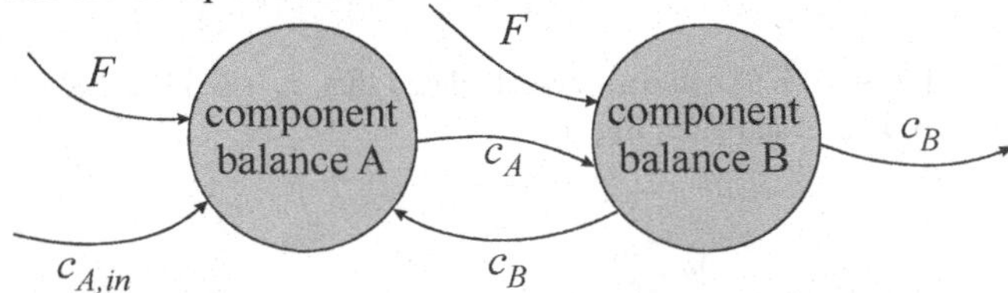

Fig. 12.5. Behavioral model of isothermal reactor with equilibrium reaction.

Also in this case, the yield of component B is directly determined by the conversion of component A, hence measuring the outlet concentration of component A would be of interest. Linearization of Eqns. (12.11) and (12.12) is straightforward and results in:

$$\delta c_A = \frac{K_1}{\tau_A s+1}\delta F + \frac{K_2}{\tau_A s+1}\delta c_{Ain} + \frac{K_3}{\tau_A s+1}\delta c_B \tag{12.13}$$

where τ_A, K_1 and K_2 were already defined in Eqn. (12.5), K_3 can now be given by:

$$K_3 = \frac{Vk_2}{F_0 + Vk_1} = \frac{k_2\tau_R}{1+k_1\tau_R} \tag{12.14}$$

and

$$\delta c_B = -\frac{K_4}{\tau_B s+1}\delta F + \frac{K_5}{\tau_B s+1}\delta c_A \tag{12.15}$$

where the time constant and gains are defined by:

$$\begin{aligned} \tau_B &= \frac{V}{F_0 + Vk_2} = \frac{\tau_R}{1+k_2\tau_R} \\ K_4 &= \frac{c_{B0}}{F_0 + Vk_2} = \frac{c_{B0}/F_0}{1+k_2\tau_R} \\ K_5 &= \frac{Vk_1}{F_0 + Vk_2} = \frac{k_1\tau_R}{1+k_2\tau_R} \end{aligned} \tag{12.16}$$

It can be seen that Eqn. (12.13) has an additional term δc_B compared with Eqn. (12.4). In addition, Eqn. (12.15) is similar to Eqn. (12.7); however, τ_R has been replaced by τ_B. If $k_2 = 0$, then $\tau_R = \tau_B$, and the gains K_4 and K_5 in Eqn. (12.16) reduce to the values given in Eqn. (12.8).

The response from δc_{Ain} to δc_A changes now, compared with the previous case, owing to the fact that c_{Ain} also affects c_B via c_A and c_B in turn affects c_A again: in other words, the component balances have become interactive. This can be seen when we combine Eqns. (12.13) and (12.15):

$$\begin{aligned} \left[(\tau_A s+1)(\tau_B s+1) - K_3K_5\right]\delta c_A &= \left[K_1(\tau_B s+1) - K_3K_4\right]\delta F \\ &\quad + K_2(\tau_B s+1)\,\delta c_{Ain} \end{aligned} \tag{12.17}$$

The relationship between c_A and c_{Ain} is no longer a first-order transfer function as shown in Eqn. (12.4): it can now be given by:

$$\frac{\delta c_A}{\delta c_{Ain}} = \frac{K_2(\tau_B s+1)}{(\tau_A s+1)(\tau_B s+1) - K_3K_5} \tag{12.18}$$

The term K_3K_5 describes the interaction between the component balances.
Similarly, the response for changes in c_B due to changes in c_{Ain} can be described as:

$$\frac{\delta c_B}{\delta c_{Ain}} = \frac{K_2K_5}{(\tau_A s+1)(\tau_B s+1) - K_3K_5} \tag{12.19}$$

Example

Assume the same parameter values as in the previous case, with k_1= 0.1 s^{-1}, k_2 = 0.01 s^{-1} and τ_R=20 s. The value of τ_B = 16.67, K_2 = 0.333, K_3 = 0.067 and K_5 = 1.667, hence Eqn. (12.18) becomes:

$$\frac{\delta c_A}{\delta c_{Ain}} = \frac{0.333(16.67s+1)}{(6.67s+1)(16.67s+1)-0.11} = \frac{0.374(16.67s+1)}{124.85s^2+26.21s+1} \tag{12.20}$$

and Eqn. (12.19) becomes:

$$\frac{\delta c_B}{\delta c_{Ain}} = \frac{0.555}{(6.67s+1)(16.67s+1)-0.11} = \frac{0.624}{124.85s^2+26.21s+1} \tag{12.21}$$

The response is shown in Fig. 12.6.

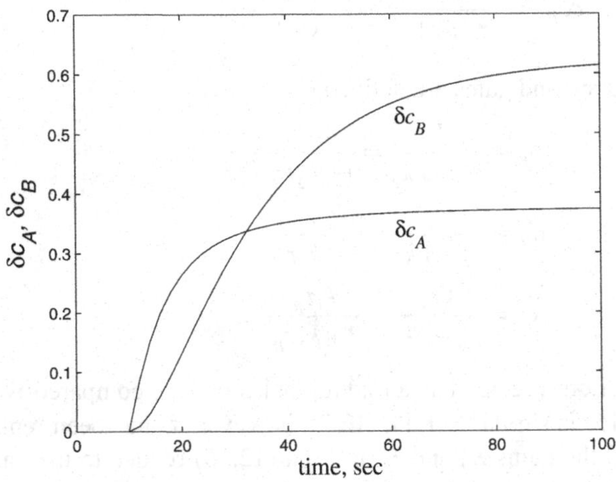

Fig. 12.6. Reactor concentration response c_A to unit step change in c_{Ain}

As can be seen, the response looks like a first-order response, it is called pseudo-first-order response. It could be approximated reasonably well by a first-order transfer function:

$$\frac{\delta c_A}{\delta c_{Ain}} \approx \frac{0.374}{(9s+1)} \tag{12.22}$$

The response of δc_A to changes in c_{Ain} has become slightly more sluggish. Owing to the interaction, the time constant increased from 6.67 s to 9 s. The response for δc_B to changes in c_{Ain} changed hardly, the damping only increased from 1.15 to 1.17, the gain decreased from 0.666 to 0.624.

The response of δc_A to changes in the flow F changed also. In the previous case it was a first-order transfer function with gain K_1 and time constant τ_A (Eqn. 12.4). From Eqn. (12.17) it can be seen that

$$\frac{\delta c_A}{\delta F} = \frac{K_1(\tau_B s+1)-K_3K_4}{(\tau_A s+1)(\tau_B s+1)-K_3K_5} \tag{12.23}$$

As can be seen, the response of c_A to changes in F could show an inverse response, when $K_1 - K_3K_4 < 0$, which will not occur in reality.

Example

Assume that the following additional values hold for this reaction, $V = 1\ m^3$, $F = 0.05\ m^3/s$, and for $k_1 \tau_R = 2$ it can be calculated that $c_A = 0.375 c_{Ain}$ and $c_B = 0.625\ c_{Ain}$. If $c_{Ain0} = 800\ kg/m^3$, then $c_{A0} = 300\ kg/m^3$ and $c_{B0} = 500\ kg/m^3$. The values of the gains can then be calculated as: $K_1 = 3333.3$, $K_3 = 0.0667$, $K_4 = 8333.3$, $K_5 = 1.667$. Substituting the values for the gains and time constants in Eqn. (12.23) results in:

$$\frac{\delta c_A}{\delta F} = \frac{3120.8(20.0s+1)}{124.85s^2 + 26.21s + 1} \tag{12.24}$$

which can also be written as a pseudo-first-order transfer function (compare Eqns. (12.20) and (12.21)).

12.4 Consecutive Reactions

In this section consecutive reactions will be investigated. In the previous two examples, maximization of the yield of component B could be achieved by maximizing the conversion of component A, which is equivalent to the minimization of the outlet concentration of component A.

It is assumed that the reaction takes place according to the following scheme:

$$A \xrightarrow{k_1} B \xrightarrow{k_2} C \tag{12.25}$$

and the reaction rates are assumed to be first order in the components:

$$r_A = -k_1 c_A,\ r_B = k_1 c_A - k_2 c_B,\ r_C = k_2 c_B \tag{12.26}$$

The component balances for all three components can be written as:

$$\begin{aligned} V\frac{dc_A}{dt} &= F(c_{Ain} - c_A) - Vk_1 c_A \\ V\frac{dc_B}{dt} &= -Fc_B + Vk_1 c_A - Vk_2 c_B \\ V\frac{dc_C}{dt} &= -Fc_C + Vk_2 c_B \end{aligned} \tag{12.27}$$

The behavioral model of the process is shown in Fig. 12.7.

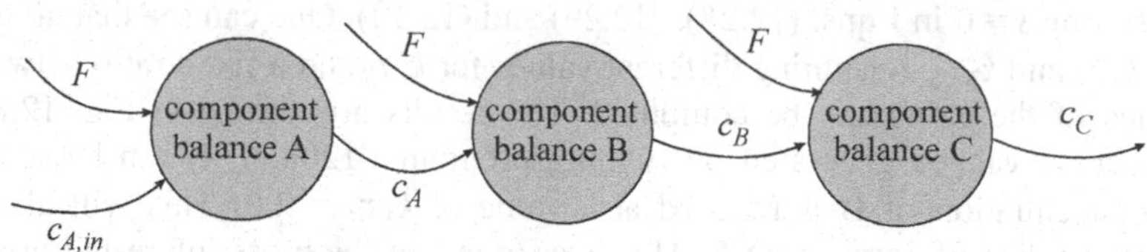

Fig. 12.7. Behavioral model for reactor in which consecutive reactions take place.

Without analyzing the model we can see two important things:

- the transfer function from c_{Ain} to c_C will be third order, since each balance constitutes a first-order transfer function.
- the transfer function from F to c_C will be a combination of a first-order, second-order and third order transfer function. The overall model will depend on the values of the gains of the individual models.

The overall transfer function between c_C and c_{Ain} can easily be obtained through linearization and Laplace transformation of Eqn. (12.27):

$$\frac{\delta c_C}{\delta c_{Ain}} = \left(\frac{\delta c_A}{\delta c_{Ain}}\right)\left(\frac{\delta c_B}{\delta c_A}\right)\left(\frac{\delta c_C}{\delta c_B}\right) = \frac{K_2}{(\tau_A s+1)}\frac{K_5}{(\tau_B s+1)}\frac{K_6}{(\tau_R s+1)} = \frac{\dfrac{k_1\tau_R}{(1+k_1\tau_R)}\dfrac{k_2\tau_R}{(1+k_2\tau_R)}}{(\tau_A s+1)(\tau_B s+1)(\tau_R s+1)} \tag{12.28}$$

where K_2, τ_A and τ_R are given by Eqn. (12.5), K_5 and τ_B are given by Eqn. (12.16), while K_6 is given by $K_6 = k_2\tau_R$.

Third-order systems are usually not difficult to control using standard-type controllers. This will be discussed in a later chapter.

If B is the preferred component, one should realize that the concentration of B has a maximum. To see this, first the responses of c_A and c_B to c_{Ain} have to be derived. The response of c_A to changes in c_{Ain} was already given in Eqn. (12.4) and is the same in this case:

$$\frac{\delta c_A}{\delta c_{Ain}} = \frac{\dfrac{1}{1+k_1\tau_R}}{\tau_A s+1} \tag{12.29}$$

The relationship between c_B and c_{Ain} is:

$$\frac{\delta c_B}{\delta c_{Ain}} = \frac{\dfrac{k_1\tau_R}{1+k_1\tau_R}\dfrac{1}{1+k_2\tau_R}}{(\tau_A s+1)(\tau_B s+1)} \tag{12.30}$$

The static relationship between the concentration changes and changes in c_{Ain} can be derived by setting $s = 0$ in Eqns. (12.28), (12.29) and (12.30). One can see that all gains are a function of $k_1\tau_R$ and $k_2\tau_R$. Assuming different values for $k_1\tau_R$ and a fixed ratio between k_1 and k_2, the values of the gains can be computed. The results are shown in Fig. 12.8 for $k_2 = 0.25k_1$. The curve can be generated by running program F1208.m. It can be seen that the maximum concentration of B is reached at a value of $k_1\tau_R = 2.0$. This will also give the maximum production of component B. However, one may be more interested in a situation where the difference between the production of component B and C is maximal. This can be achieved at a value of $k_1\tau_R = 1.0$. This situation may be desirable if component B and C have to be separated in a separation train where minimum separation energy is required.

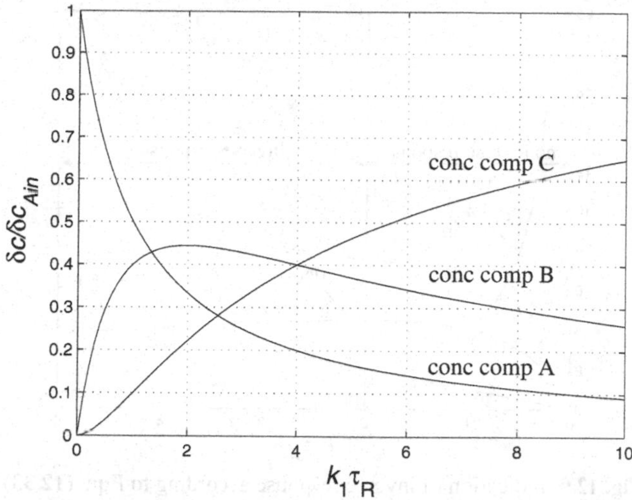

Fig. 12.8. Concentration profile in case of consecutive reactions.

It can be seen that control of the residence time τ_R is important. When the residence time changes, the concentration of component B changes. Control can be achieved by measuring the outlet concentration of component B and manipulating the reactor throughput F. As can be seen from Fig. 12.7, the relationship between the two variables will be a combination of a first- and second-order transfer function. The overall transfer function can easily be derived from Eqn. (12.27). Equation (12.27a) is the same as Eqn. (12.1), therefore linearization of Eqn. (12.27a) results in Eqn. (12.4). Equation (12.27b) is the same as Eqn. (12.12), therefore linearization of Eqn. (12.27b) results in Eqn. (12.15). Combination of Eqns. (12.4) and (12.15) results in:

$$\frac{\delta c_B}{\delta F} = \frac{-K_4(\tau_A s+1)+K_1K_5}{(\tau_A s+1)(\tau_B s+1)-K_2K_5} \tag{12.31}$$

The response of δc_B to changes in F will give an inverse response if $(-K_4+K_1K_5) > 0$. This condition can be written as:

$$-\frac{c_{B0}/F_0}{1+k_2\tau_R}+\frac{(c_{Ain}-c_A)_0/F_0}{1+k_1\tau_R}\frac{k_1\tau_R}{1+k_2\tau_R}>0 \tag{12.32}$$

which can be rewritten to:

$$k_1\tau_R(c_{Ain}-c_A-c_B)_0 > c_{B0} \quad or$$
$$k_1\tau_R > \frac{c_{B0}}{c_{C0}} \tag{12.33}$$

This condition is shown in Figure 12.9, where $k_1\tau_R$ as well as the ratio c_{B0}/c_{C0} are plotted. In addition, the concentration of component B is plotted.

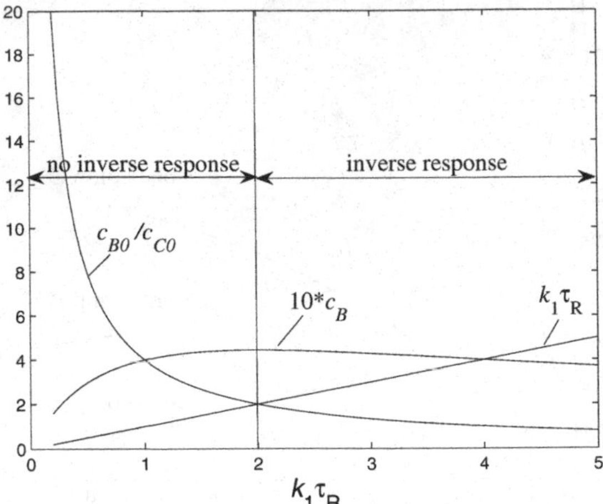

Fig. 12.9. Indication of inverse response according to Eqn. (12.33).

For values of $k_1\tau_R > 2$, the reactor can exhibit an inverse response, i.e. when flow changes are such that the value of $k_1\tau_R$ changes but remains greater than 2, there will be an inverse response of the concentration of component B to flow changes.

Whether control will give any problems still has to be determined by substituting numerical values for the process parameters. The target of the composition controller will depend on whether one is interested in maximum production of B or minimum separation energy.

Another interesting feature that Fig. 12.8 shows is the so-called state multiplicity. Two different values of the residence time, $k_1\tau_R = 1.0$ (first operating point) and $k_1\tau_R = 4.0$ (second operating point), result in the same concentration of component B. The second operating point is not of interest, since the concentration of component C is high and separation of components B and C will be costly. Moreover, the reactor feed is smaller in operating point two than in operating point one. The models $(\delta c_B/\delta F)$ in the two operating points are quite different.

12.5 Non-isothermal Reactions

Often reactions do not take place at isothermal conditions. In that case the excess amount of heat is withdrawn from the reactor by cooling or supplied to the reactor by heating. Often a jacketed reactor or reactor with cooling/heating coil or pipes is used. The component balances are now also affected by the temperature, since the rate of reaction becomes temperature dependent. Let us assume that we have to deal with an exothermic reaction with heat of reaction ΔH and that the excess heat in the reactor is withdrawn through a cooling coil with heat transfer area A. The assumptions, made in section 12.1 are still valid (with exception of isothermal conditions), in addition, the following assumptions are being made:

- the overall heat transfer coefficient U is assumed to be constant
- the temperature in the reactor is T (ideally mixed) and the average temperature in the cooling coil is T_c
- the heat capacity of the cooling coil and the cooling coil contents can be ignored since it is small compared with the heat capacity of the reactor. Note: heat capacity is defined by $C = Mc_p$, where M is the mass and c_p the specific heat
- the reaction to be considered is A → B
- the rate of reaction is first-order in component A

Let us assume that $T_{in} < T$ and that the rate of reaction can be given by an Arrhenius-type equation:

$$r = k_0 e^{-E/RT} c_A \tag{12.34}$$

in which E is the activation energy of the reaction, R the gas constant and k_0 the pre-exponential factor. T is the temperature and c_A the concentration of component A.

The energy balance for the reactor can be written as:

$$\rho V c_p \frac{dT}{dt} = F\rho c_p (T_{in} - T) + rV\Delta H - UA(T - T_c) \tag{12.35}$$

where the left-hand side of the equation represents the temporary accumulation of energy in the reactor, (J/s), the first term on the right-hand side represents the sensible heat going into the reactor, the second term on the right-hand side represents the energy that is generated due to the reaction, and the third right-hand side term is the heat withdrawn from the reactor. In some cases, especially when the physical properties are temperature dependent, it is better to use an enthalpy balance, rather than a balance for the temperature change. In Eqn. (12.35) it is assumed that the reference temperature is equal to zero.

Combining Eqn. (12.1) with Eqn. (12.34) results in:

$$V \frac{dc_A}{dt} = F(c_{Ain} - c_A) - Vr \tag{12.36}$$

A behavioral diagram based on Eqns. (12.35) and (12.36) is shown in Fig. 12.10.

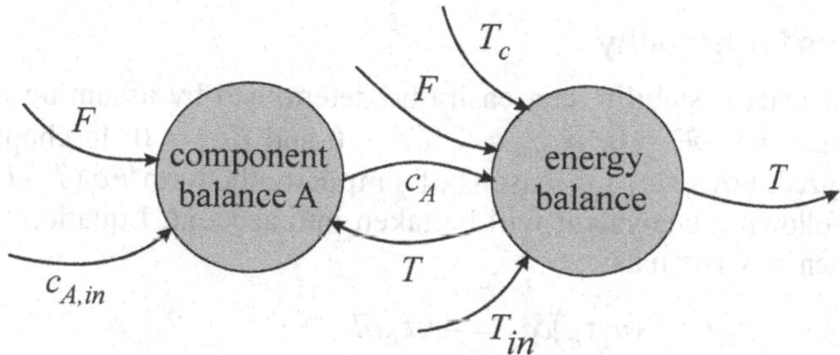

Fig. 12.10. Interaction between component and energy balance.

In general, the outlet concentration c_A is controlled by the reactor feed F and the reactor temperature T is controlled by the cooling water flow, which is this case equivalent to changing the cooling water temperature T_c. Let us therefore investigate what these models are.

Equation (12.36) can be linerarized to:

$$\delta c_A = \frac{K_1 \delta F}{\tau_R s + 1} + \frac{\delta c_{Ain}}{\tau_R s + 1} - \frac{\tau_R \delta r}{\tau_R s + 1} \tag{12.37}$$

with

$$\tau_R = V / F_0, K_1 = (c_{Ain} - c_A)_0 / F_0 \tag{12.38}$$

Equations (12.34) and (12.35) can be linearized to:

$$\delta T = \frac{-K_2 \delta F}{\tau_T s + 1} + \frac{K_3 \delta T_{in}}{\tau_T s + 1} + \frac{K_4 \delta r}{\tau_T s + 1} + \frac{K_5 \delta T_c}{\tau_T s + 1} \tag{12.39}$$

with

$$\tau_T = \tau_R / (1 + UA / F_0 \rho c_p)$$
$$K_2 = \frac{(T - T_{in})_0 / F_0}{1 + (UA / F_0 \rho c_p)}$$
$$K_3 = \frac{1}{1 + (UA / F_0 \rho c_p)} \tag{12.40}$$
$$K_4 = \frac{V \Delta H / F_0 \rho c_p}{1 + (UA / F_0 \rho c_p)}$$
$$K_5 = \frac{UA / F_0 \rho c_p}{1 + (UA / F_0 \rho c_p)}$$

Equation (12.34) can be written as:

$$\delta r = r_1 \delta c_A + r_2 \delta T$$
$$r_1 = \left(\frac{\partial r}{\partial c_A} \right)_0 = r_0 / c_{A0} \tag{12.41}$$
$$r_2 = \left(\frac{\partial r}{\partial T} \right)_0 = E r_0 / R T^2$$

in which r_0 is the rate of reaction in the steady state situation.

12.5.1 Conditions for Stability

The conditions for reactor stability can easily be determined by assuming that there are no external disturbances, i.e. $\delta F = 0$, $\delta c_{Ain} = 0$, $\delta T_c = 0$ and $\delta T_{in} = 0$. In chapter 7 the reactor stability was analyzed; however, for reasons of simplicity the term $F c_p (T_{in} - T)$ was ignored at that time. In the following analysis it will be taken into account. Equations (12.37), (12.39) and (12.41) can then be written as:

$$\begin{aligned} (\tau_R s + 1 + r_1 \tau_R) \delta c_A &= -r_2 \tau_R \delta T \\ (\tau_T s + 1 - K_4 r_2) \delta T &= K_4 r_1 \delta c_A \end{aligned} \tag{12.42}$$

which results in the so-called characteristic equation:

$$\tau_T \tau_R s^2 + (\tau_R - r_2 \tau_R K_4 + \tau_T + r_1 \tau_T \tau_R) s + (r_1 \tau_R - r_2 K_4 + 1) = 0 \tag{12.43}$$

The first condition for stability is called the static condition and requires:

$$r_1 \tau_R - r_2 K_4 + 1 > 0 \tag{12.44}$$

Substitution of parameters and combination with the static version of Eqn. (12.36) results in:

$$\frac{c_{Ain0}}{c_{A0}} > \frac{F_0 (c_{Ain0} - c_{A0}) \Delta H}{(F_0 \rho c_p + UA)} \frac{E}{R T_0^2} \tag{12.45}$$

The condition for dynamic stability is:

$$\tau_R - r_2 \tau_R K_4 + \tau_T + r_1 \tau_T \tau_R > 0 \tag{12.46}$$

Substitution of parameters and combination with the static version of Eqn. (12.36), results in:

$$\frac{c_{Ain}}{c_{A0}}\frac{F_0\rho c_p}{F_0\rho c_p+UA}+1>\frac{F_0(c_{Ain0}-c_{A0})\Delta H}{(F_0\rho c_p+UA)}\frac{E}{RT_0^{\ 2}} \tag{12.47}$$

If Eqn. (12.45) is satisfied, Eqn. (12.47) is also satisfied. If Eqn. (12.45) is not satisfied, the reactor is unstable. Its behavior can then be characterized by an increasing exponential time function.

12.5.2 Static Instability

The static behavior of the reactor can also be shown in a conversion-temperature phase plane. The static energy balance Eqn. (12.35) can be combined with the static component balance Eqn. (12.36) to give:

$$T=\frac{F_0c_{A0}V\Delta H}{F_0\rho c_p+UA}C+\frac{F_0\rho c_p}{F_0\rho c_p+UA}T_{in0}+\frac{UA}{F_0\rho c_p+UA}T_{c0} \tag{12.48}$$

which shows a linear relationship between the temperature T and the conversion C. The conversion can be calculated by combining Eqns. (12.34) and (12.36), which results in:

$$C=\frac{c_{Ain}-c_A}{c_{Ain}}=\frac{k_0\tau_R}{k_0\tau_R+e^{E/RT}} \tag{12.49}$$

Both Eqn. (12.48) and (12.49) can be shown in the C–T plane, a stable operating point is obtained when the produced heat is equal to the heat that is removed by the cooling coil (points A and B). This is shown in Fig. 12.11. Point C is an unstable operating point, when a disturbance acts upon the system, the reactor will either move to operating point A or B, depending on the direction of the disturbance.

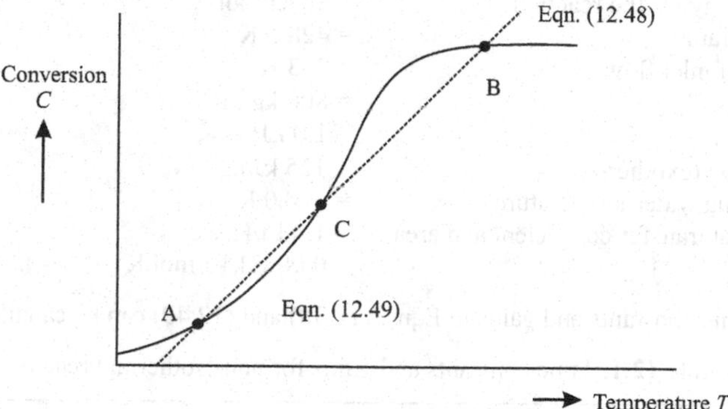

Fig. 12.11. Conversion-temperature state plane for reactor.

12.5.3 Dynamic Instability

If Eqn. (12.46) is not satisfied, dynamic instabilities can occur. This is due to the fact that there is an interaction between the component balance and the energy balance. A temperature rise will cause the reaction to proceed faster, resulting in a decrease of component A. Consequently, the temperature will decrease and the concentration of component A increases again. This cycle can continue, and under certain conditions lead to instability.

12.5.4 Relationship between Concentration and Feed Changes

Suppose that all conditions for static and dynamic stability have been met and that we would like to control the outlet concentration of component A with the feed flow F. To analyze the model, we combine Eqns. (12.37), (12.39) and (12.41) using $\delta c_{Ain} = 0$, $\delta T_{in} = 0$ and $\delta T_c = 0$. This results, after considerable rewriting, in:

$$\frac{\delta c_A}{\delta F} = \frac{K_1 \tau_T s + (K_1 - K_1 K_4 r_2 + K_2 \tau_R r_2)}{\tau_R \tau_T s^2 + (\tau_R + \tau_T - r_2 \tau_R K_4 + r_1 \tau_T \tau_R)s + (r_1 \tau_R - r_2 K_4 + 1)} \tag{12.50}$$

As can be seen, the second term of the denominator is the expression of Eqn. (12.46), the third term in the denominator is the expression of Eqn. (12.44). The shape of the final response depends on the sign of $b = K_1 - K_1K_4r_2 + K_2\tau_R r_2$. If Eqns. (12.47) and (12.49) are satisfied and $K_1 > 0$, we can distinguish two situations:

- $b>0$, this results in a pseudo-first-order response for $\delta c_A / \delta F$
- $b<0$, this will give a non-minimum phase response for $\delta c_A / \delta F$.

The relationship for $\delta T / \delta T_c$ (or between the reactor temperature and the cooling water flow) can be derived in a similar manner. The model has then to be extended with an equation, relating the cooling water temperature T_c to the cooling water flow.

Example

Suppose the reactor is operating with the following data:

V	reactor volume	$= 5$ m^3
c_A	outlet concentration of component A	$= 213.8$ kg/m^3
c_{Ain}	inlet concentration of component A	$= 800$ kg/m^3
F	total volumetric flow	$= 0.01$ m^3/s
k	pre-exponential constant	$= 25.0$ s^{-1}
E	activation energy for the reaction	$= 30$ kJ/mol
T	reactor temperature	$= 428.5$ K
T_{in}	temperature of inlet flow	$= 353$ K
ρ	density	$= 800$ kg/m^3
c_p	specific heat	$= 1.0$ kJ/kg.K
ΔH	heat of reaction (exothermic)	$= 125$ kJ/kg
T_c	average cooling water temperature	$= 300.0$ K
UA	product of heat transfer coefficient and area	$= 1.0$ kJ/K.sec
R	gas constant	$= 0.00831$ kJ/mol.K

The values of the time constants and gains in Eqns. (12.38) and (12.40) can be calculated as:

Table 12.1. Time constants and gains for non-isothermal reactor.

parameter	value	parameter	value
τ_R	500.0	K_4	69.4
K_1	58620	K_5	0.111
$UA / F\rho c_p$	0.125	r_0	1.172
τ_T	444.4	r_1	0.0055
K_2	6711.1	r_2	0.023
K_3	0.889		

The transfer function of Eqn. (12.50) can now be calculated. Equation (12.50) becomes:

$$\frac{\delta c_A}{\delta F} = \frac{29377(633.7s + 1)}{1.9E5s^2 + 808.3s + 1} \tag{12.51}$$

The non-linear model and the linearized model of Eqn. (12.51) are simulated in F1212.mdl, the concentration response for a step change of 10% in the flow is shown in Fig. 12.12. It can be seen that the responses look quite similar: the final response is somewhat different, the one model calculates a concentration change of 32.0 kg/m^3, the other model calculates a change in concentration of 29.3 kg/m^3. The response 'peaks' somewhat around $t = 600$ s; this is due to the large value of the positive time constant (633.7) in the numerator. The denominator has two time constants of the order of magnitude 450–500 seconds. If the time constant in the numerator is less than these time constants, the response would look like a first-order response.

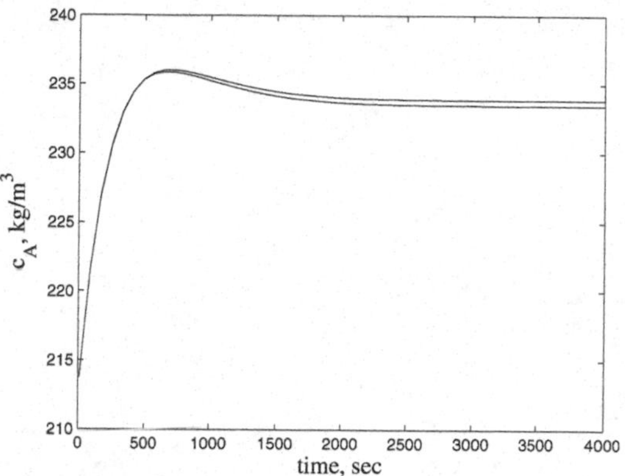

Fig. 12.12. Response of reactor outlet concentration to step change in flow.

The reactor temperature response is shown in Fig. 12.13, as can be expected this response also has a non-minimum phase character.

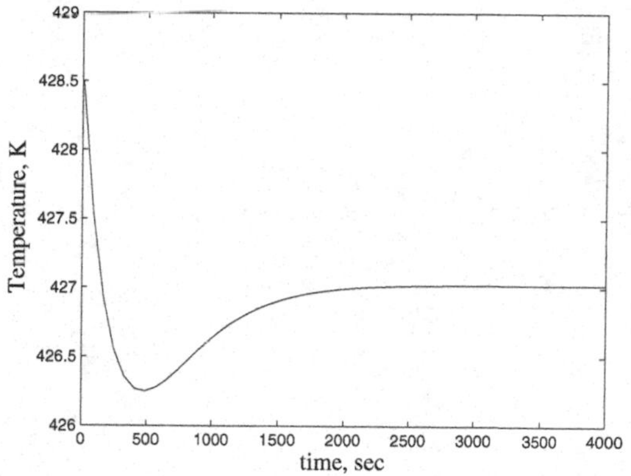

Fig. 12.13. Response of reactor temperature to step change in flow.

Files referred to in this chapter:

F1204.mdl: model to generate data for Fig. 12.4
F1204plot.m: file to plot Fig. 12.4
F1206.mdl: model to generate data for Fig. 12.6
F1206plot.m: file to plot Fig. 12.6
F1208.m: file to generate Fig. 12.8
F1212par.m: parameter definition for F1212
F1212.mdl: model to generate data for Fig. 1212
F1212plot.m: file to plot Fig. 1212 and 12.13

13 Dynamic Analysis of Tubular Reactors

In this chapter, tubular chemical reactors will be analyzed. Reaction mechanisms considered are first order in the components. As will be shown, at isothermal conditions reactor dynamics can be described by partial differential equations for the component balances. If non-isothermal conditions exist, the energy balance would also be a partial differential equation. Dynamics models are considerably different from the models for ideally stirred chemical reactors under similar operating conditions.

13.1 Introduction

The previous chapter dealt with ideally mixed reactors. Tubular reactors, as shown in Fig. 13.1, often give a higher conversion than ideally mixed reactors for the same reaction conditions because on average the reactant concentration is higher. They are therefore, theoretically, preferred over stirred tank reactors.

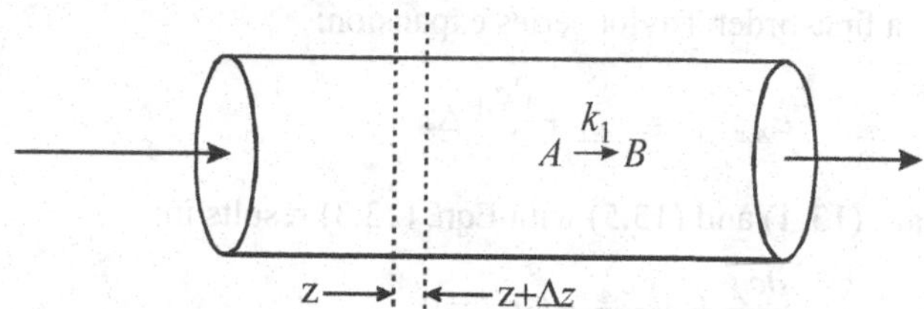

Fig. 13.1. First-order reaction under isothermal conditions in tubular reactor.

To limit complexity, let us start out by making the following assumptions:

a. the reaction conditions are isothermal
b. the rate of reaction is first order in the component A
c. the density of all components is constant and equal
d. the mixing in radial direction is ideal
e. there is no mixing in axial direction
f. the velocity of the medium is constant in axial direction
g. the dispersion in the reactor can be neglected

Assumptions d, e and g mean that the flow through the reactor can be considered to be plug flow.

Let the length of the reactor be L (m) and the cross sectional area A_c (m^2). The inlet concentration of component A at the inlet is c_{Ain}, the velocity of the medium is v (m/s).

Process Dynamics and Control: Modeling for Control and Prediction. Brian Roffel and Ben Betlem.

13.2 First-order Reaction

The component balance over a period Δt for component A in a segment with volume $A_c\Delta z$ can now be written as:

$$\begin{bmatrix} accumulation\ of \\ component\ A \\ during\ time\ \Delta t \end{bmatrix} = \begin{bmatrix} component\ in \\ during\ time\ \Delta t \end{bmatrix} - \begin{bmatrix} component\ out \\ during\ time\ \Delta t \end{bmatrix} - \begin{bmatrix} disappearance \\ of\ component \\ during\ time\ \Delta t \end{bmatrix} \tag{13.1}$$

which can be written in mathematical terms as:

$$A_c\Delta z\left[c_{A,t+\Delta t} - c_{A,t}\right] = vA_c c_{A,z}\Delta t - vA_c c_{A,z+\Delta z}\Delta t - rA_c\Delta z\Delta t \tag{13.2}$$

Equation (13.2) can be rewritten as:

$$\frac{c_{A,t+\Delta t} - c_{A,t}}{\Delta t} = v\frac{c_{A,z} - c_{A,z+\Delta z}}{\Delta z} - r \tag{13.3}$$

Since the reaction is assumed to be first order in component A, the rate of reaction r, can be written as:

$$r = k_1 c_A \tag{13.4}$$

The concentration at location $z+\Delta z$ can be written as a function of the concentration at location z, by using a first-order Taylor series expansion:

$$c_{A,z+\Delta z} = c_{A,z} + \frac{\partial c_A}{\partial z}\Delta z \tag{13.5}$$

Combination of Eqns. (13.4) and (13.5) with Eqn. (13.3) results in:

$$\frac{\partial c_A}{\partial t} + v\frac{\partial c_A}{\partial z} + k_1 c_A = 0 \tag{13.6}$$

We can now introduce deviation variables δc_A and take the Laplace transform. Equation (13.6) can then be written as:

$$\frac{d(\delta c_A)}{dz} + \left(\frac{k_1 + s}{v}\right)\delta c_A = 0 \tag{13.7}$$

The solution can be written as:

$$\delta c_A(z,s) = \delta c_A(0,s)\, e^{-\frac{k_1+s}{v}z} \tag{13.8}$$

The response for the outlet concentration at $z=L$ becomes:

$$\frac{\delta c_A(L,s)}{\delta c_A(0,s)} = \frac{\delta c_{A,out}}{\delta c_{A,in}} = e^{-k_1\tau_R}e^{-s\tau_R} \tag{13.9}$$

in which $\tau_R = L/v$, the residence time of the material in the reactor.

The term $e^{-k_1\tau_R}$ is the process gain, the term $e^{-s\tau_R}$ indicates a pure time delay. When the concentration changes at the beginning of the reactor, it takes τ_R time units before the change reaches the end of the reactor. If the residence time τ_R = 10 s and k_1 = 0.2 s^{-1}, then $k_1\tau_R = 2$, hence the process gain is 0.135, i.e. a unit change in the inlet concentration of 1.0 results in a change in the outlet concentration of 0.135. The step response for the outlet concentration is shown in Fig. 13.2.

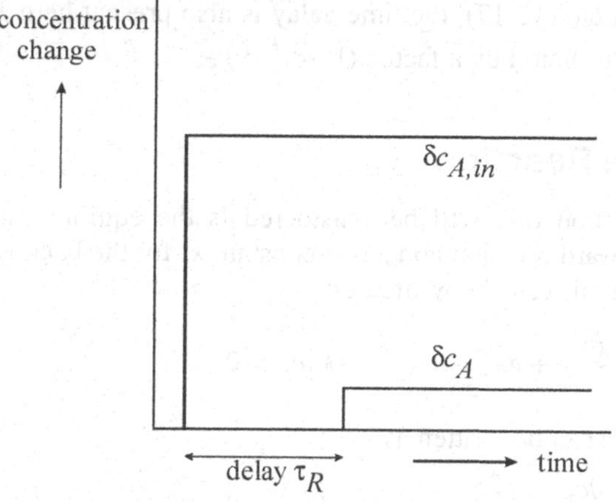

Fig. 13.2. Response of outlet concentration to step change in inlet concentration.

The component balance for the product B can be written as:

$$A_c \Delta z\left[c_{B,t+\Delta t} - c_{B,t}\right] = vA_c c_{B,z}\Delta t - vA_c c_{B,z+\Delta z}\Delta t + rA_c \Delta z \Delta t \tag{13.10}$$

which can be rewritten as:

$$\frac{\partial c_B}{\partial t} + v\frac{\partial c_B}{\partial z} - k_1 c_A = 0 \tag{13.11}$$

Linearization and Laplace transformation of Eqn. (13.11) gives:

$$s\delta c_B + v\frac{d(\delta c_B)}{dz} = k_1 \delta c_A \tag{13.12}$$

The solution of Eqn. (13.12) can be given by:

$$\delta c_B(z,s) = C_1 e^{-\frac{s}{v}z} + C_2 e^{-\frac{k_1+s}{v}z} \tag{13.13}$$

in which the first term on the right-hand side represents the solution of the homogeneous (left-hand side) part of Eqn. (13.12) and the second term represents the solution for δc_A which was obtained in Eqn. (13.8). C_1 and C_2 are integration constants to be determined from boundary conditions. At $z = 0$, we may write:

$$\delta c_B(0,s) = C_1 + C_2 = 0 \Rightarrow C_1 = -C_2 \tag{13.14}$$

Equation (13.13) can now be written as:

$$\delta c_B(z,s) = C_1 e^{-\frac{s}{v}z}\left(1 - e^{-\frac{k_1}{v}z}\right) \tag{13.15}$$

The value of C_1 can be determined from the boundary condition at $z = \infty$:

$$\delta c_B(\infty,s) = C_1 = \delta c_{A,in} \tag{13.16}$$

which gives for the change in outlet concentration of component B ($z = L$):

$$\frac{\delta c_B(L,s)}{\delta c_A(0,s)} = \frac{\delta c_{B,out}}{\delta c_{A,in}} = e^{-s\tau_R}(1 - e^{-k_1\tau_R}) \tag{13.17}$$

As can be seen from Eqn. (13.17), the time delay is also present here and incoming changes in concentration are attenuated by a factor $(1-e^{-k_1\tau_R})$.

13.3 Equilibrium Reaction

The next type of reaction that will be considered is the equilibrium reaction with a rate constant k_1 for the forward reaction and a rate constant k_2 for the backward reaction. Equation (13.6) can in this case be written as:

$$\frac{\partial c_A}{\partial t}+v\frac{\partial c_A}{\partial z}+k_1c_A-k_2c_B=0 \tag{13.18}$$

Similarly, Eqn. (13.11) can be written as:

$$\frac{\partial c_B}{\partial t}+v\frac{\partial c_B}{\partial z}+k_2c_B-k_1c_A=0 \tag{13.19}$$

Both equations have to be solved simultaneously. Taking the boundary conditions into account, the solution for the product concentration can be given as:

$$\frac{\delta c_B(L,s)}{\delta c_A(0,s)}=\frac{\delta c_{B,out}}{\delta c_{A,in}}=\frac{k_1}{k_1+k_2}\left(1-e^{-(k_1+k_2)\tau_R}\right)e^{-s\tau_R} \tag{13.20}$$

As can be seen, the response is again characterized by a gain and a time delay.

13.4 Consecutive Reactions

Let us consider a consecutive reaction:

$$A\xrightarrow{k_1}B\xrightarrow{k_2}C \tag{13.21}$$

The component balance for component A is given by Eqn. (13.6), the balance for component B is given by Eqn. (13.19). Since Eqn. (13.6) only depends on c_A it can be solved independently. The solution for the transfer function from δc_{Ain} to δc_B, taking the appropriate boundary conditions into account, can be written as:

$$\frac{\delta c_B(L,s)}{\delta c_A(0,s)}=\frac{\delta c_{B,out}}{\delta c_{A,in}}=\frac{k_1}{k_1-k_2}\left(e^{-k_2\tau_R}-e^{-k_1\tau_R}\right)e^{-s\tau_R} \tag{13.22}$$

where it is assumed that component B is the component of interest and component C is undesirable byproduct. Again, the model consists of a gain and time delay. It can also be seen from Eqns. (13.17), (13.20) and (13.22) that for different reaction mechanisms, the process dynamics (time delay) are the same, the process gains are, however, dependent on the reaction mechanism.

13.5 Tubular Reactor with Dispersion

In the previous sections it was assumed that there was no dispersion in the reactor, i.e. axial mixing due to diffusion was neglected. In reality this is never the case; however, the relevancy depends on the size of the diffusion coefficient.

Let us assume that a first-order reaction $A \xrightarrow{k_1} B$ is considered again. If there is dispersion present, the diffusion term can be described by a second-order derivative in the axial direction Dd^2c_A/dz^2. This term has to be added to Eqn. (13.6), hence

$$\frac{\partial c_A}{\partial t}+v\frac{\partial c_A}{\partial z}+k_1c_A-D\frac{\partial^2 c_A}{\partial z^2}=0 \tag{13.23}$$

in which D is the diffusion coefficient in (m^2/s).

When there is no dispersion in the section upstream of the reactor, the boundary condition at the entrance can be found from a material balance around the entrance:

$$\begin{aligned}A_c\Delta z\left(\frac{\partial c_A}{\partial t}\right)_{0^+} &= A_cvc_A\left(0^-,t\right)-A_cvc_A\left(0^+,t\right)+\\ &A_cD\left(\frac{\partial c_A}{\partial z}\right)_{0^+}-A_c\Delta z\,k_1\,c_A\left(0^+,t\right)\end{aligned} \tag{13.24}$$

where:

0^- position at the entrance outside of the reactor

0^+ position at the entrance inside the reactor

A_c cross sectional area of the reactor

$c_A\left(0^-,t\right)$ concentration at the entrance of an element with length Δz, which is the inlet concentration

$c_A\left(0^+,t\right)$ concentration at the outlet of an element with length Δz

It is assumed that there is no dispersion in the section upstream of the reactor. In the limit case when Δz approaches zero, Eqn. (13.24) can be written as:

$$vc_A(0^-,t)-vc_A(0^+,t)+D\left(\frac{\partial c_A}{\partial z}\right)_{0^+}=0 \tag{13.25}$$

Similarly, at the end of the reactor we may write:

$$vc_A\left(L^-,t\right)-vc_A\left(L^+,t\right)+D\left(\frac{\partial c_A}{\partial z}\right)_{L^+}=0 \tag{13.26}$$

In this case, the dispersion outside the reactor has again been ignored. Since $c_A\left(L^-,t\right)=c_A\left(L^+,t\right)$, Eqn. (13.26) may be reduced to:

$$D\left(\frac{\partial c_A}{\partial z}\right)_L=0 \tag{13.27}$$

The model for the reactor now consists of Eqns. (13.23), (13.25) and (13.27).
After Laplace transformation, Eqn. (13.23) can be written as:

$$s\delta c_A(z,s)-\delta c_A(z,0^+)=-v\frac{d\delta c_A(z,s)}{dz}+D\frac{d^2\delta c_A(z,s)}{dz^2}-k_1\delta c_A(z,s) \tag{13.28}$$

The general solution of Eqn. (13.28) is determined by the following characteristic equation:

$$DJ^2-vJ-(k_1+s)=0 \tag{13.29}$$

from which the following roots can be determined:

$$J = \frac{v}{2D} \pm \frac{1}{2D}\sqrt{v^2 + 4D(k_1 + s)} \tag{13.30}$$

The general solution is of the following form:

$$\delta c_A(z,s) = C_1 e^{J_1 z} + C_2 e^{J_2 z} \tag{13.31}$$

The coefficients C_1 and C_2 can be determined by differentiating Eqn. (13.31) with respect to z and equating it to the boundary conditions, Eqn. (13.25) and (13.27). The result for the solution becomes then:

$$\frac{\delta c_A(z,s)}{\delta c_{A,in}} = \frac{J_2 e^{-J_1(L-z)} - J_1 e^{-J_2(L-z)}}{J_2(1 - J_1 D/v)e^{-J_1 L} - J_1(1 - J_2 D/v)e^{-J_2 L}} \tag{13.32}$$

13.5.1 Static Analysis

The static behavior can be studied by setting $s = 0$ in Eqns. (13.30) and (13.32). The following parameters are introduced, the Péclet number:

$$Pe = vL/D \tag{13.33}$$

and a parameter a as:

$$a = \sqrt{1 + 4k_1 D/v^2} \tag{13.34}$$

which enables us to write Eqn. (13.32) as:

$$\frac{c_A(z)}{c_A(0^-)} = 2\exp\left(Pe.\frac{z}{2L}\right)\frac{(1+a)\exp\left[Pe\frac{a}{2}\left(1-\frac{z}{L}\right)\right] - (1-a)\exp\left[Pe\frac{a}{2}\left(\frac{z}{L}-1\right)\right]}{(1+a)^2\exp\left(Pe\frac{a}{2}\right) - (1-a)^2\exp\left(-Pe\frac{a}{2}\right)} \tag{13.35}$$

Figure 13.3 shows Eqn. (13.35) for $a = 1.3$ and three different values of the Péclet number. As can be seen, the outlet concentration decreases for higher Péclet numbers.

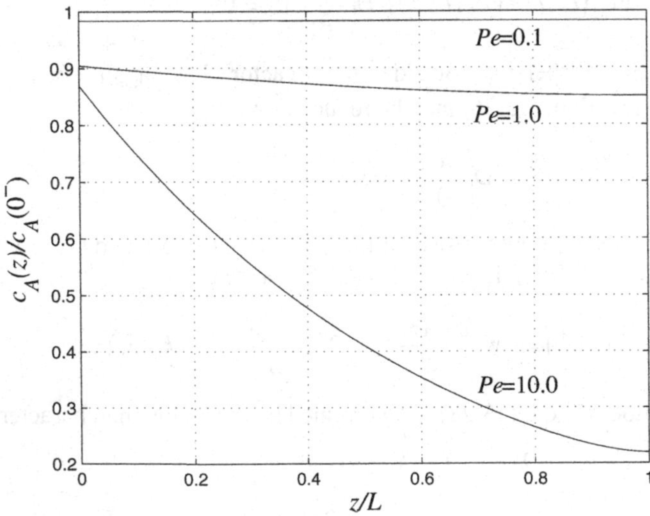

Fig. 13.3. Reactor concentration profile for $a = 1.3$ and different Péclet numbers.

13.5.2 Special Cases

Let us consider two special cases, the first one in which the dispersion becomes very small, and the second one in which the conversion becomes very large.

In the first case, when the dispersion becomes very small, *a* can be approximated by:

$$a = \sqrt{1+4k_1 D/v^2} \approx 1+2k_1 D/v^2 \tag{13.36}$$

Equation (13.35) can then be written as:

$$\frac{c_A(z)}{c_A(0^-)} = e^{-k_1 z/v} \tag{13.37}$$

The reactor conversion follows from:

$$C = 1 - c_A(L)/c_A(0^-) = 1 - e^{-k_1 L/v} = 1 - e^{-k_1 \tau_R} \tag{13.38}$$

which is the well-known equation for conversion in a plug flow reactor.

In the second case, in which the diffusion coefficient D becomes large, Pe(a/2) is proportional to $D^{-1/2}$, hence it approaches zero. If, in addition, the following approximation is made:

$$\begin{aligned} e^x &\approx 1+x \\ e^{-x} &\approx 1-x \end{aligned} \tag{13.39}$$

Eqn. (13.35) can be written as:

$$\frac{c_A(z)}{c_A(0^-)} \approx \frac{v}{v+k_1 L} \tag{13.40}$$

hence the conversion becomes:

$$C = 1 - c_A(L)/c_A(0^-) = \frac{k_1 L}{v+k_1 L} = \frac{k_1 \tau_R}{1+k_1 \tau_R} \tag{13.41}$$

which is the well-known equation for conversion in an ideally mixed reactor.

The conversion for the plug flow reactor and ideally mixed reactor are graphically shown in Fig. 13.4 for different values of $k_1 \tau_R$.

13.5.3 Dynamic Analysis

Equation (13.32) can be simplified when $z = L$. It can then be written as:

$$\frac{\delta c_A(L,s)}{\delta c_{A,in}} = exp(-Pe/2) \left\{ \begin{array}{l} exp\left[\frac{1}{2} Pe \sqrt{1+4(s+k_1)\tau_R / Pe}\right] + \\ exp\left[-\frac{1}{2} Pe \sqrt{1+4(s+k_1)\tau_R / Pe}\right] \end{array} \right\} \tag{13.42}$$

Using Laplace transform tables the function can be transformed back to the time domain, the result for the response to a pulse in the input is the so-called impulse transfer function:

$$h(t) = \frac{\tau_R}{2t}\sqrt{\frac{\tau_R Pe}{\pi t}} exp\left\{-\frac{1}{4} Pe\left(\frac{\tau_R}{t} + \frac{t}{\tau_R} - 2\right) - \frac{1}{4} k_1 t\right\} \tag{13.43}$$

Equation (13.43) is used to compute Fig. 13.5 for $k_1 = 0.1$ and for four different values of the Péclet number.

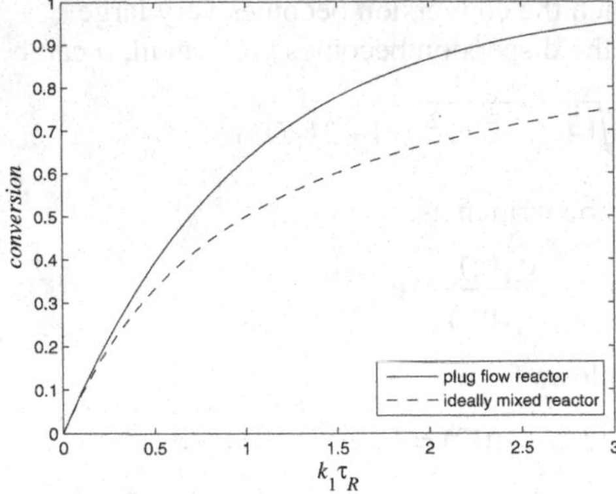

Fig. 13.4. Reactor conversion as a function of $k_1 \tau_R$.

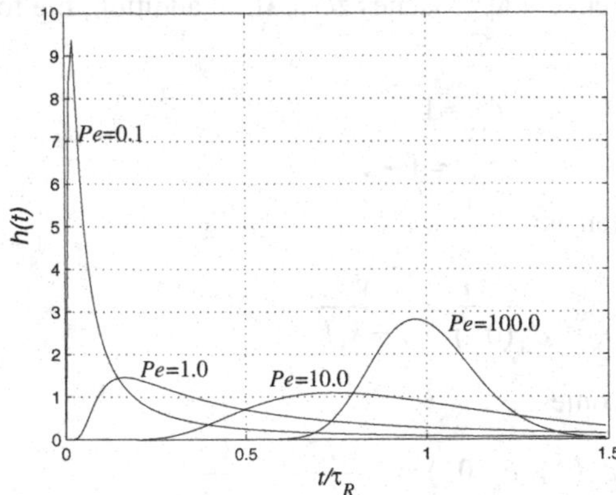

Fig. 13.5. Impulse response of a tubular reactor with back mixing, $k_1 = 0.1$.

It can be seen that for small values of the Péclet number, which corresponds to large values of the diffusion, the behavior approximates the response of an ideally mixed reactor. For large Péclet numbers, i.e. small values of the diffusion, the behavior the behavior approximates the behavior of a pure dead-time process.

13.6 Dynamics of Adiabatic Tubular Flow Reactors

In practice, temperatures in a tubular reactor mostly vary, resulting in a non-uniform temperature profile along the reactor (Fig. 13.6). The following assumptions are made:

a. the reactor is filled with catalyst with temperature T_s
b. the porosity (or fraction of the volume that is occupied by the fluid) is ε
c. radial concentration gradients can be ignored
d. diffusion can be neglected
e. the reaction is of first order in component A
f. heat development due to reaction occurs in the fluid
g. catalyst and fluid temperatures can be represented by a uniform temperature over the cross sectional area
h. the reaction is an exothermal reaction, i.e. heat is generated by the reaction
i. physical properties, such as density and heat capacity may be assumed constant

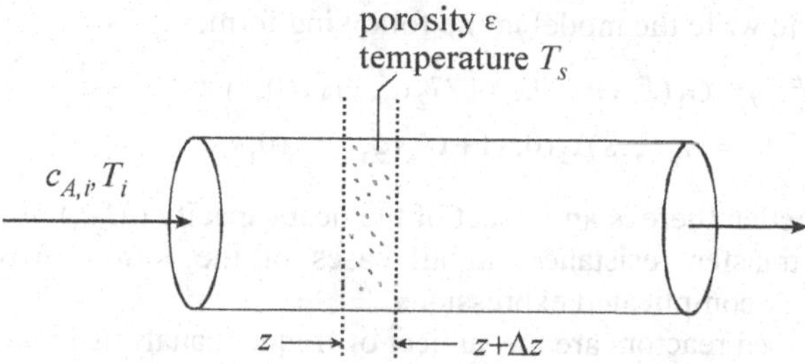

Fig. 13.6. Adiabatic tubular flow reactor.

The component balance over the length Δz for component A can still be represented by Eqn. (13.5).

An energy balance for the fluid over an element of length Δz can be written as:

$$\frac{\partial T}{\partial t}+v\frac{\partial T}{\partial z}-\frac{\alpha_s A_s}{A_c \varepsilon \rho_f c_f}(T_s-T)-\frac{\Delta H k_1 c_A}{\rho_f c_f}=0 \tag{13.44}$$

The first term in Eqn. (13.44) is the accumulation term, the second term represents the heat transport due to flow through the reactor, the third term the heat exchange from the catalyst to the fluid and the last term the heat generated by the reaction.
In Eqn. (13.44) the following nomenclature is used:

α_s heat transfer coefficient, W/m^2.K
A_s catalyst area, m^2
ρ_f fluid density, kg/m^3
c_f specific heat of the fluid, J/kg.K
T_s catalyst or solids temperature, K
T fluid temperature, K
ΔH heat of reaction, J/kg

The energy balance for the catalyst can be written as:

$$\frac{\partial T_s}{\partial t}-\frac{\alpha_s A_s}{A_c(1-\varepsilon)\rho_s c_s}(T-T_s)=0 \tag{13.45}$$

in which:

ρ_c density catalyst, kg/m^3
c_s specific heat of the catalyst, J/kg.K

The boundary and initial conditions can be defined as:

$$\begin{aligned} t=0 \quad & c_A=c_{A0}(z),\ \ T=T_0(z),\ \ T_s=T_{s0}(z) \\ z=0 \quad & c_A=c_{Ain}(t),\ \ T=T_{in}(t) \end{aligned} \tag{13.46}$$

Solution of this set of equations is not trivial. The reader is referred to the Crider and Foss (1968) for an elaborate treatment. The authors make the following substitutions:

$$\begin{aligned} x_1 &= \delta c_A / c_A \\ x_2 &= \rho_f c_f \delta T / \Delta H c_A \\ \xi &= z / L \end{aligned} \tag{13.47}$$

which enables them to write the model in the following form:

$$
\begin{aligned}
x_1(\xi,s) &= G_1(\xi,s)x_1(0,s) + G_2(\xi,s)x_2(0,s) \\
x_2(\xi,s) &= G_3(\xi,s)x_1(0,s) + G_4(\xi,s)x_2(0,s)
\end{aligned}
\tag{13.48}
$$

In every transfer function there is an impact of the heat capacity ($M_s c_s$) of the catalyst and the fluid-catalyst heat transfer resistance, in all cases of the form $as/(bs+1)$. The transfer functions become very complicated expressions.

Adiabatic packed-bed reactors are the subject of frequent analysis in the literature (see, for example, Balakotaiah 1999); non-adiabatic tubular chemical reactors are analyzed, among others, by Berezowski (2003); while non-isothermal non-adiabatic packed bed chemical reactors are discussed in, among others, Juncu (1995). The last two types of reactor show frequently state multiplicity, i.e. for different values of a reactor parameter, a so-called bifurcation parameter, there is more than one solution to the set of partial differential equations.

Files referred to in this chapter:
F1303.m: file to generate Fig.13.3
F1304.m: file to generate Fig.13.4
F1305.m: file to generate Fig.13.5

References

Balakotaiah, V., Christoforatou, E.L. and West, D.H. (1999) Transverse concentration and temperature non-uniformities in adiabatic packed-bed reactors. *Chemical Engineering Science*, **54**, 1725–34.

Berezowski, M., (2003), Fractal solutions of recirculation tubular chemical reactors. *Chaos, Solitons and Fractals*, **16**, 1–12.

Crider, J.E. and Foss, A.S. (1968) An analytic solution for the dynamics of a packed adiabatic chemical reactor. *American Institute of Chemical Engineers Journal*, **14** (1), 77–84.

Juncu, G. and Floarea, O. (1995) Multiplicity pattern in the non-isothermal non-adiabatic packed bed chemical reactor. *Computers and Chemical Enginerring*, **19** (10), 1063–68.

14 Dynamic Analysis of Heat Exchangers

In this chapter different types of heat exchanger will be analyzed for their dynamic behavior. The first type is the shell and tube type, where steam condenses inside the tubes and the contents of a well-mixed tank or reactor have to be heated. The second type is also a shell and tube type heat exchanger, where the steam condenses outside the tubes (shell-side) and the liquid to be heated flows through the tubes. The last type is the countercurrent heat exchanger, in which the liquid to be heated flows countercurrent to the heating medium.

14.1 Introduction

It would be very difficult to describe *the* model for a heat exchanger, since there are so many different types. The most common type is the cooling or heating coil in a tank or reactor, meant to transfer heat in either direction. Inside the coil the temperature varies with time and axial direction; outside the coil the temperature is usually uniform, since the tank or reactor contents is usually stirred. Therefore this type can usually be modeled fairly easily and linearization of the model can give a good estimate of the dynamics of the heat transfer process.

The second well-known type is the pipe-mounted heat exchanger in which the medium to be heated flows through the pipes and steam condenses outside the pipes. Sometimes this type is called shell and tube heat exchanger. Even though this type resembles the previous type somewhat, its dynamic behavior is different.

A third type of heat exchanger is a countercurrent heat exchanger in which one fluid flows in one direction and another fluid flows in the opposite direction. In this case both fluid temperatures are truly time and location dependent and this is the most complicated case in terms of an analytical treatment.

There are still other types of heat exchanger; however, we will limit the discussion to the three types mentioned.

14.2 Heat Transfer from a Heating Coil

In this case it is assumed that heat is transferred from the coil to the tank contents, the discussion for heat withdrawal is similar. The situation is shown in Fig. 14.1.

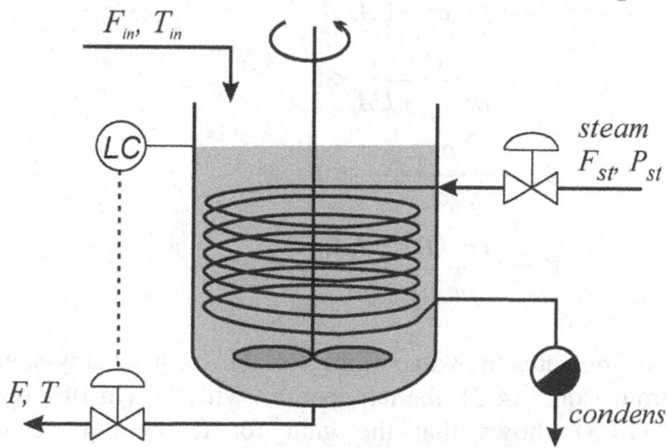

Fig. 14.1. Tank with heating coil.

Process Dynamics and Control: Modeling for Control and Prediction. Brian Roffel and Ben Betlem.

Let us make several assumptions so as not to complicate matters too much:

- there is a uniform temperature T in the tank
- the volume of fluid in the tank is constant, i.e. the inlet flow and outlet flow are the same
- steam condenses inside the coil, therefore a uniform condensation temperature T_s can be assumed
- the heat transfer coefficients inside and outside the coil are constant, i.e. the overall heat transfer coefficient is also constant
- the heat capacity of the coil ($M_{coil}c_{p,coil}$) is small compared to the heat capacity of the liquid and can therefore be ignored
- the dynamics on the steam side can be ignored, i.e. the steam temperature reacts instantaneously to changes in the steam supply

The energy balance can now be written as:

$$\rho V c_p \frac{dT}{dt} = F\rho c_p(T_{in} - T) + UA(T_s - T) \tag{14.1}$$

in which the term on the left-hand side of the equal sign is the accumulation of energy, the first term on the right-hand side is the transportation of energy by the flow of the fluid and the last term is the energy transfer from the coil. The following nomenclature is used:

ρ density of the fluid, kg/m^3
V volume of the fluid in the tank, m^3
c_p specific heat of the fluid, J/kg.K
F flow through the tank, m^3/s
T_{in} inlet temperature of the fluid, K
UA product of overall heat transfer coefficient and area, W/K

Generally, one will be interested in the response of the temperature T to changes in the flow F, the inlet temperature T_{in} and the flow F. These relationships can easily be derived from Eqn. (14.1) by linearizing it and taking the Laplace transform, the result is:

$$\delta T = \frac{K_1}{\tau_T\, s+1}\delta T_s + \frac{K_2}{\tau_T\, s+1}\delta T_{in} - \frac{K_3}{\tau_T\, s+1}\delta F \tag{14.2}$$

in which

$$\begin{aligned}
\tau_T &= \frac{\rho V c_p}{\rho F_0 c_p + UA} \\
K_1 &= \frac{UA}{\rho F_0 c_p + UA} < 1 \\
K_2 &= \frac{\rho F_0 c_p}{\rho F_0 c_p + UA} < 1 \\
K_3 &= \frac{\rho c_p (T_{in0} - T_0)}{\rho F_0 c_p + UA}
\end{aligned} \tag{14.3}$$

where the subscript '0' indicates the value of the variable in the steady-state situation.

As can be seen from Eqn. (14.2), the temperature will show a first-order response to all changes. Equation (14.3) shows that the gain for temperature changes, either inlet temperature or steam temperature, is always less than one.

The relationship between tank temperature and the input variable shown in Eqn. (14.2) will change if it is assumed that the heat capacity of the coil can no longer be ignored.

It is still assumed that the radial heat conduction through the wall is ideal and that the axial heat conduction can be ignored and that the wall temperature can be characterized by an average wall temperature T_w, which is probably a reasonable assumption given the fact that heat conduction is relatively fast compared to heat transfer (see chapter 1). Equation (14.1) can then be modified to:

$$\rho V c_p \frac{dT}{dt} = F\rho c_p (T_{in} - T) + \alpha_o A_o (T_w - T) \tag{14.4}$$

where $\alpha_o A_o$ is the product of heat transfer coefficient and area on the outside of the coil, and T_w is the wall temperature of the coil. Similar to Eqn. (14.4) we can write the energy balance for the wall:

$$M_w c_{pw} \frac{dT_w}{dt} = \alpha_i A_i (T_s - T_w) - \alpha_o A_o (T_w - T) \tag{14.5}$$

in which the subscript 'w' stands for wall.

Equations (14.4) and (14.5) form the new model of the heat exchanger with the heat capacity of the wall taken into account. Taking the Laplace transform and linearizing both equations results in:

$$\begin{aligned} \delta T &= \frac{K_1}{\tau_T\, s+1}\delta T_w + \frac{K_2}{\tau_T\, s+1}\delta T_{in} - \frac{K_3}{\tau_T\, s+1}\delta F \\ \delta T_w &= \frac{K_4}{\tau_w\, s+1}\delta T_s + \frac{K_5}{\tau_w\, s+1}\delta T \end{aligned} \tag{14.6}$$

in which the time constants and gains are defined by:

$$\begin{aligned} \tau_T &= \frac{\rho V c_p}{\rho F_0 c_p + \alpha_o A_o} & \tau_w &= \frac{M_w c_{pw}}{\alpha_i A_i + \alpha_o A_o} \\ K_1 &= \frac{\alpha_o A_o}{\rho F_0 c_p + \alpha_o A_o} & K_4 &= \frac{\alpha_i A_i}{\alpha_i A_i + \alpha_o A_o} \\ K_2 &= \frac{\rho F_0 c_p}{\rho F_0 c_p + \alpha_o A_o} & K_5 &= \frac{\alpha_o A_o}{\alpha_i A_i + \alpha_o A_o} \\ K_3 &= \frac{\rho c_p (T_{in0} - T_0)}{\rho F_0 c_p + \alpha_o A_o} \end{aligned} \tag{14.7}$$

Equation (14.6) can be rewritten by eliminating δT_w:

$$\begin{aligned} \delta T = &\frac{K_1 K_4}{(\tau_T s+1)(\tau_w s+1) - K_1 K_5}\delta T_s + \frac{K_2(\tau_w s+1)}{(\tau_T s+1)(\tau_w s+1) - K_1 K_5}\delta T_{in} \\ &- \frac{K_3(\tau_w s+1)}{(\tau_T s+1)(\tau_w s+1) - K_1 K_5}\delta F \end{aligned} \tag{14.8}$$

Equation (14.8) shows that the response of δT to changes in the steam temperature δT_s is a second-order response, the response to changes in δT_{in} and δF are pseudo-first-order responses.

14.3 Shell and Tube Heat Exchanger with Condensing Steam

Figure 14.2 shows a shell and tube heat exchanger in which steam condenses at the outside of the pipes and the fluid is flowing through the pipes.

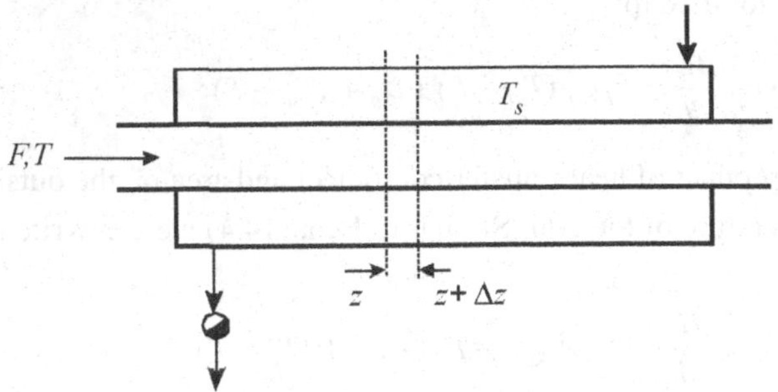

Fig. 14.2. Shell and tube heat exchanger.

The following assumptions are being made in this case:

- the steam temperature T_s outside the tubes is uniform
- the flow of the fluid through the tubes is ideal plug flow, i.e. there is an axial temperature gradient but no radial temperature gradient
- the physical properties, such as density and specific heat are constant
- the heat transfer coefficient α is constant
- the dynamics on the steam side can be ignored, i.e. the steam temperature reacts instantaneously to changes in the steam supply
- the radial heat conduction through the tube wall is ideal and the axial heat conduction can be ignored

It is assumed that the heat capacity of the wall ($M_w c_w$) cannot be ignored with respect to the heat capacity of the liquid ($M_f c_f$). If it is ignored later on, this will then be a special case of this more general case. The energy balance for a section of the wall at every point z can be written as:

$$M_w c_w \frac{\partial T_w}{\partial t} = \alpha_s A_s (T_s - T_w) - \alpha_f A_f (T_w - T) \tag{14.9}$$

in which:

A_f heat transfer area per unit length at the fluid side, m
A_s heat transfer area per unit length at the steam side, m
c_w specific heat of the wall, J/kg.K
M_w mass of the tubes per unit length, kg/m
T fluid temperature, K
T_w wall temperature, K
α_s heat transfer coefficient at the steam side, W/m^2.K
α_f heat transfer coefficient at the fluid side, W/m^2.K

The left-hand side term in Eqn. (14.9) represents the accumulation of energy, the first term at the right-hand side the heat transfer from the condensing steam to the wall and the last term the heat transfer from the wall to the fluid to be heated.

The energy balance for the fluid can be written as:

$$M_f c_f \frac{\partial T}{\partial t} + F c_f \frac{\partial T}{\partial z} = \alpha_f A_f (T_w - T) \tag{14.10}$$

in which:

M_f mass of liquid per unit length, kg/m
c_f specific heat of the fluid, J/kg.K
F mass flow of the fluid, kg/s

The first term represents the accumulation of energy, the second term the transportation of energy due to flow and the third term the flow of energy from wall to fluid.

Equations (14.9) and (14.10) can be written in a simplified form when the following time constants τ and the fluid velocity v is introduced:

$$\tau_f = \frac{M_f c_f}{\alpha_f A_f} \qquad \tau_{wf} = \frac{M_w c_w}{\alpha_f A_f}$$
$$\tau_{ws} = \frac{M_w c_w}{\alpha_s A_s} \qquad v = \frac{F}{M_f} \tag{14.11}$$

The energy balances become then:

$$\tau_{ws}\frac{\partial T_w}{\partial t} = T_s - T_w - \frac{\tau_{ws}}{\tau_{wf}}(T_w - T) \tag{14.12}$$

and

$$\tau_f \frac{\partial T}{\partial t} + v\tau_f \frac{\partial T}{\partial z} = T_w - T \tag{14.13}$$

The model description can be completed by a proper definition of boundary (for T) and initial conditions (for T and T_w).

14.3.1 Static Model Analysis

The static model can be found by setting the derivatives with respect to time equal to zero. Equation (14.9) can be written as:

$$T_{w0}(z) = \frac{\alpha_s A_s T_{s0} + \alpha_f A_f T_0(z)}{\alpha_s A_s + \alpha_f A_f} \tag{14.14}$$

Notice that a subscript '0' has been added to indicate steady-state values; T_{s0} is assumed to be uniform along the outside of the pipe, hence $T_{s0}(z) = T_{s0}$.

In steady state, Eqn. (14.10) can, after combination with Eqn. (14.14), be written as:

$$v_0 \tau_{f0} \frac{dT_0(z)}{dz} + T_0(z) = T_{s0} \tag{14.15}$$

with

$$\tau_{f0} = M_f c_f \left[\frac{1}{\alpha_f A_f} + \frac{1}{\alpha_s A_s} \right] \tag{14.16}$$

τ_{f0} is the heating time constant, which is the product of the heat capacity of the fluid times the heat resistance from steam to fluid.

The solution of this equation can be found by solving the homogeneous equation and adding the particular solution while taking into account the boundary condition $T_0 = T_{in}$ for $z = 0$:

$$T_0(z) = T_{s0} - (T_{s0} - T_{in0})\, e^{-z/v_0 \tau_{f0}} \tag{14.17}$$

The outlet temperature for $z = L$ is then given by:

$$T_0(L) = T_{out} = T_{s0} - (T_{s0} - T_{in0})\, e^{-\tau_{R0}/\tau_{f0}} \tag{14.18}$$

τ_{R0} is the transportation time for the liquid from the pipe entrance to the outlet in steady state. Figure 14.3 shows the static temperature profile along the pipe for the following conditions: $T_{s0} = 380$ K, $T_{in0} = 250$ K, $v_0 = 1$ m/s, $\tau_{f0} = 10$ s, $L = 12$ m.

The wall temperature profile in a steady-state situation can easily be calculated from Eqns. (14.14) and (14.17);

$$T_{w0}(z) = T_{s0} - \frac{\alpha_f A_f}{\alpha_s A_s + \alpha_f A_f}(T_{s0} - T_{in0})\, e^{-z/v_0\tau_{f0}} \tag{14.19}$$

14.3.2 Dynamic Model Analysis

In this section the dynamic behavior of the model for fluid temperature changes as a function of changes in the flow or steam temperature will be analyzed. Let us first assume that the heat capacity of the wall can be ignored, once the results are found, the effect of the heat capacity on the result will be analyzed.

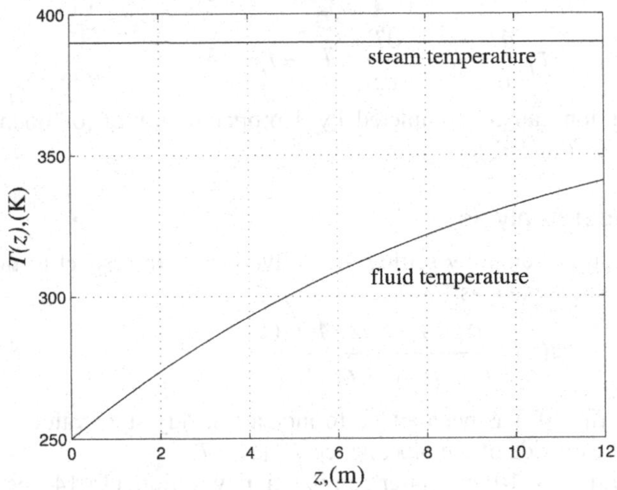

Fig. 14.3. Static fluid temperature profile along the pipe.

If the dynamics of the wall can be ignored, T_w approaches T_s and Eqn. (14.13) can be written as:

$$\tau_f \frac{\partial T}{\partial t} + v\tau_f \frac{\partial T}{\partial z} = T_s - T \tag{14.20}$$

The process variables can be represented by their steady-state values and a small variation around the steady state:

$$T_s = T_{s0} + \delta T_s, \quad T = T_0 + \delta T, \quad v = v_0 + \delta v \tag{14.21}$$

For the moment changes in the heat transfer coefficients will be neglected, it will be pointed out later what the impact is on the final result.

Linearization of Eqn. (14.20) in the operating point gives:

$$\tau_f \frac{\partial(\delta T)}{\partial t} + (v_0 + \delta v)\tau_f \frac{\partial(T_0 + \delta T)}{\partial z} = T_{s0} - T_0 + \delta T_s - \delta T \tag{14.22}$$

Equation (14.22) can be written in individual terms and combined with Eqns.(14.15) and (14.17), which results in:

$$\frac{\partial(\delta T)}{\partial t} + v_0 \frac{\partial(\delta T)}{\partial z} + \frac{1}{\tau_f}\delta T = \frac{1}{\tau_f}\delta T_s - \frac{1}{\tau_f}\left(T_{s0} - T_{in0}\right)e^{-z/v_0\tau_f}\frac{\delta v}{v_0} \tag{14.23}$$

where $\tau_f = \tau_{f0}$, since it was assumed that the dynamics of the wall could be ignored, i.e. the wall temperature was approximated by the steam temperature, which is the case if $\alpha_s A_s >> \alpha_f A_f$.

If the heat transfer coefficients are not constant, the contribution of the second right-hand side term of Eqn. (14.23) will become smaller, hence the contribution of changes in the flow (or velocity) on changes in the temperature will diminish.

Introduction of the Laplace operator s into Eqn. (14.23) results in:

$$v_0 \frac{d(\delta T)}{dz} + \left(s + \frac{1}{\tau_f}\right)\delta T = \frac{1}{\tau_f}\delta T_s - \frac{1}{\tau_f}\left(T_{s0} - T_{in0}\right)e^{-z/v_0\tau_f}\frac{\delta v}{v_0} \tag{14.24}$$

This equation is an ordinary first-order equation in δT and can easily be solved. A general solution should be assumed and a particular solution, which have the following form:

$$\begin{aligned} \delta T_{general} &= A_1 e^{-(s+\tau_f^{-1})z/v_0} \\ \delta T_{particular} &= A_2 \delta T_s + A_3 e^{-z/v_0\tau_f}\left(\delta v / v_0\right) \end{aligned} \tag{14.25}$$

Equation (14.25) should be substituted into Eqn. (14.24), combination with the boundary condition $\delta T = \delta T\,(0,s)$ at $z = 0$ gives then the final expression for δT:

$$\begin{aligned} \delta T(z,s) = {} & e^{-z/v_0\tau_f} e^{-sz/v_0}\delta T(0,s) + \frac{1}{1+\tau_f s}\left[1 - e^{-z/v_0\tau_f}e^{-sz/v_0}\right]\delta T_s \\ & - \frac{1}{\tau_f s}\left(T_{s0} - T_{in0}\right)e^{-z/v_0\tau_f}\left(1 - e^{-sz/v_0}\right)\left(\delta v / v_0\right) \end{aligned} \tag{14.26}$$

Note that T_{in0} is the inlet temperature of the fluid at steady state, $\delta T(0,s)$ is the fluid inlet temperature change which can be time dependent.

This equation is still somewhat difficult to interpret, therefore the term $e^{-z/v_0\tau_f}$ will be eliminated by using Eqn. (14.17) and evaluating the result for $z = L$. The response of the outlet temperature can then be described as:

$$\delta T_{out} = \frac{\delta T_{out}}{\delta T_{in}}\delta T_{in} + \frac{\delta T_{out}}{\delta T_s}\delta T_s + \frac{\delta T_{out}}{\left(\delta v / v_0\right)}\left(\delta v / v_0\right) \tag{14.27}$$

in which

$$\frac{\delta T_{out}}{\delta T_{in}} = \frac{T_{s0} - T_{out0}}{T_{s0} - T_{in0}} e^{-s\tau_R}$$
$$\frac{\delta T_{out}}{\delta T_s} = \frac{1}{1+\tau_f s}\left(1 - \frac{T_{s0} - T_{out0}}{T_{s0} - T_{in0}} e^{-s\tau_R}\right) \tag{14.28}$$
$$\frac{\delta T_{out}}{(\delta v / v_0)} = -\frac{1}{\tau_f}(T_{s0} - T_{out0})\frac{1 - e^{-s\tau_R}}{s}$$

The residence time τ_R is the transportation time of the fluid from the entrance of the pipe to the pipe exit for the current steady-state situation.

The ratio τ_R / τ_f can be calculated from Eqn. (14.18):

$$\frac{\tau_R}{\tau_f} = ln\frac{T_{s0} - T_{in0}}{T_{s0} - T_{out0}} \tag{14.29}$$

Using the data from the previous section and noting that $\tau_R = 12$ s and $T_{out0} = 340.84$ K, Eqn. (14.28) can be written as:

$$\frac{\delta T_{out}}{\delta T_{in}} = 0.30e^{-12s}$$
$$\frac{\delta T_{out}}{\delta T_s} = \frac{1}{1+10s}\left(1 - 0.30e^{-12s}\right) \tag{14.30}$$
$$\frac{\delta T_{out}}{(\delta v / v_0)} = -3.92\frac{1 - e^{-12s}}{s}$$

Figures 14.4–14.6 show the step responses of the transfer functions. As can be seen from Fig. 14.4, the model between fluid outlet temperature changes and fluid inlet temperature changes is a pure time delay of 12 seconds. This can be expected since the fluid has to move through the pipe, before the inlet change reaches the outlet.

The model between the fluid outlet temperature change and change in steam temperature is initially a first-order response with a time constant of 10 seconds. Upon an increase in steam temperature, the fluid temperature starts to increase over the entire length of the pipe. The fluid towards the beginning of the pipe is longer exposed to the increased steam temperature than the fluid towards the end of the pipe, hence the temperature continues to increase. After the residence time, however, the new fluid entering the pipe was only exposed to the new steam temperature; hence the fluid outlet temperature stays constant.

The transfer function between the fluid outlet temperature and changes in the fluid velocity can be considered as an integrator with the difference between an immediate and delayed response. Integration takes place over a period of 12 seconds, hence the final change in fluid temperature on a 20% change in fluid velocity will be $-3.92 \times 12 \times 0.2 = -9.4$ K.

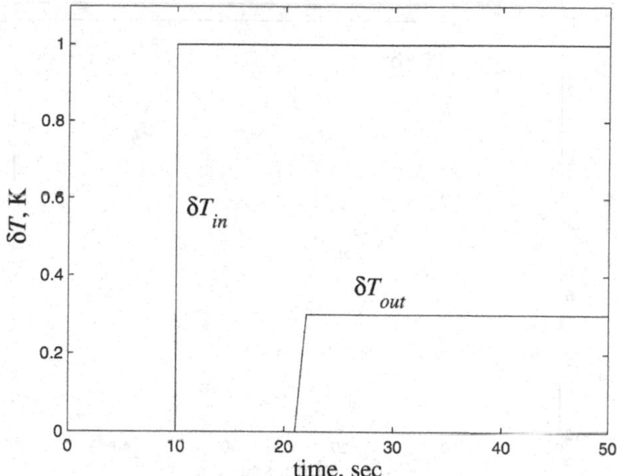

Fig. 14.4. Heat exchanger fluid outlet temperature to step change in fluid inlet temperature.

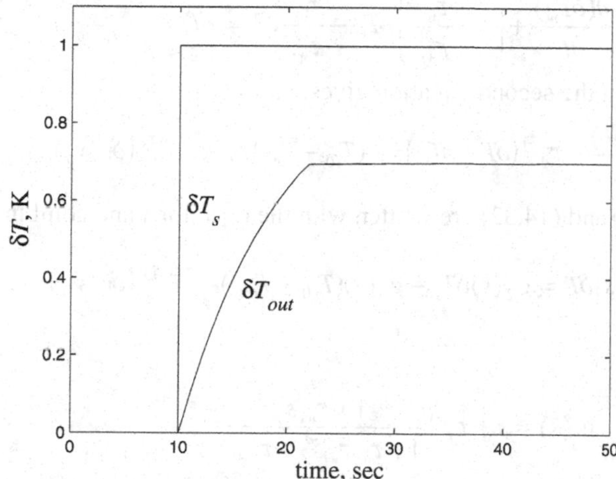

Fig. 14.5. Heat exchanger fluid outlet temperature to step change in steam temperature.

In the foregoing analysis the heat capacity of the pipe wall was ignored. Taking it into account does not result in any particular problems. Equations (14.12) and (14.13) will now have to be linearized, linearization of the first equation gives:

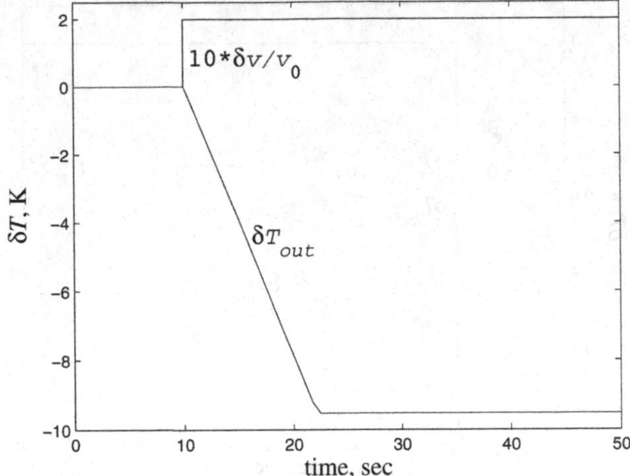

Fig. 14.6. Heat exchanger fluid outlet temperature to step change in fluid velocity.

$$\tau_{ws}\frac{\partial(\delta T_w)}{\partial t}+\left(1+\frac{\tau_{ws}}{\tau_{wf}}\right)\delta T_w-\frac{\tau_{ws}}{\tau_{wf}}\delta T=\delta T_s \tag{14.31}$$

while linearization of the second equation gives:

$$v_0\frac{\partial(\delta T)}{\partial z}+\frac{\partial(\delta T)}{\partial t}+\tau_f^{-1}(\delta T-\delta T_w)=-(T_{s0}-T_{in0})\tau_f^{-1}e^{-z/v_0\tau_f}(\delta v/v_0) \tag{14.32}$$

When Eqns. (14.31) and (14.32) are written with the operator s and combined, the result is:

$$v_0\frac{\partial(\delta T)}{\partial z}+g_1(s)\delta T=g_2(s)\delta T_s-g_3(s)(T_{s0}-T_{in0})e^{-z/v_0\tau_f}(\delta v/v_0) \tag{14.33}$$

in which

$$g_1(s)=s+\tau_f^{-1}\frac{1+\tau_{ws}s}{1+\tau_{ws}\tau_{wf}^{-1}+\tau_{ws}s}$$
$$g_2(s)=\frac{\tau_f^{-1}}{1+\tau_{ws}\tau_{wf}^{-1}+\tau_{ws}s} \tag{14.34}$$
$$g_3(s)=\tau_f^{-1}$$

Solution of Eqn. (14.33) gives:

$$\delta T=e^{-g_1z/v_0}\delta T_{in}+\frac{g_2}{g_1}\left(1-e^{-g_1z/v_0}\right)\delta T_s$$
$$-\frac{g_3}{g_1-\tau_f^{-1}}\left(T_{s0}-T_{in0}\right)\left(e^{-z/v_0\tau_f}-e^{-g_1z/v_0}\right)(\delta v/v_0) \tag{14.35}$$

If $\alpha_sA_s >> \alpha_fA_f$ then $\tau_{ws} << \tau_{wf}$ in which case the expression for $g_1(s)$ in Eqn. (14.34) can be approximated by:

$$g_1(s)=s+\tau_f^{-1} \tag{14.36}$$

Equation (14.35) can then be written as:

$$\frac{\delta T_{out}}{\delta T_{in}} \approx \frac{T_{s0} - T_{out0}}{T_{s0} - T_{in0}} e^{-s\tau_R}$$

$$\frac{\delta T_{out}}{\delta T_s} \approx \frac{1}{1 + s\left(\tau_f + \tau_{ws} + \tau_{ws}\tau_{wf}^{-1}\tau_f\right) + s^2\tau_{ws}\tau_f}\left(1 - \frac{T_{s0} - T_{out0}}{T_{s0} - T_{in0}} e^{-s\tau_R}\right) \tag{14.37}$$

$$\frac{\delta T_{out}}{\delta v / v_0} \approx -\frac{1}{\tau_f} \frac{1 + \tau_{ws}\tau_{wf}^{-1} + s\tau_{ws}}{1 + \tau_{ws}\tau_{wf}^{-1}\left(1 + \tau_{ws}\tau_f^{-1}\right) + s\tau_{ws}} \left(T_{s0} - T_{out0}\right) \frac{1 - e^{-s\tau_R}}{s}$$

The first expression in Eqn. (14.37) corresponds to Eqn. (14.28).

The major difference between the second term in Eqn. (14.37) and the second term in Eqn. (14.28) is that the term between brackets is multiplied by a first-order transfer function in Eqn. (14.28) and a second order transfer function in Eqn. (14.37).

The third term in Eqn. (14.37) is similar to the third term in Eqn. (14.28), the original transfer function is multiplied by an additional lead-lag transfer function.

Apparently, the effect of taking the heat capacity of the tube wall into account is minor, only an additional first-order effect is present in the response of the fluid outlet temperature to changes in steam temperature.

14.4 Dynamics of a Counter-current Heat Exchanger

Figure 14.7 shows a counter-current heat exchanger in which the one fluid decreases in temperature and the other fluid increases in temperature.

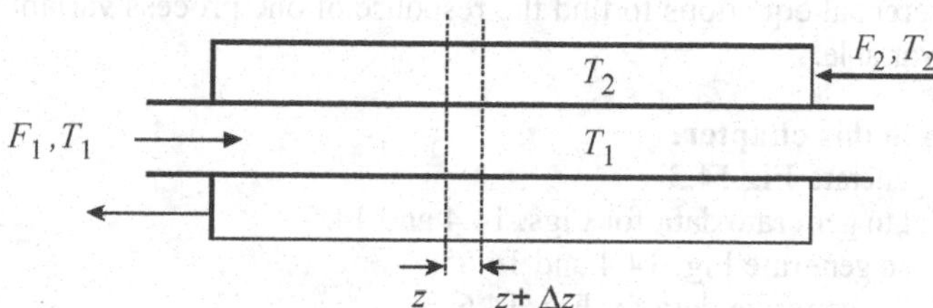

Fig. 14.7. Counter-current heat exchanger.

For each fluid we could write an equation such as Eqn. (14.10). If the heat capacity of the wall would be ignored, we would get a set of two simultaneous equations that could be solved. The outlet temperature responses of the two fluids would consist of a summation of exponential functions, which are an indication of higher order models. An approximate solution of the set of two differential equations is discussed, among others, by Friedly (1972), Harriott (1964) and Mozley (1956).

The approximation is obtained by writing the set of equations as:

$$\begin{aligned} M_{f1}c_{f1}\frac{\partial T_1}{\partial t} + F_1 c_{f1}\frac{\partial T_1}{\partial z} &= UA\left(T_2 - T_1\right) \\ M_{f2}c_{f2}\frac{\partial T_2}{\partial t} - F_2 c_{f2}\frac{\partial T_2}{\partial z} &= UA\left(T_1 - T_2\right) \end{aligned} \tag{14.38}$$

in which subscript '1' refers to the fluid inside the tubes and subscript '2' refers to the fluid outside the tubes.

Note that one equation has a term $+\partial T / \partial z$ and the other has a term $-\partial T / \partial z$ since the one fluid is warming up and the other is cooling down.

An additional approximation is made by using the arithmetic mean of the temperature:

$$T_i = \frac{T_{i,in} + T_{i,out}}{2} \quad i = 1,2 \tag{14.39}$$

From a substitution of Eqn. (14.39) into Eqn. (14.38) the transfer function between one fluid outlet temperature and the other fluid inlet temperature can then easily be derived:

$$\begin{aligned} \frac{\delta T_{2,out}}{\delta T_{1,in}} &= \frac{4\tau_1 / \tau_{R1}}{(\tau_1 s + 2\tau_1 / \tau_{R1} + 1)(\tau_2 s + 2\tau_2 / \tau_{R2} + 1) - 1} \\ \tau_1 &= M_{f1} c_{f1} / UA, \tau_2 = M_{f2} c_{f2} / UA \\ \tau_{R1} &= M_{f1} / F_1, \tau_{R2} = M_{f2} / F_2 \end{aligned} \tag{14.40}$$

Note that τ_R is the residence time of the liquid in the heat exchanger. As can be seen, the approximation is second order. When the one fluid increases in temperature, the outlet temperature of the other fluid will also start to increase. The approximation can be made more accurate by incorporating the heat capacity of the wall in equal parts in the heat capacity of the fluids.

As Harriott (1964) points out, the response depends more strongly on the residence times τ_{R1} and τ_{R2} than on the time constants τ_1 and τ_2. When all the time constants are 10 s, the effective time constants in Eqn. (14.40) become 2.5 and 5 s. Changing τ_1 and τ_2 to 20 s, gives effective time constants of 3.3 and 5 s.

Analytical solutions of partial differential equations are generally not easy to find. However, excellent and accurate numerical approximations are available today to solve this set of partial differential equations to find the response of one process variable to a change in another process variable.

Files referred to in this chapter:

F1403.m: file to generate Fig.14.3
F1404.mdl: model to generate data for Figs. 14.4 and 14.5
F1404plot.m: file to generate Fig. 14.4 and 14.05
F1406.mdl: model to generate data for Fig. 14.6
F1406plot.m: file to generate Fig. 14.6

References

Friedly, J.C. (1972) *Dynamic Behavior of Processes*, Prentice Hall.

Harriott, P. (1964) *Process control*, McGraw-Hill.

Mozley, J.M. (1956) Predicting dynamics of concentric pipe heat exchangers. *Industrial and Engineering Chemistry*, June, 1035–41.

15 Dynamics of Evaporators and Separators

Evaporators and single-stage separators are fairly similar. Both operate at the boiling point of the liquid. The main difference is that in evaporators usually pure liquids are evaporated whereas in separators usually one (light) component is separated from other components. This leads to difference in dynamic behavior. In this chapter this behavior will be analyzed for the general case where the liquid level can vary. If the liquid level is constant it is merely a simplification from the first case.

15.1 Introduction

Evaporators were discussed in chapter 2. There the environmental model was developed for the case where the heat transfer area is proportional to the height of the liquid in the vessel (Fig. 15.1).

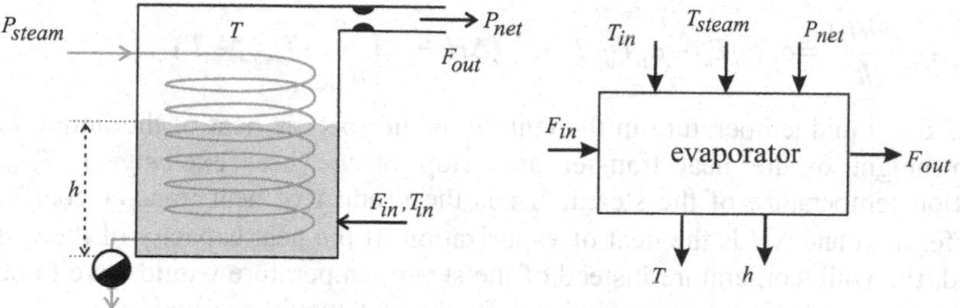

Fig. 15.1. Evaporator with variable heat transfer surface.

The goal of the model of the evaporator is to determine whether load variations are self-controlling as a function of the design variables. This means that the relationship between F_{in} and F_{out} has to be determined.

The environmental model is shown in Fig. 15.1. For the behavioral model, the level of detail has been restricted, because only the low frequency range of the disturbances is of importance. The following simplifications and assumptions are made:

- the liquid in the tank is ideally mixed
- vapor–liquid equilibrium is instantaneous
- the vapor does not exchange heat with the coil
- F_{out} depends on the square root of the pressure drop
- in the coil the same temperature exists everywhere
- the boil-up effect is negligible
- all heat capacities of the equipment can be ignored
- the mass of vapor can be ignored compared with the mass of the liquid
- all physical properties can be assumed constant in the operating range
- the cross-sectional area of the vessel is constant

The effect of some of these assumptions may be difficult to determine. The heat capacity of the coil would normally result in an additional small time constant. The capacity of the wall can be added to the capacity of the liquid. The weak point in the model is the fact that

Process Dynamics and Control: Modeling for Control and Prediction. Brian Roffel and Ben Betlem.

the boil-up effect has been ignored. The volume of the vapor bubbles can vary as a function of the pressure and temperature. However, even a simplified analysis gives a rather good impression of some of the dynamic effects that can occur.

15.2 Model Description

Since the level can vary, a mass balance is relevant and can be written as:

$$\frac{dM}{dt} = F_{in} - F_{out} \tag{15.1}$$

Because of the last two assumptions, the mass balance can be rewritten as:

$$\rho A_c \frac{dh}{dt} = F_{in} - F_{out} \tag{15.2}$$

where ρ is the liquid density, A_c the cross sectional area of the tank, h the liquid level and F the mass flow. The subscripts '*in*' and '*out*' refer to the inlet and outlet conditions respectively.

The energy balance can be written as:

$$\rho c_p A_c \frac{d(hT)}{dt} = c_p F_{in} T_{in} - c_p F_{out} T - F_{out} \Delta H + UA \frac{h}{h_{max}} (T_{steam} - T) \tag{15.3}$$

where T is the liquid temperature in the tank, c_p is the specific heat of the liquid, h_{max} is the maximum height of the heat transfer area (top of the heat exchanger), T_{steam} is the condensation temperature of the steam, UA is the product of heat transfer coefficient and heat transfer area and ΔH is the heat of vaporization. If the heat capacity of the wall cannot be ignored, the wall temperature instead of the steam temperature would have to be used in Eqn. (15.3) and an additional energy balance for the wall would be required.

Additional equations are required to complete the model description. The first one is the relationship between the outlet flow and the pressure, taking assumption four into account:

$$F_{out} = c_V \sqrt{P - P_{net}} \tag{15.4}$$

Since there is only one component present in the tank (boiling pure liquid), there is also a relationship between the pressure in the tank and the temperature of the vapor (and liquid, which is the same). This relationship can be well described by the law of Clausius-Clapeyron

$$P = P_{ref} exp\left(\frac{\Delta H}{R} \left(\frac{1}{T_{ref}} - \frac{1}{T} \right) \right) \tag{15.5}$$

As can be seen, Eqn. (15.3) contains a double derivative, it is good practice to change it to a single derivative by combining Eqn. (15.3) with Eqn. (15.2), which results in:

$$\rho c_p A_c h \frac{dT}{dt} = c_p F_{in} (T_{in} - T) - F_{out} \Delta H + UA \frac{h}{h_{max}} (T_{steam} - T) \tag{15.6}$$

The model now consists of Eqns. (15.2), and (15.4)–(15.6). The behavioral model is shown in Fig. 15.2. As can be seen, the mass balance affects the energy balance because of the varying level; the energy balance affects the pressure via the Clausius-Clapeyron equation, and a changing pressure affects the outlet flow which in turn affects the mass balance.

15.3 Linearization and Laplace Transformation

Linearization of Eqn. (15.2) results in:

$$\rho A_c s \delta h = \delta F_{in} - \delta F_{out} \tag{15.7}$$

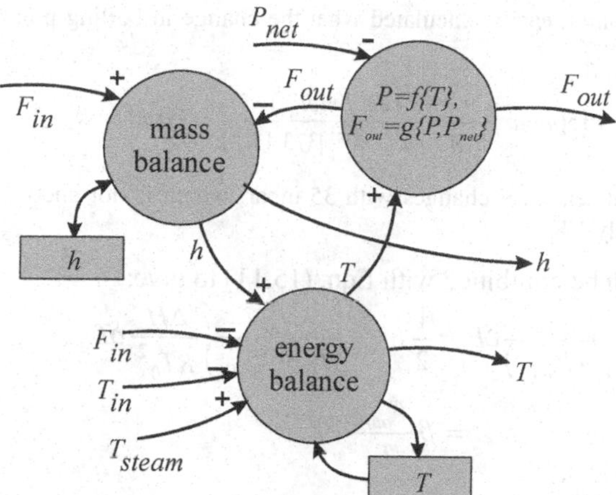

Fig. 15.2. Behavioral model for evaporator with variable heat transfer surface.

in which δ is a variation around the steady-state value.

Since we are only interested in changes of F_{out} as a result of changes to F_{in}, changes in the inputs T_{steam} and T_{in} will not be considered. Linearization of the energy balance, Eqn. (15.6) results then in:

$$\left(\rho c_p A_c h_0 s + c_p F_{in0} + \frac{UAh_0}{h_{max}}\right)\delta T = -c_p\left(T_0 - T_{in0}\right)\delta F_{in} - \Delta H \delta F_{out}$$
$$+\frac{UA}{h_{max}}\left(T_{steam0} - T_0\right)\delta h \tag{15.8}$$

where the index '0' refers to the steady state.

Linearization of Eqn. (15.4) gives:

$$\delta F_{out} = \left(\frac{\partial\left(c_v\sqrt{P-P_{net}}\right)}{\partial\left(P-P_{net}\right)}\right)\left(\delta P - \delta P_{net}\right)$$
$$\frac{\frac{1}{2}c_v}{\sqrt{P_0 - P_{net0}}} = \frac{1}{2}\frac{F_{out0}}{\left(P_0 - P_{net0}\right)}\left(\delta P - \delta P_{net}\right) \tag{15.9}$$

Linearization of Eqn. (15.5) gives:

$$\delta P = P_{ref}\frac{\partial}{\partial T}\left[exp\left(\frac{\Delta H}{R}\left(\frac{1}{T_{ref}} - \frac{1}{T}\right)\right)\right]\delta T =$$
$$P_{ref}exp\left(\frac{\Delta H}{R}\left(\frac{1}{T_{ref}} - \frac{1}{T_0}\right)\right)\frac{\partial}{\partial T}\left[\frac{\Delta H}{R}\left(\frac{1}{T_{ref}} - \frac{1}{T}\right)\right]_0 \delta T \tag{15.10}$$

Taking $T_{ref} = T_0$ and $P_{ref} = P_0$, this equation can be written as:

$$\delta P = P_0 \frac{\Delta H}{R T_0^{\,2}} \delta T \tag{15.11}$$

Example

Using this equation it can be easily calculated what the change in boiling point is of water at 100 °C and 1015 mbar:

$$\delta P = P_0 \frac{\Delta H}{R T_0^{\,2}} \delta T = 1015[mbar] \frac{40.103[J.mol^{-1}]}{8.3[J.mol^{-1}K^{-1}]373^2[K^2]} = 35\,[mbar.K^{-1}] \tag{15.12}$$

When the air pressure at sea level changes with 35 mbar, which is not uncommon, the boiling point changes by approximately 1 K.

Equation (15.9) can be combined with Eqn. (15.11) to give:

$$\begin{aligned} \delta F_{out} &= \frac{1}{2} \frac{F_{out0}}{(P_0 - P_{net0})} \delta P = \frac{1}{2} F_{out0} \frac{P_0}{(P_0 - P_{net0})} \frac{\Delta H}{R T_0^{\,2}} \delta T \\ &= \beta \frac{F_{out0}}{T_0} \delta T \end{aligned} \tag{15.13}$$

in which

$$\beta = \frac{1}{2} \frac{P_0}{(P_0 - P_{net0})} \frac{\Delta H}{R T_0} \tag{15.14}$$

which is a dimensionless constant.

15.4 Derivation of the Normalized Transfer Function

In the operating point we may write that $F_{in0} = F_{out0} = F_0$. The mass balance can then be normalized to:

$$\tau_1 s \frac{\delta h}{h_0} = \frac{\delta F_{in}}{F_0} - \frac{\delta F_{out}}{F_0} \quad with\ \tau_1 = \frac{\rho A_c h_0}{F_0} \tag{15.15}$$

The time constant τ_1 is the residence time in the tank in the steady-state situation.

Equation (15.8) can be rewritten with the help of Eqn. (15.13) to:

$$\begin{aligned} &\left(\rho c_p A_c h_0 s + c_p F_0 + \beta \Delta H \frac{F_0}{T_0} + \frac{U A h_0}{h_{max}} \right) \delta T = \\ &\qquad - c_p (T_0 - T_{in0}) \delta F_{in} + \frac{UA}{h_{max}} (T_{steam0} - T_0) \delta h \end{aligned} \tag{15.16}$$

Define the time constant for the total heating time τ_2:

$$\tau_2 = \frac{\rho c_p A_c h_0}{c_p F_0 + \beta \Delta H \frac{F_0}{T_0} + \frac{U A h_0}{h_{max}}} \tag{15.17}$$

then Eqn. (15.16) can be rewritten as:

$$(\tau_2 s + 1) \delta T = -\frac{\tau_2}{\tau_1} (T_0 - T_{in0}) \frac{\delta F_{in}}{F_0} + \tau_2 \frac{UA/h_{max}}{\rho c_p A_c h_0} (T_{steam0} - T_0) \delta h \tag{15.18}$$

Define the heating time of the vessel compared with the heating time of the coil as:

$$\tau_3 = \frac{\rho c_p A_c h_0 T_0}{\dfrac{UAh_0}{h_{max}}\left(T_{steam0} - T_0\right)} \tag{15.19}$$

then Eqn. (15.18) can be further simplified to:

$$(\tau_2 s + 1)\delta T = -\frac{\tau_2}{\tau_1}(T_0 - T_{in0})\frac{\delta F_{in}}{F_0} + \frac{\tau_2}{\tau_3} T_0 \frac{\delta h}{h_0} \tag{15.20}$$

Substitution of the mass balance and the equation for the outgoing flow finally results in:

$$\left(\tau_1 s(\tau_2 s + 1) + \frac{\tau_2}{\tau_3}\beta\right)\frac{\delta T}{T_0} = \left(-\tau_2 \frac{(T_0 - T_{in0})}{T_0} s + \frac{\tau_2}{\tau_3}\right)\frac{\delta F_{in}}{F_0} \tag{15.21}$$

The transfer function becomes now:

$$\frac{\delta T}{\delta F_{in}} = \frac{T_0}{F_0}\frac{-\tau_2 s(T_0 - T_{in0})/T_0 + \dfrac{\tau_2}{\tau_3}}{\tau_1 s(\tau_2 s + 1) + \dfrac{\tau_2}{\tau_3}\beta} = \left(\frac{T_0}{F_0}\right)\frac{-\tau_3 s(T_0 - T_{in0})/T_0 + 1}{\tau_1\tau_3 s^2 + \dfrac{\tau_1\tau_3}{\tau_2}s + \beta} \tag{15.22}$$

The desired transfer function can be obtained by substituting the equation for the outgoing flow again:

$$\frac{\delta F_{out}}{\delta F_{in}} = \frac{-\tau_3 s(T_0 - T_{in0})/T_0 + 1}{\dfrac{\tau_1\tau_3}{\beta}s^2 + \dfrac{\tau_1\tau_3}{\tau_2\beta}s + 1} \tag{15.23}$$

15.5 Response Analysis

The initial response and stationary behavior are interesting in case of a step change in the inlet flow. For a unit change in the inlet flow δF_{in} = 1/s. The initial behavior then becomes:

$$\lim_{t\to 0}\left(\frac{\mathrm{d}F_{out}}{\mathrm{d}t}\right) = \lim_{s\to\infty}\left[s^2\left[\frac{-(T_0 - T_{in0})/T_0\,\tau_3 s + 1}{\dfrac{\tau_1\tau_3}{\beta}s^2 + \dfrac{\tau_1\tau_3}{\tau_2\beta}s + 1}\right]\frac{1}{s}\right] = \frac{-\beta(T_0 - T_{in0})/T_0}{\tau_1} \tag{15.24}$$

When $T_0 > T_{in0}$, the response is initially negative, since less heat is available for evaporation. Owing to the decreasing pressure, the boiling point temperature will decrease. On an increase in inlet flow, the outlet vapor flow will therefore initially decrease.

The stationary behavior for $t\to\infty$ becomes:

$$\lim_{t\to\infty}(\delta F_{out}) = \lim_{s\to 0}\left[s\left[\frac{-(T_0 - T_{in0})/T_0\,\tau_3 s + 1}{\dfrac{\tau_1\tau_3}{\beta}s^2 + \dfrac{\tau_1\tau_3}{\tau_2\beta}s + 1}\right]\cdot\frac{1}{s}\right] = 1 \tag{15.25}$$

For a step change in inlet flow, the outlet flow will eventually become equal to the inlet flow.

15.6 General Behavior

In the stationary situation, the outlet flow becomes equal to the inlet flow. The question we could ask is: is the system stable and how does the outlet flow go to its steady-state value? To investigate this, the denominator of Eqn. (15.23) needs to be analyzed. The normalized denominator can be written as the basic equation for a second-order system:

$$Denominator = \frac{\tau_1\tau_3}{\beta}s^2 + \frac{\tau_1\tau_3}{\tau_2\beta}s + 1 = \tau^2 s^2 + 2\varsigma\tau s + 1 \tag{15.26}$$

A time constant τ and damping coefficient ζ can be defined:

$$\tau = \sqrt{\frac{\tau_1\tau_3}{\beta}}$$
$$\varsigma = \tfrac{1}{2}\frac{\tau_1\tau_3}{\tau_2\beta}\Big/\sqrt{\frac{\tau_1\tau_3}{\beta}} = \frac{1}{2\tau_2}\sqrt{\frac{\tau_1\tau_3}{\beta}} \tag{15.27}$$

To understand the mechanism of the response, the damping coefficient has to be expressed in the design variables:

$$\varsigma = \tfrac{1}{2}\frac{c_p F_0 + \Delta H\beta\frac{F_0}{T_0} + \frac{UAh_0}{h_{max}}}{\rho c_p A_c h_0}\sqrt{\frac{1}{\beta}\frac{\rho A_c h_0}{F_0}\frac{\rho c_p A_c h_0 T_0}{\frac{UAh_0}{h_{max}}(T_{steam0} - T_0)}}$$
$$= \tfrac{1}{2}\frac{c_p F_0 + \Delta H\beta\frac{F_0}{T_0} + \frac{UAh_0}{h_{max}}}{\sqrt{\beta c_p \frac{F_0}{T_0}\frac{UAh_0}{h_{max}}(T_{steam0} - T_0)}} \tag{15.28}$$

The denominator is the result of two interacting balances: the energy balance and the mass balance. This interaction is a result of the changing heat transfer surface. If the heat transfer surface is constant (level control), the interaction is eliminated. If the damping coefficient $0 < \zeta < 1$, the interaction results in oscillation. If the inlet flow increases, the level will increase so much that the outlet flow will even exceed the inlet flow. The level will oscillate for some time until a new stationary value is reached. If the damping coefficient $\zeta > 1$, the level will gradually approach a new steady-state value. The only design variables are UA/h_{max} and the outflow parameter $\beta = (T_0 / F_{out0})(\partial F_{out} / \partial T)$. Usually, the last term in the numerator of Eqn. (15.28) is much larger than the other two terms. This means that for an increase in β the damping decreases. If UA/h_{max} increases, however, the damping will increase, since the effect of UA/h_{max} in the numerator is stronger than in the denominator.

15.7 Example of Some Responses

To visualize some responses, the following parameters are assumed: $\tau_1 = 2.5$ min, $\tau_2 = 1.25$ min (assuming the effect of changes in β on τ_2 can be neglected), $\tau_3 = 5$ min, $(T_0 - T_{in0})/T_0 = 0.4$ and $\beta = 5$, 15 and 25. The responses are shown in Fig. 15.3.

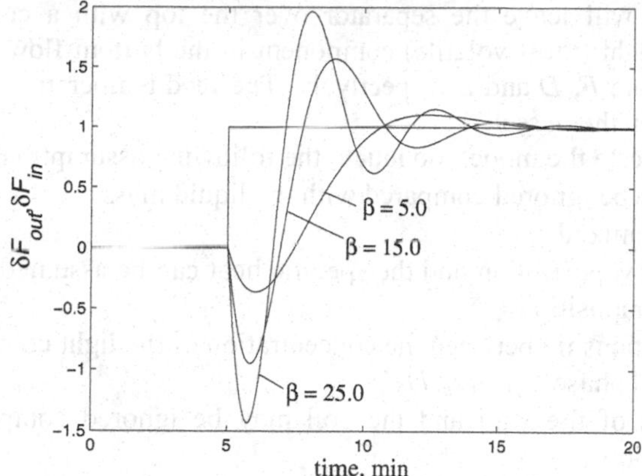

Fig. 15.3. Response of δF_{out} to a step change in δF_{in} for different values of β.

It can clearly be seen that when the value of β increases, the response gets an inverse character initially and starts to oscillate. The shape of the response depends strongly on the values of the parameters (τ and β).

Physically this phenomenon can be explained as follows. Since the inlet temperature is lower than the temperature in the tank, an increase in inlet flow will lead to a decrease in temperature. However, the increased inlet flow will eventually lead to an increase in outlet flow since a new equilibrium will be found and outlet flow will become equal again to the increased inlet flow. This means that the temperature will increase as well as the level. In addition, the pressure will increase, resulting in the increased outlet vapor flow. The oscillatory behavior is a result of the interaction between mass balance and energy balance.

For most industrial evaporators, τ_2 is small compared with τ_1, hence the response of F_{out} to F_{in} will approach a first-order response; however, in small-scale evaporators the situation may be different and τ_2 can be significant compared with τ_1.

15.8 Separation of Multi-phase Systems

When more than one component is present in the liquid mixture, the components can be separated since they have different relative volatility. Let us assume that a binary mixture is separated as shown in Fig. 15.4.

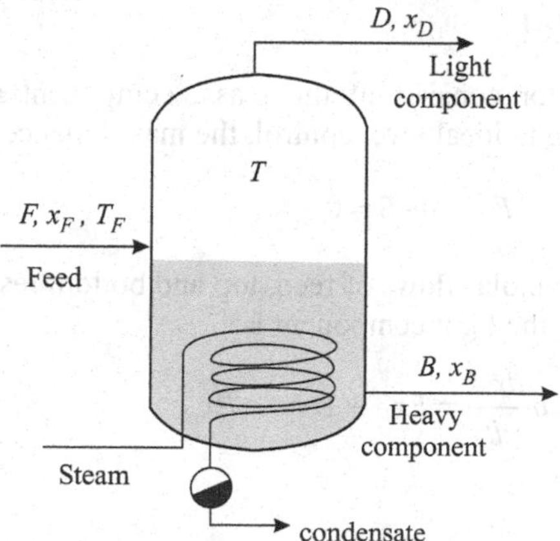

Fig. 15.4. Binary mixture separation.

The light component will leave the separator over the top with a concentration x_D, the concentration of the light (most volatile) component in the bottom flow is x_B. The flows of feed, top and bottom are F, D and B respectively. The feed temperature T_F can be different from the temperature in the vessel T.

So as not to complicate the model too much, the following assumptions will be made:

- the vapor mass may be ignored compared with the liquid mass
- the liquid is ideally mixed
- the density, heat of vaporization and the specific heat can be assumed independent of the temperature and composition
- there is a fixed relationship between the concentration of the light component in the vapor phase and the liquid phase: $x_D=f(x_B, T)$
- the heat capacities of the wall and the coil may be ignored compared with the heat capacity of the liquid
- heat losses can be ignored
- the heating coil is always covered by the liquid
- to realize the previous assumption, the level is ideally controlled by manipulating the outlet flow B
- the heat of mixing can be neglected

The environmental diagram for the separator is shown in Fig. 15.5

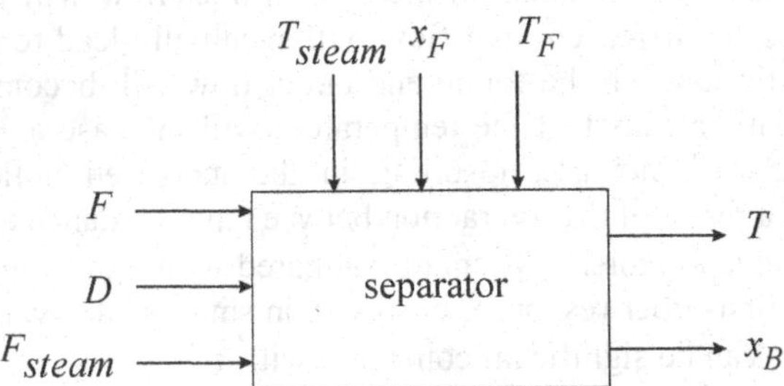

Fig. 15.5. Environmental diagram for the separator.

As can be seen, it is assumed that the steam temperature (or pressure) and the feed conditions (composition and temperature) are assumed to be disturbance variables.

15.9 Separator Model

The model for the separator consists of the mass, component and energy balance and additional equations. Owing to ideal level control, the mass balance can be written as:

$$F - D - B = 0 \tag{15.29}$$

in which F, D and B are the molar flows of feed, top and bottom respectively (mol/s).
The component balance for the light component is:

$$\rho_L A_c h \frac{dx_B}{dt} = Fx_F - Dx_D - Bx_B \tag{15.30}$$

in which x is the concentration of the light component, ρ_L = density (mol/m^3), A_c = cross sectional area of the vessel (m^2) and h is the liquid level (m),

The energy balance for this case can be written as:

$$\rho_L c_p A_c h \frac{dT}{dt} = F c_p T_F - B c_p T - D(c_p T + \Delta H) + UA(T_{steam} - T) \quad (15.31)$$

in which c_p is the specific heat of the liquid (J/mol.K), ΔH the heat of vaporization (J/mol), and UA the product of heat transfer coefficient and heat transfer area (J/K.s).

The left-hand side of the equation represents the change in energy of the liquid, the first term on the right-hand side represents the sensible heat coming into the vessel with the feed flow, the second term is the sensible heat leaving the tank with the bottom flow, the third term is the sensible heat leaving the vessel with the top flow, and the fourth term is the heat transferred from the steam to the liquid. Owing to the assumption that the heat capacity of the wall can be ignored, the last term contains the steam temperature rather than the coil temperature.

Equation (15.30) can, after combination with the mass balance, be written as:

$$\rho_L A_c h \frac{dx_B}{dt} = F(x_F - x_B) - D(x_D - x_B) \quad (15.32)$$

Similarly, energy balance (15.31) can be rewritten as:

$$\rho_L c_p A_c h \frac{dT}{dt} = F c_p (T_F - T) - D\Delta H + UA(T_{steam} - T) \quad (15.33)$$

Since there is a fixed equilibrium relationship between x_D, x_B and T we may write:

$$x_D = f(x_B, T) \quad (15.34)$$

The behavioral model for this set of equations is shown in Fig. 15.6.

15.10 Model Analysis

It is possible to analyze various situations in which disturbance variables change. We could, for example, analyze the change in bottom composition as a result of a change in feed flow. If it can be assumed that there are no changes in feed and steam temperature and bottom and distillate flow, how would the bottom composition x_B respond to feed changes? To analyze this, the model will be linearized with the condition that $\delta T_{steam} = \delta T_F = \delta x_F = 0$.

Linearization of the mass balance, Eqn. (15.29) gives then:

$$\delta F - \delta D - \delta B = 0 \quad (15.35)$$

Linearization of the component balance, Eqn. (15.32) gives:

$$(\rho_L A_c h_0 s + B_0)\delta x_B = (x_{F0} - x_{B0})\delta F - D_0 \delta x_D - (x_{D0} - x_{B0})\delta D \quad (15.36)$$

Linearization of the energy balance, Eqn. (15.33) gives:

$$(\rho_L c_p A_c h_0 s + F_0 c_p + UA)\delta T = c_p (T_{F0} - T_0)\delta F - \Delta H \delta D \quad (15.37)$$

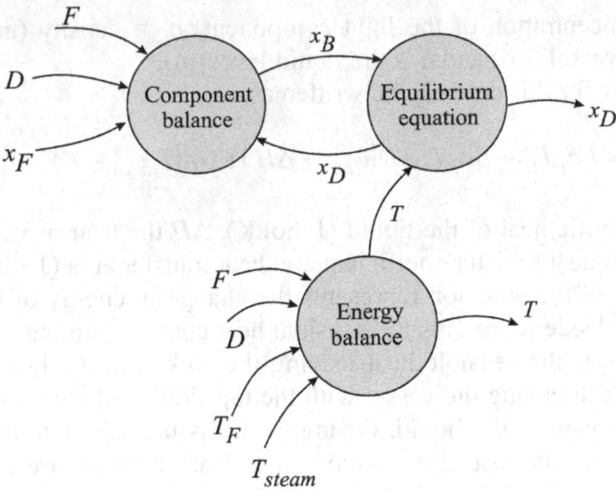

Fig. 15.6. Behavioral model for the separator.

Linearization of Eqn. (15.34) gives:

$$\delta x_D = \beta \delta x_B + \gamma \delta T$$
$$\kappa = \left(\frac{\partial f}{\partial x_B}\right), \; \lambda = \left(\frac{\partial f}{\partial T}\right) \tag{15.38}$$

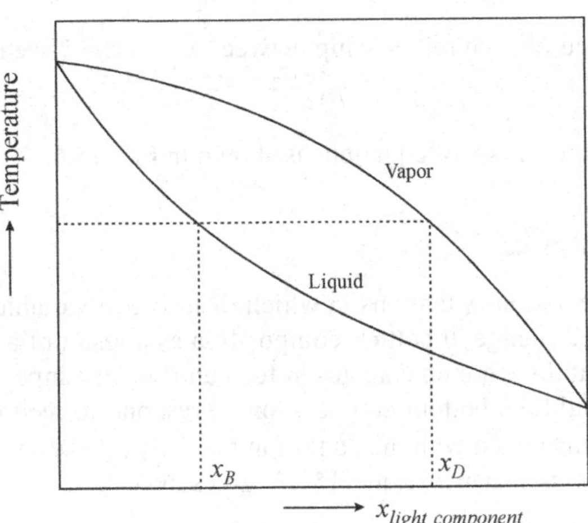

Fig. 15.7. Isobaric vapor-liquid equilibrium curves.

Typical vapor–liquid equilibrium curves at constant pressure are shown in Fig. 15.7. As can be seen, $\kappa > 0$ and $\lambda < 0$. This can easily be explained: when the concentration of a component in the liquid increases, its concentration in the vapor will also increase, hence κ is positive. Figure 15.7 shows that, if the temperature increases, x_D decreases, thus λ is negative.

The model for this situation in which we want to make this specific analysis, consists of Eqns. (15.35)–(15.38).

15.11 Derivation of the Transfer Function

To investigate how the bottom concentration x_B changes as a result of a change in the feed flow F, the component balance, Eqn. (15.36), is combined with Eqn. (15.38) to give:

$$\begin{aligned} \delta x_B &= \frac{\tau_C / M_0}{\tau_C s + 1}\left[(x_{F0} - x_{B0})\delta F - \lambda D_0 \delta T - (x_{D0} - x_{B0})\delta D\right] \\ \tau_C &= \frac{M_0}{B_0 + \kappa D_0} \\ M_0 &= \rho_L A_c h_0 \end{aligned} \tag{15.39}$$

The energy balance, Eqn. (15.37), can be written as:

$$\begin{aligned} \delta T &= \frac{\tau_T / c_p M_0}{\tau_T s + 1}\left[c_p (T_{F0} - T_0)\delta F - \Delta H \delta D\right] \\ \tau_T &= \frac{c_p M_0}{F_0 c_p + UA} \end{aligned} \tag{15.40}$$

Combination of Eqns. (15.39) and (15.40) results in:

$$\begin{aligned} \delta x_B &= \frac{\tau_C M_0}{\tau_C s + 1}\left(x_{F0} - x_{B0} - \frac{\tau_T (T_{F0} - T_0)\lambda D_0}{M_0(\tau_T s + 1)}\right)\delta F + \\ &- \frac{\tau_C M_0}{\tau_C s + 1}\left(x_{D0} - x_{B0} - \frac{\tau_T \Delta H \lambda D_0}{c_p M_0 (\tau_T s + 1)}\right)\delta D \end{aligned} \tag{15.41}$$

The response of x_B to changes in the feed F can easily be obtained from Eqn. (15.41). To get more insight, it is rewritten to the following equation:

$$\begin{aligned} \frac{\delta x_B}{\delta F} &= \frac{\tau_C\left[\tau_T M_0 (x_{F0} - x_{B0}) s + M_0 (x_{F0} - x_{B0}) - \tau_T (T_{F0} - T_0)\lambda D_0\right]}{(\tau_C s + 1)(\tau_T s + 1)} \\ &= K \frac{\tau s + 1}{(\tau_C s + 1)(\tau_T s + 1)} \end{aligned} \tag{15.42}$$

The overall response of x_B to changes in F depends on the sign of the term $M_0(x_{F0} - x_{B0}) - \tau_T (T_{F0} - T_0)\gamma D_0$. If $T_{F0} - T_0 > 0$ then the term is positive and the response of x_B to changes in F will be pseudo-first-order, since τ is positive. If, however, $T_{F0} - T_0 < 0$, the term could become negative; consequently τ would become negative, resulting in an inverse response. Figure 15.8 shows some responses for $\tau_C = \tau_T = 2$, $K = 1$ and $\tau = 1, 0, -1, -2$.

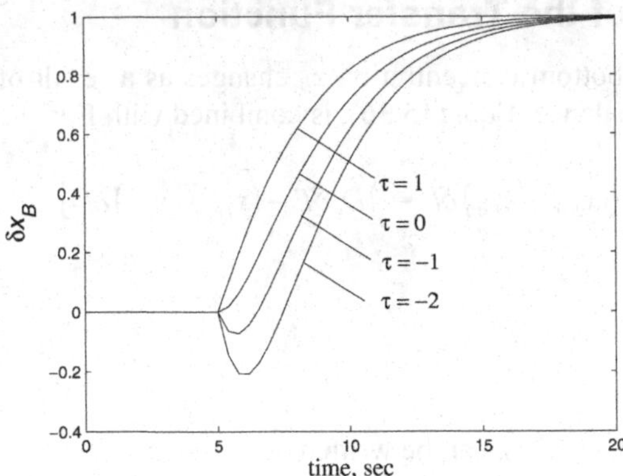

Fig. 15.8. Change in bottom concentration to change in feed according to Eqn. (15.42) for different values of the time constant τ.

As can be seen, the response increasingly gets an inverse character when the time constant τ becomes more negative.

In the hypothetical case when $\tau_T(T_{F0}-T_0)\lambda D_0 = M_0(x_{F0}-x_{B0})$, the bottom concentration only shows a dynamic response; however, there is no static impact of feed changes on the bottom concentration. The process transfer function can now be written as:

$$\frac{\delta x_B}{\delta F} = \frac{Ks}{(\tau_C s+1)(\tau_T s+1)} \tag{15.43}$$

The response is shown in Fig. 15.9, for a value of $\tau_C = 2$, $\tau_T = 2$ and $K=1$. It is obvious that for different values of K, responses with a different peak height will be obtained.

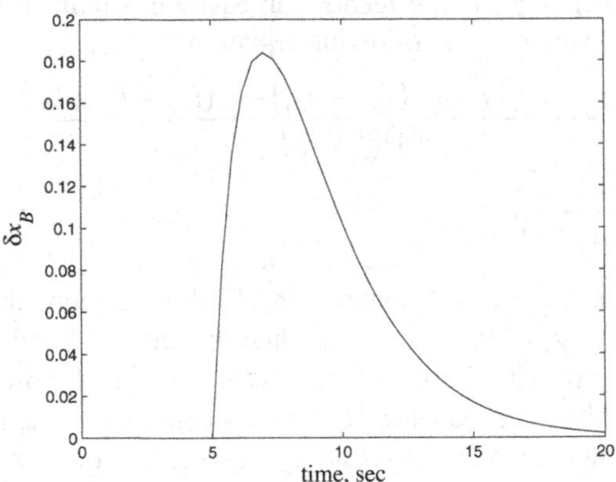

Fig. 15.9. Change in bottom concentration to change in feed according to Eqn. (15.43).

Files referred to in this chapter:

F1503.mdl: model to generate data for Fig. 15.3
F1503plot.m: file to generate Fig. 15.3
F1508.mdl: model to generate data for Fig. 15.8
F1508plot.m: file to generate Fig. 15.8
F1509.mdl: model to generate data for Fig. 15.9
F1509plot.m: file to generate Fig. 15.9

16 Dynamic Modeling of Distillation Columns

In this chapter the dynamics of distillation columns will be discussed. Vapor flow, as well as liquid flow, responses will be discussed. First, an environmental model will be defined, which specifies the model inputs and outputs of a continuous column. Next, a first-principles behavioral model is presented consisting of mass, component and energy balances for each tray. The tray molar mass depends on the liquid and the vapor load, as well as on the tray composition. Two tray load regimes are distinguished: an aeration regime and an obstruction regime. The energy balance is strongly simplified. Finally, dynamic models are derived to describe liquid, vapor and composition responses for a single tray and for an entire distillation column.

16.1 Column Environmental Model

The *environmental model* defines the system boundaries and the relations between the system and its surroundings. In the environmental diagram (Fig. 16.1), it is assumed that the feed flow and conditions are determined externally.

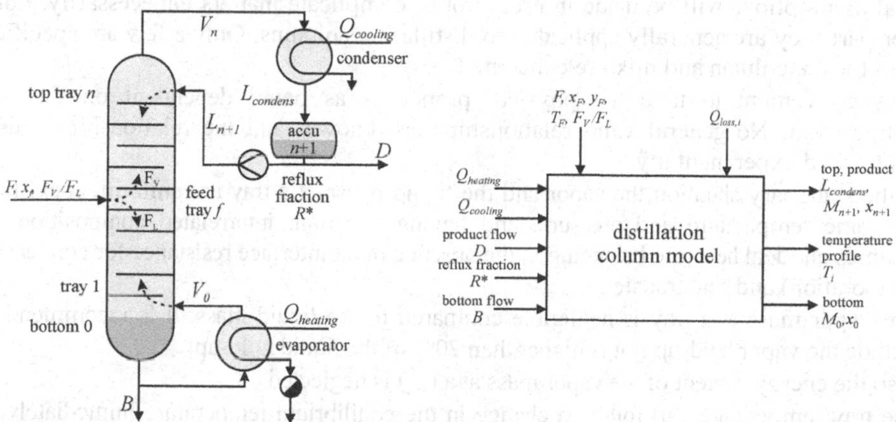

Fig. 16.1. Sketch and environmental model of distillation column with floating pressure. $L_{condens}$ = condensate flow, D = top product flow, R^*= reflux fraction ($L_{n+1}/L_{condens}$), B = top product flow, M = molar mass, x = liquid mol fraction, T = temperature.

The most important manipulated variables are the bottom energy supply $Q_{heating}$, the top energy removal $Q_{cooling}$, and the reflux ratio R, which influence the column operating pressure, the tray load and the degree of separation. Later in this chapter the reflux fraction R^* will be used, instead of the reflux ratio R. R^* is the ratio between the reflux volumetric flow and the vapor volumetric flow. This fraction directly determines the slope of the operating line in the well-known McThiele diagram.

Concerning the system outputs, a distinction can be made between the controlled and the uncontrolled variables. If the purpose of the column is to produce a required product quality, then the top and bottom qualities are the most important controlled variables. At a tray only the temperature can be continuously measured. It gives a good indication of the condition of the column.

Process Dynamics and Control: Modeling for Control and Prediction. Brian Roffel and Ben Betlem.

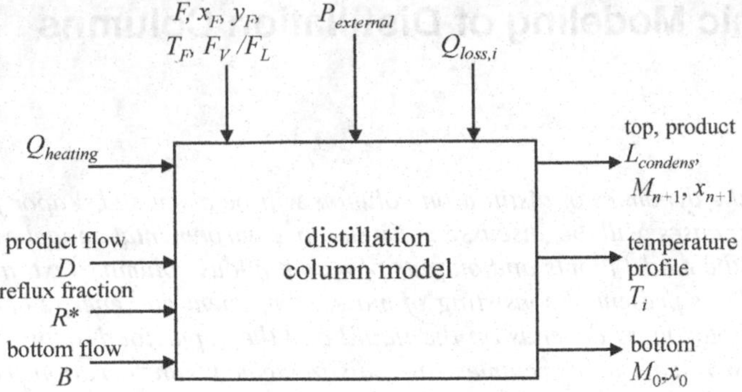

Fig. 16.2. Alternative environmental model of a distillation column with controlled pressure.

Figure 16.2 shows an alternative environmental diagram, in which the condenser cooling is determined by the external (atmospheric) pressure. This configuration is also considered in this chapter.

16.2 Assumptions and Simplifications

Several assumptions will be made in order not to complicate matters unnecessarily. For the greater part they are generally applicable to distillation columns. Only a few are specifically defined for the column and mixture concerned.

- It is convenient to take the physical properties as being dependent on the molar composition. No general valid relationships are known, and the relationships must be established experimentally.
- In the stationary situation the vapor and the liquid phase at a tray are uniform, coexisting at the same temperature and pressure, and having a certain interrelated composition. This assumes an ideal heat rate balancing in the absence of an interface resistance for condensation (evaporation) and heat transfer.
- The vapor mass at a tray is negligible compared to the liquid mass. It is recommended to include the vapor hold-up if it is higher than 20% of the liquid hold-up.
- Also the energy content of the vapor mass at a tray is neglected.
- The tray temperature can follow a change in the equilibrium temperature immediately. The equilibrium temperature is considered to be a dependent variable that is determined by the tray pressure and composition.
- The deviation from the ideal vapor–liquid equilibrium of a tray can be described with the Murphree tray efficiency. For binary distillation this concept works reasonably well when the mass transfer resistance is concentrated in the vapor phase.
- The tray vapor rate and the liquid hold-up have no effect on the heat transfer to the environment.
- The heat of mixing is negligible.
- The component dynamics of condenser and evaporator are neglected. The condenser composition is set equal to the top vapor composition and the evaporator composition is set equal to the bottom vapor composition.
- The reflux consists of liquid approximately at boiling point.

16.3 Column Behavioral Model

The *behavioral model* describes the internal system behavior. The inputs and outputs of the environmental diagram in Fig. 16.2 have to correspond to the inputs and outputs of the column behavioral diagram shown in Fig. 16.3. The distillation column is a multi-stage chain process, with interactive links between the stages or trays as indicated in the data flow diagram. The liquid flows from the top to the bottom through a sequence of forwardly coupled first-order processes. Therefore, for many trays the flow dynamics can be described by a pure time delay. The liquid flow determines mainly the liquid hold-up behavior. This differs basically from the composition and pressure behavior determined by the strong mutual influence between the trays. The pressure reacts mainly on vapor flow changes and settles quickly. The vapor propagates relatively rapidly through the column. However, the composition settles slowly. Of interest is the influence between the vapor and the liquid flow.

The tray behavior can be decomposed further to show the sub-processes (Fig. 16.4). For the top and bottom similar data flow diagrams apply. The inputs and outputs of this data flow diagram are consistent with the inputs and outputs of the tray process in the column model. The diagram can be read clockwise from the bottom left to the bottom right:

- The pressure drop together with the vapor flow settles virtually instantaneously. The pressure drop consists of the dry and wet pressure drop. The vapor flow depends on the energy balance. Roughly, the leaving vapor flow equals the arriving vapor flow minus condensation due to heat losses.
- The mass balance together with the tray hydraulics determines the liquid flow.
- The component balance together with the equilibrium equation and the tray efficiency equation deliver the liquid and vapor composition.
- The tray temperature depends on the pressure and composition.

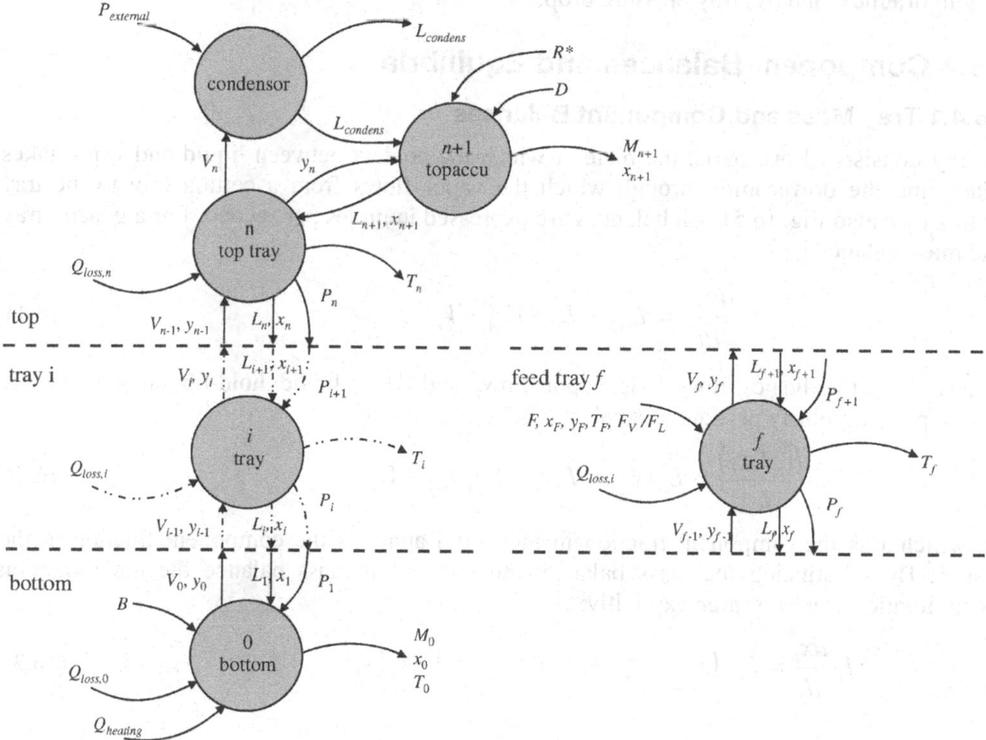

Fig. 16.3. Column behavioral model, → data flow, ⋯→ data flow of tray i, $1 \leq i \leq n-1$.

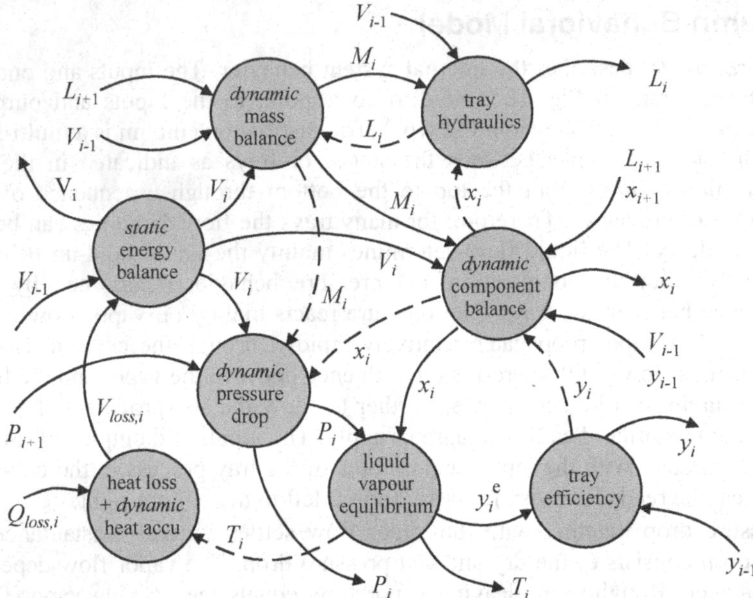

Fig. 16.4. Behavioral model of tray i, → data flow, --→ feedback data flow.

The model equations are described in the subsequent sections. The sequence of the sections is: (i) material balances including equilibrium relations; (ii) energy balances; (iii) tray hydraulics; and (iv) tray pressure drop.

16.4 Component Balances and Equilibria

16.4.1 Tray Mass and Component Balances

A tray consists of two parts: the plate at which the contact between liquid and vapor takes place and the downcomer through which the vapor flows from a certain tray to the tray below (see also Fig. 16.5). All balances are expressed in moles per second. For a general tray the mass balance is:

$$\frac{dM_i}{dt} = L_{i+1} - L_i + V_{i+1} - V_i \tag{16.1}$$

where L is the liquid flow, V is vapor flow, and M is liquid hold-up on a tray. The accompanying binary partial mass balance is:

$$\frac{d(M_i x_i)}{dt} = L_{i+1}x_{i+1} - L_i x_i + V_{i-1}y_{i-1} - V_i y_i \tag{16.2}$$

in which x is the component fraction in the liquid and y is the component fraction in the vapor. By substituting the mass balance into the partial mass balance the mole fraction accumulation can be written explicitly:

$$M_i \frac{dx_i}{dt} = L_{i+1}(x_{i+1} - x_i) + V_{i-1}(y_{i-1} - x_i) - V_i(y_i - x_i) \tag{16.3}$$

16.4.2 Feed, Top and Bottom Mass and Component Balances

For the feed tray, the bottom and the top accumulator, slightly different mass and component balances are necessary considering the side flows:

$$\frac{dM_f}{dt} = L_{f+1} + F - L_f + V_{f+1} - V_f \tag{16.4}$$

$$\frac{dM_0}{dt} = L_1 - V_0 - B \tag{16.5}$$

$$\frac{dM_{n+1}}{dt} = L_{condens} - L_{n+1} - D \tag{16.6}$$

where the subscript f refers to the feed tray, the subscript 0 to the bottom tray and n+1 to the top tray.

By substituting the mass balance into the partial mass balance the mole fraction accumulation for the feed tray, bottom and top accumulator can be written explicitly:

$$M_f \frac{dx_f}{dt} = L_{f+1}(x_{f+1} - x_f) + F_L(x_F - x_f) + F_V(y_F - x_f) + V_{f-1}(y_{f-1} - x_f) - V_f(y_f - x_f) \tag{16.7}$$

$$M_0 \frac{dx_0}{dt} = L_1(x_1 - x_0) - V_0(y_0 - x_0) \tag{16.8}$$

$$M_{n+1} \frac{dx_{n+1}}{dt} = L_{condens}(y_n - x_{n+1}) \tag{16.9}$$

Since the reflux L_{n+1} is set by the reflux ratio, the reflux can be directly calculated from the condensed vapor flow. The molar density ratio corrects for the difference between molar flows and volumetric flows. Usually, no correction is required, as the difference in composition is small:

$$L_{n+1} = R^* . \frac{\rho_m(y_n)}{\rho_m(x_{n+1})} L_{cond} \approx R^* L_{cond} \tag{16.10}$$

in which $\rho_m(y_n)$ = liquid molar density at composition y_n (mol/m^3), and R^* = the reflux fraction.

16.4.3 Phase Equilibrium

At equilibrium the temperature of the liquid, its composition and the pressure are interrelated. Hence any one of these variables can be expressed as a function of the other. In the description of the column, for the equilibrium relationships and the energy balances the composition and the pressure are taken as the independent variables and the temperature as the dependent variable.

The liquid concentration and the pressure are taken as the independent thermodynamic tray state variables of tray i. Then the equations of Dalton, Raoult and Antoine determine the temperature and the vapor composition. For a non-ideal mixture the adapted Raoult equation can be used to compute the partial pressure of component j (P_j) from the pure component vapor pressure (P_j^0), which only depends on the temperature. The vapor fraction at equilibrium y_j^e is the ratio between the partial pressure and the total pressure. As stated before, the liquid and vapor phase tray pressure and temperature, are assumed to be equal.

$$P_j = x_j \beta_j P_j^o, \quad y_j^e = P_j / P, \quad j = A, B \tag{16.11}$$

The sum of the vapor fractions should be equal to one:

$$y_A^e + y_B^e = 1 \tag{16.12}$$

The Antoine equation gives the vapor pressure P_j^0 of a pure component j:

$$P_j^o = exp\left(A_j - \frac{B_j}{C_j + T}\right), \quad j = A,B \tag{16.13}$$

The liquid activity coefficients β_j can be determined from several equilibrium equations known in the literature. Usually, the Wilson equation gives good results.

$$\begin{aligned} ln\beta_A &= -ln(x_A + \Lambda_{AB} x_B) + x_B D \\ ln\beta_B &= -ln(x_B + \Lambda_{BA} x_A) - x_A D \\ D &= \frac{\Lambda_{AB}}{x_A + \Lambda_{AB} x_B} - \frac{\Lambda_{BA}}{\Lambda_{BA} x_A + x_B} \end{aligned} \tag{16.14}$$

The interaction coefficients Λ_{AB} and Λ_{BA} between the components A and B in a binary mixture, can be derived from formulas with the free molar volume and the absolute temperatures as variables. However, fixed values approximate the boiling points better. With the state equations, a set of relationships is specified, in which two variables are unknown. If the liquid composition x and the pressure P are given, then the vapor equilibrium composition y^e and the temperature T can be calculated.

For ideal binary systems, the relation can be simplified to:

$$y_i = \frac{\alpha(P_i)\, x_i}{(\alpha(P_i) - 1)\, x_i + 1} \tag{16.15}$$

in which α is the relative volatility, which is a function of the local pressure P_i. When the pressure drop is much smaller than the absolute pressure, then α may be assumed constant.

At a tray the liquid and vapor have a limited contacting period. When the mass transfer is governed by the vapor-film resistance, the static Murphree vapor efficiency can be used to describe the deviation from equilibrium. For binary distillation this concept works reasonably well. It is assumed that the efficiency is constant throughout the column.

$$\varepsilon_V = \frac{y_i - y_{i-1}}{y_i^e - y_{i-1}} \tag{16.16}$$

The equation assumes that a part (1-ε_V) of the vapor flow bypasses the tray. The mean tray efficiency ε_V is applied as a fitting parameter for the column dynamics and derived from measurements. Usually, the bottom has a Murphree efficiency of 100%. The bottom is well mixed and the liquid and vapor are at equilibrium. In the boiling liquid in the bottom no bypassing of vapor occurs.

16.4.4 Component Tray Dynamics

For small variations, the relationship between the liquid and vapor phase can be reduced to:

$$\delta y_i = \varepsilon_i \delta x_i \tag{16.17}$$

in which ε_i represents the slope of the equilibrium curve at tray i. This relationship may be adjusted for the tray efficiency. For small variations, the component balance Eqn. 16.3 can be linearized and simplified to:

$$\begin{aligned}\overline{M}\frac{d\delta x_i}{dt} &= \overline{L}(\delta x_{i+1} - \delta x_i) + \overline{V}(\varepsilon_{i-1}\delta x_{i-1} - \varepsilon_i \delta x) + \\ &(\overline{x}_{i+1} - \overline{x}_i)\,\delta L + (\overline{y}_{i-1} - \overline{y}_i)\,\delta V\end{aligned} \tag{16.18}$$

In this equation the operating value of the molar mass $\overline{M}$ and the liquid and vapor flows $\overline{L}$ and $\overline{V}$ are the mean values for a column section. Using the steady state of the component balance (Eqn. 16.3):

$$(\overline{y}_{i-1} - \overline{y}_i) = -(\overline{x}_{i+1} - \overline{x}_i)\frac{\overline{L}}{\overline{V}} \tag{16.19}$$

and rearrangement gives:

$$\begin{aligned}-\delta x_{i+1} + \left((1+\Sigma_i)\cdot \delta x_i + \tau_x \frac{d\delta x_i}{dt}\right) - \Sigma_i \cdot \delta x_{i-1} &= \\ (\overline{x}_{i+1} - \overline{x}_i)\cdot \left[\frac{\delta L_{i+1}}{\overline{L}} - \frac{\delta V_{i-1}}{\overline{V}}\right]&\end{aligned} \tag{16.20}$$

in which

$$\tau_x = \frac{\overline{M}}{\overline{L}}, \quad \text{and } \Sigma_i = \left(\frac{\delta y_i}{\delta x_i}\right)\cdot\left(\frac{V_i}{L_i}\right) = \varepsilon_i \left(\frac{V_i}{L_i}\right) \tag{16.21}$$

In the equations δx indicates a perturbation around the mean value $\overline{x}$, τ_x is the tray mixing time, and Σ is the stripping factor incorporating the tray efficiency. The relationship between the composition of the sequential trays is described by the stripping factor. The difference between the terms $(\delta x_i + \Sigma_i \cdot \delta x_i)$ and $(\delta x_{i+1} + \Sigma_i \cdot \delta x_{i-1})$ can be considered to be a measure for the composition interaction between adjoining trays.

16.5 Energy Balances

16.5.1 Tray Energy Balance

The liquid specific enthalpy changes H_i are determined by the enthalpy of the entering and outgoing tray flows (L_{i+1}, V_{i-1}, L_i, and V_i), the tray material heat capacity $C_{m,i}$ and the heat loss through the column wall $Q_{loss,i}$. The mechanical work term is usually small and has therefore been ignored. As the vapor mass of the tray has been neglected, also the internal energy of the vapor is not included. The energy balance can be written as:

$$\frac{dM_i H_{L,i}}{dt} = L_{i+1}H_{L,i+1} + V_{i-1}H_{V,i-1} - L_i H_{L,i} - V_i H_{V,i} - C_{m,i}\frac{dT_i}{dt} - Q_{loss,i} \tag{16.22}$$

Combining the mass balance with the energy balance and assuming that the density $\rho_{L,i}$ is constant, results in:

$$M_i \frac{dH_{L,i}}{dt} + C_{M,i}\frac{dT_i}{dt} = \\ L_{i+1}\left(H_{L,i+1} - H_{L,i}\right) + V_{i-1}\left(H_{V,i-1} - H_{L,i}\right) - V_i\left(H_{V,i} - H_{L,i}\right) - Q_{loss,i} \tag{16.23}$$

The accumulation terms on the left-hand side need further specification. The temperature and the enthalpy changes have to be transformed into relationships of the independent variables, i.e. pressure and composition. At the right-hand side the vapor enthalpy can be written as the sum of the liquid enthalpy H_L and the heat of evaporation H_{vap}, which results in:

$$C_{P,i}\frac{dP_i}{dt} + C_{x,i}\frac{dx_i}{dt} = \\ L_{i+1}\left(H_{L,i+1} - H_{L,i}\right) + V_{i-1}\left(H_{vap}(y_{i-1}) + H_L(y_{i-1}) - H_{L,i}\right) \\ -V_i\left(H_{vap}(y_i) + H_L(y_i) - H_{L,i}\right) - Q_{loss,i}$$

with (16.24)

$$C_{P,i} = \left[M_i\left(\frac{\partial H_{L,i}}{\partial P_i}\right)_x + C_{M,i}\left(\frac{\partial T_i}{\partial P_i}\right)_x\right]$$

$$C_{x,i} = \left[M_i\left(\frac{\partial H_{L,i}}{\partial x_i}\right)_P + C_{M,i}\left(\frac{\partial T_i}{\partial x_i}\right)_P\right]$$

Note that $H_{L,i}$ refers to the liquid enthalpy based on the liquid composition x_i, when the liquid enthalpy is based on the vapor composition y_{i-1} it is written as $H_L(y_{i-1})$.

Multiplying both sides of the component balance (16.3) with the term $\partial H/\partial x$, results in the following equation:

$$M_i\left(\frac{\partial H_{L,i}}{\partial x_i}\right)_P \frac{dx_i}{dt} = L_{i+1}\left(\frac{\partial H_{L,i}}{\partial x_i}\right)_P (x_{i+1} - x_i) + \\ V_{i-1}\left(\frac{\partial H_{L,i}}{\partial x_i}\right)_P (y_{i-1} - x_i) - V_i\left(\frac{\partial H_{L,i}}{\partial x_i}\right)_P (y_i - x_i) \\ = \underbrace{L_{i+1}\left(H_{L,i+1} - H_{L,i}\right)}_{\text{heating liquid}} \underbrace{- V_i\left(H_L(y_i) - H_{L,i}\right) + V_{i-1}\left(H_L(y_{i-1}) - H_{L,i}\right)}_{\text{cooling vapour}} \tag{16.25}$$

In the stationary state, the cooling down of the rising vapor flow balances the heating up of the descending liquid flow. Assuming that the molar enthalpy of a liquid at boiling point is linear in the molar composition, it becomes possible to remove the liquid enthalpy change due to composition by a substitution. Subtracting Eqn. (16.25) from the energy balance Eqn. (16.24) gives the general tray heat balance of Eqn. (16.26):

$$C_{P,i}\frac{dP_i}{dt} + C_{M,i}\left(\frac{\partial T_i}{\partial x_i}\right)_P \frac{dx_i}{dt} = V_{i-1}H_{vap}(y_{i-1}) - V_i H_{vap}(y_i) - Q_{loss,i} \tag{16.26}$$

with

$$C_{P,i} = \left[\left(M_i c_p(x_i) + C_{M,i}\right)\left(\frac{\partial T_i}{\partial P_i}\right)_x\right] \tag{16.27}$$

in which $H_{vap}(y_i)$ is heat of vaporization, $Q_{loss,i}$ is heat loss to the environment, M_i is the liquid molar mass, $c_p(x_i)$ is the molar heat capacity and $C_{M,i}$ is the tray material heat capacity.

The energy equation (16.26) shows that the heat supplied by the vapor from the tray below is used to compensate for the sum of the tray heat changes, heat losses to the surroundings, and the vapor released to the tray above. If the column pressure rises or the concentration of the most volatile component decreases, each tray will retain some vapor by condensation. In the opposite situation liquid will evaporate, if the heat losses are not small enough. The resulting condensation or evaporation flow $V_{e,i}$ can be written explicitly:

$$V_{e,i} = C_{P,i}\frac{dP_i}{dt} + C_{M,i}\left(\frac{\partial T_i}{\partial x_i}\right)_P \frac{dx_i}{dt} + Q_{loss,i} \tag{16.28}$$

When the composition changes are relatively small, for small perturbations the energy balance Eqn. (16.26) directly determines the vapor flow from the heat input decreased by heat accumulation and the heat loss to the environment.

$$C_{P,i}\frac{d\delta P_i}{dt} = H_{vap}(y_{i-1})\,\delta V_{i-1} - H_{vap}(y_i)\,\delta V_i - \delta Q_{loss,i} \tag{16.29}$$

Although the vapor mass has been neglected, it is simple to adapt this equation to additional vapor lags by increasing the hold-up fitting parameter $C_{p,i}$. When condensation or extra evaporation can be neglected, then the equation can be simplified further and consequently the pressure drop settles instantaneously.

$$\delta V_i = \delta V_{i-1}\frac{H_{vap}(y_{i-1})}{H_{vap}(y_i)} \tag{16.30}$$

This is called pseudo-molar overflow. Only when the heats of evaporation of the components are equal, the equation can be simplified further to molar overflow. This does not happen very often.

16.5.2 Feed, Top and Bottom Energy Balance

For the feed tray this equation should be modified to consider the feed energy supply depending on its temperature and liquid vapor ratio.

$$\begin{aligned} &C_{Pf}H_{vap}(y_f)\frac{dP_f}{dt} + C_{Mf}\left(\frac{\partial T_f}{\partial x_f}\right)_P \frac{dx_f}{dt} = \\ &c_p(x_F)\,(T_F - T_f)\,F_L + \left(c_p(y_F)\,(T_F - T_f) + H_{vap}(y_F)\right)F_V + \\ &V_{f-1}H_{vap}(y_{f-1}) - V_i H_{vap}(y_f) - Q_{loss,f} \end{aligned} \tag{16.31}$$

The energy balance for the top depends strongly on the equipment configuration of the condenser. Two situations will be considered.

- The top pressure is determined by the ambient pressure (Fig. 16.2).
  In case of a total condenser, the pressure is coupled with the ambient pressure. The required cooling is provided directly by covering more or less cooling surface. Thus, the cooling flow and temperature have no influence on the condensation and consequently the condenser dynamics play no role. The hold-up of the condenser can be settled with the top-accumulator

$$L_{condens} = V_n \tag{16.32}$$

- The top pressure is determined by vapor and condensation flows (Fig. 16.1). In case of a limited condenser, the cooling flow determines the condensate flow. Then the condenser heat capacity should be considered, and:

$$\tau_{condens} \frac{dL_{condens}}{dt} = Q_{cooling} / H_{vap}(y_n) - L_{condens} \tag{16.33}$$

The starting point for the bottom energy model is the general tray heat balance Eqn. (16.26), adjusted for the bottom. The heat is not supplied by the vapor flow from the tray underneath but by a heating source, which can be mounted internally or externally.

$$C_{P,0} \frac{dP_0}{dt} = Q_{heating,0} - V_0 H_{vap}(y_0) - Q_{loss,0} \tag{16.34}$$

When the heat requirement by the tray capacity and the heat losses are negligible with respect to the heat supply of the vapor flow, then the equation for the vapor flow V_0 for a tray (Eqn. (16.34)) can be simplified further as indicated in the sequel. When the heat supply has a heat capacity, this can often be described by a first-order lag behavior with a time constant $\tau_{reboiler}$.

$$\tau_{reboiler} \frac{dQ_{heating,0}}{dt} = Q_{heating} - Q_{heating,0} \tag{16.35}$$

When the heat is supplied by steam, then the supply can be described by:

$$Q_{heating} = F_{steam} H_{vap,steam} \tag{16.36}$$

16.6 Tray Hydraulics

16.6.1 Liquid Hold-up

The hold-up of a tray M_i consists of the liquid mass on the tray M_{cl}, downcomer M_{dc}, and the vapor hold-up M_V (Fig. 16.5).

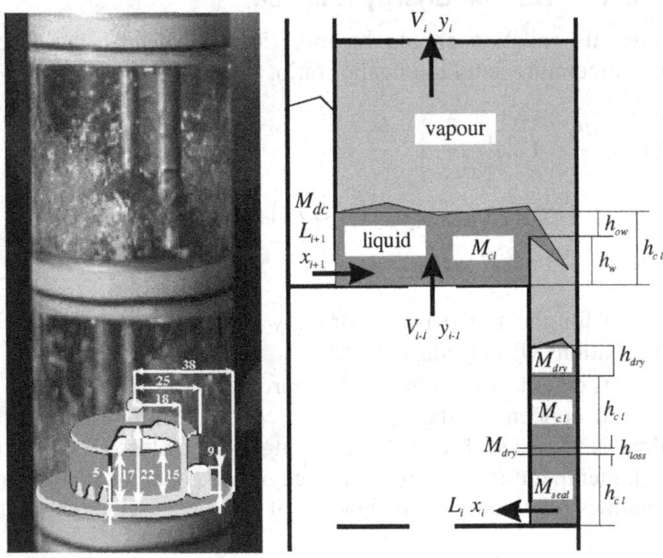

Fig. 16.5. Photograph of two trays with dimensions in millimetres and sketch of tray compartment. The crossbar right in the upper tray of the photograph is the downcomer. The tray distance is 80 mm. h_{cl}, h_{ow} , h_{dry}, and h_{loss} concern clear liquid heights, only h_w is the real height.

As stated before the vapor hold-up has been neglected.

$$M_i = M_{cl} + M_{dc} + M_V \approx M_{cl} + M_{dc} \tag{16.37}$$

The molar tray hold-up can be given by:

$$M_{cl} = h_{cl} . A_{active} \cdot \frac{\rho_L}{N_{m,L}} \tag{16.38}$$

In this equation h_{cl} is the clear liquid height, implying the net liquid height when the vapor would be omitted. $N_{m,L}$ is the liquid molar mass.

The liquid height in the downcomer equals the fixed volume of the downcomer seal W_{seal} increased by the liquid equivalent of the dry and wet pressure drop across the tray. The molar downcomer hold-up is (Fig. 16.5):

$$M_{dc} = \left(\left(h_{dry} + h_{cl} + h_{loss} \right) A_{dc} + W_{seal} \right) \frac{\rho_L}{N_{m,L}} \tag{16.39}$$

The molar hold-up of the plate and downcomer relates directly to its liquid volume, which firstly is the result of tray design and secondly depends on the settings of two manipulated variables L and V. The following assumptions can be made:

- h_{loss} is due to the pressure drop under the downcomer apron. Usually this term is negligible compared to h_{cl} and h_{dry}
- no liquid overhang occurs (liquid height difference between inlet and outlet)
- the tray specific mass is equal to the downcomer specific mass and denoted by ρ_L
- the downcomer is not aerated
- it holds that $W_{seal} = A_{dc} \cdot h_{cl}$
- the molar weight of the vapor and liquid phase are about equal: $N_{m,L} \approx N_{m,V} = \overline{N}_m$.

The total mass on a tray is:

$$M_i = M_{cl} + M_{dc} = \left[h_{cl} \left(A_{active} + 2A_{dc} \right) + h_{dry} A_{dc} \right] \frac{\rho_L}{\overline{N}_m} \tag{16.40}$$

with

$$\overline{N}_m = xN_{m,light} + (1-x)N_{m,heavy} \tag{16.41}$$

The photograph in Fig. 16.5 clearly shows the dispersion having no distinct upper surface. The liquid behaves like froth and spray. The clear liquid height equals the net liquid height beneath the weir increased by the clear liquid height above the weir.

$$h_{cl} = \alpha h_{weir} + h_{ow} \tag{16.42}$$

The three variables α, h_{ow}, and h_{dry} and Eqn. (16.42) will now be discussed in more detail.

The partial liquid hold-up α is a mean value for the froth and the spray regime based on the relative vapor load F/F_{max} (Stichlmair, 1978):

$$\alpha = 1 - \left(F/F_{max} \right)^a \tag{16.43}$$

in which a is a constant.

At high gas loads, the gas forms the continuous phase. The liquid is dispersed in single drops which can float along with the gas. F_{max} is the value of F at critical gas load. In this situation the tray is just at the transition point of being blown empty. In this case the particles are floating. Development of the F-factor gives:

$$F \equiv v_V \sqrt{\rho_V} = \frac{V}{A_{column}\,\rho_{m,V}}\sqrt{\rho_V} = \frac{\sqrt{\overline{N}_m}}{A_{column}}\frac{V}{\sqrt{\rho_{m,V}}} \tag{16.44}$$

F_{max} is considered to be constant. Then the partial liquid hold-up equation can be derived after substitution of F in Eqn. (16.43):

$$\alpha = 1 - \left(\frac{\sqrt{\overline{N}_m}}{F_{max}A_{column}}\frac{V}{\sqrt{\rho_{m,V}}}\right)^a = 1 - \left(C_\alpha\sqrt{\overline{N}_m}\frac{V}{\sqrt{\rho_{m,V}}}\right)^a \tag{16.45}$$

For the liquid height above the weir h_{ow}, usually a modified Francis weir equation is used. This classical equation stemming from 1883 is restricted to conditions where the weir height is very large compared to the crest height. This equation ignores the influence of the vapor load, however.

From more recent measurements (Betlem, 2000), it appears that the partial liquid hold-up has a strong influence on the liquid volume above the weir. Therefore, the equation has been corrected for the partial liquid hold-up.

$$h_{ow} = C^*_{ow}\left(\frac{\overline{N}_m.L}{\rho_L}\right)^b \Big/ \alpha \tag{16.46}$$

For $b = 2/3$ the modified Francis weir equation is obtained.

For the calculation of the liquid height, also the dry liquid height is required:

$$h_{dry} = \frac{\Delta P_{dry}}{\rho_L \cdot g} \tag{16.47}$$

The dry pressure drop ΔP_{dry} can be given by:

$$\Delta P_{dry} = \tfrac{1}{2}\xi_{dry}F^2 \tag{16.48}$$

After substitution of the F-factor from Eqn. (16.44):

$$h_{dry} = \frac{\frac{1}{2}\xi_{dry}}{\rho_L \cdot g}\frac{\overline{N}_m}{A^2_{column}}\frac{V^2}{\rho_{m,V}} = C^*_{dry}\frac{V^2}{\rho_{m,V}}\frac{\overline{N}_m}{\rho_L} \tag{16.49}$$

Then the molar mass of a tray becomes:

$$M_i = M_{cl} + M_{dc} = \left(h_{cl}\left(A_{active} + 2A_{dc}\right) + h_{dry}\,A_{dc}\right)\frac{\rho_L}{\overline{N}_m} \tag{16.50}$$

which can be written as:

$$M_i = \left[\begin{array}{l}\left(\alpha.h_{weir} + C^*_{ow}\left(\frac{\overline{N}_m.L}{\rho_L}\right)^b \Big/ \alpha\right)\left(A_{active} + 2A_{dc}\right) \\ \quad + \left(C^*_{dry}\frac{V^2}{\rho_{m,V}}\frac{\overline{N}_m}{\rho_L}\right)A_{dc}\end{array}\right]\frac{\rho_L}{\overline{N}_m} \tag{16.51}$$

Simplification by grouping of constants leads to:

$$M_i = \alpha \frac{\left(h_{weir}\left(A_{active} + 2A_{dc}\right)\right)\rho_L}{\overline{N}_m} + \frac{C_{ow}}{\overline{N}_m}\frac{\left(\overline{N}_m L\right)^b}{\alpha} + C_{dry}\frac{V^2}{\rho_{m,V}} \tag{16.52}$$

while for the partial liquid hold-up α, Eqn. (16.45) can be written.

16.6.2 Liquid Dynamics

Time constants can be derived directly from the mass hold-up equation of tray i by linearization. The tray mass hold-up is defined in Eqn. (16.52) in which the liquid hold-up fraction is defined according to Eqn. (16.45) and the mean molar mass by Eqn. (16.41). Often, the molar vapor densities $\rho_{m,V}$ of both components are equal (ideal gas law), but are still a function of the absolute pressure. Hence the tray hold-up according to the Eqn. (16.52) can be written as function of its independent variables:

$$M_i = M_i\left\{L, V\sqrt{\rho_{mV}}, x\right\} \tag{16.53}$$

This means that the liquid leaving the tray is determined by the liquid on the tray, the impulse of the vapor flow and by the molar composition. For small variations one can write:

$$\begin{aligned}\delta M_i = &\left(\frac{\partial M_i}{\partial L}\right)_{V,P,x} \delta L_i + \\ &\left(\frac{\partial M_i}{\partial V}\right)_{L,P,x}\left(\delta V_i - \frac{\overline{V}}{2\overline{\rho}_{m,V}}\left(\frac{\partial \rho_{m,V}}{\partial P}\right)_x \delta P_i\right) + \\ &\left(\frac{\partial M_i}{\partial \overline{N}_m}\right)_{L,V,P}\left(\frac{d\overline{N}_m}{dx}\right)\delta x_i\end{aligned} \tag{16.54}$$

As the molar vapor density depends on the pressure, according to $\rho_{m,V} = P/(RT)$ and T is a function of the variables P and x, this equation can be simplified to:

$$\begin{aligned}\delta M_i = &\left(\frac{\partial M_i}{\partial L}\right)_{V,P,x} \delta L_i + \left(\frac{\partial M_i}{\partial V}\right)_{L,P,x}\left(\delta V_i - \frac{\overline{V}}{2\overline{P}}\delta P_i\right) + \\ &\left(\frac{\partial M_i}{\partial \overline{N}_m}\right)_{L,V,P}\left(\frac{d\overline{N}_m}{dx}\right)\delta x_i\end{aligned} \tag{16.55}$$

In a limited operating area the derivatives can be considered to be constant. The first two derivatives on the right hand site are time constants: τ_L and τ_V. Now the equation can be rewritten as the liquid flowing from the tray being the dependent variable of the tray state and vapor flow:

$$\begin{aligned}&\delta L_i = \frac{1}{\tau_L}\delta M_i + \lambda\left(\delta V_i - \frac{V}{2P}\delta P_i\right) - \frac{\kappa_L}{\tau_L}\delta x_i \\ &\text{with } \lambda = -\frac{\tau_V}{\tau_L} = \left(\frac{\partial L}{\partial V}\right)_{M,P,x}\end{aligned} \tag{16.56}$$

In this equation τ_L can be considered to be a liquid hydraulic time constant. λ is a dimensionless combination of τ_L and τ_V : $\lambda = -\left(\tau_V / \tau_L\right)$ and describes the influence of the vapor flow on the liquid flow ($\partial L / \partial V$) in order to give the same hold-up. κ_L represents the sensitivity of the liquid hold-up to the mean molar weight. A composition shift will result in molecular contents shift. Increasing the light component fraction, the number of moles will rise as the tray liquid volume will nearly remain the same. κ_L cannot always be neglected owing component shifts caused by the vapor or liquid flow changes. The equation for τ_L can be given by (using Eqn. (16.52)):

$$\tau_L = \left(\frac{\partial M_i}{\partial L}\right)_{V,P,x} = b\,\frac{C_{ow}}{\overline{N}_m}\cdot\frac{\left(\overline{N}_m L\right)^b}{\alpha.}\cdot\frac{1}{L} > 0 \qquad (16.57)$$

As expected, the hydraulic time constant is completely determined by the term from Eqn. (16.52) which describes the mass above the weir. Thus it depends largely on the liquid flow and the hold-up fraction. The equation for τ_V becomes:

$$\tau_V = \left(\frac{\partial M_i}{\partial V}\right)_{L,P,x} = \left(\underbrace{-\frac{\left(h_{weir}\left(A_{active}+2A_{dc}\right)\right)\rho_L}{\overline{N}_m}}_{<0} + \underbrace{\frac{C_{ow}}{\overline{N}_m}\frac{\left(\overline{N}_m L\right)^b}{\alpha^2}}_{>0}\right)\cdot \frac{a(1-\alpha)}{V} + \underbrace{2C_{dry,h}\frac{V}{\rho_{m,V}}}_{>0} \qquad (16.58)$$

The sensitivity of the molar mass to the vapor flow can be positive or negative depending on the dominant term on the right-hand side of Eqn. (16.58). The first term in Eqn. (16.58) expresses mainly the change of the liquid hold-up below the weir edge and the second term above the weir edge. The third term describes the change in the downcomer hold-up when the vapor flow changes. This term is usually relatively small as changes only affect the downcomer hold-up. At low vapor load the first term dominates and the time constant will have a negative value. At higher vapor loads, α decreases. Consequently, the tray mass below the weir decreases by driving out liquid, whereas the mass above the weir and in the downcomer increases by the driving up of liquid. Thus, if V increases in the equation, the first term will decrease, whereas the second and third term will increase hence the sign of the result may be positive or negative.

The explanation of the tray behavior is implied in the equation for τ_V (Eqn. (16.58)). If the liquid height on a tray is approximately equal to the weir height, a vapor flow increase will lower the liquid hold-up fraction. In a sense the gas pushes the liquid aside by higher aeration. The negative term in Eqn. (16.58) determines the driven out liquid mass. If the liquid height is much larger than the weir height, increasing the vapor flow will obstruct the liquid from flowing away and the tray liquid mass rises. This is represented by the second positive term in Eqn. (16.58). In summary, for the dynamic tray behavior at higher reflux loads two different tray load regimes can be distinguished:

- *aeration regime*
  At low tray load, the weir height determines the liquid hold-up. At increasing tray loads the liquid hold-up is only marginally higher, because the liquid flow mainly determines the magnitude of the hold-up.
- *obstruction regime*
  At high tray load the liquid is driven up by the vapor impulse. Downcomer dimensions, limited stilling zone, wall effects, and other design limitations cause this behaviour. The liquid mass on the tray strongly increases with higher tray load.

16.7 Tray Pressure Drop

16.7.1 Vapor Hold-up

The total pressure drop ΔP across a tray is defined by the general equation:

$$\Delta P_{tray} = \Delta P_{dry} + \Delta P_{wet} + \Delta P_{rest} \tag{16.59}$$

The third term is caused by the surface tension of the liquid. Usually this term can be neglected. The first term is the dry pressure drop due to the vapor flow through empty bubble caps. This term is defined in Eqn. (16.48). Similar to h_{dry} we find:

$$\Delta P_{dry} = h_{dry} \rho_L g \tag{16.60}$$

The second term takes into account the pressure drop through the aerated liquid over and around the dispenser correspond to the clear liquid height. This pressure drop has a relationship with the liquid height on a tray. The method, commonly used, defines an aeration factor β, such that,

$$\Delta P_{wet} = \beta . h_{cl} \rho_L g \tag{16.61}$$

where h_{cl} (Eqn. (16.42)) equals the liquid height on the tray when the liquid is non-aerated and no entrainment occurs. β is the weighted mean between one and the partial liquid hold-up α.

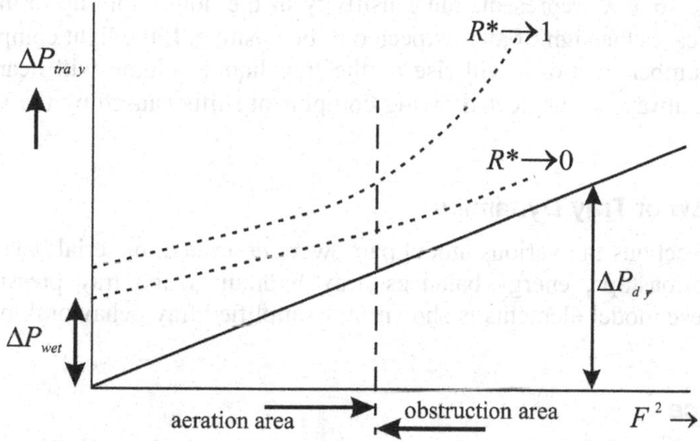

Fig. 16.6. Pressure drop as a function of the vapor and liquid load.

Figure 16.6 shows the dry and wet pressure drop. Both terms make significant contributions to the total pressure drop. The dry pressure drop depends on the square of the vapor load. The wet pressure drop is the result of the liquid height on the tray. At low tray load ($R^* \rightarrow 0$), a higher vapor load decreases the tray liquid mass. Even though there is a reduction in the tray liquid mass, the overall tray pressure drop still increases at higher vapor loads due to the strong increase of the dry pressure drop. At high tray loads ($R^* \rightarrow 1$) and high vapor rates (F^2 above a certain value), a strong increase of the pressure drop result, owing the increasing tray liquid mass.

16.7.2 Vapor Dynamics

The equation for ΔP_{tray} is strongly related to the equation for M_i. Time constants can be derived directly from the pressure drop equation of tray i by linearization. The pressure drop is defined in Eqn. (16.59), it can be written as function of its independent variables:

$$\Delta P_i = P_{i-1} - P_i = \Delta P_i\,(M_L\,, V/\sqrt{\rho_{m,V}}\,, x) \tag{16.62}$$

This means that the pressure drop is determined by the mass on the tray, the impulse of the vapor flow and by the molar composition. For small variations one can write:

$$\begin{aligned}\delta\Delta P_i = &\left(\frac{\partial \Delta P_i}{\partial L}\right)_{V,P,x} \delta M_i + \left(\frac{\partial \Delta P_i}{\partial V}\right)_{M,P,x} \left(\delta V_i - \frac{\overline{V}}{2\overline{\rho}_{m,V}}\left(\frac{\partial \rho_{m,V}}{\partial P}\right)_x \delta P_i\right) \\ &+ \left(\frac{\partial \Delta P_i}{\partial \overline{N}_m}\right)_{M,V,P} \left(\frac{d\overline{N}_m}{dx}\right) \delta x_i\end{aligned} \tag{16.63}$$

or

$$\delta\Delta P_i = \pi\ \delta M_i + R_V\left(\delta V_i - \frac{V}{2P}\delta P_i\right) + \kappa_P \delta x_i \tag{16.64}$$

In this equation π is always positive, as the pressure drop increases with a rise of the molar mass. R_V is a resistance to the vapor flow, similar to Ohms' law. This term is always positive as shown in Fig. 16.6. κ_P represents the sensitivity of the liquid hold-up to the mean molar weight. In most cases the sign of κ_P is expected to be positive. If the light component fraction increases, the number of moles will rise as the tray liquid volume will nearly remain the same. κ_P cannot always be neglected owing component shifts caused by the vapor or liquid flow changes.

16.7.3 Overview of Tray Dynamics

In the previous sections the various model parts were described: material balances including equilibrium relationships, energy balances, tray hydraulics and tray pressure drop. The coherence of these model elements is shown in a simplified tray behavioral model shown in Fig. 16.7.

Energy Balance

In the left most circle of Fig. 16.7, the vapor and pressure formulae are taken together. These calculations are strongly linked as the pressure drop depends mostly on the vapor flow. The energy balance, Eqn. (16.26), directly determines the vapor flow from the heat input, decreased by heat accumulation and the heat loss to the environment. The changes in absolute pressure and pressure drop on tray i can be described by Eqns. (16.35) and (16.64).

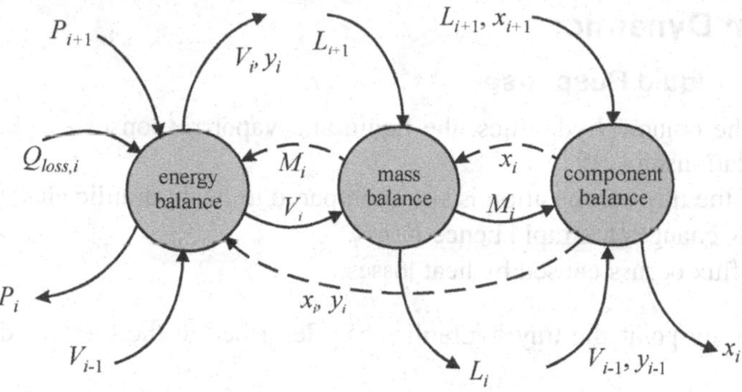

Fig. 16.7. Simplified representation of tray behavioural model, ⟶ data flow, - - - -▸ feedback data flow

Mass Balance

The mass balance together with the mass equation determines the liquid mass on a tray and the liquid flow from the tray. For small perturbations the mass balance Eqn. (16.1) can be written as:

$$\frac{d\delta M_i}{dt} = \delta L_{i+1} - \delta L_i + \delta V_{i+1} - \delta V_i \tag{16.65}$$

in which δL_i is given by Eqn. (16.56).

Component Balance

From the liquid fraction and the tray pressure the equilibrium vapor fraction and the temperature can be determined by a combination of the Dalton, Raoult, and Antoine equations. For small variations Eqn. (16.17) holds:

$$\delta y_i^e = \varepsilon_i \delta x_i \tag{16.17}$$

The liquid composition is determined by the component balance. In a two component mixture only one component balance for fraction x is required. The other fraction is 1-x. The changes in component concentration can be described by Eqn. (16.20).
Rademaker *et al.* (1975 p. 150) concludes that it serves no purpose to account for incomplete tray efficiency unless one has intimate knowledge of the transfer processes on the tray. In column sections with more than approximately 12 trays deviations from the equation are negligible.

The temperature on a tray can be calculated from the pressure and component changes:

$$\delta T_i = \left(\frac{\partial T_i}{\partial P_i}\right)_x \delta P_i + \left(\frac{\partial T_i}{\partial x_i}\right)_P \delta x_i \tag{16.66}$$

16.8 Column Dynamics

16.8.1 Column Liquid Response

To understand the column hydraulics, the liquid and vapor responses can be derived using the basic tray relationships. If:

- the change of the tray composition is slow compared to the hydraulic changes,
- the vapor flow changes are rapid hence $V = V_i$,
- no internal reflux occurs caused by heat losses,

then in an operating point the tray hydraulics can described in the Laplace-domain by first-order systems.

$$s\delta M_i = \delta L_{i+1} - \delta L_i \quad with \quad \delta L_i = \lambda\, \delta V + \frac{1}{\tau_L}\delta M_i \tag{16.67}$$

Let us consider the liquid propagation. A combination of the two equations in (16.67) results in:

$$\delta L_i = \lambda \frac{s\tau_L}{s\tau_L + 1}\delta V + \frac{1}{s\tau_L + 1}\delta L_{i+1} \tag{16.68}$$

The response of the liquid flow δL from tray j to the liquid from tray i ($i > j$) becomes:

$$\frac{\delta L_j}{\delta L_i} = \frac{1}{(s\tau_L + 1)^{i-j}} \tag{16.69}$$

This is a chain of first-order delays in series. When i-j is large, this response tends to become a time delay $(i\text{-}j)\tau_L s$:

$$\frac{\delta L_j}{\delta L_i} = e^{-(i-j)\tau_L s} \tag{16.70}$$

The response of the liquid flow δL from tray i due to a vapor change to δV can be found by a repeated application of Eqn. (16.68) and (16.69) starting from $\delta L_{n+1} = 0$ (constant reflux).

$$\begin{aligned}\frac{\delta L_j}{\delta V} &= \lambda \cdot s\tau_L \cdot \left[\frac{1}{(s\tau_L + 1)} + \frac{1}{(s\tau_L + 1)^2} \cdots\cdots \frac{1}{(s\tau_L + 1)^{n+1-j}}\right] \\ &= \lambda \cdot \left[1 - \frac{1}{(s\tau_L + 1)^{n+1-j}}\right]\end{aligned} \tag{16.71}$$

For the bottom hold-up one can write:

$$\frac{\delta M_0}{\delta V} = -\frac{1}{s} + \frac{1}{s} \cdot \frac{\delta L_1}{\delta V} = -\frac{1}{s} \cdot \left[1 - \lambda + \frac{\lambda}{(s\tau_L + 1)^n}\right] \tag{16.72}$$

For many trays, the response will become:

$$\frac{\delta M_0}{\delta V} = -\frac{1}{s} \cdot \left[1 - \lambda + \lambda e^{n\tau_L s}\right] \tag{16.73}$$

The vapor change results in a liquid mass change $\lambda.\delta V$ at all trays simultaneously (removal or addition). Consequently, the change of the mass δL_i will be compensated by the change of the mass δL_{i+1}. However, since $\delta L_{n+1} = 0$, at the top this compensation ends. Initially, at the bottom the net flow change in vapor and liquid is $(1-\lambda)\delta V$. After a period of approximately $n.\tau_L$, the new liquid flow has been settled through the entire column and the net flow change becomes δV. For $\lambda = 0$, 1 and 2, this is shown in Fig. 16.8.

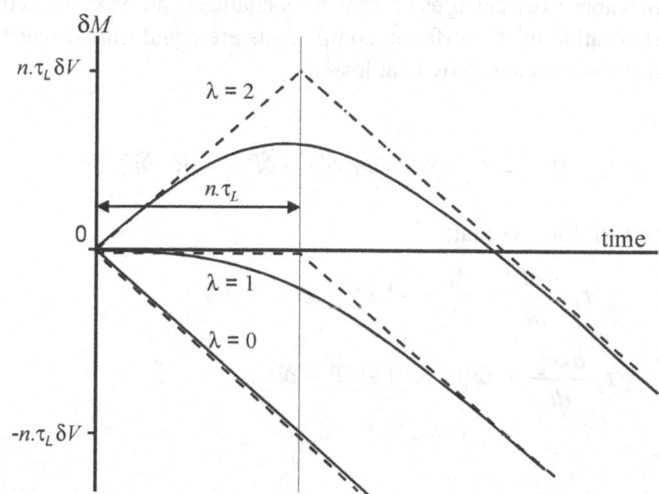

Fig. 16.8. Bottom mass response to a step change in vapor flow for $\lambda = 0$, 1, and 2.

- If $\lambda = 0$ then:

$$\frac{\delta M_0}{\delta V} = -\frac{1}{s} \tag{16.74}$$

The vapor flow has no influence on the liquid mass on the trays and the bottom mass will gradually decrease owing the step change in the vapor flow.

- If $\lambda = 1$ then:

$$\frac{\delta M_0}{\delta V} = -\frac{1}{s} \cdot \frac{1}{(s\tau_L + 1)^n} \tag{16.75}$$

Initially, an increase of the vapor flow will be completely compensated by an equal increase of the liquid flow. This will continue until the liquid mass from the top tray has arrived at the bottom. Because of the delay time between the time of the vapor change and the time the response becomes measurable at the bottom, it is difficult to control the bottom level with the vapor flow.

- If $\lambda = 2$ then:

$$\frac{\delta M_0}{\delta V} = -\frac{1}{s} \cdot \left(-1 + \frac{2}{(s\tau_L + 1)^n} \right) \tag{16.76}$$

Initially, the bottom level will rise, as more liquid will be pushed from the trays than the change in the vapor flow. This will continue until the liquid mass from the top tray arrives at the bottom. Then, the bottom level starts to fall. The response of the bottom shows an inverse response in this case.

For a column with a large number of trays, it can be proven that the controllability of the bottom level by the vapor flow can only be assured if $\lambda < 0.5$ (Betlem 1998).

16.8.2 Column Pressure Response

To understand column-pressure drop dynamics, the vapor responses can be derived using the basic tray relationships. If:

- the change of the tray composition is slow compared to the pressure changes,
- the influence of vapor flow changes on tray mass changes can be neglected,
- the heats of evaporation of the different components are equal (molar overflow),
- no internal reflux occurs caused by heat losses,

then:

$$C_{P,i}\frac{d\delta P_i}{dt} = H_{vap}\delta V_{i-1} - H_{vap}\delta V_i \text{ with } \delta P_i - \delta P_{i+1} = R_V \delta V_i \tag{16.77}$$

For the different trays it follows that:

$$\begin{aligned}\tau_P\frac{d\delta P_1}{dt} &= \frac{\delta V_0}{H_{vap}} - (\delta P_0 - \delta P_1)\\ \tau_P\frac{d\delta P_2}{dt} &= (\delta P_1 - \delta P_2) - (\delta P_1 - \delta P_2)\\ &\text{etc....}\end{aligned} \tag{16.78}$$

in which

$$\tau_P = \frac{R_V C_P}{H_{vap}} \tag{16.79}$$

The settling time is determined by all tray pressure interactions in the column. Although the pressure changes rapidly, all tray compositions react in the same way when the time approaches to large values, as the interactions ($\delta P_{i-1} - \delta P_i$) and ($\delta P_i - \delta P_{i+1}$) are nearly equal. The overall response of this type of process can be described by a first-order system:

$$\tau_{P,column}\frac{d\delta P_i}{dt} + P_i = \mu_{P,i}\frac{\delta V_{i-1}}{\overline{V}} \tag{16.80}$$

Since all tray time constants are equal, the time constant becomes:

$$\tau_{P,column} = \tau_P(1+2+3.....n) = \tau_P\frac{1+n}{2}n \approx \tau_P n^2 \tag{16.81}$$

Usually, neglecting the vapor mass can be justified. The induced time constant is very small. Vapor flow variations propagate very rapidly through the column. Therefore, all trays will simultaneously condensate vapor or evaporate liquid to bring their vapor capacities at equilibrium.

The mechanism above describes the pressure increase or decrease without other interactions. However, increasing the pressure by raising the vapor flow or reducing the condensate flow, invokes two negative feedback influences. At a higher column pressure, the bottom and top temperature will be higher. Consequently, the temperature difference across the reboiler will fall whereas the temperature difference across the condenser will rise. This will decrease the evaporation and simultaneously increase the condensation, as a result of which the pressure tends to rise. This is a counter reaction and works as a form of self-regulation. The interaction between vapor flow and pressure is shown in Fig. 16.9; the mechanism makes the column less sensitive to flow variations.

In addition to this mechanism, a composition shift will take place (x-shift). Because of the higher vapor flow, the components with a higher boiling point will shift to the top. The new stationary pressure fully depends on the sensitivity of partial pressure to the temperature and composition. When the partial pressure of the heavy components is much lower, even an inverse response may be possible. The interactions in this case are shown in Fig. 16.10.

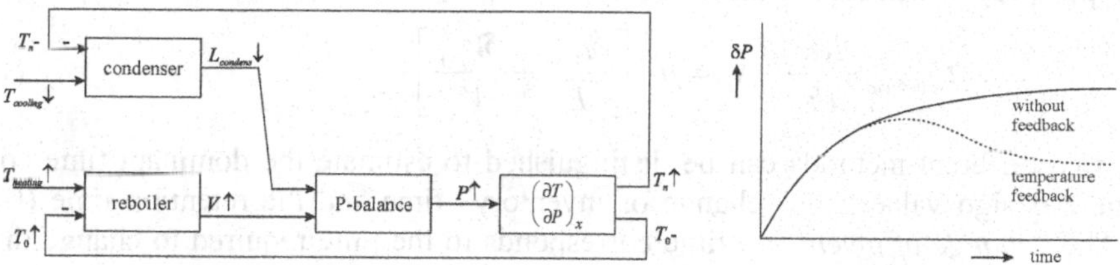

Fig. 16.9. Self-regulating control of the column pressure in case of a vapor step change.

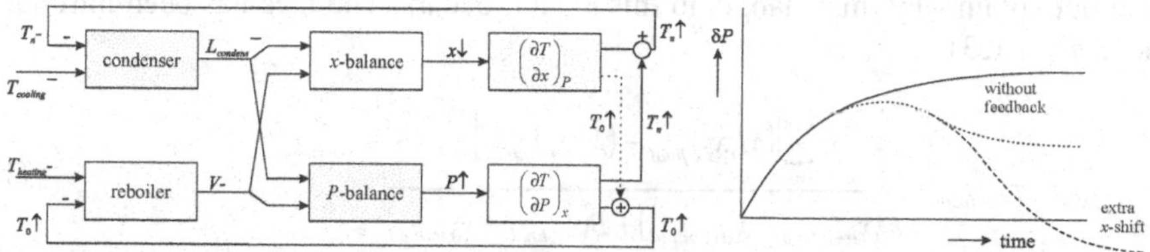

Fig. 16.10. Column pressure response in case of a vapor step change, owing temperature feedback and composition shift.

16.8.3 Column Section Composition Response

The short-term response of the tray composition δx_i depends on δL_i and δV_i, which behavior is determined by the relatively fast liquid and vapor dynamics. Since the compositions change slowly and the terms $(\delta x_i + \Sigma_i \delta x_i)$ and $(\delta x_{i+1} + \Sigma_i \delta x_{i-1})$ in Eqn. (16.20) tend to cancel each other, all tray compositions react equally for a considerable long time and Eqn. (16.20) can be written as:

$$\tau_x \frac{d\delta x_i}{dt} = (\bar{x}_{i+1} - \bar{x}_i)\left[\frac{\delta L_{i+1}}{\bar{L}} - \frac{\delta V_{i-1}}{\bar{V}}\right] \qquad (16.82)$$

The initial response is linear and correlates closely with the local gradient $(\bar{x}_{i+1} - \bar{x}_i)$. The initial part of the concentration response on tray j depends on the *relative* changes in the local vapor and liquid flows. For the response in the vapor flow when $\delta V = \delta V_i$ and $\delta L_{n+1} = 0$:

$$\begin{aligned} \frac{\delta x_i}{\delta V} &\approx \frac{(\bar{x}_{i+1} - \bar{x}_i)}{s\tau_x} \cdot \left[-\frac{1}{\bar{V}} + \frac{1}{\bar{L}} \cdot \frac{\delta L_{i+1}}{\delta V}\right] \\ &\approx -\frac{\lambda^* (\bar{x}_{i+1} - \bar{x}_i)}{s\tau_x V} \cdot \left[\frac{(1-\lambda^*)}{\lambda^*} + \frac{1}{(s\tau_L + 1)^{n-i}}\right] \quad when\ \lambda^* = \lambda \cdot \frac{V}{L} \end{aligned} \qquad (16.83)$$

Evidently, the behaviour is comparable to that for the bottom hold-up. For trays in the stripping section $V < L$, thus $\lambda^* < \lambda$. The opposite holds for trays in the rectifying section.

The long-term response results from the relatively slow column section settling time for a new composition distribution pertaining to the new stationary flows. This can be described by the left-hand side of Eqn. (16.20):

$$-\delta x_{i-1} + \left((1 + \Sigma_i)\, \delta x_i + \tau_x \frac{d\delta x_i}{dt}\right) - \Sigma_i \delta x_{i+1} = 0 \qquad (16.84)$$

The settling time is determined by all tray composition interactions in the column. When the flow dynamic effects are neglected, the overall behaviour becomes first order (Rademaker *et al.* 1975 pp. 140), which can be described by the dominant time constant:

$$\tau_{x,column}\frac{d\delta x_i}{dt}+x_i=\mu_{x,i}\left[\frac{\delta L_{i+1}}{\overline{L}}-\frac{\delta V_{i-1}}{\overline{V}}\right] \tag{16.85}$$

In fact, two different methods can be distinguished to estimate the dominant time constant from static design values: the "change of inventory" time and the retention time (Betlem, 2000). The "*change of inventory*" time corresponds to the time required to change the total column hold-up of each component from an initial to a final composition distribution. This global method uses the observed result of a composition distribution change in the column, but does not explain why this change, in this extent, occurs. The idea has been introduced by Moczek *et al.* (1963):

$$\tau_{x,column}=\frac{\sum_{i=1}^{n+1}\left[(M_ix_i)_{final}-(M_ix_i)_{initial}\right]}{D\left(x_{n+1,final}-x_{n+1,initial}\right)+B\left(x_{0,final}-x_{0,initial}\right)} \tag{16.86}$$

The *retention time* corresponds to the average delay time of a response. In this approach, all mutual tray interactions are considered. The time constant is calculated from the mean column-stripping factor that determines the internal column composition gain. This detailed method starts from the tray driving forces causing the composition distribution change. The analytical solution is found by frequency or Laplace transformation. Armstrong and Wood (1961), and Edwards and Jassim (1977) determined experimentally the transient response for reflux and feed disturbances respectively. The responses appeared to be almost entirely dominated by the major time constant and the agreement between theory and experiment was good.

$$\tau_{x,column}=\tau_x\frac{n}{\left(\overline{\Sigma}-1\right)\left(\overline{\Sigma}^{n+1}-1\right)}\cdot\left[\overline{\Sigma}^{n+1}+1-\frac{2\overline{\Sigma}\left(\overline{\Sigma}^{n}-1\right)}{n\left(\overline{\Sigma}-1\right)}\right] \tag{16.87}$$

The approximation assumes that the average column operating values are a good approximation of the static column behaviour. Notice that if $\overline{\boldsymbol{\Sigma}}$ approaches 1 and the number of trays is large, then the time constant is approximately proportional to the square of the trays number times the tray residence time:

$$\tau_{x,column}\cong\tau_x n^2 \tag{16.88}$$

Notation

A_{active}	plate area with active mass transfer, m^2
A_{column}	cross-sectional area column, m^2
A_{dc}	downcomer area, m^2
A_j	constant in Antoine equation concerning component j, —
B	bottom quantity, mol
B_j	constant in Antoine equation concerning component j, °C
b	exponent (fitting parameter) in weir equation (16.46), —
C^*_{dry}	preliminary C_{dry}, s^2/m^5
C_j	constant in Antoine equation concerning component j, °C
$C_{M,i}$	heat capacity of tray material of tray i, joule/°C
C_{ow}	coefficient (fitting pararameter) weir equation (16.46), s^b.kg^{1-b}

C^*_{ow}	preliminary C_{ow}
$C_{P,i}$	effective vapor capacity of tray i, Eqn. (16.27), mol/bar
c_p	molar heat capacity of liquid, J.C^{-1}mol^{-1}
$C_{x,i}$	effective composition capacity of tray i, Eqn. (16.24), mol
C_α	coefficient (fitting parameter) in equation (16.45), s.kg$^{-1/2}$.m$^{-3/2}$
D	distillate quantity, mol
F	feed quantity [mol] or F-factor: $v_v\rho^{1/2}$, (kg/m)$^{1/2}$.s^{-1}
F_{max}	F-factor (F) at maximum vapor load, (kg/m)$^{1/2}$.s^{-1}
g	gravitational constant, 9.8, m/s^2
H	enthalpy, J/mol
H_L	enthalpy of liquid, J/mol
$H_{L,i}$	liquid enthalpy of tray i, J/mol
H_V	enthalpy of vapor, J/mol
$H_{V,i}$	vapor enthalpy of tray i, J/mol
H_{vap}	heat (enthalpy) of evaporation, J/mol
$H_{vap,i}$	heat of evaporation at tray i, J/mol
h_{cl}	clear liquid height = net (non-aerated) height of liquid, m
h_{dry}	clear liquid height caused by vapor flow pressure drop over dry tray, m
h_{loss}	clear liquid height caused by pressure drop over apron, m
h_{ow}	clear liquid height over weir, m
h_{weir}	height of weir; aerated liquid height below weir, m
L	liquid flow, mol/s
$\overline{L}$	mean liquid flow over tray section, mol/s
L_{cond}	liquid flow from condenser to top accumulator, mol/s
L_D	distillate product liquid flow, mol/s
L_F	feed flow (in case of continuous distillation), mol/s
L_i	liquid flow from tray i, mol/s
L_{n+1}	reflux flow, mol/s
M	molar mass, mol
$\overline{M}$	mean molar liquid mass over tray section, mol/s
M_0	bottom molar liquid mass, mol
M_{cl}	clear liquid mass on tray, mol
M_{dc}	molar liquid mass in downcomer, mol
M_i	mol mass at tray i, mol
$M_{L,i}$	molar liquid tray mass of tray i, mol
M_{n+1}	top accumulator molar mass, mol
$M_{V,i}$	molar vapor tray mass of tray i, mol
N_m	molar weight, kg/mol
$N_{m,L}$	liquid molar weight, kg/mol
$N_{m,V}$	vapor molar weight, kg/mol
$\overline{N}_m$	mean molar weight, kg/mol
$N_{m,heavy}$	molar weight less volatile component, kg/mol
$N_{m,light}$	molar weight most volatile component, kg/mol
n	number of trays, —
P	pressure, bar
P_i	pressure of tray i, bar
P_j	partial pressure of component j, bar
P_j^0	pure vapor pressure of component j, bar
Q_{loss}	heat loss to surroundings, watts
$Q_{loss,i}$	heat loss to surroundings of tray i, watts

R	reflux ratio — or gas constant, 8.31, joule.mol^{-1}.°C^{-1}
R^*	reflux fraction L/V, —
R_v	change in tray pressure drop with vapor flow, Eqn (16.64), bar.s.mol^{-1}
T	temperature, C
T_i	temperature of tray i, C
t	time [s] or, min
V	vapor flow, mol/s
$\overline{V}$	mean vapor flow over tray section, mol/s
V_0	bottom vapor flow, mol/s
V_i	vapor flow from tray i, mol/s
v_V	superficial vapor velocity, m/s
$W_{L,i}$	liquid volume tray i, m^3
W_{seal}	(liquid) volume downcomer seal, m^3
x	mol fraction light component in liquid phase, —
x_A	mol fraction component A in liquid phase, —
x_B	mol-fraction component B in liquid phase, —
x_B	bottom purity, —
x_D	distillate purity, —
x_F	feed purity, —
x_i	mol fraction in liquid phase at tray i, —
x_j	mol fraction in liquid phase of component j, —
y	mol fraction light component in vapor phase, —
y_i	mol fraction in vapor phase at tray i, —
y_j	mol fraction in vapor phase of component j, —
y^e	vapor mol fraction at equilibrium, —
y_i^e	y^e at tray i, —
y_j^e	vapor mol fraction at equilibrium for component j, —

Greek symbols

α	relative volatility, —
α_j	relative volatility of component j, —
α	partial liquid hold-up (froth density), —
β	aeration factor, —
β_j	activity coefficient component j, —
ΔP	pressure drop, bar
ΔP_{dry}	tray pressure drop when tray contains no liquid, bar
ΔP_i	tray total pressure drop, bar
ε_V	mean Murphree vapor tray efficiency, —
$\varepsilon_{V,j}$	ε_V of component j, —
κ	change of tray molar mass due to composition change $(\partial M_t/\partial x)_{P,L,V}$, mol
$\Lambda_{j1,j2}$	interaction coefficient between two components (Wilson equation), —
λ	change of tray liquid flow due vapor changes $(\partial L/\partial V)_{P,x,M} = -\tau_V/\tau_L$, —
ξ_{dry}	vapor resistance coefficient of dry column, —
ξ^*_{dry}	ξ_{dry} related to hole area, —
π	change in tray pressure drop with liquid flow, Eqn. (16.64), bar.s.mol^{-1}
ρ	mass density, kg/m^3
ρ_L	liquid mass density, kg/m^3
$\rho_{L,i}$	liquid mass density of tray i, kg/m^3
ρ_m	molar density, mol/m^3

$\rho_{m,L}$ liquid molar density, mol/m^3
$\rho_{m,V}$ vapor molar density, mol/m^3
ρ_V vapor mass density, kg/m^3
Σ stripping factor: ratio slope equilibrium line to slope operating line, —
Σ_i stripping factor at tray i, —
τ_L hydraulic liquid time constant $(\partial M_i/\partial L)_{P,x,V}$, s
τ_V hydraulic vapor time constant $(\partial M_i/\partial V)_{P,x,L}$, s
τ_x residence time constant $(M_t/L)_{P,x,V}$, s] = time constant of composition responses

References

Armstrong,W.D. and Wood, R.M. (1961) The dynamic response of a distillation column to changes in the reflux and vapour flow rates. *Transactions of the Institution of Chemical Engineers*, **39**, 65–90.

Betlem, B.H.L. (1998) Influence of tray hydraulics on tray column dynamics. *Chemical Engineering Science*, **53**, 3991–4003.

Betlem, B.H.L. (2000) Batch distillation column low-order models for quality control. *Chemical Engineering Science*, 55, 3187–94.

Edwards, J.B. and Jassim, H.J. (1977) An analytical study of the dynamics of binary distillation columns. *Transactions of the Institution of Chemical Engineers*, **55**, 17–28.

Moczek,J.S., Otto, R.E. and Williams, T.J. (1963) Approximation models for the dynamic responses of large distillation columns. Proceedings second IFAC world conference, Basel. pp. 238–46.

Rademaker, O, Rijnsdorp, J.E. and Maarleveld, A. (1975) *Dynamics and Control of Continuous Distillation Columns*, Elsevier, Amsterdam.

Stichlmair, J., (1878) *Grundlagen der Dimensionierung des gas/flüssigkeit-kontakt apparatus bodenkolonne*, Verlag Chemie, Weinheim

17 Dynamic Analysis of Fermentation Reactors

In this chapter the dynamics of fermentation reactors for the production of penicillin will be discussed. Since these reactors are usually operated in a fed-batch mode, this mode will be discussed in addition to a continuous operation mode. For the continuous reactor, the non-linear dynamic model will be linearized to explain the dynamic responses of the reactor concentrations to a change in feed rate.

17.1 Introduction

Continuous operation is the preferred mode of operation for many processes. For fermentation reactors, however, often a batch or fed-batch mode of operation is preferred to be able to exhaust the feed sufficiently. If continuous operation is used, feed rates that can be realized are rather low, to achieve a large residence time. This means that the typical residence times of seconds to minutes for chemical reactors cannot be realized for continuous fermentation reactors.

In this chapter fed-batch as well as continuous operation will be discussed and some of the differences will be explained.

Numerous models for penicillin production are available in the literature. Some of them are very detailed (Paul *et al.*, 1998); some are less detailed but do take aeration into account (Rodrigues and Filho, 1999; Birol *et al.*, 2002); others ignore the aeration dynamics (van Impe and Bastin, 1995; Betlem *et al.*, 2002). If aeration is sufficient and the oxygen concentration is close to saturation, the oxygen dynamics can indeed be ignored. If aeration is not sufficient, it has a relevant impact on the kinetic equations, and thus on the biomass and penicillin dynamics.

In this chapter the influence of temperature and pH on the penicillin production will be ignored, i.e. it is assumed that temperature as well as pH are constant. Birol *et al.* (2002) give detailed models for the impact of temperature and pH on the kinetic equations.

The model developed in this section can be arranged in the group of global models, since all physiological information is gathered into a single biomass term and no detailed physical information can be obtained from the model. More structured models could include effects of cell physiology on penicillin production.

The reactor is generally fed with substrate (glucose), which is used for producing new biomass and for the maintenance of the biomass. Biomass is used in the penicillin production, part of the penicillin is hydrolyzed to acid.

17.2 Kinetic Equations

There are numerous kinetic equations for substrate to biomass conversion and biomass to penicillin conversion. Some of the equations will be discussed in this section.

Van Impe and Bastin (1995) describe the substrate to biomass conversion by the Contois equation (biomass growth rate μ):

$$\mu = \mu_{c1} \frac{c_s}{K_x c_x + c_s} \tag{17.1}$$

Process Dynamics and Control: Modeling for Control and Prediction. Brian Roffel and Ben Betlem.

in which μ_{c1} is the maximum specific growth rate for Contois kinetics (h^{-1}), K_x the Contois saturation constant for substrate limitation of biomass production (g/g), c_s the substrate concentration (g/l) and c_x the biomass concentration (g/l).

Sometimes, authors propose the Monod equation for substrate to biomass conversion, which is merely a simplification of the Contois equation:

$$\mu = \mu_m \frac{c_s}{K_s + c_s} \tag{17.2}$$

with μ_m the maximum specific growth rate for Monod kinetics (h^{-1}), and K_s the Monod saturation constant substrate limitation of biomass production (g/l).

Birol *et al.* (2002) suggest also an impact of the oxygen concentration on the substrate to biomass conversion:

$$\mu = \mu_{c2} \frac{c_s}{K_x c_x + c_s} \frac{c_o}{K_{ox} c_x + c_o} \tag{17.3}$$

where μ_{c2} is the maximum specific growth rate for Contois kinetics (h^{-1}), K_{ox} is the oxygen limitation constant (g/g) and c_o is the oxygen concentration in the reactor (g/l).

For the substrate to penicillin conversion, van Impe and Bastin (1995) suggest the Haldane equation (penicillin specific production rate π):

$$\pi = \pi_{m1} \frac{c_s}{K_p + c_s + c_s^2 / K_i} \tag{17.4}$$

in which π_{m1} is the specific production constant (h^{-1}), K_p the saturation constant for substrate limitation of product formation (g/l) and K_i the substrate inhibition constant for product formation (g/l). Figure 1 shows the different substrate conversion paths.

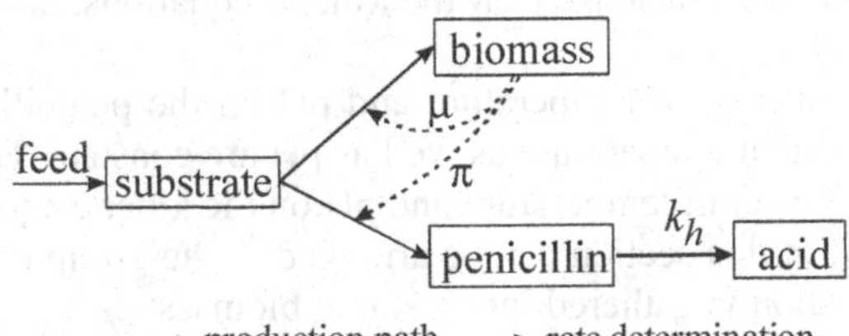

Fig. 17.1. Conversion paths for substrate.

If the penicillin hydrolysis is ignored for the moment, the maximum achievable penicillin production rate can be determined by setting $d\pi / dc_s = 0$, which results in:

$$c_{s,optimal} = \sqrt{K_p K_i} \tag{17.5}$$

Birol *et al.* (2002) modified Eqn. (17.4) by including an oxygen dependency:

$$\pi = \pi_{m2} \frac{c_s}{K_p + c_s + c_s^2 / K_i} \frac{c_o^3}{K_{op} c_x + c_o^3} \tag{17.6}$$

where K_{op} is the oxygen limitation constant (g/g), c_o is the oxygen concentration (g/l) and π_{m2} the specific production rate constant (h^{-1}).

The specific substrate consumption rate σ can be described by the following equation (van Impe and Bastin, 1995; Birol *et al.*, 2002):

$$\sigma = \frac{\mu}{Y_{XS}} + \frac{\pi}{Y_{PS}} + m_x \tag{17.7}$$

with Y_{XS} the biomass on substrate yield coefficient (g/g), Y_{PS} the penicillin on substrate yield coefficient (g/g) and m_x the overall specific maintenance demand (h^{-1}).

Birol (2002) describes the specific oxygen consumption by the following equation:

$$\phi = \frac{\mu}{Y_{XO}} + \frac{\pi}{Y_{PO}} + m_o \tag{17.8}$$

in which Y_{XO} the biomass on oxygen yield coefficient (g/g), Y_{PO} the penicillin on oxygen yield coefficient (g/g) and m_o the maintenance coefficient on oxygen (h^{-1}).

The rate of penicillin hydrolysis is assumed to be first order in the penicillin concentration:

$$r_p = -k_h c_p \tag{17.9}$$

in which k_h the penicillin hydrolysis rate constant (h^{-1}).

17.3 Reactor Models

First we will describe the model of a fed-batch reactor. The purpose of this model is to describe the penicillin production as a function of the feed rate and substrate feed concentration. The reactor is usually operated in such a way that the first period is a batch phase or growing period during which the feed is zero. The second phase is the production phase, during which a specific feeding strategy is used. In this case, the feed is increase to an initial value and ramped subsequently to a final value.

Assuming that no density changes take place, the volumetric flow balance can be written as:

$$\frac{dV}{dt} = F_{in} \tag{17.10}$$

in which V is the reactor volume (l) and F the reactor feed (l/h). There is some addition of acid or caustic, but these flows are usually small compared with the feed rate. There is also some water production and entrainment of water as a result of aeration; however, these flows will also be ignored.

The component balance for the biomass can be written as:

$$\frac{d(Vc_x)}{dt} = \mu V c_x \tag{17.11}$$

in which the right-hand side term is the growth of biomass. Equation (17.10) can be rewritten by combining it with Eqn. (17.11) to:

$$\frac{dc_x}{dt} = \left(\mu - \frac{F_{in}}{V}\right) c_x \tag{17.12}$$

Like the component balance for the biomass, the component balance for the substrate can be written as:

$$\frac{dc_s}{dt} = (c_{sin} - c_s)\frac{F_{in}}{V} - \sigma c_x \tag{17.13}$$

in which c_{sin} the substrate concentration in the feed (g/l).

The component balance for the penicillin production can be given by:

$$\frac{dc_p}{dt} = -\left(\frac{F_{in}}{V} + k_h\right)c_p + \pi c_x \tag{17.14}$$

If aeration is taken into account, the oxygen balance can be given by:

$$\frac{dc_o}{dt} = -\phi c_x + K_{La}(c_o^* - c_o) - \frac{F_{in}}{V}c_o \tag{17.15}$$

in which c_o^* is the dissolved oxygen concentration at saturation (g/l). The second right-hand side term is the overall mass transfer of oxygen to the reactor contents, the mass transfer coefficient K_{La} is usually a function of the speed of rotation of the mixer and other factors that affect the mass transfer.

The model for a continuously operated reactor with inlet flow F_{in} and outlet flow F_{out} is similar to the model for a fed-batch reactor. In this case, the volumetric flow balance can be written as:

$$\frac{dV}{dt} = F_{in} - F_{out} \tag{17.16}$$

The component balance for the biomass, Eqn. (17.11) becomes in this case:

$$\frac{d(Vc_x)}{dt} = \mu V c_x - F_{out} c_x \tag{17.17}$$

Equation (17.17) can be combined with Eqn. (17.16) to give equation (17.12) again. Similarly, Eqns. (17.13) to (17.15) also hold for the continuous reactor.

17.4 Dynamics of the Fed-batch Reactor

The non-linear model for the fed-batch operation was simulated in Simulink and its behavior studied. Some results will be given in this section. For the substrate to biomass conversion, Eqn. (17.1) is used, for substrate to penicillin conversion, Eqn. (17.4) is used. Equations (17.7) and (17.8) complete the kinetic model. Table 17.1 shows the model parameters that are used. Note that the dynamic response of the process variables is dependent on the parameter values.

Table 17.1. Simulation parameters for fed-batch reactor model.

parameter	value	parameter	value	parameter	initial value
μ_{cl}	0.07 h^{-1}	Y_{PO}	0.20 g/g	V	7.5 l
K_x	0.10 g/g	m_o	0.467 h^{-1}	c_x	1.4 g/l
π_m	0.005 h^{-1}	m_x	0.014 h^{-1}	c_s	43.0 g/l
K_p	0.10 g/l	k_h	0.02 h^{-1}	c_p	0.0 g/l
K_i	0.10 g/l	K_{op}	5×10^{-4} g/g	c_o	1.0 g/l
Y_{XS}	0.45 g/g	K_{ox}	0.02 g/g	c_o^*	1.16 g/l
Y_{PS}	0.90 g/g	K_{La}	100.0 h^{-1}		
Y_{XO}	0.04 g/g				

Figure 17.2 shows the reactor feed rate as a function of time, Fig. 17.3 the corresponding response of the reactor volume.

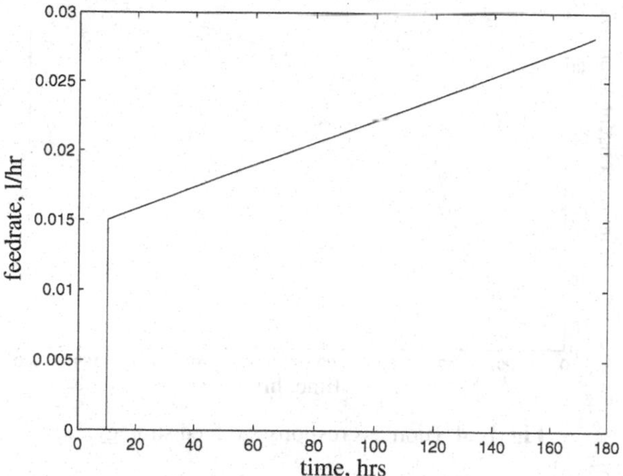

Fig. 17.2. Reactor feed strategy.

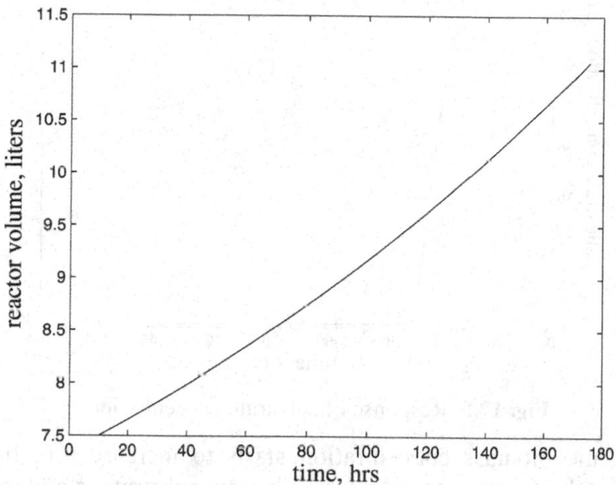

Fig. 17.3. Response of reactor volume to feed increase.

As can be seen, the reactor volume is initially constant, since no feed enters the reactor, then it starts to increase at an increasing rate due to the ramp in feed rate.

Figures 17.4 and 17.5 show the response of the biomass and substrate concentrations respectively.

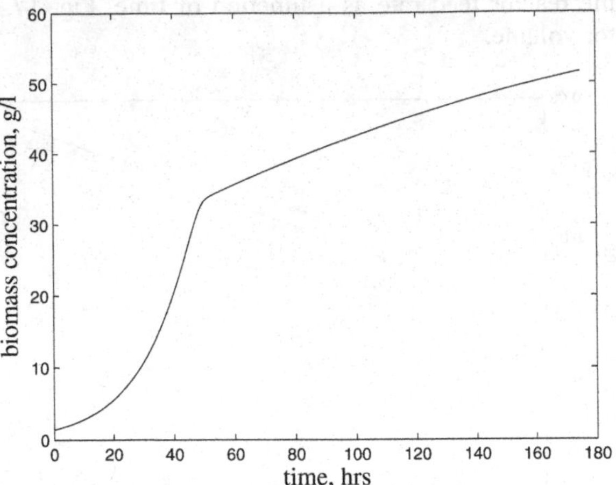

Fig. 17.4. Biomass response to feed strategy.

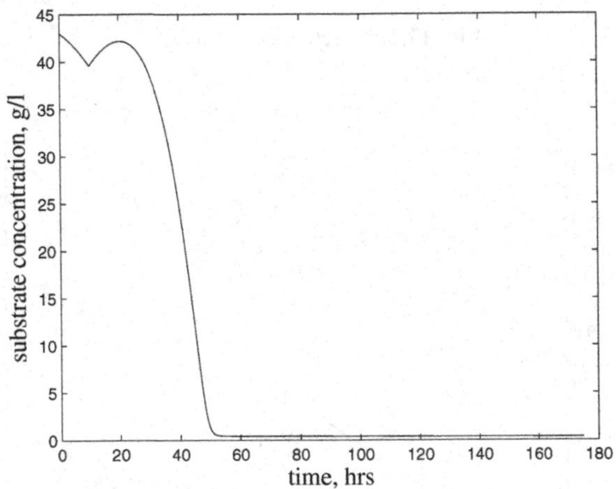

Fig. 17.5. Response of substrate concentration.

As can be seen, the biomass concentration starts to increase initially, since sufficient substrate is present in the reactor. At time $t = 12$ h, new substrate is added to the reactor and since the substrate concentration is not depleted yet, the biomass concentration continues to grow. Eventually, the growth of biomass slows down, owing to the growth of penicillin.

The substrate concentration starts to fall initially, since it is consumed by the biomass. Then, at $t = 12$ h, new substrate enters the reactor since the reactor feed is started.
As the biomass concentration continues to increase, the substrate concentration will fall. At $t = 50$ h, a situation is reached, where all incoming substrate is immediately consumed by the biomass.

Figure 17.6 shows the penicillin concentration in response to the feed strategy. It can be seen that the penicillin production starts when the substrate concentration has reached low levels, in the proximity of the substrate concentration for maximum production rate $c_{s,optimal} = \sqrt{K_p K_i}$.

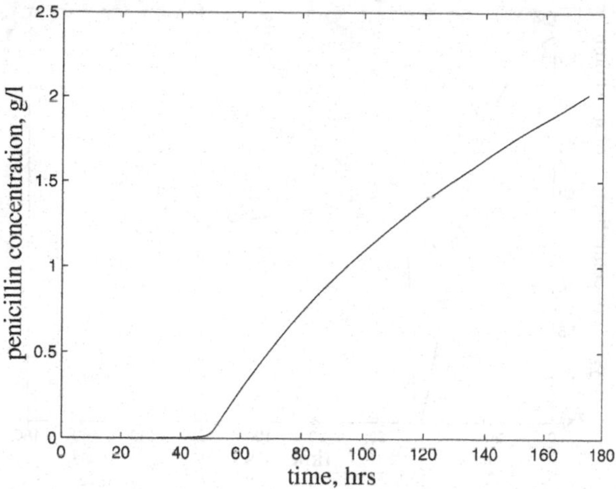

Fig. 17.6. Response of penicillin concentration for fed-batch reactor.

The reason for this behavior is that the penicillin specific production rate π is very small when the substrate concentration is high, this is shown in Fig. 17.7. Only at low level of substrate concentration is there a reasonable penicillin production. Figure 17.5 shows that the substrate concentration is high for the first 50 hours, hence hardly any penicillin will be produced during that period.

Figure 17.8 shows the response of the oxygen concentration. As can be seen, initially, the concentration falls sharply owing to the feed addition.

Many authors have studied optimal feeding strategies for fed-batch reactors (Rodrigues and Filho, 1999; Paul *et al.*, 1998; van Impe and Bastin, 1995). They do usually only include the feed rate F_{in}, however, the substrate feed concentration and initial period of batch operation should also be taken into account (Betlem *et al.*, 2002).

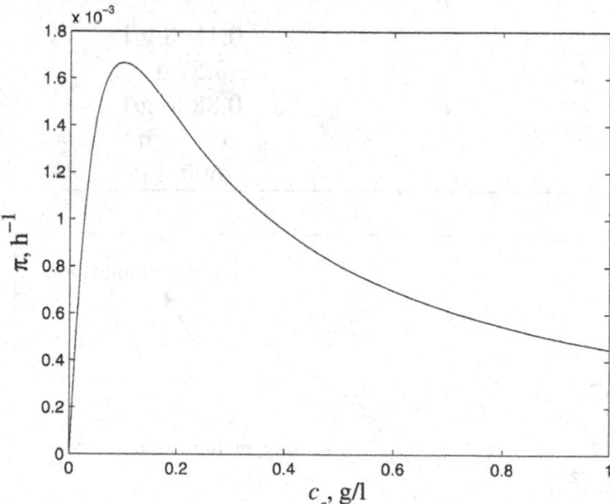

Fig. 17.7. Penicillin specific production rate π as a function of the substrate concentration.

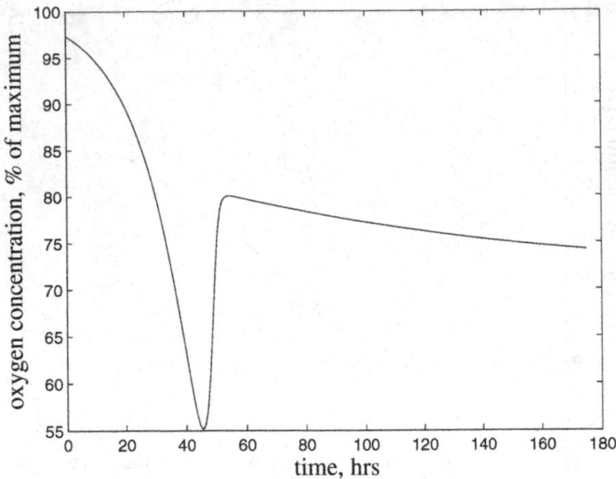

Fig. 17.8. Oxygen concentration as a function of time.

17.5 Dynamics of Ideally Mixed Fermentation Reactor

In this case, the model consists of Eqns. (17.16), (17.12)–(17.15) and the equations for the kinetics, Eqns. (17.1), (17.4), (17.7) and (17.8). The feed rate is 0.17 l/h and the initial conditions are given in Table 17.2. Figure 17.9 shows the reactor penicillin concentration for various values of the reactor substrate feed rate F_{in}, using the initial conditions from Table 17.2.

Table 17.2. Initial values for continuous fermentation reactor.

parameter	initial value
V	100 l
c_x	52.52 g/l
c_s	0.1848 g/l
c_p	3.57 g/l
c_o	0.8886 g/l
F_{in}	0.17 l/h
$c_{s,in}$	600 g/l

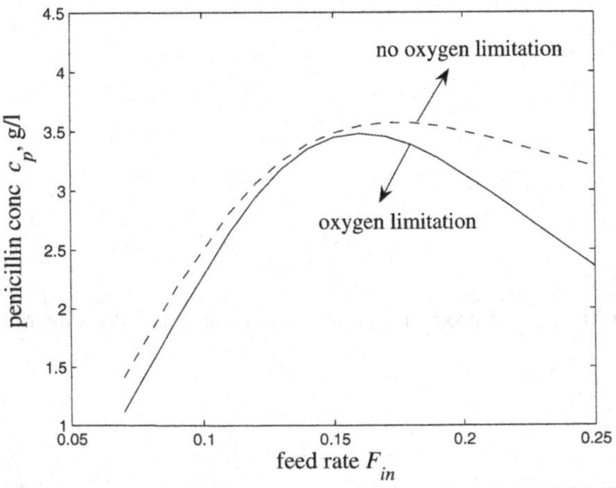

Fig. 17.9. Reactor penicillin concentration as a function of the feed rate.

Parameter values in the kinetic models have been chosen in such a way that the maximum penicillin production is achieved at approximately the same feed rate.

Starting with a steady-state feed rate of 0.17 l/h, the dynamic responses for a small change in the feed of –10% are shown in Figs. (17.10)–(17.12). Figure (17.10) shows the response of the biomass concentration, Fig. (17.11) shows the response of the substrate concentration and Fig. (17.12) the response of the penicillin concentration.

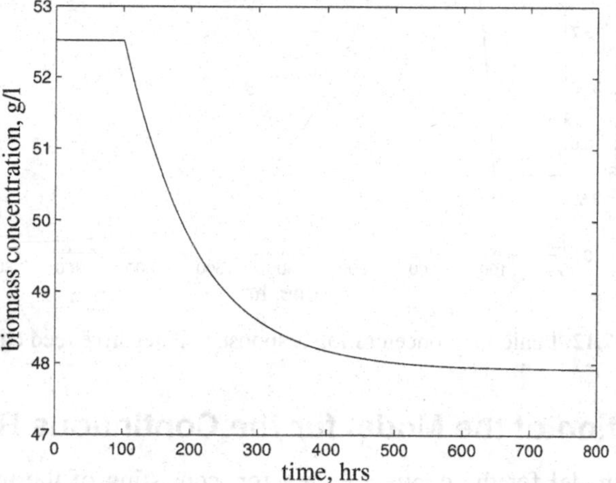

Fig. 17.10. Biomass concentration response to negative feed change.

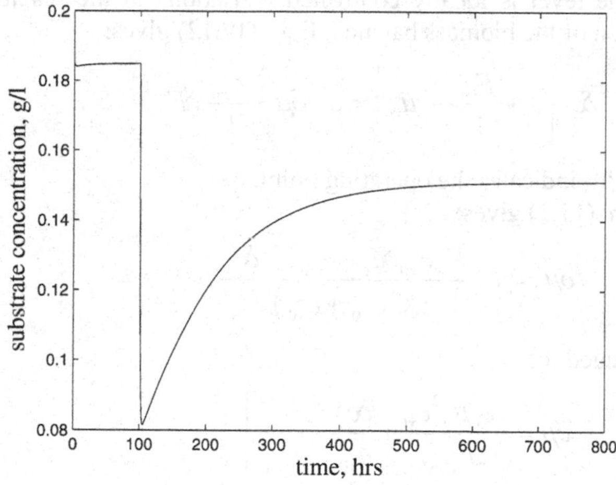

Fig. 17.11. Substrate concentration response to negative feed change.

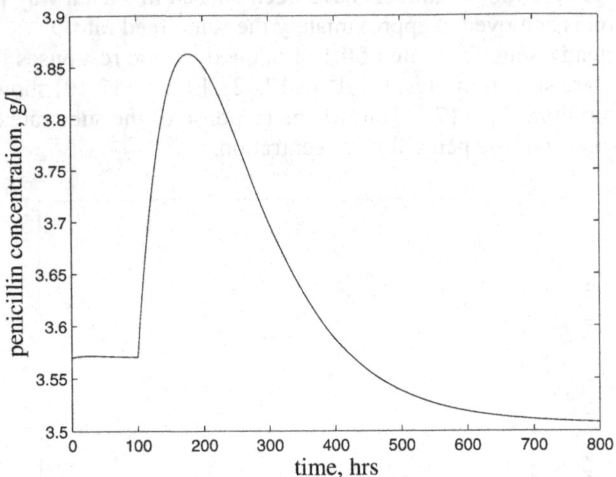

Fig. 17.12. Penicillin concentration response to a negative feed change.

17.6 Linearization of the Model for the Continuous Reactor

In this section the model for the continuous reactor, consisting of the three balances Eqns. (17.12), (17.13) and (17.14) and additional kinetic equations, will be Laplace transformed and linearized to get some insight in the responses to feed changes.

Assuming that the level is ideally controlled, variations in the reactor volume will be ignored. Linearization of the biomass balance, Eqn. (17.12) gives:

$$\delta c_x \left[s + \frac{F_{in0}}{V} - \mu_0 \right] = c_{x0} \delta\mu - \frac{c_{x0}}{V} \delta F_{in} \tag{17.18}$$

where the subscript '0' indicates the operating point.
Linearization of Eqn. (17.1) gives:

$$\delta\mu = \mu_{c1} \frac{K_x c_{x0} \delta c_s - K_x c_{s0} \delta c_x}{(K_x c_{x0} + c_{s0})^2} \tag{17.19}$$

which can be rearranged to:

$$\delta\mu = \frac{K_x \mu_0^2 c_{x0}}{\mu_{c1} c_{s0}} \left[\frac{\delta c_s}{c_{s0}} - \frac{\delta c_x}{c_{x0}} \right] \tag{17.20}$$

Substitution of Eqn. (17.20) into Eqn. (17.18) results in:

$$\delta c_x \left[s + \frac{F_{in0}}{V} - \mu_0 + \frac{K_x \mu_0^2 c_{x0}}{\mu_{c1} c_{s0}} \right] = \frac{K_x \mu_0^2 c_{x0}^2}{\mu_{c1} c_{s0}^2} \delta c_s - \frac{c_{x0}}{V} \delta F_{in} \tag{17.21}$$

Define the following time constant and gains:

$$\tau_x = 1/\left[\frac{F_{in0}}{V} - \mu_0 + \frac{K_x \mu_0^2 c_{x0}}{\mu_{c1} c_{s0}}\right]$$

$$K_1 = \frac{K_x \mu_0^2 c_{x0}^2}{\mu_{c1} c_{s0}^2} / \left[\frac{F_{in0}}{V} - \mu_0 + \frac{K_x \mu_0^2 c_{x0}}{\mu_{c1} c_{s0}}\right] \tag{17.22}$$

$$K_2 = (c_{x0}/V) / \left[\frac{F_{in0}}{V} - \mu_0 + \frac{K_x \mu_0^2 c_{x0}}{\mu_{c1} c_{s0}}\right]$$

then Eqn. (17.21) can be written as:

$$\delta c_x = \frac{K_1}{(\tau_x s + 1)} \delta c_s - \frac{K_2}{(\tau_x s + 1)} \delta F_{in} \tag{17.23}$$

Linearizing the substrate balance, Eqn. (17.13) and assuming that there are no changes in the inlet substrate concentration, results in:

$$\delta c_s \left[s + \frac{F_{in0}}{V}\right] = -c_{x0} \delta\sigma - \sigma_0 \delta c_x + \frac{c_{sin} - c_{s0}}{V} \delta F_{in} \tag{17.24}$$

Linearization of Eqn.(17.7) gives:

$$\delta\sigma = \frac{\delta\mu}{Y_{XS}} + \frac{\delta\pi}{Y_{PS}} \tag{17.25}$$

Linearization of Eqn.(17.4) gives:

$$\delta\pi = \frac{\pi_0^2}{\pi_{m1} c_{s0}^2} \left(K_p - \frac{c_{s0}^2}{K_i}\right) \delta c_s \tag{17.26}$$

Substitution of Eqns.(17.25), (17.26) and (17.19) into Eqn. (17.24) results in:

$$\delta c_s \left[s + \frac{F_{in0}}{V} + \frac{K_x \mu_0^2 c_{x0}^2}{Y_{xs} \mu_{c1} c_{s0}^2} + \frac{\pi_0^2 c_{x0}}{Y_{ps} \pi_{m1} c_{s0}^2} \left(K_p - \frac{c_{s0}^2}{K_i}\right)\right] =$$
$$\left(\frac{K_x \mu_0^2 c_{x0}}{Y_{xs} \mu_{c1} c_{s0}} - \sigma_0\right) \delta c_x + \frac{c_{sin} - c_{s0}}{V} \delta F_{in} \tag{17.27}$$

Define the following time constant and gains:

$$\tau_s = 1/\left[\frac{F_{in0}}{V} + \frac{K_x \mu_0^2 c_{x0}^2}{Y_{xs} \mu_{c1} c_{s0}^2} + \frac{\pi_0^2 c_{x0}}{Y_{ps} \pi_{m1} c_{s0}^2} \left(K_p - \frac{c_{s0}^2}{K_i}\right)\right]$$

$$K_3 = \left(\frac{K_x \mu_0^2 c_{x0}}{Y_{xs} \mu_{c1} c_{s0}} - \sigma_0\right) / \left[\frac{F_{in0}}{V} + \frac{K_x \mu_0^2 c_{x0}^2}{Y_{xs} \mu_{c1} c_{s0}^2} + \frac{\pi_0^2 c_{x0}}{Y_{ps} \pi_{m1} c_{s0}^2} \left(K_p - \frac{c_{s0}^2}{K_i}\right)\right] \tag{17.28}$$

$$K_4 = \left(\frac{c_{sin} - c_{s0}}{V}\right) / \left[\frac{F_{in0}}{V} + \frac{K_x \mu_0^2 c_{x0}^2}{Y_{xs} \mu_{c1} c_{s0}^2} + \frac{\pi_0^2 c_{x0}}{Y_{ps} \pi_{m1} c_{s0}^2} \left(K_p - \frac{c_{s0}^2}{K_i}\right)\right]$$

then Eqn. (17.27) can be written as:

$$\delta c_s = \frac{K_3}{(\tau_s s+1)}\delta c_x + \frac{K_4}{(\tau_s s+1)}\delta F_{in} \tag{17.29}$$

Note that $K_3 < 0$ as more biomass ($\delta c_x > 0$) will consume more substrate, hence the substrate concentration will decrease ($\delta c_s < 0$).
Linearization the penicillin balance, Eqn.(17.14) results in:

$$\delta c_p\left[s+\frac{F_{in0}}{V}+k_h\right] = -\frac{c_{p0}}{V}\delta F_{in} + \pi_0 \delta c_x + c_{x0}\delta\pi \tag{17.30}$$

Combination with Eqn. (17.26) results in:

$$\delta c_p\left[s+\frac{F_{in0}}{V}+k_h\right] = -\frac{c_{p0}}{V}\delta F_{in} + \pi_0 \delta c_x + \frac{\pi_0^2 c_{x0}}{\pi_{m1} c_{s0}^2}\left(K_p - \frac{c_{s0}^2}{K_i}\right)\delta c_s \tag{17.31}$$

Define the following time constant and gains:

$$\begin{aligned}
\tau_p &= 1/\left(\frac{F_{in0}}{V}+k_h\right)\\
K_5 &= \pi_0 /\left(\frac{F_{in0}}{V}+k_h\right)\\
K_6 &= \frac{\pi_0^2 c_{x0}}{\pi_{m1} c_{s0}^2}\left(K_p - \frac{c_{s0}^2}{K_i}\right)/\left(\frac{F_{in0}}{V}+k_h\right)\\
K_7 &= \left(c_{p0}/V\right)/\left(\frac{F_{in0}}{V}+k_h\right)
\end{aligned} \tag{17.32}$$

then Eqn. (17.31) can be written as:

$$\delta c_p = \frac{K_5}{(\tau_p s+1)}\delta c_x + \frac{K_6}{(\tau_p s+1)}\delta c_s - \frac{K_7}{(\tau_p s+1)}\delta F_{in} \tag{17.33}$$

As can be seen from Eqns. (17.33), (17.29) and (17.23), the response of the penicillin concentration c_p to a change in the feed rate F_{in} is a combination of a first-order response (Eqn. (17.33)), followed by the effects of second and third-order response. Equation (17.33) also shows that the immediate effect of F_{in} will be in opposite direction, which means that c_p will initially decrease when F_{in} increases. The long-term effect is determined by a combination of the process gains that were defined.

Since the process operates at a feed rate where the penicillin concentration is maximal (Fig. 17.9), any feed change will lead to a decrease in penicillin concentration. When the feed rate is decreased, the dilution effect decreases, hence the concentration will initially increase. Thus in this case an inverse response would result. If the feed rate increases, the dilution increases, leading to an initially lower penicillin concentration. Also the long-term effect is a lower concentration, hence an inverse response is not expected.

Time constants and process gains in the operating point are given in Table 17.3.

Table 17.3. Time constants and process gains for the linearized model.

parameter	value	parameter	value
τ_x	608.9 h	K_3	–0.0184
τ_s	1.16 h	K_4	6.98
τ_p	46.1 h	K_5	0.068
K_1	284.1	K_6	–7.45
K_2	319.8	K_7	1.65

It can be seen that there is only a small time constant from the inlet flow to substrate concentration (Eqn. (17.29)), hence on a feed change the substrate concentration will respond immediately, which can also be seen in Fig. 17.11. Then there is another path from the feed flow via the biomass concentration to the substrate concentration (Eqn. (17.29) via Eqn. (17.23)). This path has a negative contribution and major time constant, hence from its peak value the substrate concentration will now change slowly.

Another interesting feature that can be seen from the linearized model is, that there is only one major time constant in the system, even though there are two interacting balances, the interaction between the biomass and substrate. The response of the penicillin concentration to the feed flow is primarily determined by the first-order path from F_{in} to c_p, which has a time constant of 46.1 h. This time constant is determined primarily by the kinetic constant k_h for the hydrolysis reaction, as can be seen from Eqn. (17.31), since $V/F_{in} = 588.2$ and $1/k_h = 50.0$.

Figure 17.13 shows the response of the penicillin concentration to a +10% change in the feed rate.

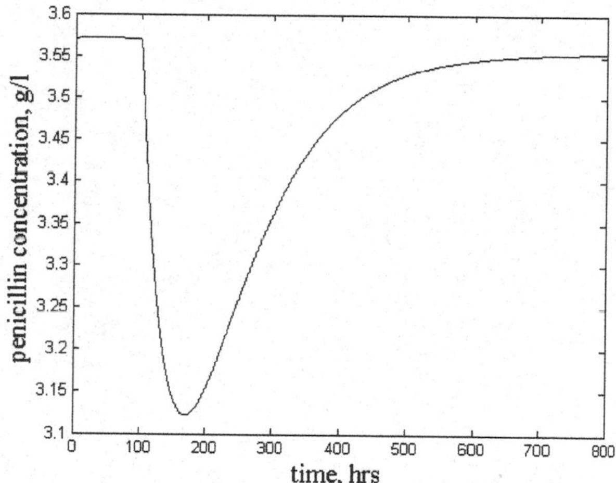

Fig. 17.13. Penicillin concentration response to a negative feed change.

It can be seen that there is indeed no inverse response as was expected. However, owing to the fact that the optimum in Fig. 17.9 is rather flat, the final penicillin concentration is only slightly lower than its starting value. This makes the feed rate not a suitable candidate for control of the penicillin concentration. Also the large initial response would have a negative impact on a control scheme if the feed rate were to be for control of the penicillin concentration.

Files referred to in this chapter:
T1703.m: calculation of model parameters in Table 17.3
F1702plot.m: plot files for Figs. 17.02 to 17.08
F1709.m: plot files for Fig. 17.09
F1710plot.m: plot files for Figs. 17.10 to 17.12
F1713plot.m: plot files for Figs. 17.13
PenFedBatch.mdl: model for fed batch production of penicillin
Pennicilinb.m: model file used by PenFedBatch.mdl
PenCSTR.mdl: model for continuous production of penicillin
Pennicillinc.m: model file used by PenCSTR.mdl
PenLinMod.mdl: linear model for continuous penicillin production

References

Betlem, B.H.L., Mulder, P. and Roffel, B. (2002) Optimal mode of operation for biomass production. *Chemical Engineering Science*, **57**, 2799–2809.

Birol, G., Undey, C. and Cinar, A. (2002) A modular simulation package for fed-batch fermentation: penicillin production. *Computers and Chemical Engineering*, **26**, 1553–65.

Paul, G.C., Syddall, M.T., Kent, C.A. and Thomas, C.R. (1998) A structured model for penicillin production on mixed substrates. *Biochemical Engineering Journal*, **2**, 11–21.

Rodrigues, J.A.D. and Filho, R.M. (1999) Production optimization with operating constraints for a fed-batch reactor with DMC predictive control. *Chemical Engineering Science*, **54**, 2745–51.

van Impe, J.F. and Bastin, G. (1995) Optimal adaptive control of fed-batch fermentation processes. *Control Engineering Practice*, **3** (7), 939–54.

18 Physiological Modeling: Glucose–Insulin Dynamics and Cardiovascular Modeling

So far, chemical processes have been discussed as well as fermentation reactors. These systems can be characterized by the fact that there are usually no discontinuities in the model equations. In physiological systems, however, there is often a threshold value that has to be succeeded before certain phenomena take place. This causes a threshold non-linearity in the model and therefore these systems can be analyzed well through simulation but theoretically they are difficult to analyze. It is not the purpose of this chapter to discuss different physiological models: the two systems that will be selected for illustration of system dynamics are the modeling of glucose–insulin dynamics and cardiovascular modeling approaches.

18.1 Introduction to Physiological Models

Modeling starts to play an increasingly important role in medicine. For insulin infusion (Parker, 2001; Andersen and Højbjerre, 2003; Bergman *et al.*, 1981), respiratory systems (Hämäläinen, 1992, 2000), heart rate fluctuations (McSharry *et al.*, 2003), blood-pressure regulation (Isaka, 1993), prediction of epileptic seizures (McSharry *et al.*, 2003) and the study of metabolic disorders, to name just a few, models have been reported.

It is not the intent of this chapter to discuss physiological systems modeling in detail, but rather to show some differences with the modeling of chemical and biological systems.

The reader who is unfamiliar with the field of physiological modeling is referred to the books of Baura (2002) and Khoo (2000). These two books give an excellent impression of physiological systems modeling and analysis.

The PHYSBE software package (Mathworks) is also interesting. It is a physiological simulation benchmark experiment using Simulink. In the software, the body's circulatory system is divided into three parts, the lungs (pulmonary circulation), the heart (coronary circulation) and everything else (systemic circulation). Simulations can easily be performed to display pressures, blood-flows, volumes, temperatures and heat flows as a function of time. It should be stressed that this is a simple representation of reality; however, it gives the reader an insight into some aspects of the circulatory system.

18.1.1. Introduction to glucose–insulin modeling

In the first part of this chapter, a simple model will be developed which describes the glucose and insulin concentrations in the blood. This will enable us to study the glucose–insulin equilibrium and determine the optimal insulin addition for different types of diabetes patients. Maintenance of this equilibrium is an example of a self-regulating system. When the glucose level is increased, insulin secretion is stimulated, consequently the insulin level increases leading to increased uptake of glucose by the tissues. When the glucose level falls, insulin secretion also decreases, and the levels of glucose and insulin return to normal values.

There are numerous articles on glucose–insulin modeling owing to its relevance for patients with diabetes mellitus. Diabetes is a chronic syndrome, characterized by insufficient insulin production in the pancreas or decreased insulin activity, resulting in increased blood

Process Dynamics and Control: Modeling for Control and Prediction. Brian Roffel and Ben Betlem.

sugar levels. There are two types: type I which usually starts at an early age, around age 4–5 or 12–14. It is an insulin-dependent diabetes, since the patient needs to supply insulin in order to control the disease. Type II usually occurs around age 65 and the use of insulin is usually not required.

Several well-known articles will be mentioned here. The first one is by Bergman, Phillips and Cobelli (1981). The model presented in this article is often referred to as the Bergman minimal model, it describes the measurement and modeling of insulin sensitivity and β-cell glucose sensitivity from the response to intravenous glucose. The second article is that by Nucci and Cobelli (2000). It gives an overview of models of subcutaneous insulin kinetics. The models reviewed are modifications of the Bergman minimal model. The Berman minimal model includes one equation to describe the insulin dynamics and one equation to describe the glucose dynamics. Since the dynamics usually do not fit experimental data, often an additional separate insulin compartment is introduced to describe the "delayed" effect of insulin on glucose absorption.

A more detailed model is given by Topp *et al.* (2000). In addition to the insulin and glucose dynamics it includes an additional balance for the pancreatic β-cell mass dynamics.

Other models, such as neuro-fuzzy models (Dazzi, 2001) and models using artificial intelligence techniques (Liszka-Hackzell, 1999), have also been reported.

The Automated Insulin Dosage Advisor (AIDA) is combining a rule-based system and a non-linear dynamic model of the glucose–insulin system (Lehman, 1994).

In the next section a non-linear model for glucose–insulin dynamics will be discussed.

18.2 Modeling of Glucose and Insulin Levels

To model the glucose and insulin dynamics, an understanding of the processes that take place is essential. Lehman (1994) and Khoo (2000) suggest interactions between glucose and insulin as shown in Fig. 18.1.

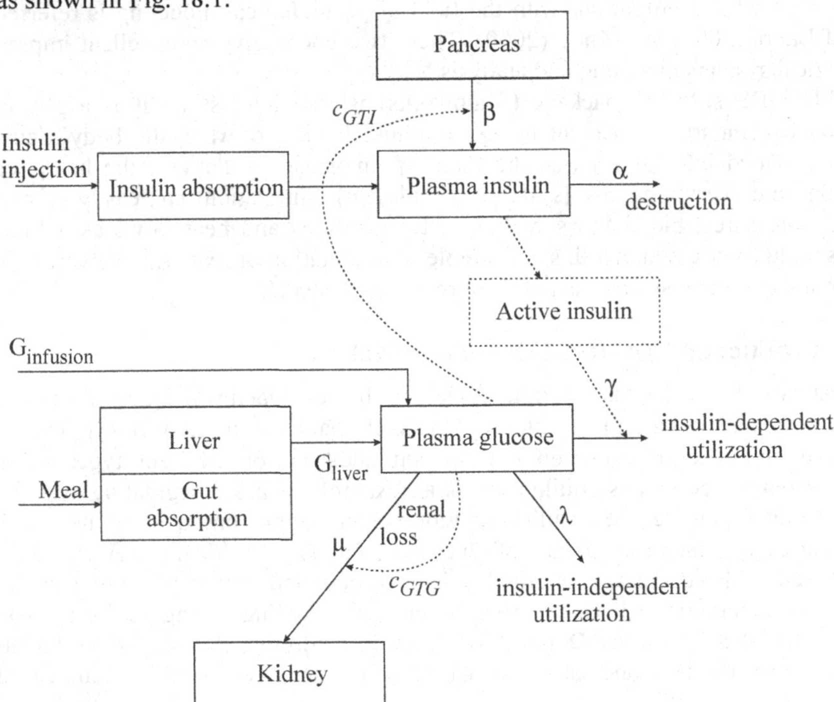

Fig. 18.1. Model of the glucose–insulin interaction proposed by Lehmann (1994) and Khoo (2000).

Active or remote insulin is used by various authors to indicate that the insulin does not immediately stimulate the glucose production by the liver and glucose utilization by the tissue. This will not be considered at this point but discussed after initial model development. In addition, many authors assume that glucose and insulin dynamics can be modeled by assuming a one compartment model, this is equivalent to the assumption that the glucose and insulin concentrations are the same everywhere in the blood volume.

The component balance for the insulin concentration (mU/ml) in the blood plasma can then be written as:

$$V\frac{dc_I}{dt} = -\alpha V c_I + \beta V(c_G - c_{GTI}) \tag{18.1}$$

in which the left-hand side term represents the accumulation of insulin, the first right-hand side term is the insulin destruction through a reaction involving the insulinase enzyme, α is a rate constant (h^{-1}), the second right-hand side term is the insulin produced by the pancreas with rate constant β (mU/mg.h) and the production is dependent on the plasma glucose level (mg/ml). If the glucose concentration falls below a certain level, c_{GTI}, the threshold glucose concentration, insulin production by the pancreas ceases and this term becomes zero. V is the volume of blood in a person, it is assumed to be 5500 ml in this case. Typical values for α and β and the threshold level are:

Table 18.1. Typical values for component balance.

parameter	value
α	1.38 h^{-1}
β	0.26 mU/mg.h
c_{GTI}	0.5 mg/ml

It is evident that this equation contains a threshold non-linearity, the second right-hand side term.

The component balance that describes the glucose concentration in the blood plasma can be written as:

$$V\frac{dc_G}{dt} = -\lambda V c_G - \gamma V c_I c_G - \mu V(c_G - c_{GTG}) + G_{liver} + G_{infusion} \tag{18.2}$$

The left-hand side term represents the accumulation of glucose in the blood. The first right-hand side term represents the tissue utilization of glucose, λ is the rate constant (h^{-1}). The mechanism for glucose utilization by the tissue is not the same for all cells, there are some cells in which insulin acts as a stimulus for glucose utilization. This is represented by the second right-hand side term of Eqn. (18.2) in which γ is the rate constant (ml/mU.h).

The third term of Eqn. (18.2) represents the glucose removal by the kidneys, the so-called renal loss, this term is proportional to the glucose concentration above a threshold value c_{GTG}. For glucose concentrations below this threshold value, this term is zero.

The fourth right-hand side term of Eqn. (18.2) is the normal glucose addition from the liver or gastrointestinal tract. It is assumed that this flow is constant. The last term of Eqn. (18.2) represents a glucose injection, if present.

Typical values for the rate constants and loads are:

Table 18.2. Typical values of the rate constants and loads.

parameter	value
λ	0.45 h^{-1}
γ	22.27 ml/mU.h
μ	1.31 h^{-1}
c_{GTG}	2.2 mg/ml
G_{liver}	6000 mg/h
$G_{infusion}$	80000 mg/h

It should be noted that the component balances are not any different than in the case of chemical or biological systems, one has to know the mechanisms of production and disappearance. In this case, however, we have to deal with two balances that have terms that depend on threshold levels. Consequently, the system description becomes non-linear. Also, the second right-hand side term in Eqn. (18.2) introduces a non-linearity. This actually makes the system bi-linear.

18.3 Steady-state Analysis

The steady-state situation can be modeled by setting the left-hand side terms in Eqns. (18.1) and (18.2) equal to zero. Assuming $G_{infusion}$ is zero, the steady-state equations can be written as:

$$c_I = \frac{\beta}{\alpha}.max(0, c_G - c_{GTI}) \tag{18.3}$$

and

$$c_I = \frac{-\lambda c_G - \mu.max(c_G - c_{GTG}) + G_{liver}/V}{\gamma c_G} \tag{18.4}$$

The insulin concentration and glucose concentrations are plotted in Fig. 18.2. The straight line represents Eqn. (18.3), the other curve Eqn. (18.4), the intersection of the two lines yields the steady-state situation, which is $c_I = 0.0457$, $c_G = 0.7427$.

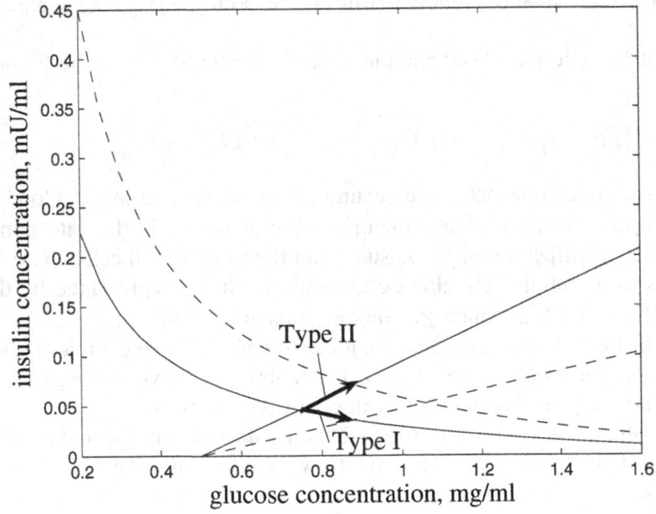

Fig. 18.2. Steady-state situation for a normal person.

For a patient with type I diabetes, the pancreas fails to produce sufficient insulin. This could be modeled by decreasing the value of the rate constant β. A lower value of β decreases the slope of the straight line, resulting in a decreased value of the insulin concentration and increased value of the glucose concentration. If β is reduced by 50%, the new steady-state values for the insulin and glucose concentrations are: $c_I = 0.0356$, $c_G = 0.8779$. As can be seen from Fig. 18.2, the insulin concentration is too low and the glucose level too high, hence insulin treatment would be useful.

For a patient with type II diabetes, there is a reduction in the ability of the insulin to stimulate the glucose uptake by the tissues. This can be simulated by decreasing the value of γ. Decreasing γ by 50% results in the steady-state values of $c_I = 0.0712$, $c_G = 0.8779$. In Fig. 18.2 this can be explained as follows. Equation (18.3) is not affected by changing the value of γ. The straight line does therefore not shift. However, the curved line is affected by a lower value of γ, in fact, it moves upward. This results in a higher value of the insulin concentration as well as the glucose concentration.

18.4 Dynamic Analysis

It is interesting to see how the glucose and insulin concentrations react to an infusion with glucose. It demonstrates the ability of the system of a healthy person to maintain desired glucose and insulin levels without external intervention. A disturbance is given of 80,000 mg/h glucose during 12 minutes.

Figure 18.3 shows the glucose response and Fig. 18.4 the insulin response.

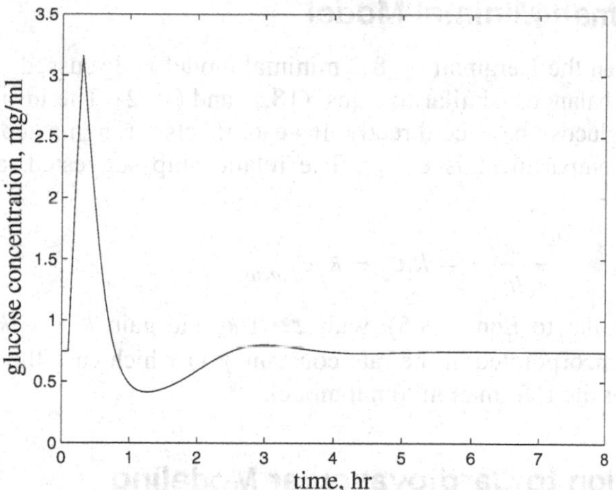

Fig. 18.3. Glucose transient after a pulse disturbance in the glucose supply.

As can be seen both concentrations fall below the steady-state values, before returning to normal. This is also observed in a real situation. From Eqn. (18.2) it can be seen that the insulin impacts on the glucose level.

In the simulation a first-order transfer function has been added between the plasma insulin concentration and the so-called active insulin concentration. The active insulin concentration affects the glucose balance. The relationship between insulin concentration and active insulin concentration is therefore:

$$\frac{\delta c_{I,active}}{\delta c_I} = \frac{1}{\tau s + 1} \tag{18.5}$$

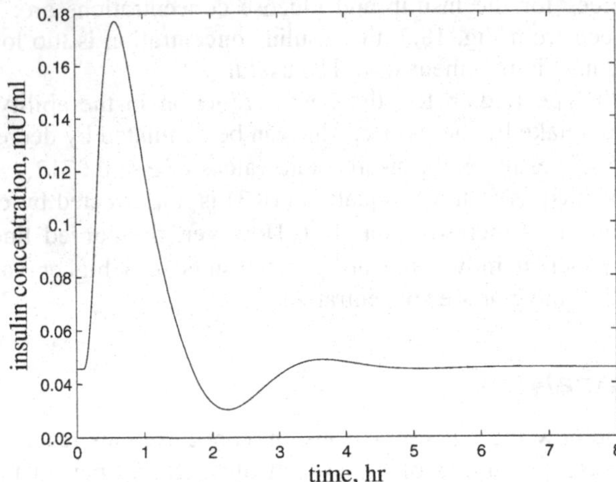

Fig. 18.4. Insulin transient after a pulse disturbance in the glucose supply.

18.5 The Bergman Minimal Model

In the literature, often the Bergman (1981) minimal model is discussed. This starts with the glucose and insulin balances, similar to Eqns. (18.1) and (18.2). The insulin concentration c_I does not affect the glucose balance directly. Instead, it acts through a separate compartment, the output of the compartment is $c_{I,active}$. The relationship suggested, amongst others, by Khoo (2000) is:

$$\frac{dc_{I,active}}{dt} = k_1 c_I - k_2 c_{I,active} \tag{18.6}$$

This equation is similar to Eqn. (18.5), with $\tau = 1/k_2$ and gain $K = k_1/k_2$. The gain K can, however, easily be incorporated in the rate constant γ, in which case the model discussed in this chapter becomes the Bergman minimal model.

18.6 Introduction to Cardiovascular Modeling

The human cardiovascular system is very complex and the possibilities for model validation are limited, even though blood flow and pressures at many locations in the body can be measured. The purpose of a model is usually to study the blood circulation, in particular flow and pressure effects.

In cardiovascular modeling often flow elements are used and it is assumed that the heart pulsates with a particular frequency. This pulsation is called the ‘driving function’ (Sacca, 2003).

In another approach in which heart rate variability, variations in cardiac cycle and arterial blood pressure are modeled, also flow elements are used, in addition, a baroreflex model is introduced, affecting the heart rate and the stroke volume of the heart. By introducing a time delay between the baroreflex input and output, an unstable system is created that continues to oscillate and explains the heart rate variability and variations in the cardiac cycle. Both modeling approaches will be briefly discussed in this chapter.

18.7 Simple Model using Aorta Compliance and Peripheral Resistance

The simplest possible model of the cardiovascular system takes into account the compliance of the aorta and the arterial system. The resistance that the blood flow encounters is assumed to be represented by one so-called peripheral resistance. This situation is shown in Fig. 18.5.

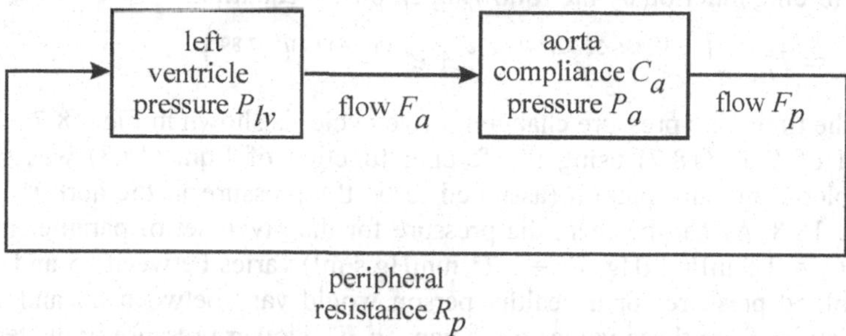

Fig. 18.5. Simple model of circulatory system.

It is assumed that the heart delivers a flow F_a with a pressure P_{lv}.

This situation can easily be represented by a flow analogy as shown in Fig. 18.6. Here a pump delivers a pressure P_{lv}, the liquid flows through a resistance R_a with flow F_a and subsequently through a resistance R_p with flow F_p. Attached is a piston with capacitance C_a which is equal to A^2/k, A being the piston area and k the spring constant. The following equations hold for this analog:

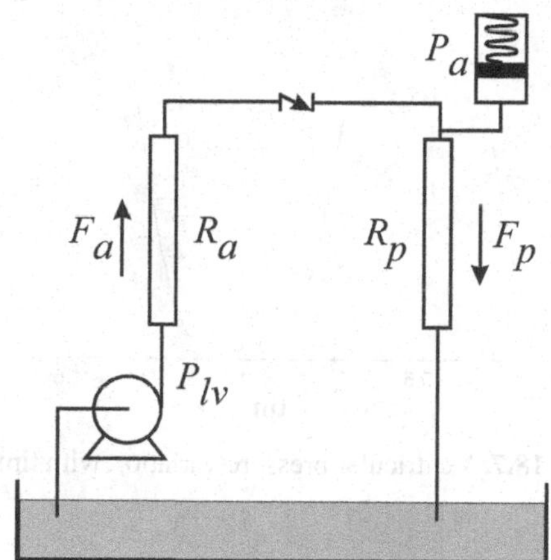

Fig. 18.6. Flow analogy of circulatory system.

$$
\begin{aligned}
P_{lv} - P_a &= R_a F_a \\
C_a \frac{dP_a}{dt} &= F_a - F_p \\
P_a &= R_p F_p
\end{aligned}
\tag{18.7}
$$

in which P_{lv} = ventricular pressure (mmHg), P_a is the pressure in the aorta (mmHg), R_a = resistance for flow in the aorta and through the heart valve (mmHg.s/ml), F_a =blood flow into the aorta (ml/s), C_a = aorta compliance (ml/mmHg), F_p = blood flow through the peripheral system (ml/s) and R_p = peripheral resistance (mmHg.s/ml).

It is assumed that the ventricular pressure changes P_{lv} are associated with the heart beat. This is often modeled by a 'forcing function' (Smith *et al.*, 2003; Sacca, 2003). Sacca describes the forcing function by the following empirical equation:

$$P_{lv} = P_{lv,max}\left(e^{37.83t-32.32t^2-11.232} + e^{38.10t-49.14t^2-7.89} \right) \tag{18.8}$$

in which t is the time. The pressure change for two cycles is shown in Fig. 18.7.

The model of Eqn. (18.7) using the forcing function of Eqn. (18.8) was simulated to produce the blood pressure pattern (assumed to be the pressure in the aorta). The result is shown in Fig. 18.8. As can be seen, the pressure for the given set of parameters (R_a = 0.03 mmHg.s/ml, C_a = 1.0 ml/mmHg, R_p = 1.25 mmHg.s/ml) varies between 65 and 120 mmHg. The normal blood pressure for a healthy person would vary between 80 and 120 mmHg, although deviations from these values often happen. The lower pressure limit stems from the fact that blood flow is only possible in one direction, from pressure P_{lv} to P_a, if $P_a > P_{lv}$ the flow is zero.

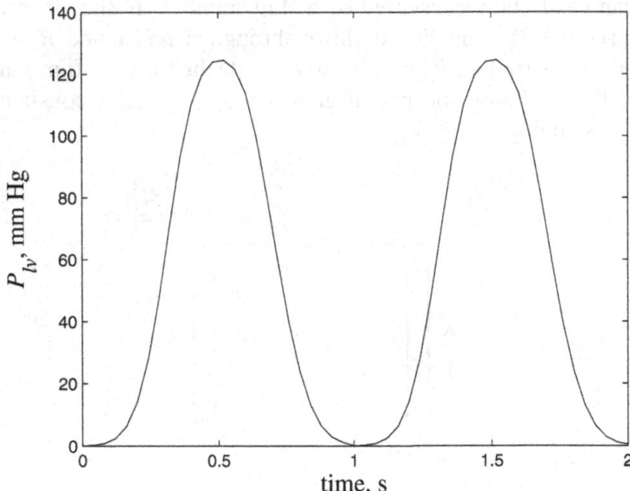

Fig. 18.7. Ventricular pressure variation with time.

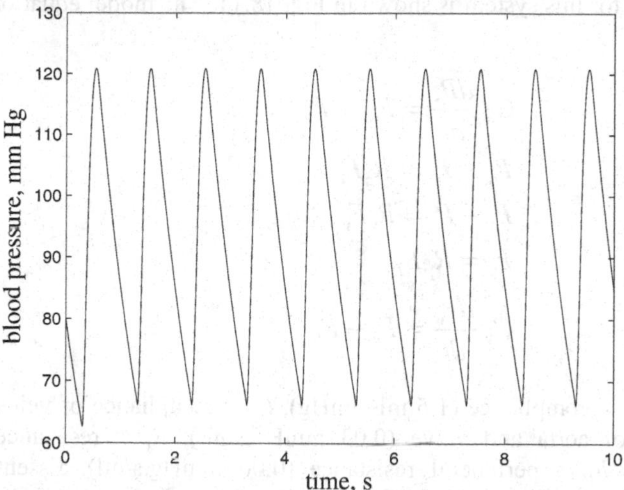

Fig. 18.8. Variation of blood pressure with time.

Figure 18.9 shows the blood pressure for one heart beat. It can be seen that the pressure rises faster than it decreases; this is also observed in reality.

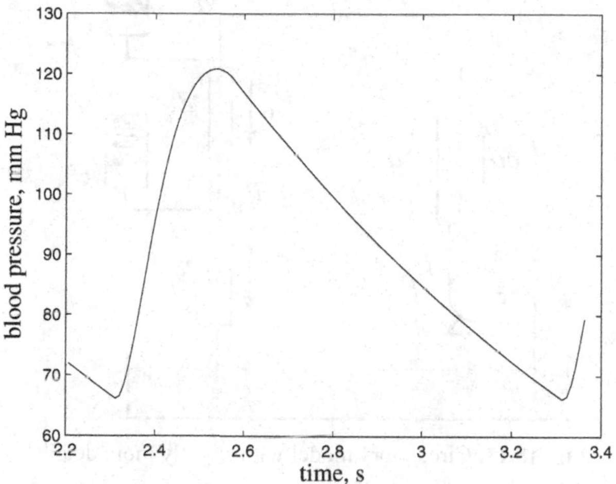

Fig. 18.9. Variation of blood pressure during one heart beat.

It was subsequently investigated what the impact is of increasing the level of detail in the model. It was assumed that the veins also have a certain compliance and the extended model is shown in Fig. 18.10.

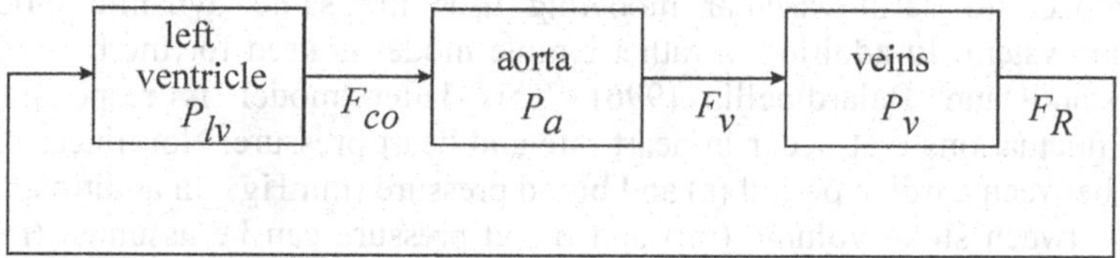

Fig. 18.10. Flow analog of circulatory model.

The flow analogy for this system is shown in Fig. 18.11. The model equations can be written as:

$$
\begin{aligned}
&C_a \frac{dP_a}{dt} = F_{co} - F_v \\
&P_{lv} - P_a = R_a F_{co} \\
&P_a - P_v = R_v F_v \\
&P_v = R_p F_R \\
&C_v \frac{dP_v}{dt} = F_v - F_R
\end{aligned}
\qquad (18.9)
$$

in which C_a = aorta compliance (1.5 ml/mmHg), C_v = compliance of veins (80 ml/mmHg), R_a = resistance of aorta and valve (0.03 mmHg.s/ml), R_v = resistance of veins (0.90 mmHg.s/ml) and R_p = peripheral resistance (0.06 mmHg.s/ml). Essentially, one more differential equation has been added. The model could be further detailed by distinguishing more 'compartments', such as systemic and pulmonary sections (arterial and venous), left and right heart and so forth (Sacca, 2003; Geertsema *et al.*, 1997).

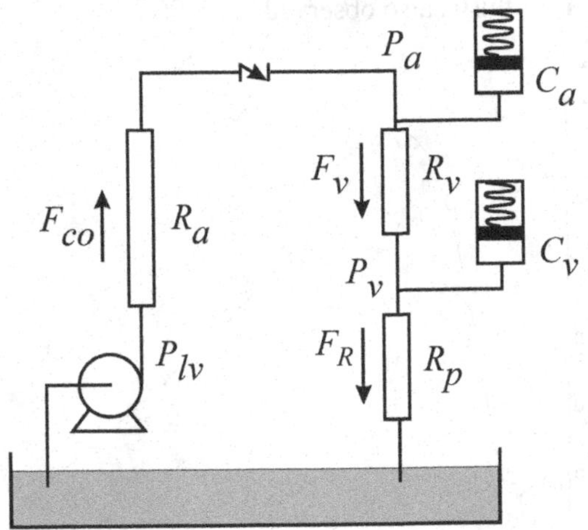

Fig. 18.11. Circulatory model with slightly more detail.

The simulation results of the slightly more detailed model are not much different from the ones already shown in Figs. 18.8 and 18.9.

18.8 Modeling Heart Rate Variability using a Baroreflex Model

Another approach to cardiovascular modeling uses the same dynamic model for the cardiovascular system. In addition, a rather simple model is used for the baroreflex (Akay, 2001; Cavalcanti and Belardinelli, 1996). This latter model is responsible for the spontaneous fluctuations that occur in heart rate and heart pressure. Non-linear relationships are assumed between cardiac period (s) and blood pressure (mmHg). In addition, a non-linear relationship between stoke volume (ml) and blood pressure can be assumed (mmHg). The relationship between cardiac period T and blood pressure P can be given by (Cavalcanti and Belardinelli, 1996):

$$T(s) = \frac{0.54}{1 + e^{-0.4(P-68)}} + 0.66 \tag{18.10}$$

For the relationship between stoke volume SV and blood pressure P, the authors proposed the following relationship:

$$SV(ml) = \frac{86}{1 + 72\left(\dfrac{P}{25} - 1\right)^{-7}} \tag{18.11}$$

The non-linear relationship between cardiac period and blood pressure is shown in Fig. 18.12.

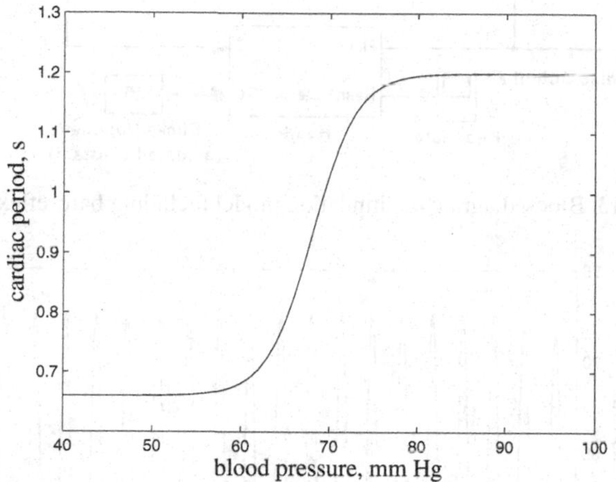

Fig. 18.12. Non-linear relationship between cardiac period and blood pressure.

The behavior of the cardiovascular system is approximated by a simple flow model as given by Eqn. (18.7). From this equation, it can easily be derived that:

$$\frac{\delta P_{lv}}{\delta F_a} = \frac{R_a R_p C_a s + (R_a + R_p)}{R_p C_a s + 1} \tag{18.12}$$

in which s is the Laplace operator. Assuming the following values: $R_a = 0.03$ mmHg.s/ml, $R_p = 0.90$ mmHg.s/ml and $C_a = 1.5$ ml/mmHg, Eqn. (18.12) can be written as:

$$\frac{\delta P_{lv}}{\delta F_a} = \frac{0.04s + 0.93}{1.35s + 1} \tag{18.13}$$

It is further assumed that the baroreflex output does not immediately affect the heart cycle and the stroke volume. A time delay between 1.0 and 2.0 seconds is assumed, and it can be shown that with increasing delay, the blood pressure pattern becomes more chaotic. This is not difficult to understand, the time delay introduces a phase shift in the closed loop system and when it is large enough, the phase shift will become larger than –180°, and oscillation will be sustained.

The model structure is shown in Fig. 18.13, for reasons of simplicity a constant stroke volume of 75 ml is assumed in the simulation. The variation in heart rate is shown in Fig. 18.14. It can be seen that the average heart rate is about 62 beats/min, fluctuations are in the order of ±10 beats/min.

When the time delay is decreased, the period becomes shorter, however, the pressure cycle disappears and the pressure will become a steady signal. Both types of model give some insight into phenomena that takes place in the cardiovascular system.

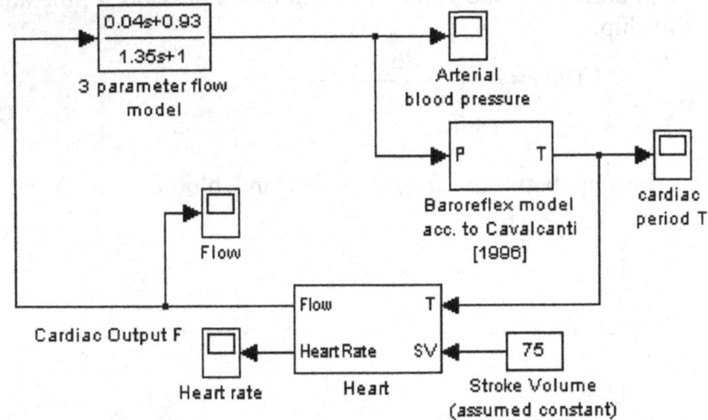

Fig. 18.13. Block diagram of simulation model including baroreflex model.

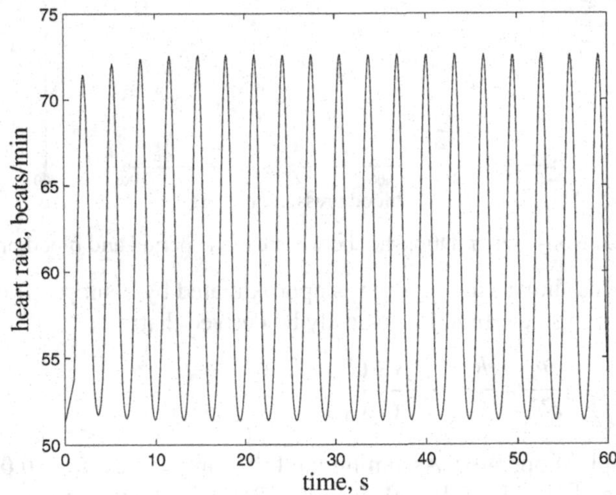

Fig. 18.14. Heart rate variability versus time using baroreflex model.

More detailed models that also take the dependency of the stroke volume on the arterial pressure into account and describe their effect on the behaviour of the cardiac heart rate and cardiac output (heart rate multiplied by stroke volume) are described by Cavalcanti and Belardinelli (1996).

Files referred to in this chapter:
F1802.m: steady-state analysis glucose–insulin equilibrium
F1803par.m: parameters for glucose–insulin model
F1803.mdl: model glucose–insulin simulation
F1803plot.m: plot file for Fig. 18.03
F1807.m: forcing function for cardiovascular dynamics
F1808.mdl: model for cardiovascular dynamics
F1808plot.m: plot file for Fig. 18.08

F1811.mdl: more detailed model for cardiovascular dynamics
F1812.m: generation cardiac period – blood pressure relationship
F1813.mdl: cardiovascular simulation using baroreflex model
F1814plot.m: plot file for Fig. 18.14

References

Akay, M. (ed.) (2001) *Non-Linear Biomedical Signal Processing, Vol. II, Dynamic Analysis and Modeling*, IEEE Press, New York.

Andersen, K.E. and Højbjerre, M. (2003) A Bayesian approach to Bergman's minimal model. A MaPhySto Workshop on computational aspects of graphical models, 17–20 Sept. Aalborg University, Denmark.

Baura, G.D. (2002) *System Theory and Practical Applications of Biomedical Signals*, IEEE Press.

Bergman, R.N., Phillips, L.S. and Cobelli, C. (1981) Physiologic evaluation of factors controlling glucose tolerance in man. *Journal of Clinical Investigation*, **68**, 1456–67.

Dazzi, D., Taddei, F., Gavarini, A., Uggeri, E., Negro, R. and Pezzarossa, A. (2001) The control of blood glucose in the critical diabetic patient: a neuro-fuzzy method. *Journal of Diabetes and its Complications*, **15** (2), 80–7.

Geertsema, A.A., Rakhorst, G., Mihaylov, D., Blanksma, P.K. and Verkerke, G.J. (1997) Development of a numerical simulation model of the cardiovascular system, *International Journal of Artificial Organs*, l. **21** (12), 1297–1311.

Hämäläinen, R.P. and Kettunen, A. (2000) Stability of fourier coefficients in relation to changes in respiratory air flow patterns. *Medical Engineering and Physics*, **22** (10), 733–9.

Hämäläinen, R.P. (1992) Optimal control analysis of the breathing pattern by using the WBREPAT computer software, in *Control of Breathing and its Modelling Perspective* (eds Y. Honda, Y. Miyamoto, K. Konno and J.G. Widdicombe), Plenum Press, New York, pp. 393–6.

Isaka, S. (1993) Control strategies for arterial blood pressure regulation. *IEEE Transactions on Biomedical Engineering*, **40** (4), 353–63.

Khoo, M.C.K. (2000) *Physiological Control Systems, Analysis, Simulation and Estimation*, Wiley Interscience, IEEE.

Lehmann, E.D., Deutsch, T., Carson, E.R. and Sonksen, P.H. (1994) AIDA: an interactive diabetes advisor. *Computer Methods and Programs in Biomedicine*, **41**, 183–203.

Liszka-Hackzell, J.J. (1999) Prediction of blood glucose levels in diabetic patients using a hybrid AI technique. *Computers and Biomedical Research*, **32** (2), 132–45.

The Mathworks: http://www.mathworks.com/products/demos/simulink/physbe/index.shtml.

McSharry, P.E., Smith, L.A. and Tarassenko, L. (2003) Comparison of predictability of epileptic seizures by a linear and a nonlinear method. *IEEE Transactions on Biomedical Engineering* **50** (5), 628–33.

McSharry, P.E., Clifford, G., Tarassenko, L. and Smith, L.A. (2003) A dynamic model for generating synthetic electrocardiogram signals. *IEEE Transactions on Biomedical Engineering* **50** (3), 289–94.

Nucci, G. and Cobelli, C. (2000) Models of subcutaneous insulin kinetics, a critical review. *Computer Methods and Programs in Biomedicine*, **62**, 249–57.

Parker, S.P., Doyle III, F.J. and Peppas, N.A. (2001) The intravenous route to blood glucose control. *IEEE Engineering in Medicine and Biology*, Jan./Feb., 65–73.

Sacca, L.F. (2003) A numerical model of the cardiovascular system, M.Sc. Thesis, University of Rome.

Smith, B.W., Chase, J.G., Nokes, R.I., Shaw, G.M. and Wake, G. (2003) Minimal haemodynamic system model including ventricular interaction and valve dynamics. *Medical Engineering and Physics*, **26**, 131–9.

Topp, B., Promislow, K., de Vries, G., Miura, R.M. and Finegood, D.T. (2000) A model of β-cell mass, insluin and glucose kinetics: pathways to diabetes. *Journal of Theoretical Biology* **206**, 605–19.

19 Introduction to Black Box Modeling

Black box or experimental modeling is a method for the development of models based on process data. Since physical modeling is usually very time consuming, black box modeling is a popular method for gaining insight into the overall (input–output) process behavior. The developed models are usually used for prediction of future process values or in process control applications. Aspects of empirical modeling that will be discussed in this chapter are data pre-conditioning, model complexity, model linearity and model extrapolation. The various black box or empirical modeling techniques addressed in the following chapters are: partial least squares modeling, time series modeling, neural network modeling, fuzzy modeling and neuro-fuzzy modeling.

19.1 Need for Different Model Types

In the previous chapters deterministic models were derived. They were designed based on the chemical and physical balances and mechanisms of the process and consequently the model described the internal functional behavior. Black-box models, on the other hand, are designed based on the input–output behavior of the process and consequently the model describes the overall behavior. A black-box model consists of a certain structure of which the parameters are determined by means of experimental results. Therefore, they often are called experimental models. The main properties of black-box models are the structure characteristics, which are: level of detail, degree of non-linearity and the structural way in which dynamics are composed.

As already discussed in the previous chapters, process behavior is usually non-linear. Whether or not the empirical model to be developed should also be non-linear depends on the operating range in which the model will be used. If the process is controlled and the operating range is small, a linear process model may be an adequate approximation of reality. The application of the model will determine whether the model needs to be dynamic or static. For control and prediction type applications, models are usually dynamic.

If the process conditions vary over a wide range, there may be a need for a non-linear empirical model. In case of a dynamic non-linear model there are a few possibilities for developing such a model, for example a dynamic neural network or a dynamic fuzzy model. One could also develop a Wiener model, in which the process dynamics are represented by a linear model, such as a state space model. The static characteristics of the process are then modeled by a polynomial, able to represent the non-linearity.

If the empirical dynamic model is linear, one could use for example a time series model. Different model types are available and will be discussed in a subsequent chapter.

If the model is linear and static one could for example use partial least squares modeling. State space modeling can be used in its linear or non-linear form, depending on the situation.

In all cases the availability of a sufficient number of data points and the quality of the data determines how good the model is that will be developed. If industrial process data is collected, one can almost be certain that there will be sections of data missing or faulty data may be included in the data set. In that case one needs to pay considerable attention to data pre-conditioning.

Process Dynamics and Control: Modeling for Control and Prediction. Brian Roffel and Ben Betlem.

19.2 Modeling steps

In chapter 1, section 1.3, the general steps for process modeling were introduced. The main phases in modeling are: system analysis, model design and model analysis. Fig 19.1 shows the modeling steps more dedicated to black box model design.

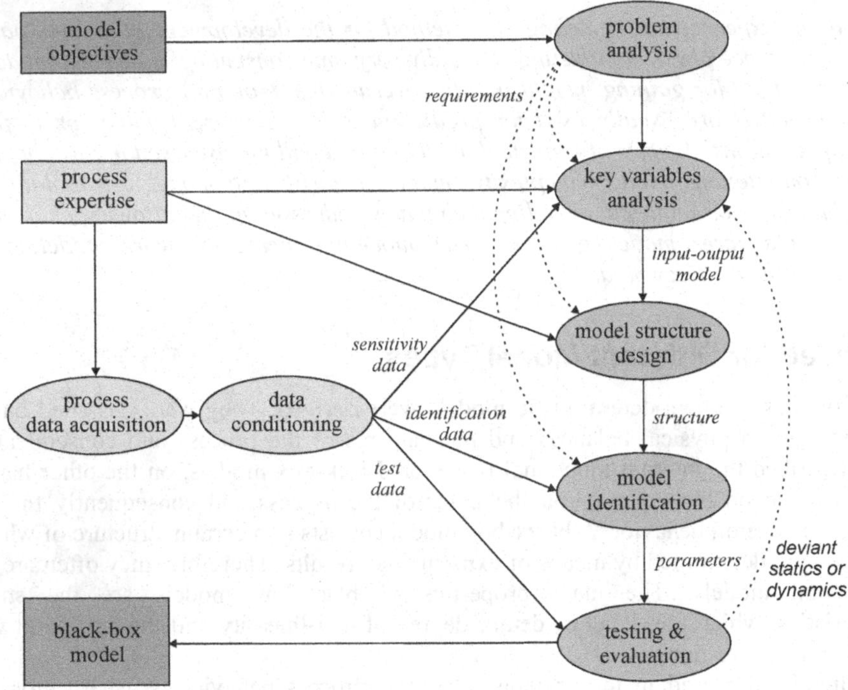

Fig. 19.1. Modeling steps for black-box modeling.

The phases in back-box modeling are:

- System analysis.
  This phase is similar to deterministic modeling, as described in section 1.3.1. During this phase the goals and the requirements of the model are formulated and the boundaries are determined. Often, it is already necessary to consider the three most important characteristics of a black-box model: level of detail, degree of non-linearity of the process and the order of the dynamics.
- Data conditioning.
  Black-box modeling is also called experimental modeling. The models are based on the data available and form the starting point of the design. The data mining and conditioning is therefore a crucial initial step.
- Key variable analysis.
  The choice of the input and output variables can be based on process insight, but can also be determined by a sensitivity analysis. Important is that the input space is not too large and the variables as much as possible mutually independent. By othogonalization the number of variables can be reduced.
- Model structure design.
  When a selection for a certain type of black-box model has been made, choices have to be made about handling model-order and non-linearity. The number of parameters is a good measure of the level of detail that will be obtained. The level of detail should be balanced with the amount and the quality of the data available.

- Model identification.
  In this step the model is fitted to the measured data. Usually, the error between model and reality is minimized.
- Model evaluation.
  The model is tested by means of special test data sets to determine whether the model has sufficient capacity to predict stationary and dynamic behavior. Most black-box models show good interpolation properties. However, their extrapolation qualities are mostly limited and should be tested if required.

In this chapter the subjects of data conditioning, key variable selection, model-order non-linearity and extrapolation will be discussed in more detail.

19.3 Data Preconditioning

After process analysis, data preconditioning is the first important step in the modeling process. If faulty data is used or data is insufficiently rich in information, a poor model will result. An important tool that can be used is principal component analysis. This tool gives a good indication whether there are points in the data sets which are abnormal. Bad data points should be removed from the data set. One should be careful if a dynamic model has to be developed. If data is removed from a dynamic data set, the continuity in the data is lost and this has consequences for the model. The best thing would be to initialize the model at the start of each new section of the data set. In case of static data, points can be arbitrarily removed without consequences for the model to be developed.

It often happens that collected data is bad, for example in case of analyzer readings it may happen that the gas chromatograph was out of service for some period of time. This situation is similar to the situation where bad data points have to be removed. Rather than removing bad data points, one could also fill the bad data section with good data points. This is a difficult issue. One could use linear interpolation between the last and first good process value. One could also look at larger data sections that look similar to the data section in which the bad data points are present and then copy the 'similar' data section to the bad data section. This is probably a better method than linear interpolation.

In case of dynamic modeling, the data collection frequency is also an important issue. What will be the execution frequency of the model? If this frequency is for example 1 minute, it would make sense to collect data at a frequency higher than 1 minute, for example, every 10 to 20 seconds. If the data collection frequency is high and one wants to use every data point while prediction of the process measurement is only required at a much lower frequency, than the process data will be highly correlated, i.e. the current process reading will depend heavily on the previous process reading. This is often not a desirable situation for good prediction and/or control.

19.4 Selection of Independent Model Variables

Process operators often have the feeling that a monitored process variable depends on many input variables. The authors have come across a situation where the process operators were under the impression that a polymer quality was affected by 33 process input variables. This is not often the case. The first thing to do is to make a distinction between process input variables and state variables. This also determines the hierarchy of the variables in the model. Figure 19.2 show the hierarchy in the modeling exercise.

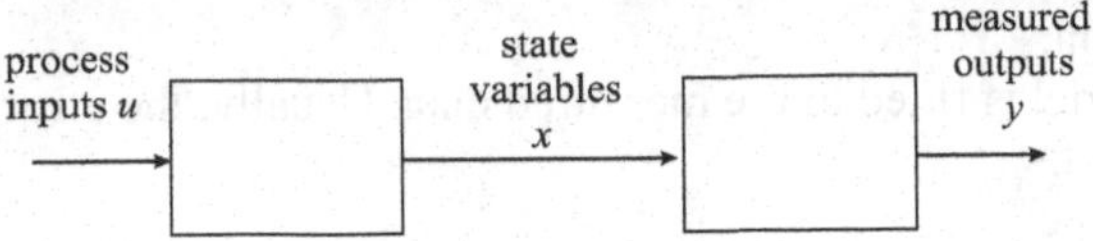

Fig. 19.2. Measure process output in the context of process inputs and state variables.

The process input variables (usually flows) affect the process state variables (usually pressure, level, temperature, concentration) which in turn have an impact on the measured process output such as a quality (for example viscosity). One should then ask oneself the question: what part of the process do I want to model: the relationship between the process inputs and the quality variable or the relationship between the state variables and the measured quality variable or the relationship between the process inputs and the state variables? It would not be logical to model the quality variable as a function of both the process inputs and the state variables. The latter exercise would be one that is not likely to be successful, since the process inputs affect the state variables and the state variables affect the process quality variable. Furthermore, the state variables are usually not mutually independent, for example temperature, pressure and composition. This means that a model that includes the process inputs as well as the state variables as additional inputs, models the effect of the true process inputs multiple times. Therefore a proper selection of the inputs and output(s) of the model is very important.

An effective way to reduce the number the number of inputs is to transform the input space of mutually dependent variables into a space of independent variables by principal component analysis (PCA). This space is usually much smaller. This transformation forms the basis of the construction of partial least square (PLS) models. Also in many packages for neural-network design a PCA toolbox is present.

19.5 Model Order Selection

It is difficult to fix the order of the model before starting the modeling exercise. In case of a dynamic model, the autocorrelation of the signal will give a good indication, how many past process values should be taken into account. If, for example, the current process output reading depends on the previous two process output readings and the previous two input readings, then a second-order model would be appropriate. However, model order is a concept that is strongly related to time series models. If, however, a fuzzy model or neural network model is used, it is difficult to speak about model order. In the case of a neural network one could speak about the size of the hidden layer in the neural network and in case of a fuzzy model one could speak about the number of fuzzy rules in the model. One general recommendation could be made: start always with a low model order or a simple network or a model with only a few rules. If the model fit is poor, try to expand the model in order to improve the fit. One should always monitor the model complexity. When a second-order time series model ($y_k = f(y_{k-1}, u_{k-1}, u_{k-2})$ is expanded to a third-order model ($y_k = f(y_{k-1}, y_{k-2}, u_{k-1}, u_{k-2}, u_{k-3})$, the number of model parameter increases from 3 to 5. In one develops a neural network model for a process in which the output depends on the previous two inputs and outputs ($y_k = f(y_{k-1}, y_{k-2}, u_{k-1}, u_{k-2})$, then the neural network already has four inputs. Using three hidden neurons and one output neuron now yields a model with 23 model parameters. Adding two additional inputs in the model ($y_k = f(y_{k-1}, y_{k-2}, y_{k-3}, u_{k-1}, u_{k-2}, u_{k-3})$, increases the number of model parameters to 31. As can be seen, a neural network model is easily over-parameterized. This will result in an excellent fit to the training data set, but a poor fit to a validation data set. A similar situation occurs when using fuzzy models. A fuzzy model with only three rules and of second-order type, has between 18 and 24 parameters, depending on the complexity of each rule.

It is also important to realize that a fifth-order state space model has considerably more than five model parameters!

One often speaks in this context about the curse of dimensionality, by making the model slightly more complex, one increases the number of model parameters significantly. One should therefore watch out for models that produce a good model fit due to a high model dimensionality!

19.6 Model Linearity

If possible, develop a linear model for any particular application with a low number of model parameters. Only if this is not possible, one should revert to other options. It is good to realize that a fuzzy model that is used for non-linear modeling is a combination of several linear models. The non-linear relationship between the process inputs and the process output is fitted by a piecewise approximation of a series of linear models. This is attractive if one still would like to retain some feel for the actual meaning of the developed model. Often, for fuzzy models, one linear model is valid for a particular operating region. The overall fuzzy model combines the individual linear models in an intelligent way to describe the nonlinearity in the process. This is one of the reasons that fuzzy models generally provide more insight into the process than neural networks, which we consider to be true black box models.

Another efficient way of developing non-linear models is to include non-linear terms in the model. In order to do this in an effective way, one must have some understanding of the relationship between the input and output variables. If, for example, one knows that a variable y depends on x and x^2, then one could use x and x^2 as basis variables and develop a linear relationship between y, x and x^2. This is much more efficient than using y as an output variable and x as an input variable and develop a non-linear model by using, for example, fuzzy logic or neural network modeling.

19.7 Model Extrapolation

Whether a model is good enough for the purpose it has been developed for, can be judged by testing the model on a validation data set that contains data that falls within the range of data the model was developed for. Another important issue is model extrapolation, the use of the model for a data set that contains data points that fall outside the region of the data test set. If the process behaves in a linear fashion and a linear model is developed one may expect the model to posses good extrapolation capabilities. If process behavior is non-linear and a fuzzy model was developed, one (linear) rule will be used for extrapolation of the model values outside the operating range. A linear prediction of a strongly non-linear phenomenon will never be exactly correct. A neural network, which is a true non-linear model description *may* provide better results, although this is not guaranteed. This will depend on many factors, such as the number of data points near the operating region boundary for which the model was developed and the non-linearity at the boundary which the model tries to fit versus the non-linearity of the phenomenon that is modeled.

19.8 Model Evaluation

Model structure as well as the model parameters can be empirically identified from plant data. Structural considerations concern the order of the parametric model. The model is usually identified by minimizing the error between process data and model prediction. For the process industries, in many instances low order models will suffice, especially when the process is kept in a certain operating point. In most cases, part of the available data is used for model development, this data set is called the test set or training set, another part is used for model validation and is called the validation set. A criterion should be defined which indicates how good the model fit is. For a process, operating around a steady-state value, one could use:

$$fit = 100 \times \left[1 - \frac{norm(y - \hat{y})}{norm(y - \bar{y})} \right] \tag{19.1}$$

in which $\hat{y}$ is the model prediction and $\bar{y}$ the mean or average value of the data series.

For a process not operating at steady-state value, such as a batch process, it would be better to use another criterion for the model fit, such as a quadratic error criterion or the percentage variance (*vae*) explained:

$$vae = 100 \times \left[1 - \frac{\sigma^2(y - \hat{y})}{\sigma^2(y)} \right] \tag{19.2}$$

in which $\sigma^2(y)$= the variance in y.

In the following chapters different approaches to black box modeling will be discussed. In addition, principal component analysis will be discussed as a tool for data conditioning and data reduction. However, first an introduction to linear algebra will be given, since some modeling tools rely heavily on it.

20 Basics of Linear Algebra

To understand the concepts of the methods described in the following chapters, one should have an understanding of the basics of linear algebra. This chapter will serve as a short refresher. Linear algebra essentially deals with vector and matrix manipulations, all of which can easily be performed by using the MATLAB software. However, an insight into some of the concepts behind the operations may be helpful.

20.1 Introduction

There are many good references that give a brief overview of Linear Algebra (Lay, 1996; Bretscher, 1997; and Wise *et al.*, 2004).

The building blocks of linear algebra are scalars, vectors and matrices. An example of a scalar is a single number, for example $a = 3$. There are two types of vector: column vectors and row vectors. An example of a column vector is:

$$b = \begin{bmatrix} 2 \\ 1 \\ 4 \end{bmatrix} \tag{20.1}$$

in MATLAB defined as $b = [2;1;4]$;

An example of a row vector is:

$$c = \begin{bmatrix} 2 & 1 & 4 \end{bmatrix} \tag{20.2}$$

in MATLAB defined as $c = [2\ 1\ 4]$;

A column vector can be transformed into a row vector by taking the transpose:

$$c = b^T \tag{20.3}$$

in MATLAB this expression is written as: $c = b'$.

A matrix is an array of vectors, for example:

$$A = \begin{bmatrix} 2 & 5 & 1 \\ 3 & 4 & 2 \end{bmatrix} \tag{20.4}$$

In MATLAB this matrix can be defined as $A = [2\ 5\ 1;\ 3\ 4\ 2]$;

Matlab uses the colon notation, if you would like to address all columns or all rows, for example, $A1 = A(:,1{:}2)$ means that matrix $A1$ has the same rows than matrix A, the number of colums of $A1$ are only columns one to two of matrix A.

The transpose of a matrix is obtained by inverting rows and columns, i.e. the first column becomes the first row, and so forth:

$$A^T = \begin{bmatrix} 2 & 3 \\ 5 & 4 \\ 1 & 2 \end{bmatrix} \tag{20.5}$$

Matrices (and also vectors) can be added and subtracted, as long as the dimensions agree. Matrix and vector addition is associative and commutative:

Process Dynamics and Control: Modeling for Control and Prediction. Brian Roffel and Ben Betlem.

$$A + B = B + A$$
$$A + (B + C) = (A + B) + C \tag{20.6}$$

Two matrices $A(m,n)$ and $B(n,p)$ can be multiplied as long as the number of columns in A is the same than the number of rows in B. The result of the multiplication is a matrix $C(m,p)$.

$$A = \begin{bmatrix} 4 & 3 & 2 \\ 1 & 3 & 5 \end{bmatrix}, B = \begin{bmatrix} 2 & 5 \\ 3 & 1 \\ 4 & 2 \end{bmatrix} \tag{20.7}$$

$$A * B = \begin{bmatrix} 4*2+3*3+2*4 & 4*5+3*1+2*2 \\ 1*2+3*3+5*4 & 1*5+3*1+5*2 \end{bmatrix} = \begin{bmatrix} 25 & 27 \\ 31 & 18 \end{bmatrix} \tag{20.8}$$

20.2 Inner and Outer Product

Vectors can be multiplied in two ways: one can calculate the inner product, which is a scalar, and one can take the outer product, the result being a matrix.

The inner product of two vectors is the sum of the products of the corresponding elements. For example:

$$a = \begin{bmatrix} 3 \\ 2 \\ 5 \end{bmatrix}, \quad b = \begin{bmatrix} 1 \\ 3 \\ 2 \end{bmatrix} \tag{20.9}$$

then the inner product is

$$c = a^T b = \begin{bmatrix} 3 & 2 & 5 \end{bmatrix} \begin{bmatrix} 1 \\ 3 \\ 2 \end{bmatrix} = [3*1+2*3+5*2] = 19 \tag{20.10}$$

In MATLAB this would be calculated as follows: a = [3;2;5]; b=[1;3;2]; c=a'*b;

The length of a vector or the 2-norm of a vector, denoted by $||a||$, also uses the inner product. The length of vector a is defined as:

$$\| a \| = \sqrt{a^T a} = \sqrt{\begin{bmatrix} 3 & 2 & 5 \end{bmatrix} \begin{bmatrix} 3 \\ 2 \\ 5 \end{bmatrix}} = \sqrt{38} = 6.1644 \tag{20.11}$$

in MATLAB sqrt(a'*a);

The outer product of two vectors is obtained by multiplying the first element of a column vector by a row vector, thus forming the first row of a matrix. Subsequently, the second element of the column vector is multiplied by the row vector, and so forth. In the example of Eq. (20.7) this would lead to:

$$C = ab^T = \begin{bmatrix} 3 \\ 2 \\ 5 \end{bmatrix} \begin{bmatrix} 1 & 3 & 2 \end{bmatrix} = \begin{bmatrix} 3*1 & 3*3 & 3*2 \\ 2*1 & 2*3 & 2*2 \\ 5*1 & 5*3 & 5*2 \end{bmatrix} = \begin{bmatrix} 3 & 9 & 6 \\ 2 & 6 & 4 \\ 5 & 15 & 10 \end{bmatrix} \tag{20.12}$$

Both vectors do not necesarily have to be of the same length.

20.3 Special Matrices and Vectors

20.3.1 The Identity Matrix

The identity matrix is a special matrix with all elements being zero, except for the diagonal elements, which are equal to one.

$$I = \begin{bmatrix} 1 & 0 & 0 \\ 0 & 1 & 0 \\ 0 & 0 & 1 \end{bmatrix} \tag{20.13}$$

In MATLAB this matrix can easily be created by the eye command: I = eye(3);

Any matrix that is multiplied by the identity matrix yields the original matrix again, i.e. $A*I=A$.

20.3.2 Orthogonal and Orthonormal Vectors

Vectors are orthogonal if their inner product is zero. Geometrically, this can be interpreted that the vectors have right angles to each other or perpendicular. Vectors are orthonormal if they are orthogonal and of unit length, in other words, the inner product with themselves is unity. For an orthonormal set of column vectors v_i, $i = 1,\ldots,n$, it should hold that:

$$\begin{aligned} v_i^T v_j &= 0 \ \ for\ i \neq j \\ v_i^T v_i &= 1 \end{aligned} \tag{20.14}$$

An example of an orthonormal set is our coordinate system, consisting of three vectors at right angles of each other. In principal component analysis it will be shown that there are other coordinate systems that are more convenient to use.

Example: The set of column vectors $(0.6\ \ 0.8)^T$ and $(-0.8\ \ 0.6)^T$ are orthonormal. Applying (20.14) to this set, we obtain:

$$\begin{aligned} [0.6 \ \ 0.8]\begin{bmatrix} 0.6 \\ 0.8 \end{bmatrix} &= 1 \\ [-0.8 \ \ 0.6]\begin{bmatrix} -0.8 \\ 0.6 \end{bmatrix} &= 1 \\ [0.6 \ \ 0.8]\begin{bmatrix} -0.8 \\ 0.6 \end{bmatrix} &= 0 \end{aligned} \tag{20.15}$$

20.4 Gauss–Jordan Elimination, Rank and Singularity

Often it occurs that a system of linear equations has to be solved. A simple way to accomplish this is using Gauss–Jordan elimination, which makes use of the property of equations that a linear combination of the equations still gives the same solution. Consider the following example:

$$\begin{bmatrix} 2y_1 + y_2 + 2y_3 \\ 2y_1 + 2y_2 + 0 \\ -4y_1 + 2y_2 + y_3 \end{bmatrix} = \begin{bmatrix} 9 \\ 10 \\ -1 \end{bmatrix} \tag{20.16}$$

This system can be written in the following form:

$$Ay = b \tag{20.17}$$

or:

$$\begin{bmatrix} 2 & 1 & 2 \\ 2 & 2 & 0 \\ -4 & 2 & 1 \end{bmatrix} \begin{bmatrix} y_1 \\ y_2 \\ y_3 \end{bmatrix} = \begin{bmatrix} 9 \\ 10 \\ -1 \end{bmatrix} \tag{20.18}$$

We would like to start with the first row and eliminate elements of the rows below. The first coefficient in the first row, 2, is called the *pivot*, this is the element that we would like the elements below to be zero.

The Gauss–Jordan elimination starts with subtracting the first row from the second row, and subsequently adding twice the first row to the third row in order to eliminate y_1. This gives:

$$\begin{bmatrix} 2 & 1 & 2 \\ 0 & 1 & -2 \\ 0 & 4 & 5 \end{bmatrix} \begin{bmatrix} y_1 \\ y_2 \\ y_3 \end{bmatrix} = \begin{bmatrix} 9 \\ 1 \\ 17 \end{bmatrix} \tag{20.19}$$

The first nonzero coefficient in the second row, 1, is called the *pivot*, since we would like the element below it to be zero.

The Gauss–Jordan elimination proceeds by subtracting four times the second row from the third row. The result is:

$$\begin{bmatrix} 2 & 1 & 2 \\ 0 & 1 & -2 \\ 0 & 0 & 13 \end{bmatrix} \begin{bmatrix} y_1 \\ y_2 \\ y_3 \end{bmatrix} = \begin{bmatrix} 9 \\ 1 \\ 13 \end{bmatrix} \tag{20.20}$$

from which it can be seen that $y_3 = 1$. From the second equation then $y_2 = 3$ and from the first equation it follows that $y_1 = 2$. This form of the matrix is called the *echolon* form, all non-zero entries are on or above the main diagonal.

In MATLAB this system can be solved by using the backslash operator. Define: A = [2 1 2; 2 2 0; -4 2 1]; b = [9; 10; –1]; $y = A\backslash b$.

One should realize that $X = A\backslash B$ denotes the solution to the matrix $AX = B$ while $X = B/A$ denotes the solution to the matrix $XA = B$.

One could also write the solution of Eqn. (20.17) as $y = A^{-1}.b$ and could then write in Matlab y = inv(A)∗b which gives the same solution as using the backslash operator. It is obvious that the latter solution is only found as long as the inverse of A exists.

By elimination, the number of independent equations can be determined. Only when the number of variables equals the number of equations, while the equations are independent, can a non-singular solution be found. In that case the "eliminated" matrix is a full square matrix. In this context the concept rank is useful. The rank of a matrix is the maximum number of independent rows (or, the maximum number of independent columns). A square matrix $A(n,n)$ is non-singular only if its rank is equal to n. The rank can easily be found from the number of non-zero rows obtained by reducing the matrix to echelon form.

If for a non-square matrix $A(m,n)$ the rank is equal to the minimum of (m,n), then the matrix is of full rank. If the rank is less than $min(m,n)$, then the matrix is rank-deficient or collinear (not all column are linearly independent). In MATLAB the rank is calculated by using *rank*(A).

20.5 Determinant of a matrix

Determinants indicate whether a matrix is singular or not, they are very useful in the analysis of linear equations and can, amongst others, be used to calculate the inverse of a matrix. The determinant is usually denoted by *det*(*A*) or |*A*|.

Suppose we have the following 2 × 2 matrix:

$$A = \begin{bmatrix} a_{11} & a_{12} \\ a_{21} & a_{22} \end{bmatrix} \tag{20.21}$$

then the determinant of this matrix is:

$$det(A) = |A| = \begin{vmatrix} a_{11} & a_{12} \\ a_{21} & a_{22} \end{vmatrix} = a_{11}a_{22} - a_{21}a_{12} \tag{20.22}$$

As can be seen, the determinant is a scalar.

The determinant for a 3 × 3 matrix is slightly more complicated. Suppose we have the following matrix:

$$A = \begin{bmatrix} a_{11} & a_{12} & a_{13} \\ a_{21} & a_{22} & a_{23} \\ a_{31} & a_{32} & a_{33} \end{bmatrix} \tag{20.23}$$

then the determinant of this matrix is:

$$det(A) = |A| = \begin{vmatrix} a_{11} & a_{12} & a_{13} \\ a_{21} & a_{22} & a_{23} \\ a_{31} & a_{32} & a_{33} \end{vmatrix} = a_{11}\begin{vmatrix} a_{22} & a_{23} \\ a_{32} & a_{33} \end{vmatrix} - a_{12}\begin{vmatrix} a_{21} & a_{23} \\ a_{31} & a_{33} \end{vmatrix} + a_{13}\begin{vmatrix} a_{21} & a_{22} \\ a_{31} & a_{32} \end{vmatrix} \tag{20.24}$$

In MATLAB the determinant of a matrix can easily be calculated by using the expression *det*(*A*).

Example: The determinant of the matrix:

$$A = \begin{bmatrix} 2 & 1 & 2 \\ 2 & 2 & 0 \\ -4 & 2 & 1 \end{bmatrix} \tag{20.25}$$

is

$$det(A) = 2\begin{vmatrix} 2 & 0 \\ 2 & 1 \end{vmatrix} - 1\begin{vmatrix} 2 & 0 \\ -4 & 1 \end{vmatrix} + 2\begin{vmatrix} 2 & 2 \\ -4 & 2 \end{vmatrix} = \\ = 2*(2-0) - 1*(2-0) + 2*(4+8) = 26 \tag{20.26}$$

Systems with a singular matrix can not be solved. Wise (2004) considered the following example:

$$\begin{bmatrix} 1 & 3 & 2 \\ 2 & 6 & 9 \\ 3 & 9 & 8 \end{bmatrix} \begin{bmatrix} y_1 \\ y_2 \\ y_3 \end{bmatrix} = \begin{bmatrix} 1 \\ -4 \\ -4 \end{bmatrix} \tag{20.27}$$

which produces after elimination:

$$\begin{bmatrix} 1 & 3 & 2 \\ 0 & 0 & 5 \\ 0 & 0 & 2 \end{bmatrix} \begin{bmatrix} y_1 \\ y_2 \\ y_3 \end{bmatrix} = \begin{bmatrix} 1 \\ -6 \\ -7 \end{bmatrix} \tag{20.28}$$

The last two equations yield conflicting values for y_3 and therefore the solution can not be found.

If we calculate the determinant of the A-matrix we find:

$$det(A) = \begin{vmatrix} 1 & 3 & 2 \\ 2 & 6 & 9 \\ 3 & 9 & 8 \end{vmatrix} = 0 \tag{20.29}$$

As can be seen, there is no solution to the set of linear equations when the determinant of the A-matrix is equal to zero.

Equation (20.27) shows that the A-matrix contains two columns (one and two) that are multiples of each other. In that case the system $Ay = b$ will not have a unique solution, since there are two sets of equations and three unknows. This can easily be seen if we carry out one more elimination step, by subtracting 0.4 times the second equation from the third equation, Eqn. (20.28) becomes:

$$\begin{bmatrix} 1 & 3 & 2 \\ 0 & 0 & 5 \\ 0 & 0 & 0 \end{bmatrix} \begin{bmatrix} y_1 \\ y_2 \\ y_3 \end{bmatrix} = \begin{bmatrix} 1 \\ -6 \\ -4.6 \end{bmatrix} \tag{20.30}$$

20.6 The Inverse of a Matrix

Before discussing how to calculate the inverse of a matrix, we will introduce the concept of "adjoint of a matrix". The adjoint of a matrix is the matrix of cofactors of the transpose of the matrix. The adjoint of the matrix can be found by using the following steps:

- determine the transpose of matrix A, i.e. interchange the columns and rows of A
- determine the cofactors for every element. Let M_{ij} denote the (n–1) square sub-matrix of A, obtained by deleting the i-th row and the j-th column. The determinant $|M_{ij}|$ is called the minor of the element a_{ij} of A. Then we define a cofactor, denoted by A_{ij}, which is the signed minor:

$$A_{ij} = (-1)^{i+j} |M_{ij}| \tag{20.31}$$

Example: Determine the adjoint matrix of matrix A, given in Eqn. (20.25).
The first step is determining the transpose:

$$A^T = \begin{bmatrix} 2 & 2 & -4 \\ 1 & 2 & 2 \\ 2 & 0 & 1 \end{bmatrix} \tag{20.32}$$

The cofactors can be written as:

$$adj(A)=\begin{bmatrix} +\begin{vmatrix}2&2\\0&1\end{vmatrix} & -\begin{vmatrix}1&2\\2&1\end{vmatrix} & +\begin{vmatrix}1&2\\2&0\end{vmatrix} \\ -\begin{vmatrix}2&-4\\0&1\end{vmatrix} & +\begin{vmatrix}2&-4\\2&1\end{vmatrix} & -\begin{vmatrix}2&2\\2&0\end{vmatrix} \\ +\begin{vmatrix}2&-4\\2&2\end{vmatrix} & -\begin{vmatrix}2&-4\\1&2\end{vmatrix} & +\begin{vmatrix}2&2\\1&2\end{vmatrix}\end{bmatrix}=\begin{bmatrix}2&3&-4\\-2&10&4\\12&-8&2\end{bmatrix} \tag{20.33}$$

The inverse of matrix A, denoted by A^{-1} can be calculated according to:

$$A^{-1}=\frac{1}{|A|}adj(A) \tag{20.34}$$

Example: The inverse of matrix A as given in Eqn. (20.25) can be calculated by combining Eqns. (20.26) and (20.33):

$$A^{-1}=\frac{1}{26}\begin{bmatrix}2&3&-4\\-2&10&4\\12&-8&2\end{bmatrix}=\begin{bmatrix}0.077&0.115&-0.154\\-0.077&0.385&0.154\\0.462&-0.308&0.077\end{bmatrix} \tag{20.35}$$

For the inverse of a matrix A the following properties hold: $AA^{-1}=I$ and $A^{-1}A=I$. Using the Gauss–Jordan elimination method, one can easily calculate the inverse of a matrix. It can be shown that

$$\begin{aligned}(AB)^{-1}&=B^{-1}A^{-1}\\(A^T)^{-1}&=(A^{-1})^T\end{aligned} \tag{20.36}$$

In MATLAB the inverse can be calculated by the *inv* function. The inverse of matrix A in Eq. (20.25) can be calculated by defining: *A*=[2 1 2;2 2 0;-4 2 1]; *B*=inv(*A*); resulting in Eqn. (20.35).

In some cases the inverse of the matrix cannot be calculated since the matrix becomes ill-conditioned, then one could calculate the so-called pseudo-inverse, which will be explained in the following section.

20.7 Inverse of a Singular Matrix

When the columns of a matrix A are not linearly independent, the inverse of the matrix can not be found, since there are many solutions to $Ay=b$. In this case it is still possible, however, to find a least squares solution to the problem. The matrix that solves this problem is called pseudo-inverse A^+, it can be calculated by using singular value decomposition.
When the columns of A are independent, it can be shown that

$$A^+=(A^TA)^{-1}A^T \tag{20.37}$$

In case the columns of matrix A are linearly dependent (A is collinear), one can use singular value decomposition to calculate the pseudo inverse. A matrix $A(m,n)$ is factored into:

$$A=U\,S\,V^T \tag{20.38}$$

where $U(m,n)$ is orthogonal, $V(n,n)$ is orthogonal and $S(n,n)$ is diagonal. The nonzero entries of S are called singular values: they decrease monotonically from upper left to lower right of S.

Suppose one has the following matrix:

$$A = \begin{bmatrix} 2 & 1 & 4 \\ 2 & 2 & 6 \\ -4 & 2 & 0 \end{bmatrix} \tag{20.39}$$

This matrix is singular; since the third column is linearly dependent on the other two, it is equal to the first column plus two times the second column. If one uses Matlab and one enters A = [2 1 4;2 2 6;-4 2 0]; B = inv(A), then Matlab returns that the matrix is singular and the result is undetermined.

If one types [U,S,V] = *svd*(A) then the result is:

```
U =
   -0.5655     0.0315     0.8242
   -0.8118    -0.1976    -0.5494
    0.1456    -0.9798     0.1374

S =
    8.1002          0          0
         0     4.4030          0
         0          0          0

V =
   -0.4119     0.8146    -0.4082
   -0.2343    -0.5277    -0.8165
   -0.8806    -0.2407     0.4082
```

and then B=U∗S∗V', the result is

```
B =
    2.0      1.0      4.0
    2.0      2.0      6.0
   -4.0      2.0     -0.0
```

from which it can be seen that B is the original A matrix that was entered.

As one can see, matrix S is diagonal and its elements are decreasing from upper left to lower right. There are only two nonzero elements, since there are two independent equations.

This means that in reconstructing matrix A, two columns of U, two columns of V and two rows and columns of S would suffice. Type therefore in Matlab B = U(:,1:2)∗S(1:2,1:2)∗V(:,1:2)′, the result is

```
B =
    2.0      1.0      4.0
    2.0      2.0      6.0
   -4.0      2.0     -0.0
```

which is the same as before.

Using the special properties of orthogonal matrices:

$$\begin{aligned} A^T A &= I \\ A^{-1} &= A^T \end{aligned} \tag{20.40}$$

and the fact that the inverse of a product is the product of the inverses in reverse order, one can write for the inverse of matrix A:

$$A^{-1} = V\, S^{-1}\, U^T \tag{20.41}$$

taking into account that the rank of the matrix S is two (two independent equations). The Matlab command for calculation of the inverse therefore becomes:

```
A1 = V(:,1:2)*inv(S(1:2,1:2))*U(:,1:2)'
```

resulting in:

```
A1 =
      0.0346    0.0047   -0.1887
      0.0126    0.0472    0.1132
      0.0597    0.0991    0.0377
```

One would get the same result with the command pinv(A), which in fact uses the SVD algorithm to calculate the matrix inverse.

20.8 Generalized Least Squares

In Eq. (20.16) we had a system with three equations and three unknowns. In this case the solution was obtained through matrix inversion. However, it can happen that there are more equations than unknowns, in which case a number of solutions can be found, all with an associated error or residual.

Assume there is a system of equations:

$$X\,b - y = e \tag{20.42}$$

in which e is an error vector. (Note that the equation $Ax=b$ is now written in different nomenclature, namely $Xb=y$.)

If we want to find the optimal solution we will minimize the errors, i.e.

$$\min_x J = ee^T \tag{20.43}$$

The solution of this equation is

$$b = (X^T X)^{-1} X^T y \tag{20.44}$$

The vector b is called the regression vector that relates X to y.

Suppose there are two variables x_1 and x_2 with the values arranged in a matrix A and another variable y with the values arranged in a vector (note that there are 5 equations and 2 unknowns):

$$X = \begin{bmatrix} 4 & 1 \\ 3 & 2 \\ 4 & 3 \\ 5 & 1 \\ 3 & 2 \end{bmatrix},\ y = \begin{bmatrix} 5 \\ 8 \\ 9 \\ 8 \\ 7 \end{bmatrix} \tag{20.45}$$

then the vector that relates x to y can be computed from Eq. (20.44). In Matlab that is:

```
b = inv(X'*X)*X'*y
```

which computes to $b = [1.0982\ \ 1.7768]^T$.

Therefore y can be estimated from x_1 and x_2 by the following equation:

$y = 1.0982x_1 + 1.7768x_2$.

Rather than using the previous Matlab command, one could also use the backslash operator

```
b = X\y
```

which is a simplification of this equation.

The projection of y into X (or calculating the predicted value of y) is done via the equation:

```
p = X*b
```

resulting in:

```
p =
     6.1696
     6.8482
     9.7232
     7.2679
     6.8482
```

The prediction error is $y–p$:

```
e =
    -1.1696
     1.1518
    -0.7232
     0.7321
     0.1518
```

The error has the property that it is orthogonal to the columns of X. To verify this, compute $X^T e$, which results in

```
0.1421 e-13
0.0799 e-13
```

which is zero (apart from the computer precision). Thus the inner product of each column of X with e is zero, which is precisely the definition of orthogonality (Eq. (20.14)).

In the previous example the projection of y onto the columns of X (or the calculation of the predicted value of y) was computed from:

$$p = X * b = X * (X^T X)^{-1} X^T y = P\,y \tag{20.46}$$

20.9 Eigen Values and Eigen Vectors

An eigenvector of a square nxn matrix A is a nonzero vector x such that $(A - \lambda I)\,x = 0$ for some scalar λ. A scalar λ is called an eigenvalue of A if there is a non-trivial solution x of $Ax = \lambda x$, such an x is called an eigenvector corresponding to λ.

It is straightforward to determine if a vector is an eigenvector of a matrix. In addition, it is easy to determine if a specified scalar is an eigenvalue.

Lay [1996] gives an illustrative example of eigenvalue and eigenvector. Let

$$A = \begin{bmatrix} 1 & 6 \\ 5 & 2 \end{bmatrix}, u = \begin{bmatrix} 6 \\ -5 \end{bmatrix}, v = \begin{bmatrix} 3 \\ -2 \end{bmatrix} \tag{20.47}$$

then show that u is an eigenvector of A and v is not.

$$Au = \begin{bmatrix} 1 & 6 \\ 5 & 2 \end{bmatrix} \begin{bmatrix} 6 \\ -5 \end{bmatrix} = \begin{bmatrix} -24 \\ 20 \end{bmatrix} = -4 \begin{bmatrix} 6 \\ -5 \end{bmatrix} \tag{20.48}$$

$$Av = \begin{bmatrix} 1 & 6 \\ 5 & 2 \end{bmatrix} \begin{bmatrix} 3 \\ -2 \end{bmatrix} = \begin{bmatrix} -9 \\ 11 \end{bmatrix} \neq \lambda \begin{bmatrix} 3 \\ -2 \end{bmatrix} \tag{20.49}$$

thus u is an eigenvector of A, corresponding to an eigenvalue of –4, but v is not an eigenvector of A.

The eigenvalue can easily be obtained using MATLAB by typing: *A*=[1 6;5 2]; eig(*A*), the result is –4 and 7.

The eigenvalues can also be computed by using the determinant. For matrix A one computes:

$$\begin{aligned} det(A,\lambda) = \begin{vmatrix} 1-\lambda & 6 \\ 5 & 2-\lambda \end{vmatrix} &= (1-\lambda)(2-\lambda) - 30 = \\ &= (\lambda - 7)(\lambda + 4) \end{aligned} \tag{20.50}$$

from which the eigenvalues follow: $\lambda = 7$ and $\lambda = -4$.

Singular value decomposition is a convenient way to compute the eigenvalues of a non-square matrix. Suppose the A matrix is:

$$A = \begin{bmatrix} 1 & 2 \\ 3 & 4 \\ 0 & 0 \end{bmatrix} \tag{20.51}$$

In order to compute the eigenvalues and eigenvectors, the matrix is made square by multiplying it with its transpose:

$$W = A * A^T = \begin{bmatrix} 1 & 2 \\ 3 & 4 \\ 0 & 0 \end{bmatrix} \begin{bmatrix} 1 & 3 & 0 \\ 2 & 4 & 0 \end{bmatrix} = \begin{bmatrix} 5 & 11 & 0 \\ 11 & 25 & 0 \\ 0 & 0 & 0 \end{bmatrix} \tag{20.52}$$

Calculating SVD consists of finding the eigenvalues and eigenvectors of AA^T and A^TA. The eigenvectors of A^TA make up the columns of V, the eigenvectors of AA^T make up the columns of U. The singular values in S are the square roots of eigenvalues from A^TA and AA^T. The singular values are the entries on the diagonal of the S-matrix, arranged in descending order. U, S and V are defined in Eqn. (20.38).

The eigenvalues of W can be calculated from the condition $det(W) = 0$, or:

$$\begin{aligned} \begin{vmatrix} 5-\lambda & 11 & 0 \\ 11 & 25-\lambda & 0 \\ 0 & 0 & -\lambda \end{vmatrix} = (5-\lambda)\begin{vmatrix} 25-\lambda & 0 \\ 0 & -\lambda \end{vmatrix} - 11 \begin{vmatrix} 11 & 0 \\ 0 & -\lambda \end{vmatrix} = \\ -\lambda\left(\lambda^2 - 30\lambda + 4\right) = 0 \end{aligned} \tag{20.53}$$

from which the following eigenvalues can be computed:

$$\begin{aligned} \lambda_1 &= 29.8661 \\ \lambda_2 &= 0.1339 \\ \lambda_3 &= 0.0 \end{aligned} \tag{20.54}$$

The S-matrix becomes then:

$$S = \begin{bmatrix} \sqrt{\lambda_1} & 0 & 0 \\ 0 & \sqrt{\lambda_2} & 0 \\ 0 & 0 & \sqrt{\lambda_3} \end{bmatrix} = \begin{bmatrix} 5.465 & 0 & 0 \\ 0 & 0.366 & 0 \\ 0 & 0 & 0 \end{bmatrix} \tag{20.55}$$

For $\lambda = 29.8661$ we may write (see Eqn. (20.53)):

$$\begin{bmatrix} -24.8661 & 11 & 0 \\ 11 & -4.8661 & 0 \\ 0 & 0 & -29.8661 \end{bmatrix} \begin{bmatrix} v_1 \\ v_2 \\ v_3 \end{bmatrix} = \begin{bmatrix} 0 \\ 0 \\ 0 \end{bmatrix} \tag{20.56}$$

in which the vector v is the eigenvector, i.e. the first column of U. The eigenvector can be found through inverse iteration (Press *et al.* 1988). Similarly the eigenvector for $\lambda = 0.1339$ and $\lambda = 0.0$ can be found. Arranging the three eigenvectors in the U-matrix gives:

$$U = \begin{bmatrix} -0.4046 & -0.9145 & 0 \\ -0.9145 & 0.4046 & 0 \\ 0 & 0 & 1 \end{bmatrix} \tag{20.57}$$

The columns of V can be found by repeating the analysis for $A^T A$, it is then found that:

$$V = \begin{bmatrix} -0.5760 & 0.8174 \\ -0.8174 & -0.5760 \end{bmatrix} \tag{20.58}$$

References

Lay, D.C. (1996) *Linear Algebra and its Applications*, 2nd edn, Addison-Wesley.

Bretscher, O., (1997), *Linear algebra with applications*, Prentice Hall.

Press, W.H., Flannery, B.P., Teukolsky, S.A. and Vetterling, W.T. (1988) *Numerical Recipes in C – The Art of Scientific Computing, (Improving Eigenvalues and/or Finding Eigenvector by Inverse Iteration)*, Cambridge University Press.

Wise, B.M., Callagher, N.B., Bro, R., Shaver, J.M., Windig, W. and Koch, R.S. (2004) *PLS_Toolbox 3.5*, Eigenvector Research Inc.

21 Data Conditioning

Process data, which is used for the identification of a black box model, such as training of a neural network, has to be a good representation of the process considered. Therefore the data set has to be examined carefully and bad data will have to be removed from it. The data set should be further examined to determine the degree of correlation within and between the outputs and inputs of the process. This is needed to determine the structure of the model to be developed. In this chapter some aspects of the analysis of process measurements will be discussed.

21.1 Examining the Data

When data from a process or experiment are used for the development of a black box model, it first has to undergo a thorough examination. Much attention should be given to the data set, because it will determine how the final model will look like. Erroneous data could lead to a model that has poor prediction capability and bad generalization properties. Therefore the data set has to meet the following conditions:

- it has to be a good representation of the process
- it should have a high signal to noise ratio
- no bad data should be present in the set
- the data should preferably be continuous without missing data.

This means that the data set needs to be examined carefully and calculations should be made for suspicious data points that might be the result of bad measurements. This is explained further in the following sections.

To ensure that the data in the set gives an accurate representation of the process, the entire set has to be examined. This examination has to be made in person, because no software is available to achieve this. The data set should comply with the following guidelines:

- *The data set has to consist of a sufficient number of data points*
  For developing black box models, many data points are required. Depending on the number of parameters in the model, one could state that the number of data points should preferably be at least 10–50 times the number of model parameters. For low order models, the number could probably be at the low end, for higher order models the number of points should be at the high end. However, the more data points are available, the more accurate the process model will be. Especially for the purpose of training a neural network, the number of measurements in the data set must be large. To train a neural network (see chapter 27) a training set as well as a test set of data is needed. An absolute minimum for a training-set of process data for a neural network of a reasonable size is 1000 measurements.
- *The data within the set has to have sufficient range*
  The entire range of values that the process measurement can assume, should be present within the data set. Although some black box models have reasonable generalization capabilities, no risk should be taken. The maximum as well as the minimum values of the process measurements should be present, together with a wide variety of other process states.

Process Dynamics and Control: Modeling for Control and Prediction. Brian Roffel and Ben Betlem.

- *The data set has to contain sufficient dynamic information*
  If a dynamic process model is required, the data set should contain a large number of changes. It is difficult to give an exact number, because it depends of many factors such as the size of the data set, the process dynamics, etc. The changes should be diverse in size and sign, thus generating an information rich signal.
- *The data set has to contain sufficient information about the low frequency and gain of the process*
  In case a dynamic model has to be developed, the measurement changes in the data set should reach their steady-state values. If the measurement is constantly changing and never reaches a steady-state value, the static behavior of the model will be poor and the process gain of the model will be underestimated. It is best that the entire range of frequencies the process can experience, is present. A frequency analysis in the form of an amplitude plot can be made to verify this.
- *The data should be sampled at a sufficiently high frequency*
  If the data sampling frequency is not correctly chosen, the true process behaviour will not be captured. A reasonable sampling frequency is of the order of $\Delta t = (1/3 \ldots 1/5)\tau$, where τ is the main process time constant.

Ensuring that the data contains sufficient information and has an adequate range is sometimes difficult to realize for controlled processes. The controlled variable will often not be represented in the model, since it contains little variance because it is controlled. However, the controlled variable may be one we are interested in; hence, some data should be collected with the controller in the manual mode (i.e. with an uncontrolled process).

21.2 Detecting and Removing Bad Data

When the data set is being examined visually, several measurements can already be marked as potentially bad data, especially the measurements that lie considerably out of the expected range of the process. But there are also bad data that are not directly visible. Therefore several methods have been developed to recognize this. One of these methods, the principal component analysis, will be discussed in the following chapter, but its merits will already be shown here.

In principal component analysis, the process input data can easily be analyzed: outliers can be detected and redundant measurements identified. Process inputs and output data can be combined in one data set.

A matrix $A(I*J)$ is constructed containing all the process input data. This matrix is decomposed into a set of scores $T(I*K)$ and loadings $P(J*K)$, where K is the number of principal components chosen so as to explain the important variation in the data using as few orthogonal components as possible.

In a principal component model, each principal component is a linear combination of the original process variables defined in the data set. For a process with 10 process variables (file pv.mat), a principal components analysis was made using the PLS toolbox (Eigen vector research, 2004) and the result is shown in Table 21.1:

Table 21.1. Rercent variance captured by PCA model.

principal component number	eigenvalue of cov(x)	% variance captured by this principal component	total % variance captured
1	3.81	38.14	38.14
2	2.38	23.78	61.92
3	1.13	11.30	73.21
4	0.94	9.41	82.63
5	0.70	7.01	89.64

Almost 90% of the variance in the data set can be explained by the first five principal components (PC).

A scores plot of PC1 versus PC2, using a 5 PC model, is shown in Fig. 21.1. The plot shows that most of the batches are within the 95% confidence limits for the model. It can clearly be seen that three batches, namely the batches with numbers 4, 8 and 84 fall far outside the 95% confidence limit. There are some other batches which fall just outside the confidence limits. This way outliers can easily be identified and removed from the data set.

Also from the so-called Q-residuals plot (error plot), it can be seen (Fig. 21.2) that some batches behave in an abnormal way and should be removed from the data set, because they have an exceptionally large error contribution to the PC model.

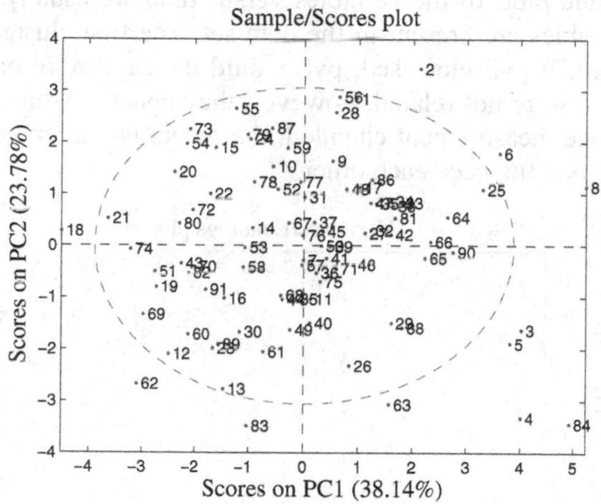

Fig. 21.1. Scores plot from the principal component analysis.

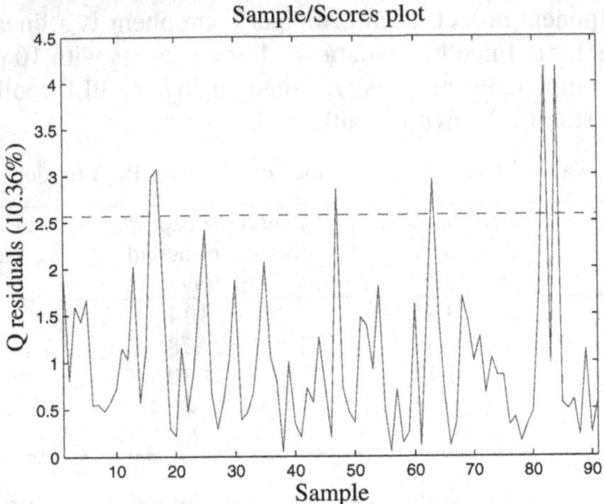

Fig. 21.2. Q-residuals plot.

Combining the information from both plots, one would start with removing batch 84 and create new figures for the reduced data set and repeat the elimination process.

Figure 21.3 shows a loadings plot from the same principal component analysis, principal component two is plotted versus principal component three. One could also plot other principal component combinations. As can be seen, there are two clusters and each cluster consists of more than one measurement. The first cluster comprises the measurements pv2 and pv8, the second cluster consists of the measurements pv4, pv6, pv7 and pv9. Using physical insight and knowledge and plots of the variables versus time are usually sufficient to identify whether duplicate variables are present in the data set. The first cluster contained measurements that were related, if pv2 increased, pv8 would increase more or less linearly. Measurements in cluster two were not related; however, they constituted measurements in one of the raw materials. If one measurement changed, the others would change as well and therefore these measurements influenced each other.

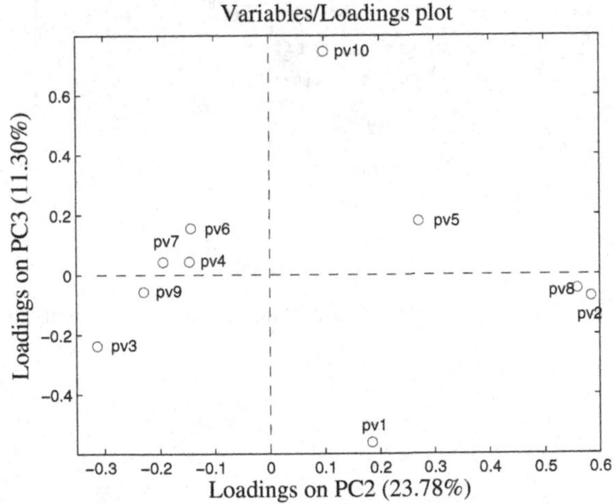

Fig. 21.3. Loadings plot from the principal component analysis.

Imagine a chemical reactor where several variables in and around the reactor are measured. When the reactor is ideally mixed and the top and bottom temperature in the reactor are measured, the two measurements will fall together in one cluster. One could easily eliminate one measurement because the second measurement would not contribute anything to the model and only hamper the generalization capability of the model.

Sometimes measurements are evenly distributed in the scores plot but one or more measurements fall distinctly outside the envelope of the other measurements. Also in this case, the measurements that fall outside the envelope should receive special attention. It might be that these measurements were faulty, for example, owing to problems with the sensor. It is best to eliminate such a measurement from the data set, at least initially.

The details of the scores and loadings plot will be explained in chapter 21; however, as can already be seen, a principal component analysis can be useful in detecting outliers, abnormal patterns and redundant variables.

21.3 Filling in Missing Data

It could be that the measurement set that is collected is not complete. This is often the case if quality measurements are included in the data set. Quality indicators, such as chromatographs, are prone to failure. In that case usually measurement values of "zero" or "no good" are stored. These values should be removed from the data set. However, the gap that is created in the data set, now has to be filled with good values. Since no values are available, one could use one of the following three approaches:

- Use data interpolation techniques, such as smoothing splines (Friedman *et al*., 1983) or time series modeling (Box *et al*., 1994). In case of time series modeling, a time series model for the variable in question is developed and the missing data during the time frame that no measurements are available are estimated with the help of the model. This is a complicated approach, since it is difficult to develop a reasonable time series model for the variable in the data set for which the measurements are missing.
- Use linear interpolation between the last good measurement and the next first good measurement. This is often a procedure that gives acceptable results in the modeling process.
- Use a matching pattern approach. In this case, one looks for a similar pattern in the data set and copies that data to the missing time frame.

In general one could say that the second approach works reasonably well in many cases.

21.4 Scaling of Variables

The process variables that make up the data set usually relate to a variety of measurements, often with different physical meaning. Process measurements can often be divided into two groups: measurements that are indicative of or related to quality and measurements that are not directly related to quality. Process measurements that fall into the first category are usually compositions, pressures and temperatures; process measurements that fall into the latter group are usually flows, weights and volumes. Much of the variance in the latter group of variables can be explained by the variance of the process load. It is therefore useful to "normalize" these measurements by dividing them by the value of the process load.

Some method of scaling might be useful in order to avoid giving undue significance to the variables with high absolute values. The most common method is to autoscale the data by subtracting the mean and dividing by the standard deviation for each column:

$$X_{ij} = \frac{x_{ij} - \bar{x}_j}{\sqrt{\sigma_j}} \tag{21.1}$$

where X is an autoscaled matrix of dimension $I \times J$ and $\bar{x}_j$ and σ_j are the mean and the variance of the j-th column of the original matrix X.

21.5 Identification of Time Lags

It could be that a change in process input does not have an immediate impact on the change in process output in a certain operating point. If one is collecting steady-state data, this is not relevant, but if one is collecting dynamic data, this could become an important issue. In most modeling approaches, an estimate is required of the time lag between a process input and the corresponding output. This means that an estimate has to be made of the time that it takes before a change in the input reaches its largest impact on the output. This could be determined by calculating the cross correlation between the process input and output. It could also be estimated from process knowledge.

Suppose, a liquid flows through a pipeline and a valve position is changed at the beginning of the pipeline. If the temperature is measured at the end of the pipeline, the time lag, which is equal to the transportation time of the liquid from valve to measurement location, can easily be calculated. Figure 21.4 shows a typical cross correlation plot between a process input and a process output.

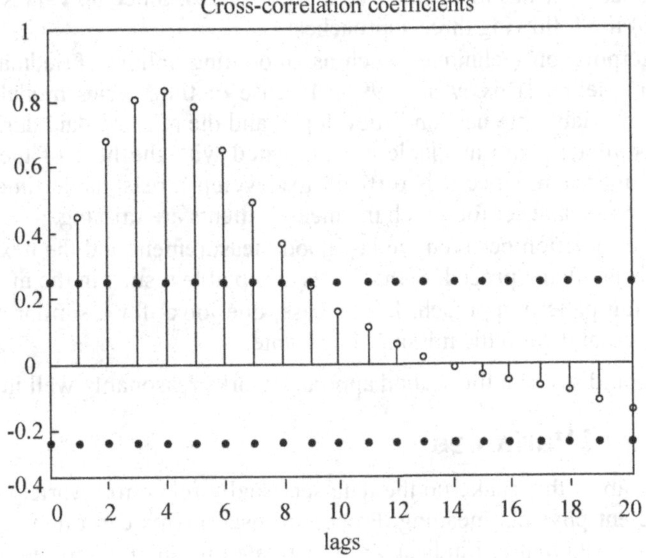

Fig. 21.4. Cross correlation between a process input and output.

As can be seen, there is a peak at $k = 4$, meaning that there is an estimated time lag between process input and process output of 4 sampling intervals. The process output should then be shifted four sampling intervals backward in time to make a static principal component analysis meaningful.

As mentioned, the cross correlation function is used to determine the correlation between two series of data. The most common use is to determine the correlation between the input and output of a process. It indicates how a measurement of the process output is related to the process input (also likely to be measurements), several sampling periods ago. It can also be used to determine the maximum number of neurons in the input layer of a neural network used for the historical input data (neural network with delayed inputs).

Given an input series $u_1, u_2, \ldots, u_n$ and an output series $y_1, y_2, \ldots, y_n$, the cross covariance coefficients c_{uy} can be calculated as follows:

$$c_{uy}(k) = \frac{1}{N}\sum_{t=1}^{N-k}(u_t - \overline{u})(y_{t+k} - \overline{y}) \qquad k = 0,1,2,\ldots \tag{21.2}$$

where $\overline{y}$ and $\overline{u}$ refer again to the means of the output and input series and N to the number of measurements. The dimensionless cross correlation $r(k)$ is calculated from:

$$r_{uy}(k) = \frac{c_{uy}(k)}{\sqrt{c_{uu}(0)\, c_{yy}(0)}} \tag{21.3}$$

In this equation $c_{uu}(0)$ and $c_{yy}(0)$ are the values of the auto covariance at lag 0 for the input and the output series.

Values of $r_{uy}(k)$ near to 1 or –1, imply high dependency, low values imply independence. In examining the cross correlation of a certain data set a boundary can be chosen below which independence is assumed. The value of the confidence limit for the cross correlation of process data is usually in the order of 0.15 to 0.3, depending on the number of points in the data set.

The calculation of the cross correlation is a linear technique. Sometimes it is also used for a non-linear process. This means that applying it to the entire range of process data at once would also give a distorted result. The cross correlation function can only be applied to single operating points of the process. Dividing the process range into several small parts, in which it can be assumed that the process behaves linear, will give better results.

21.6 Smoothing and Filtering a Signal

To develop an accurate model, the signal to noise ratio should be more than five. This means that changes in a particular process variable should be at least five times the value of the process noise for that variable. If this is not the case, one could use a smoothing or filtering technique to suppress the noise.

Processes and measurement devices often show a kind of stochastic behavior. Usually, we may assume that the process and sensor can be described by an original signal y' that is corrupted by an independent noise signal n. This is indicated in Figure 21.5. The measured signals y can be filtered or smoothed in order to eliminate the noise in a certain frequency range to retrieve a deduced signal $\hat{y}$, that should represent the original signal as much as possible.

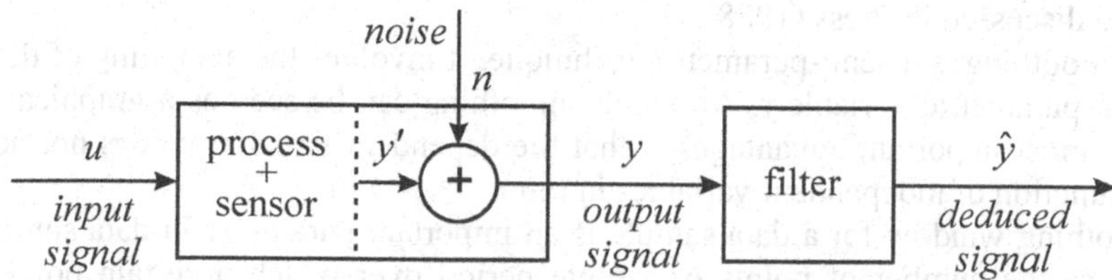

Fig. 21.5. Filtering of an output signal.

Smoothing and filtering are related techniques to reconcile a signal or a sampled data sequence to compensate for random errors or fluctuations. Filtering is the reconstruction of a signal by removing unwanted components; smoothing can be defined as reconciliation of data in space or time of a scale significant smaller than the problem at hand. Both techniques are related to data modeling by polynomial least squares fitting or optimal filtering.

The following classifications concerning signal smoothing and filtering can be made:

- real-time versus off-line
- causal versus acausal
- recursive versus non-recursive
- time domain versus frequency domain

Simple techniques in the time domain are first-order filtering or advanced averaging. These methods are usually applied in real-time applications, where the rate of filtered data should follow the rate of sampled data. Off-line operation offers the possibility to try out different kinds of filters or to adapt a certain filter parameter setting to attain the most suited result. The data set considered can be the complete set or a broad window including future data.

A filter is causal if its output at a certain time moment depends only on the input data at that moment or earlier. If the output also relies on future input data, then the filtering is acausal.

The techniques are called recursive when the output depends on the recent input data points as well as on the recently calculated outputs. Conversely, the technique is non-recursive when the result only depends on the data points. An example of off-line smoothing is averaging over a time window or number of data points. This is a non-recursive technique as the result only depends on the measured data.

For calculations in the frequency domain, the sampled data should be converted, for instance, by fast Fourier transform (FFT). The advantage is that a specific frequency range can be excluded. Examples are low-pass, notch or high-pass filtering to eliminate noise in the high, specific, or low frequency range respectively. For optimal filtering in the frequency domain, it is convenient to consider the entire data record. Usually, filtering in the frequency domain is off-line, non-recursive and acausal.

21.6.1 Data Smoothing

Data smoothing is perhaps most justified when it is used as a graphical technique, to guide the eye through a forest of data points. Therefore one or more criteria should be set, to obtain a data set that is satisfying. In this chapter a brief explanation of data smoothing is given, without making a full mathematical justification of the technique. Several smoothing techniques are discussed in Press (1988).

Data smoothing is a semi-parametric technique, it involves the averaging of the measured dependent parametric variable y. Although smoothing can be seen as a graphical technique, one of its most important advantages is that the dependent variable y does not need to be a specific function of independent variables in time.

The smoothing window for a data sample is an important parameter in data smoothing. It is specified as the number of points or sample period over which a certain point or sample should be reconciled, also called "the amount of smoothing". The point that is adapted can either be located in the middle of the window (off-line smoothing) or at the end of the window (on-line smoothing).

A window-size of zero gives no smoothing at all, whereas any number larger than half the number of data points will render the data virtually featureless. Therefore the smoothing window should be optimized according to a pre-determined criterion. When the undisturbed data are known, an error can be calculated between the undisturbed data and the smoothed data. Also the difference between the original signal with noise and the smoothed line can be calculated. The smoothing window can be optimized by minimizing one of these errors.

21.6.2 Time Domain Filtering

The most general filter is a linear recursive algorithm that calculates a deduced sampled signal $\hat{y}_n$ based on the samples y_n:

$$\hat{y}_n = \sum_{i=0}^{M} c_i y_{n-i} + \sum_{j=1}^{N} d_j \hat{y}_{n-j}$$

while: (21.4)

$$\sum_{i=0}^{M} c_i + \sum_{j=1}^{N} d_j = 1$$

Assume a simple first-order filter:

$$\hat{y}_n = a y_n + (1-a)\hat{y}_{n-1} \tag{21.5}$$

in which a is a weighing of the last data point. This equation can be rewritten into a discrete differential equation:

$$\frac{\Delta t}{a}\frac{\hat{y}_n - \hat{y}_{n-1}}{\Delta t} = (y_n - \hat{y}_{n-1}) \tag{21.6}$$

The time constant of this discrete filter is $\Delta t/a$. If more recent data points have to be considered, a should be smaller. Points from the past receive an exponentially lower weight.

21.6.3 Frequency Domain Filtering

When a signal y is corrupted with noise, filtering in the frequency domain can partly restore this signal into the deduced signal $\hat{y}$:

$$\hat{Y}(f) = F(f)Y(f) \tag{21.7}$$

in which $F(f)$ is the filter equation. In the most common situation we have an evenly spaced sampled signal. The sample period Δt is measured in seconds and f in cycles per second. Up to now, the angular frequency ω has been used in this book. In case of the discrete transform, it is common to use f, where:

$$\omega = 2\pi f \tag{21.8}$$

A simple low-pass filter that suppresses the higher frequency can be given by:

$$F(f) = \frac{1}{1 + 2\pi j f \tau} \tag{21.9}$$

The turn-over frequency of this filter is $1/\tau$. This filter resembles the filter in the time domain mentioned earlier in Eqn.(21.6).

The filtering technique in the frequency domain that is most frequently used performs the following actions:

- remove the linear trend from the data;
- use a Fast Fourier Transform on the data;

- use the filter algorithm $F(\omega)$ on the entire data set or in case of smoothing use a certain smoothing window,
- perform inverse Fourier transformation;
- reinsert the linear trend.

With this technique, it is possible to remove a certain part of the signal in the frequency domain. This enables us to remove the high frequencies from the data, while keeping the lower frequencies. Also a combination between filtering and smoothing can be applied. First specific frequencies can be filtered considering the entire signal, followed by smoothing of the signal by considering a suited window.

The fast Fourier transform will be explained briefly. Subsequently, an explanation is given on how data is filtered, passing only the low frequencies.

21.6.4 Fast Fourier Transforms

For a sampled signal in the time domain $y(t_k)$ with the sample interval Δt, the discrete Fourier transform of a finite number of samples results in a discrete counterpart signal $Y(f_j)$ in the frequency domain. The discrete Fourier transform can directly be derived from the continuous transformation (Eqn. (5.1)):

$$Y_i = \sum_{k=0}^{n-1} y_k e^{-2\pi j f_i t_k} \tag{21.10}$$

The discrete Fourier transform represents a signal in the frequency domain and consists of n frequency values. The inverse discrete Fourier transform of the frequency domain signal is :

$$y_k = \frac{1}{n} \sum_{i=0}^{n-1} Y_i e^{2\pi j f_i t_k} \tag{21.11}$$

The definitions used, are:

$$Y_i \approx \frac{Y(f_i)}{\Delta t}, \; f_i \equiv \frac{i}{n\Delta t} \text{ with } i = -\frac{1}{2}, \ldots, \frac{1}{2} \tag{21.12}$$

and:

$$y_k = y(t_k), \; t_k \equiv k\Delta t \text{ with } k = 0,1,2,\ldots,n-1 \tag{21.13}$$

The discrete Fourier transform maps n complex numbers y_k into n complex numbers Y_i. The mapping of Eqns. (21.10) and (21.11) is only slightly different. By using Eqn. (21.12), Eqn. (21.10) is made independent of the sample interval.

When the sample interval is Δt, then the frequency range covered is $-f_c < f < f_c$, where f_c is the Nyquist critical frequency:

$$f_c \equiv \frac{1}{2\Delta t} \tag{21.14}$$

One should be aware that when the original signal is not limited to this band-width, all the power of the signal which lies outside this band-width is shifted into this range. This phenomenon is called aliasing.

With FFT it is possible to compute the discrete Fourier transfer in the order of $\log_2 N$ in stead of N^2 complex multiplications. FFT is based on the idea that a discrete Fourier transform of length n can be rewritten as the sum of two discrete Fourier transforms, one with even and one with odd subscripts, each of length $n/2$. In case n is not only even, but is expressed in a power of two, the process can be repeated and the number of required calculations is minimal. If the length of the data set is not a power of two, it is recommended to pad the data up with zeros to the next power of two. There are several variants of FFT. The most common one rearranges the input elements into bit-reverse order (Press *et al.*, 1988).

21.6.5 Optimal Filtering

An application of filtering, frequently used, is the removal of noise n of a measured signal y to reconstruct the uncorrupted signal y' (see Fig. 21.5).

$$y(t) = y'(t) + n(t) \quad \text{or} \quad Y(f) = Y'(f) + N(f) \tag{21.15}$$

where $Y(f)$, $Y'(f)$ and $N(f)$ are the discrete Fourier transform of the measured signal, the original signal and the noise respectively. We like to design an optimal filter F in the frequency domain, which when applied to the measured signal Y, can produce a signal $\widehat{Y}$, that is close to the uncorrupted signal Y'.

$$\hat{Y}(f) = F(f)\,Y(f) \tag{21.16}$$

In a least-square sense this means that:

$$\int_{-\infty}^{\infty} \left|\hat{Y}(f) - Y'(f)\right| df \text{ is minimal} \tag{21.17}$$

From substitution of Eqns. (21.15) and (21.16) into (21.17) it follows that:

$$\int_{-\infty}^{\infty} \left|(Y'(f) + N(f))F(f) - Y'(f)\right| df \quad \text{is minimal} \tag{21.18}$$

When the noise N is uncorrelated with signal Y', which is actually the definition of noise, then by differentiating with respect to the independent variable F, it can be derived (Economou *et al.*, 1996) that an ideal (Wiener) filter can be written as:

$$F(f) = \frac{|Y'(f)|^2 - |N(f)|^2}{|Y'(f)|^2} \tag{21.19}$$

Functions like $|Y(f)|^2$ are actually power spectral density functions. An estimation by means of FFT usually suffices. To calculate the FFT, a rather long period of a signal $y(t)$ has to be considered. Next, the power can be calculated according to Eqns. (21.10) and (21.12):

$$|Y(f)|^2 = |\Delta t\, Y_i|^2 \tag{21.20}$$

One has to consider that in case of FFT, the calculated power is not a continuous function but is only defined at specific frequencies. More sophisticated methods to determine the density power spectrum from FFT, by means of data windowing or pole approximation (maximal entropy method), can be found in Press *et al.* (1988).

To determine the optimal filter, $|N(f)|^2$ should be estimated. $|N(f)|^2$ can be obtained by plotting $|Y(f)|^2$ versus f. It generally will look like the curve shown in Fig. 21.6 (Press, 1988).

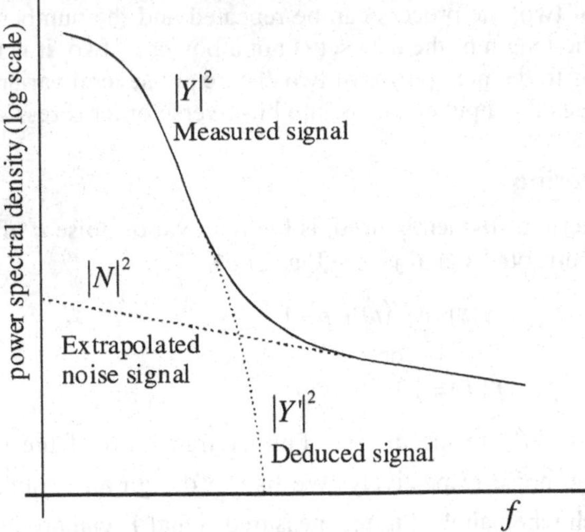

Fig. 21.6. Estimation of noise component.

This is just an example, the noise spectrum may have different forms. Usually, the original signal $Y'(f)$ will show a peak for low frequencies. The tail end of the curve represents mainly the noise component, it is extrapolated back to the signal region to give the noise model $N(f)$. According to Press *et al.* (1988), the model does not need to be accurate for the method to be useful.

21.7 Initial Model Structure

There are several well-known model structures that are often used in non-linear modeling. Some that are used frequently are:

- Non-linear Finite Impulse Response (NFIR) models, which use u_{t-k} as regressors, $k = 1, \ldots N$
- Non-linear AutoRegressive with eXogenous inputs (NARX) models which use u_{t-k} and $\hat{y}_{t-k}$ as regressors
- Non-linear Output Error (NOE) models which use u_{t-k} and $\hat{y}_{t-k} - \hat{e}_{t-k}$ as regressors
- Non-linear AutoRegressive Moving Average with eXogenous inputs (NARMAX) models which use u_{t-k}, $\hat{y}_{t-k}$ and $\hat{e}_{t-k}$ as regressors

The regressors NOE and NARMAX are recurrent structures since the prediction at the next time step is dependent on the prediction on the previous time step. It is generally more difficult to work with recurrent structures.

Although selection of an initial model structure is not really part of the data conditioning exercise, it will be dealt with here, since bad data can have a profound effect on the apparent model structure and the model performance.

Some tools that are useful for determining an initial model structure are:

- calculation of the autocorrelation of the output(s) of the process;
- residuals analysis

The auto correlation function is used to determine the degree of correlation of a data set $y_1, y_2, ..., y_n$ within itself. It indicates how a measurement at a certain time is related to measurements taken several sampling periods ago. It is used to determine whether the process output at time k depends only on the process inputs or also at the process output at time $(k–1)$ and possibly at time $(k–2)$.

The theoretical autocovariance, $c_{yy}(k)$, at lag k can be estimated by:

$$c_{yy}(k) = \frac{1}{N}\sum_{t=1}^{N-k}(y_t - \bar{y})(y_{t+k} - \bar{y}) \qquad k = 0,1,2,... \tag{21.21}$$

where $\bar{y}$ is the average of the data set, and N the number of measurements.

It is more common however to calculate the normalized autocorrelation $r_{yy}(k)$:

$$r_{yy}(k) = \frac{c_{yy}(k)}{c_{yy}(0)} \tag{21.22}$$

The auto correlation function of a data set is always symmetrical around $k = 0$, hence $r_{yy}(k) = r_{yy}(-k)$. From Eqn. (21.22) it is easy to see that $r_{yy}(0) = 1$, but it can also be proven that $|r_{yy}(k)| < 1$ for any $k \neq 0$.

Values of $r_{yy}(k)$ near to 1 or –1, imply high dependency, low values imply independence. In examining the autocorrelation of a certain data set a boundary can be chosen below which independence is assumed. Several different values for these so-called confidence limits are used, depending on the application for which it is used. A very common value for the confidence limit, for the auto correlation of process data is a value between 0.15 and 0.3. This is a rough estimation of the confidence limit $C(k)$ function defined by Box and Jenkins (1994) which states:

$$C(k) = \frac{2}{\sqrt{N}} f\left(r_{yy}(k)\right) \tag{21.23}$$

The calculation of the autocorrelation is a linear technique that is often also applied to a non-linear process. This means that applying it to the entire range of process data at once would give a distorted result. For a non-linear process it should only be applied to single operating points or regions.

After an initial model has been developed, one should check the autocorrelation of the residuals (error between predicted variable and measured variable). For a good model, the autocorrelation of the residuals should be small. In addition, there should be no cross correlation between the process input and residuals. If either the autocorrelation or cross correlation of the residuals is not small, the model structure should be expanded, usually resulting in a more complex model.

However, it could also be that there is still some bad data present in the data set. It will then not be possible to describe it with the selected model and thus it will lead to auto and cross correlation of the residuals. It should be noted that bad data is not always data that represents exceptional readings of the process variable. It could also be that during a certain period of time a disturbance entered the process. During the time that the disturbance is active, data that is collected during the process upset, will be hard to describe with a model that does not have a disturbance model. The data collected during the process upset should either be removed from the data set, or a disturbance model should be added to the process model.

Files referred to in this chapter:
pv.mat: data for 10 process variables

References

Box, G.E.P. and Jenkins, G.M. (1994) *Time Series Analysis: Forecasting and Control*, 3rd edn, Holden-Day.

Economou, A., Fielden, P.R. and Packham, A.J. (1996) Wiener filtering of electro-analytical data by means of fast Fourier transform. *Analytica Chimica Acta*, **319** (1–2), 3–12.

Friedman, J.H., Grosse, E. and Stuetzle, W. Multidimensional additive spline approximation. *SIAM Journal of Scientific and Statistical Computing*, **4**, 291–301.

Press, W.H., Flannary, B.P. Teukolsky, S.S. and Vetterling, W.T. (1988) *Numerical Recipes: Example Book (C)*, Cambridge University Press.

Wise, B.M., Callagher, N.B., Bro, R., Shaver, J.M., Windig, W. and Koch, R.S. (2004) *PLS_Toolbox 3.5*, Eigenvector Research Inc.

22 Principal Component Analysis

From a statistical perspective, principal component analysis (PCA) is a method for reducing the dimensionality of data sets by transforming correlated variables into a smaller set of uncorrelated variables and finding linear combinations of the original variables with maximum variability.

As chemical processes become increasingly instrumented and data is recorded more frequently, both data compression and extraction become important. However, the information should be compressed in such a way that the essential features of the data are safeguarded and the new data is more easily displayed than the original data.

PCA is a tool, initially used by chemometricians for data compression and extraction of analytical instruments. Nowadays, it is a tool that is also frequently used to analyze data from process plants, where process data is often correlated. PCA offers the advantage that it finds a set of new uncorrelated variables that are a linear combination of the original variables. The new variables describe the maximum variance in the original data set in ascending order.

22.1 Introduction

Principal component analysis (PCA) can be performed on data matrices, consisting of raw data. Analysis of raw data can be used to estimate the importance and characteristics of the individual variables. The data matrix for n variables measured by m data points is given by:

$$X = \begin{bmatrix} x_{11} \; x_{12} \; \ldots \; x_{1n} \\ x_{12} \; x_{22} \; \ldots \; x_{2n} \\ \ldots \\ x_{m1} \; x_{m2} \; \ldots \; x_{mn} \end{bmatrix} \tag{22.1}$$

where each column represents a variable and each row a time slot or measurement value. x_{mn} is therefore the n-th variable measured by the m-th sample point.

PCA entails transforming n original variables $x_1, x_2, \ldots, x_n$ into an ordered sequence of n new variables $y_1, y_2, \ldots, y_n$, such that the new variables are uncorrelated with each other and account for decreasing portions of the variance of the original variables. The new variables are called "principal components" and they are estimated from the eigenvectors of the covariance or correlation matrix of the original variables.

If the first few principal components capture most of the original variability, then they can be used to represent the original variables in a lower dimension.

PCA is often used to show relationships between variables and to identify outliers of data sets (see chapter 21).

Process Dynamics and Control: Modeling for Control and Prediction. Brian Roffel and Ben Betlem.

22.2 PCA Decomposition

Mathematically, PCA relies upon eigenvector decomposition of the covariance or correlation matrix of the original process variables. For a given data matrix X with m rows (data points) and n columns (variables), the covariance or correlation matrix of X is defined as :

$$R = \mathrm{cov}(X) = \frac{X^T X}{m-1} \tag{22.2}$$

where the columns of X have been mean centered, that is, the mean has been subtracted from each column.

If the columns of X have been autoscaled, i.e. the mean subtracted from each column and divided by the standard deviation, then Eqn. (22.2) is the correlation matrix of X.

In PCA, X is always mean centered or autoscaled.

PCA now decomposes the data matrix X into the sum of the outer product of a so-called score vector t_i and a so-called loading vector p_i with a residual error E:

$$X = t_1 p_1^T + p_2 t_2^T + \ldots + t_k p_k^T + E = TP^T + E \tag{22.3}$$

where $k<\min(m,n)$. The score vector t_i contains information on how samples relate to each other. The loading vector p_i contains information on how variables relate to each other.

Using singular value decomposition, the covariance matrix R can be decomposed into

$$R = \sum_{j=1}^{n} p_j \lambda_j p_j^T = P \Lambda P^T \tag{22.4}$$

where $P=[p_1,p_2,\ldots p_n]$ is an orthogonal nxn matrix of characteristic vectors or eigenvectors (i.e. the first column is the first eigenvector of R) and Λ is a diagonal matrix of characteristic roots or eigenvalues associated with the eigenvectors p_i:

$$\Lambda = diag(\lambda_1, \lambda_2, \ldots \lambda_n) = \begin{Bmatrix} \lambda_1 & 0 & \ldots & 0 \\ 0 & \lambda_2 & 0 & 0 \\ \ldots & \ldots & \ldots & \ldots \\ 0 & 0 & 0 & \lambda_n \end{Bmatrix} \tag{22.5}$$

The eigenvalues are arranged in decreasing order:

$$\lambda_1 \geq \lambda_2 \geq \ldots \geq \lambda_n \tag{22.6}$$

and because P is orthogonal, one can also write:

$$P^T R P = \Lambda \tag{22.7}$$

Because the p_i vectors are the orthonormal eigenvectors of the covariance matrix R, it can also be written that:

$$R\, p_i = \lambda_i\, p_i \tag{22.8}$$

The t_i vectors form an orthogonal set ($t_i^T t_j$=0 for $i \# j$), while the p_i are orthonormal ($p_i^T p_j$=0 for $i \# j$, $p_i^T p_j$=1 for $i = j$). Because of these special properties, the equation for the scores t_i becomes:

$$t_i = Xp_i \tag{22.9}$$

where the score vector is a linear combination of the original X data defined by p_i.

Assume that the n measured variables form an m-dimensional space. The first step is to find the direction of the largest variance in the data set, this is the first loading vector and becomes the major axis (Fig. 22.1). The projection of the measurements on this axis is the first score vector. The next loading vector is perpendicular to the first one and determines the direction of the second largest variation. An associated score vector can then be determined and the process is terminated until the number of directions is equal to the number of measured variables. However, usually one does not proceed to the end, since the last principal components merely describe the noise in the data set.

Example

Assume that 500 data points for 5 measurements x_1 ... x_5 were collected, the first column of $X5$ is the measurement x_1, and so forth. Matrix $X5$ is first autoscaled and subsequently the matrix P and Λ can be calculated using MATLAB: [P,L] = eigs(R), where R is calculated from Eqn (22.2).

```
load X5;
X5as = ascale(X5.data);
aX   = X5as.data;
[nrows,ncols]=size(aX);
R = aX'*aX/(nrows-1)
[P,L] = eigs(R)
```

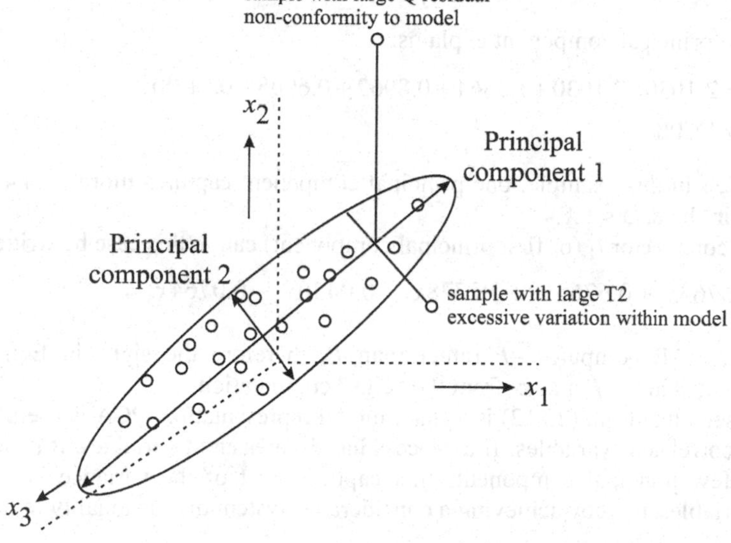

Fig. 22.1. Principal components for a data set (Wise *et al.*, 2004).

The result is (see program pcacalc.m):

```
P = -0.6763     0.0723     0.0625    -0.1844     0.7067
    -0.6765     0.0781     0.0229    -0.1896    -0.7069
    -0.2778    -0.5989    -0.0739     0.7474    -0.0031
     0.0430    -0.6167    -0.5756    -0.5350     0.0156
     0.0764    -0.4997     0.8116    -0.2918    -0.0237

L = 2.1030          0          0          0          0
         0     1.3864          0          0          0
         0          0     0.8962          0          0
         0          0          0     0.6005          0
         0          0          0          0     0.0138
```

One can easily verify Eqns. (22.4), (22.7) and (22.8).

Another way of looking at PCA is that the score vector t_i are the projections of X onto the p_i. Because the t_i,p_i pairs are arranged in descending order of λ_i, the first pair capture the largest amount of information of any pair in the decomposition. The λ_i can be seen as a measure of the amount of variance described by the t_i,p_i pair.

As can be seen from the values of P and L, the characteristic roots or eigenvalues of R are the diagonal elements of L, the first column of P is the first characteristic vector or eigenvector of R, the second column is the second eigenvector of R, etc.

22.3 Explained Variance

The portion of the total variance associated with the j-th characteristic root or eigenvalue is:

$$v_j = \lambda_j \Big/ \sum_j \lambda_j \tag{22.10}$$

Thus, the first principal component explains:

$$\begin{aligned} v_1 &= 2.1030/(2.1030+1.3864+0.8962+0.6005+0.0138) \\ &= 42.1\% \end{aligned} \tag{22.11}$$

As can be seen in this example, one principal component captures more than 40% of all of the variance in the data set X.

The first score vector t_1 (or first principal component) can in this case be written as:

$$t_1 = 0.6763x_1 + 0.6765x_2 + 0.2778x_3 - 0.0430x_4 - 0.0764x_5 \tag{22.12}$$

Note: MATLAB computes $-P$ rather than P, therefore the signs in Eqn (22.12) are opposite the signs in the P matrix from the "eigs" computation).

It can be seen that Eqn. (22.12) is a static model representation. PCA is useful for systems with many (correlated) variables. If a process has 40 measured variables, it is often possible to define a few principal components that capture most of the variance in the originally measured variables, thereby achieving a considerable system dimensionality reduction.

22.4 PCA Graphical User Interface

The PLS toolbox by Eigenvector Research (2004) is a very useful tool to perform a principal component analysis. When the toolbox is installed under Matlab, it can be called by typing pca. The following graphical user interface appears:

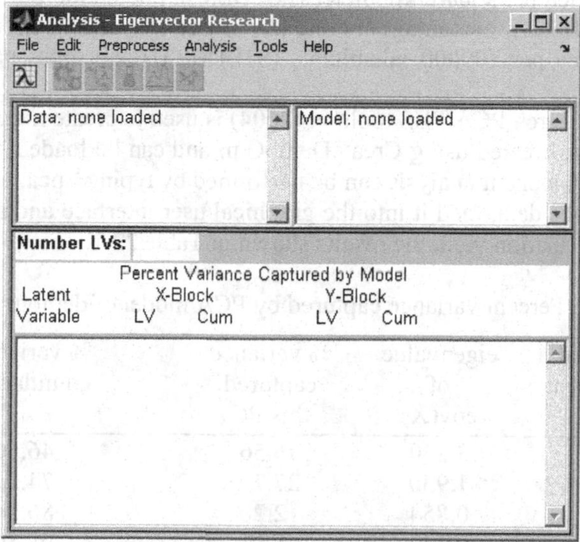

Fig. 22.2. PCA graphical user interface.

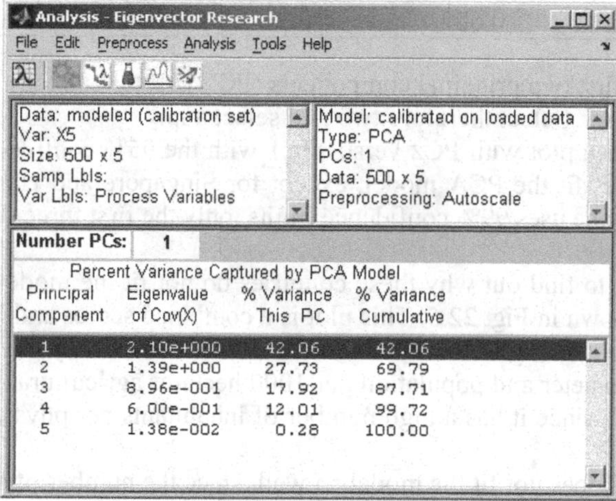

Fig. 22.3. PCA calculations for data set X.

When the data set X5 is loaded into the workspace (using load X5), go to "File", select "Load Data", "load X-block" and load X5. Data auto-scaling should be performed. You could select Preprocess to change this. Now calculate the principal components by selecting "calc", the window as shown in Fig. 22.3 will appear.

As can be seen, the first eigenvalue of the covariance matrix is 2.1 and the percentage explained variance by the first principal component is 42.06%, which is in agreement with Eqn (22.11).

22.5 Case Study: Demographic Data

The case study in this section is taken from Gunst and Mason (1980). The authors used a data set from Loether *et al* (1974), in which for 49 countries the following 7 items were collected: infant death per 1000 live births (INFDPTB), number of inhabitants per physician (INHPHYS), population per square kilometer (POPDENS), population per 1000 hectares agricultural land (AGRIDENS), percentage literate above age 14 (LITPERC), number of students with higher education per 100000 inhabitants (STUDHIED) and gross national product per capita (GNP).
The Eigenvector Research PCA/PLS toolbox (2004) is used to create a data set object named demoG. This file was created using CreateDemoG.m and can be loaded by typing >load demoG. A principal component analysis can be performed by typing >pca.

Let us autoscale the data, load it into the graphical user interface and selecting calc. Principal component calculation yield the results shown in Table 22.1.

Table 22.1. Percent variance captured by PCA model of demographic data.

principal component number	eigenvalue of cov(X)	% variance captured this PC	% variance cumulative
1	3.260	46.56	46.56
2	1.950	27.79	74.34
3	0.854	12.20	86.54
4	0.442	6.32	92.86
5	0.288	4.11	96.97
6	0.165	2.65	99.62
7	0.027	0.38	100.00

As can be seen, the first two principal components (PC's) capture already close to 75% of the variation in the data set. Select therefore two and select apply.

Figure 22.4 shows a plot with PC2 versus PC1 with the 95% confidence limits. It can be seen that all countries fit the PCA model, except for Singapore and Hong Kong, India and the United States. If one uses 99% confidence limits, only the first three countries fall outside the limits.

If one would like to find out why these countries do not fit the model so well, one could look at the biplot, shown in Fig. 22.5. This plot is a combined scores and load plot.

Singapore and Hong Kong do not fit the model so well, since they have a very high population per square kilometer and population per 1000 hectares agricultural land. India does not fit the model so well, since it has a high number of inhabitants per physician and high infant death.

The United States does not fit the model so well, since the number of students with higher education is high, the GNP is high and the percentage literacy is high.

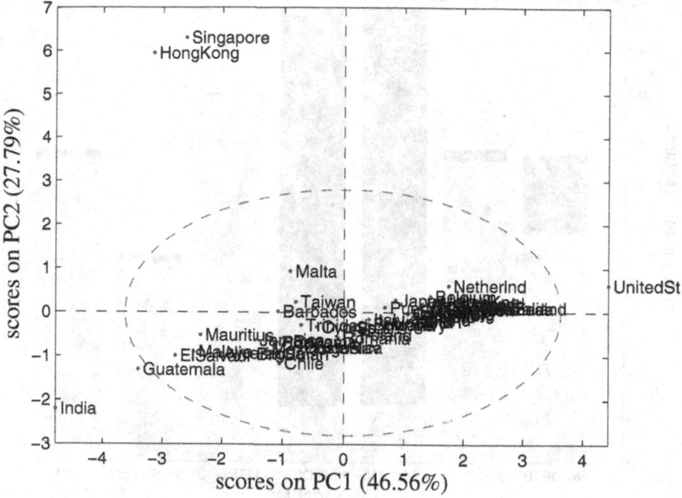

Fig. 22.4. Scores plot from PCA analysis.

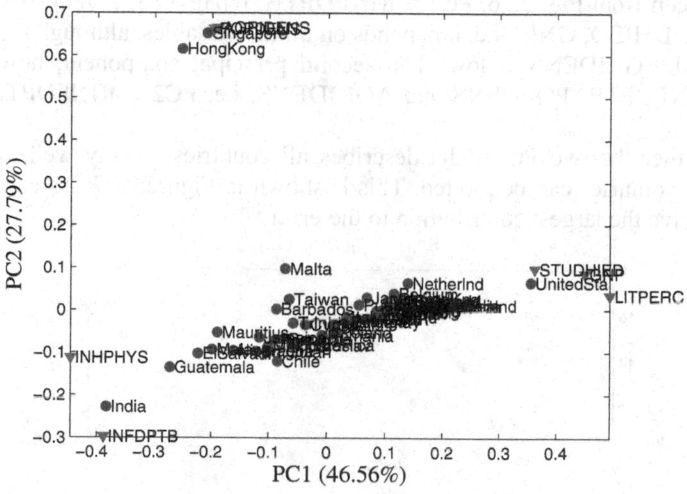

Fig. 22.5. Biplot from PCA analysis.

The scores plot gives us an idea of the relationship between samples; the loads plot gives us an idea of the relationship between variables.

It is also interesting to look at which variables make up principal component one and two. A two PC model is created and the variance captured is plotted versus the variable number as shown in Fig. 22.6.

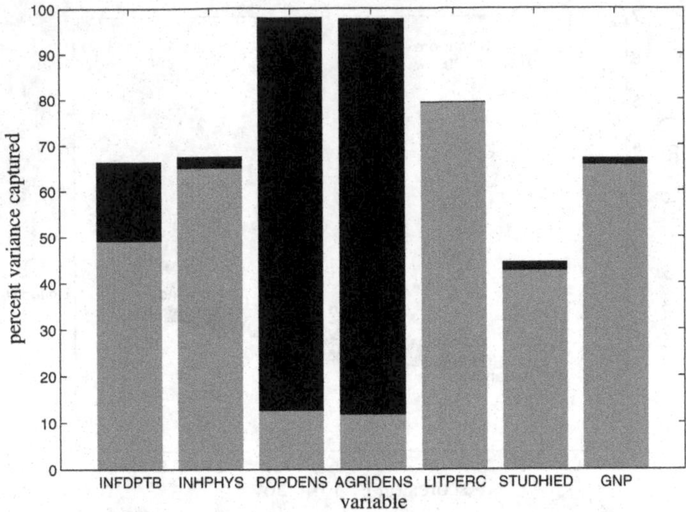

Fig. 22.6. Variance captured by a two PC model.

As can be seen from Fig. 22.6, PC1 = f(INFDPTB, INHPHYS, POPDENS, AGRIDENS, LITPERC, STUDHIED, GNP), i.e. it depends on all the variables, although the weighting for POPDENS and AGRIDENS is low. The second principal component, however, depends primarily on INFDPTB, POPDENS and AGRIDENS, i.e. PC2 = f(INFDPTB, POPDENS, AGRIDENS).

To see whether the two PC model describes all countries equally well, the error or Q-residual for all countries can be plotted. This is shown in Figure 22.7. As can be seen, India and the USA have the largest contribution to the error.

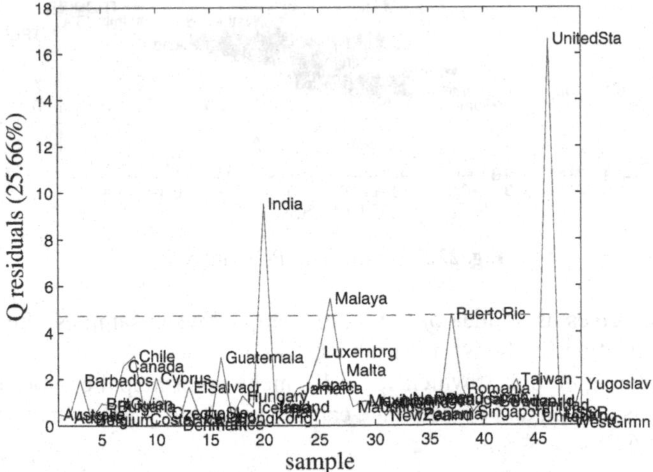

Fig. 22.7. Q-residuals for two PC model.

22.6 Case Study: Reactor Data

From a chemical reactor, agitator current, temperatures and some analytical data was collected. The example was briefly discussed in the example of section 22.2. In that section only the mechanism of calculating the PCA model was explained. Let us look at the model of Eqn. (22.12).

It is apparent that measurement x_1 and measurement x_2 have approximately the same coefficient in calculating the first score t_1. It can also be seen in Fig. 22.3 that two principal components capture close to 70% of the variation in the data set. Let us therefore create a PCA model with two principal components. The loading plot for this model is shown in Fig. 22.8.

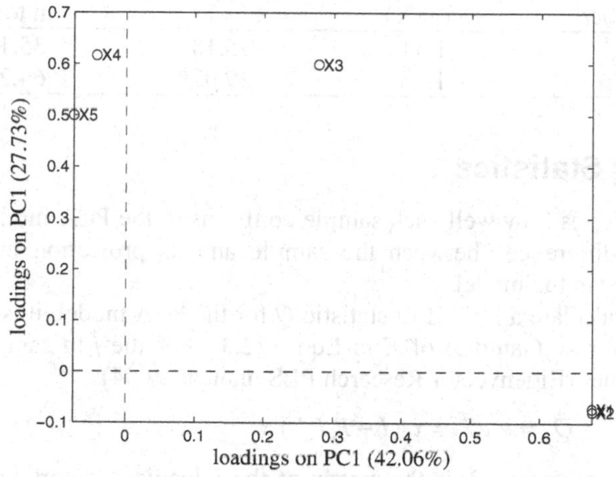

Fig. 22.8. Loading plot for two principal component model.

The variable numbers are also shown, they are numbered from $X1$ to $X5$. As can be seen, the loadings for variable $X1$ and $X2$ are virtually the same for both principal components (in the right lower corner of the figure). This might point to similarity in the measurements. Looking at the information flow diagram of the reactor, it was found that variable $X1$ was a temperature in the top of the reactor and variable $X2$ a temperature in the bottom of the reactor. They behaved very similar, in fact, they even showed the same dynamics but differed in actual reading by about 0.8 °C due to instrument offset and non-ideal mixing.

Therefore measurement $X1$ was deleted from the data set, because it did not provide additional information that was not captured by variable $X2$. Subsequently, a new PCA model was created and in this case 2 principal components explained more than 64% of the variance in the data set (Table 22.2).

Although it might seem that the latter PCA model is not as good as the first one, the model between the reactor output variable (the conversion) and the reactor input variables (stored in the X5 matrix) will be better.

Table 22.2. Percent variance captured by the PCA model of reactor dator.

principal component number	eigenvalue of cov(X)	% variance captured this PC	% variance captured in total
1	1.41	35.18	35.18
2	1.16	29.07	64.25

22.7 Modeling Statistics

The Q-statistic indicates how well each sample conforms to the PCA model (Fig. 22.1). It is a measure of the difference between the sample and its projection into the k principal components retained in the model.

One can easily calculate a lack of fit statistic Q for the PCA model, it is simply the sum of the squares of each row (sample) of E in Eqn. (22.3). For the j-th sample in X, it can be shown that x_j becomes (Eigenvector Research PLS manual, 2004):

$$Q_j = e_j e_j^T = x_j (I - P_k P_k^T) x_j^T \tag{22.13}$$

where e_j is the j-th row of E, P_k is the matrix of the k loading vectors retained in the PCA model (where each vector is a column of P_k) and I is the (nxn) identity matrix.

The sum of normalized squared scores, known as the Hotelling's T^2 statistic, is a measure of the variation in each sample within the PCA model. The T^2 statistic is defined as:

$$T_j^2 = t_j \lambda^{-1} t_j^T = x_j P \lambda^{-1} P^T x_j^T \tag{22.14}$$

where t_j refers to the j-th row of T_k, the matrix of k score vectors from the PCA model and λ^{-1} is a diagonal matrix of the inverse of the eigenvalues associated with the k principal componets retained in the model.

It is also possible to calculate confidence limits for the overall residual Q and T^2 (Jackson and Mudholkar, 1979).

Let us go back to the matrix of X data stored in X5.mat. The first column will be deleted and the last four columns will be retained. The function spca can be used to perform a principal component analysis and calculate the model statistics, or one can use the pca graphical interface.

```
load X5;
X4 = X5.data(:,2:5);
model = spca(X4,2);
```

The variance captured is:

Table 22.3. Percent variance captured by the PCA model of reactor data.

PC number	eigenvalue of Cov(X)	% variance this PC	total % var captured
1	1.410	35.18	35.18
2	1.160	29.07	64.25
3	0.884	22.09	86.34
4	0.546	13.66	100.00

The statistics based on 2 PC model are: 95% Q limit is 4.3478 and the 95% T^2 limit is 6.0418. The Q-residual is plotted in Fig. 22.9 and the Hotelling T^2 statistic in Fig. 22.10.

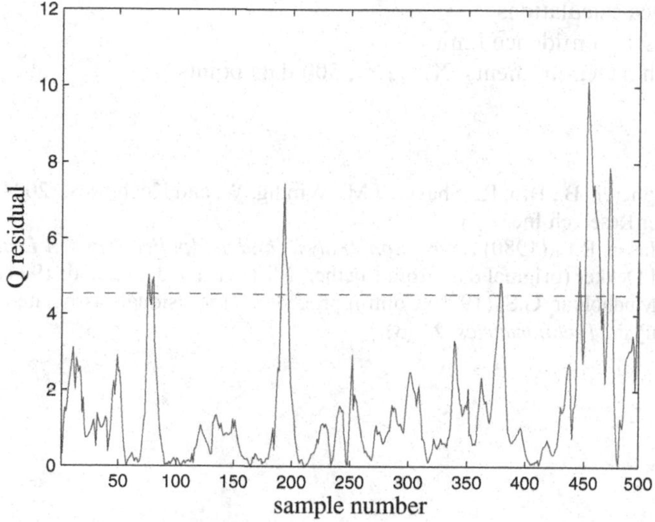

Fig. 22.9. Q-residual for 2 PC model.

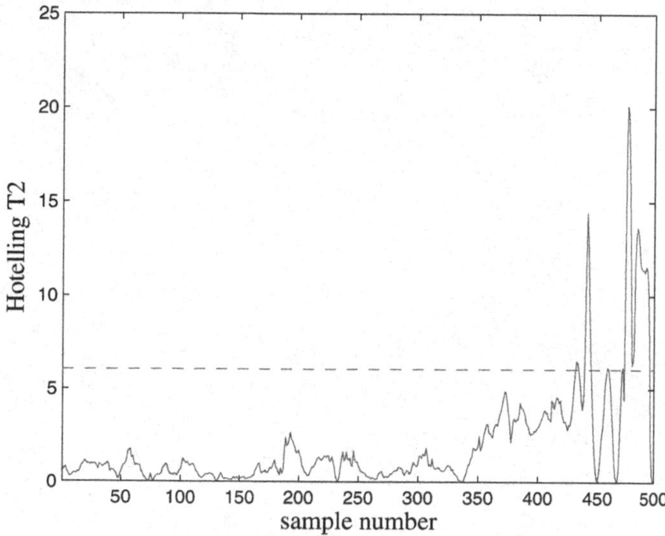

Fig. 22.10. Hotelling T^2 for 2 PC model.

As can be seen, there are many data points that are located outside the confidence limit, especially towards the end of the data section. If we look at the raw process data then it can also be seen that the process was apparently hit by a disturbance. It is better to exclude the bad data points from the analysis.

Files referred to in this chapter:
ascale: autoscaling of data
CreateDemoG: file to create data object DemoG
demoG: data object file containing demographic data
ftest: performs F-test
mcentre: mean centers data
pcacalc.m: principal component engine
Qclim: calculates Q-limit
spca: performs pca calculations
T2lim: calculates T^2 confidence limit
X5: data file with 5 measurements, $X_1 \ldots X_5$, 500 data points

References

Wise, B.M., Callagher, N.B., Bro, R., Shaver, J.M., Windig, W. and Koch, R.S. (2004) *PLS Toolbox 3.5*, Eigenvector Research Inc.

Gunst, R.F. and Mason, R.L. (1980) *Regression Analysis and its Applications – A Data-Oriented Approach*, Marcel Dekker (original data from Loether, McTavish and Voxland, 1974).

Jackson, J.E. and Mudholkar, G.S. (1979) Control procedures for residuals associated with principal component analysis. *Technometrics*, **21** (3).

23 Partial Least Squares

Partial Least Squares (PLS) attempts to find so-called latent variables that capture the variance in the data and at the same time achieves maximum correlation between predicted variables Y and predictor variables X.

Originally, PLS was a technique that would produce a static linear model, although also non-linear and dynamic versions have been published in the literature. As principal components in principal component analysis, the use of latent variables in PLS can reduce the dimensionality of the problem considerably.

23.1 Problem Definition

An excellent article, explaining the basics of Partial Least Squares (PLS), is given by Geladi and Kowalski (1986).

As in PCA, the data matrix is represented by a score matrix. The idea of PLS is to develop a model that relates the scores of the X data to the scores of the Y data. The PLS model consists of outer relations (X and Y data individually) and an inner relation that links the X data to the Y data.

The outer relationship for the input matrix or matrix with predictor variables is written as:

$$X = t_1 p_1^T + t_2 p_2^T + ... + t_n p_n^T + E = T P^T + E \qquad (23.1)$$

Similarly, the outer relationship for the output matrix or matrix with predicted variables can be written as:

$$Y = u_1 q_1^T + u_2 q_2^T + ... + u_n q_n^T + F = U Q^T + F \qquad (23.2)$$

where T and U represent the matrices of scores, P and Q represent the loading matrices for the X and Y data. If all components are described, E and F will become zero.

For the inner relationship, the idea is to describe Y as best as possible. Also, a good relationship between X and Y should be developed, therefore $||F||$ should be minimized.

The simplest model relating X to Y is one that relates the scores T to the scores U:

$$U = T B \qquad (23.3)$$

in which B is the regression matrix.

The prediction of Y can now be obtained by combining (23.3) with (23.2):

$$Y = TBQ^T + F \qquad (23.4)$$

To determine the dominant directions in which the data should be projected, the maximization of the covariance within X and Y is used as a criterium. The first set of loading vectors p_1 and q_1, representing the dominant direction, is obtained by maximizing the covariance between X and Y. Projection of the X data on p_1 and Y data on q_1 results in the first set of score vectors t_1 and u_1. This procedure is known as establishing the outer relation (Lakshminarayanan et al, 1997), shown in Fig. 23.1. The matrices X and Y can now be related through their respective scores, which is called the inner model, representing a linear regression between t_1 and u_1: $\hat{u}_1 = t_1 b_1$. The calculation of the first two dimensions is shown in Fig. 23.1.

Process Dynamics and Control: Modeling for Control and Prediction. Brian Roffel and Ben Betlem.

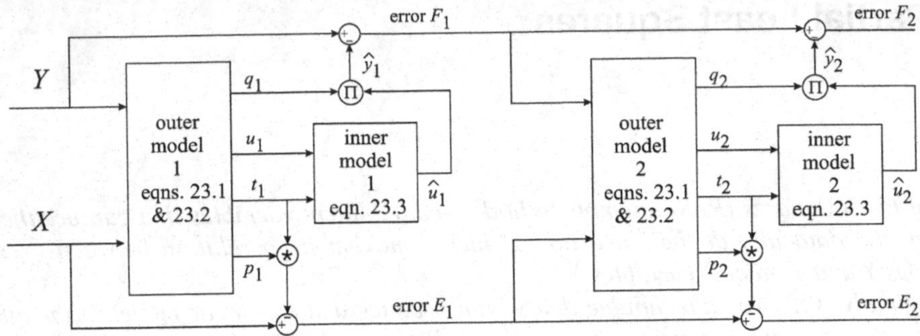

Fig. 23.1. Basic PLS calculations.

$\hat{u}_1 q_1^{\mathrm{T}}$ can be seen as that part of the Y data that has been predicted already by the first so-called latent variable, similarly the part of the X data that has already been used is $t_1 p_1^T$.

The residuals can then be computed as:

$$\begin{aligned} E_1 &= X - t_1 p_1^T \\ F_1 &= Y - \hat{u}_1 q_1^T = Y - b_1 t_1 q_1^T \end{aligned} \tag{23.6}$$

Using the newly computed residuals, the procedure of determining scores and loading vectors is repeated, until the residuals have become small enough. In practice, the number of PLS dimensions (number of latent variables) is determined based on the percentage variance explained. Irrelevant directions originating from noise and redundancy are left as errors E and F.

As is clear from the foregoing explanation, PLS is in fact a technique that breaks a multivariate regression problem up into a series of univariate regression problems.

23.2 The PLS Algorithm

The procedure for calculating the PLS model is given by several authors, amongst others, Geladi and Kowalski (1986). The procedure starts with assuming a score vector told that is any column of matrix X, for example x_1, and a score vector u which is any column of matrix Y, for example y_1. Assuming that X and Y are auto-scaled, the following steps are then carried out:

```
for each component take the following starting values:
   1. u = y1
   2. t_old = x1
calculations for the X matrix:
   3. w = (u^T * X / (u^T * u))^T
   4. w = w / ||w||
   5. t = X * w / (w^T * w)
calculations for the Y matrix:
      if number of Y variables > 1
   6.          q = (t^T * Y / (t^T * t))^T
   7.          q = q / ||q||
   8.          u = Y * q / (q^T * q)
      else
               q = 1
      end if
improvement last step > threshold
   9. if ||t-t_told|| > threshold
```

t_old = t
return to step 3
end if
calculate the X loadings and rescale the scores and weights:
10. $p = (t^T * X / (t^T * t))^T$
11. $p = p / ||p||$
12. $w = w * ||p||$
13. $t = t / ||p||$
calculate for each component i:
14. $b_i = u_i^T t_i / (t_i^T t_i)$
calculate the residuals.
15. $X = X - t_i p_i^T$
16. $Y = Y - b_i t_i q_i^T$

23.3 Dealing with Non-linearities

It could be that there are non-linearities in the data. Wold (1989) suggested replacing the inner relationship

$$\hat{u}_i = t_i b_i \tag{23.7}$$

by a quadratic relationship., i.e. t_i and u_i are then related by a quadratic model. Although this may be adequate in many cases, it will not function well when there are non-linearities that cannot be described by a quadratic relationship.

Qin and McAvoy (1992) suggested to replace the inner relationship model by a neural network model. This approach keeps the outer relationship of the PLS but employs the neural network as the inner regressor, the structure is shown in Fig. 23.2.

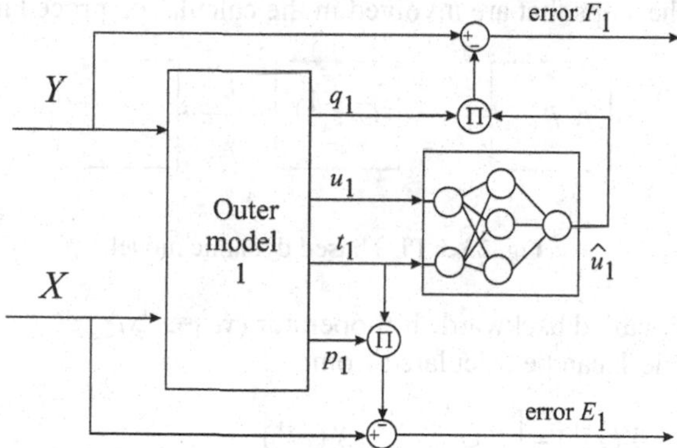

Fig. 23.2. PLS model with neural net inner relationship.

The algorithm that uses a neural network for the inner relationship is called NNPLS by the authors.

23.4 Dynamic Extensions of PLS

If the U block contains past values of the process inputs and outputs, the PLS model would be a dynamic model of the form:

$$Y = G_1(t_1)q_1^T + G_2(t_2)q_2^T + ... + F = GQ^T + F \tag{23.8}$$

Kaspar and Ray (1992, 1993) developed dynamic PLS models by filtering the process inputs and subsequent application of the standard PLS algorithm. Lakshminarayanan (1997) and Patwardhan (1998) proposed the modification of the PLS inner relation to be able to identify dynamic models. Instead of using the input scores t_i and output scores u_i, a dynamic ARX model G_i (see chapter 24) was used. The general ARX model structure can be described by:

$$A(z^{-1})y(k) = B(z-1)x(k-f) + e(k) \tag{23.9}$$

where y is the process output and x the process input. (Note that in the process control literature x is often replaced by u.)

An example of this structure is the following model:

$$y(k) + a_1 y(k-1) + a_2 y(k-2) = b_1 x(k-4) + b_2 x(k-5) \tag{23.10}$$

$G_i(t_i)q_i^{\mathrm{T}}$ is the measure of the Y space that is explained by the i-th PLS dimension. The matrix G is a diagonal matrix, comprising the dynamic elements identified at each of the n PLS dimensions:

$$G = \begin{bmatrix} G_1 & 0 & 0 & 0 & \dots & 0 \\ 0 & G_2 & 0 & 0 & \dots & 0 \\ \dots & & & & & \\ 0 & 0 & 0 & 0 & \dots & G_n \end{bmatrix} \tag{23.11}$$

Figure 23.3 shows the steps that are involved in the calculation procedure.

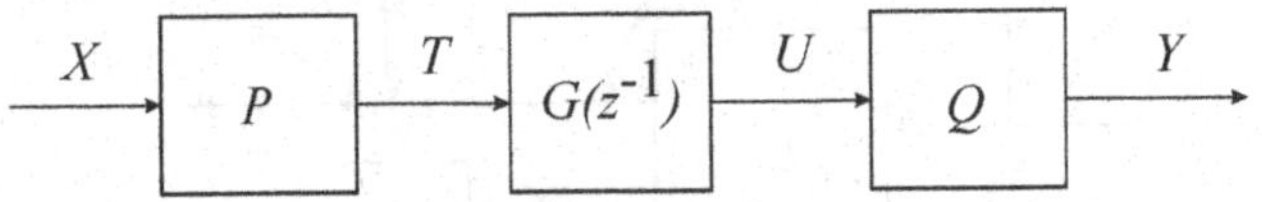

Fig. 23.3. PLS based dynamic model.

in which z^{-1} is the so-called backward shift operator ($x_{k-1}=z^{-1}x_k$).

The output variable Y can be calculated from:

$$Y(z^{-1}) = \left[QG(z^{-1})P\right]X(z^{-1}) \tag{23.12}$$

Lakshminarayanan (1997) also shows the procedure to apply PLS to Hammerstein models, which describe a process in terms of a non-linear static element and a linear dynamic element.

23.5 Modeling Examples

This section describes two examples of PLS modeling. The first one describes the model for the reactor already introduced in the previous chapter; the second example describes the model for a non-linear process.

23.5.1 Reactor Model

In the previous chapter, the input data of a chemical reactor was analyzed. It was shown that the first variable contained essentially the same information than the second variable. The first variable was therefore deleted and the remaining four input variables can be correlated to the output variable *Y*, in this case the fraction unconverted reactant. The program is saved in file F2304.m, a partial listing is shown below:

```
load X4Y;
%
% PLS model development; note: X4Y is an object data set
% The model will be stored in a structured array
%
X = X4Y.data(1:300,1:4);
Y = X4Y.data(1:300,5);
%
% one could also load X4Ya, which is a 500x5 double array
%
% load X4Ya
% X = X4Ya(:,1:4);
% Y = X4Ya(:,5);
%
model = npls(X,Y);
%
% model verification, data is not auto scaled anymore!!
%
X1 = X4Y.data(301:500,1:4);
Y1 = X4Y.data(301:500,5);
[nrx,ncx] = size(X1);
b  = model.regcoef;
Ypred = X1*b(1:ncx)'+b(ncx+1)';
plot([Y1 Ypred]);
```

The results of the model are:

Table 23.1. Percentage variance captured by PLS model.

	X-Block		Y-Block	
LV #	This LV	Total	This LV	Total
1	38.30	38.30	89.07	89.07
2	11.35	49.65	5.78	94.85
3	23.51	73.17	0.25	95.10
4	26.83	100.00	0.03	95.13

The program npls auto-scales the data first, i.e. from each variable the mean is subtracted and divided by the standard deviation. Then the first 300 data points are used to develop a linear PLS model, the modeling results are shown in Fig. 23.4.

As can be seen, the model performs reasonably well, as is usually the case during the model development stage. When we now test the model on the remaining 200 data points in the data set, the results as shown in Fig. 23.5 are obtained. The model predicts well, especially taking into account that even the four latent variables (PLS components) could explain approximately 95% of the variance in the process output data. This is caused, firstly, by the fact that the process output measurement is noisy, even when it is constant and secondly, as was explained in the previous chapter, the data contained the effects of a process upset towards the end of the data. If better data would have been used for the test, the result would even have been better.

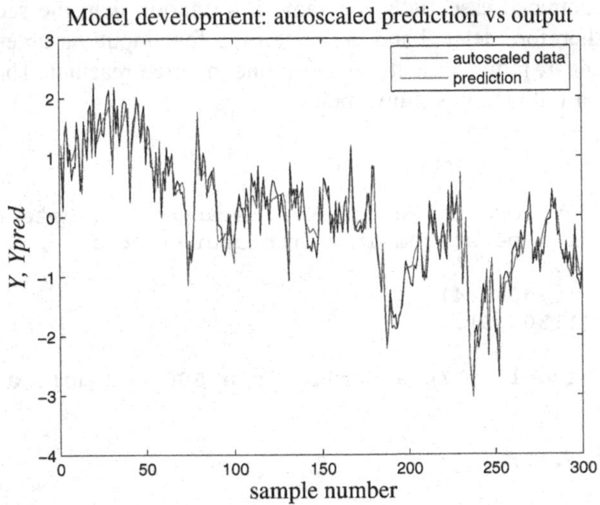

Fig. 23.4. Reactor model development.

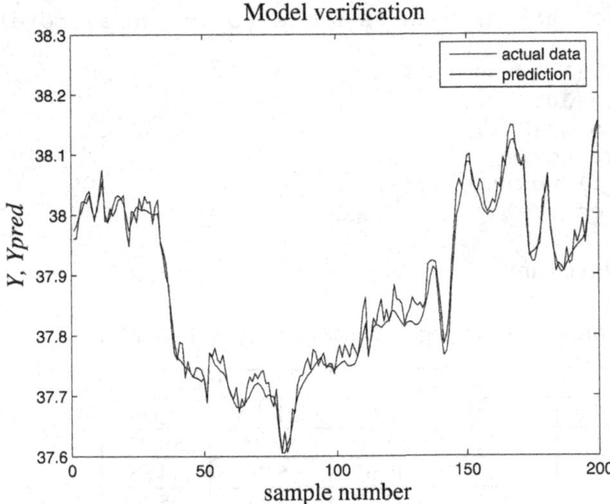

Fig. 23.5. Model validation for the reactor data.

One could also use the PLS model user interface (Wise *et al.*, 2004):

```
> X=X4Y.data(1:300,1:4);
> Y=X4Y.data(1:300,5);
> pls
```

and load the X and Y data by choosing the load options under File. Then choose "calc" to calculate the PLS model. The results are the same as shown before. You can save the PLS model by choosing the "save model" option from the File menu again, let us say you named it PLSmodel. The Eigenvector function (Wise et al., 2004) regcon could be used to convert the model parameters (PLSmodel.reg), taking the means and standard deviations into account (see also model results from reactorpls.m). The resulting model coefficients are:

```
[-0.0799 -0.0593 0.2105 0.0140 0.1250]
```

which means that the final PLS model is:

$$Y = -0.0799X_1 - 0.0593X_2 + 0.2105X_3 + 0.0140X_4 + 0.1250 \tag{23.13}$$

where X and Y are now the unscaled variables. As can be seen from Table 23.1, it is not necessary to include all latent variables in the model, two latent variables already explain 95% in the process output.

23.5.2 Non-linear Dynamic Model

Zhao (1994) described the following non-linear dynamic process:

$$y_k = y^3{}_{k-1} - 0.2\,|\,y_{k-1}\,|\,u_{k-1} + 0.08u^2{}_{k-1} \tag{23.14}$$

where y_k is the value of the process output at time k, u_{k-1} the value of the process input at time k–1.

As can be seen, the process is non-linear, this is also apparent from Fig. 23.6, where the response of the process output is shown for a positive unit step change at t = 5 in the input and a negative unit change in input.

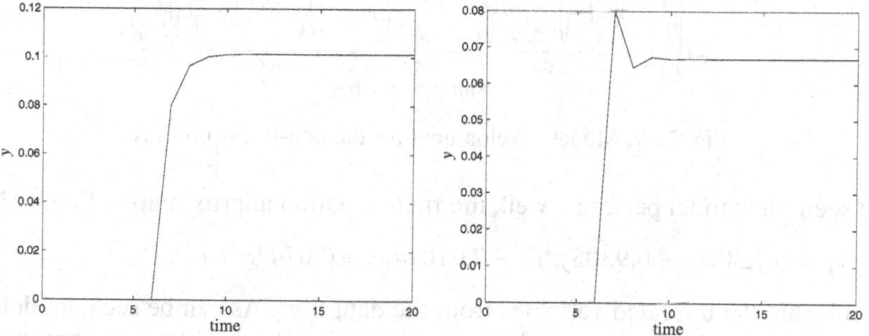

Fig. 23.6. System response for unit step change in input.

Using Eq. (23.14), data was generated for input changes between –1.5 and +1.5.

In order to model the system, the following X-matrix (input matrix) was defined:

$$X_k{}^T = \left[y_{k-1}, y_{k-1}^2, u_{k-1}, u_{k-1}^2\right] \tag{23.15}$$

The data is stored in file XYZhao.mat, the input (X) data is stored in columns 1:4, the Y data in column 5. 300 data points were generated, the first 200 were used to build the model, the last 100 points were used to validate the model. The code is similar to that of the previous example, and is stored in zhaopls.m.

The results of the model identification are:

Table 23.2. Percent variance captured by PLS non-linear model.

	X-Block		Y-Block	
LV #	This LV	Total	This LV	Total
1	67.91	67.91	74.21	74.21
2	9.35	77.26	23.88	98.09
3	16.78	94.04	0.10	98.19
4	5.96	100.00	0.19	98.38

and as can be seen, two latent variables are sufficient to predict the process output.

The results of the model development are shown in Fig. 23.7, the results of the model verification in Fig. 23.8.

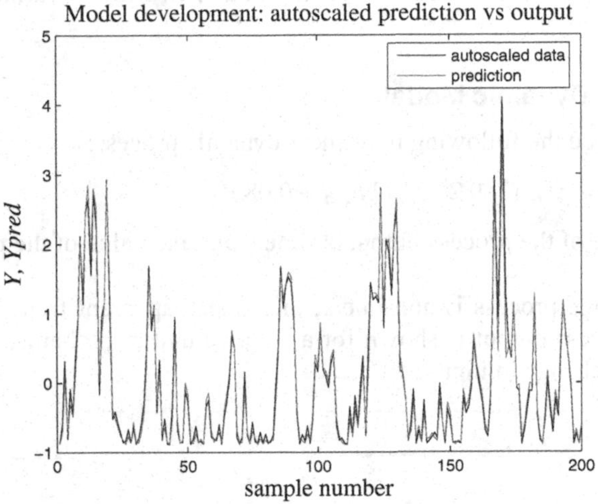

Fig. 23.7. Model development for the non-linear process.

As can be seen, the model performs well, the final equation approximating Eqn. (23.14) is:

$$y_k = 0.1249y_k - 0.9305{y^2}_{k-1} - 0.0101u_{k-1} + 0.0743{u^2}_{k-1} \tag{23.16}$$

where u and y are the unscaled variables from the data files. As can be seen, modeling a non-linear process is not a problem, as long as proper input variables are chosen. It is also possible to use a neural network to model the inner relationship. The PLS toolbox software offers the user to either model this relationship by a polynomial function or a neural network.

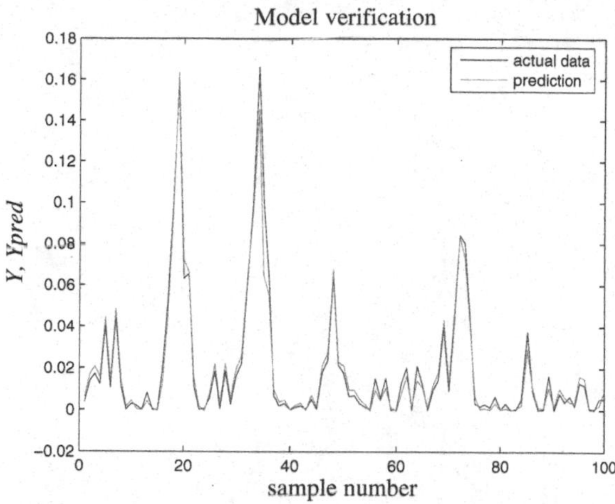

Fig. 23.8. Model verification for the non-linear process.

Files referred to in this chapter:
ascale.m: autoscales a matrix to zero mean and unit variance
bcalc.m: calculation of the contribution of each latent variable to the regression factor
bconv.m: calculates PLS model regression parameters
F2304.m: PLS model identification for reactor data (Fig. 23.4 and Fig. 23.5)
F2306.m: file to generate Zhao step response model (Fig. 23.6)
F2307.m: PLS model identification for Zhao's data (Fig. 23.7 and Fig. 23.8)
nipals.m: PLS engine for partial least squares regression
npls.m: calculates PLS model using nipals algorithm
X4Y.mat: data file with 4 process inputs and 1 process output, 500 data points, data set object structure
X4Ya.mat: data file with 4 process inputs and 1 process output, 500 data points, double array
XYzhao.mat: input (X) data and output (Y) data for Zhao process, data set object

References

Geladi, P. and Kowalski, B.R. (1986) Partial least squares regression: a tutorial. *Analytica Chimica Acta*, **185**, 1–17.

Kaspar, M.H. and Ray, W.H. (1992) Chemometric method for process monitoring. *American Institute of Chemical Engineers Journal*, **38**, 1593.

Kaspar, M.H. and Ray, W.H. (1993) Dynamic PLS modeling for process control. *Chemical Engineering Science*, **48**, 3447.

Lakshminarayanan, S., Shah, S.L. and Nandakumar, K. (1997) Modeling and control of multivariable processes: dynamic PLS approach, *American Institute of Chemical Engineers Journal*, **43** (9), 2307–22.

Patwardhan, R.S., Lakshminarayanan, S. and Shah, S.L. (1998) Constrained non-linear MPC using Hammerstein and Wiener models: PLS framework. *American Institute of Chemical Engineers Journal*, **44** (7), 1611–22.

Wise, B.M., Callagher, N.B., Bro, R., Shaver, J.M., Windig, W. and Koch, R.S. (2004) *PLS_Toolbox 3.5*, Eigenvector Research Inc.

Zhao J., Wertz, V. and Gorez, R. (1994) A fuzzy clustering method for the identification of fuzzy models for dynamic systems. Proceedings of the 9th IEEE International Symposium on Intelligent Control, Columbus, Ohio, USA, pp. 172–7.

24 Time-series Identification

Dynamic process models are useful for detailed dynamic process analysis and in process control applications. They are also increasingly used for predicting process variables, such as quality. Especially when the quality variable is difficult to measure or can be measured infrequently, dynamic models can provide an inferential measurement, based on easily and frequently measurable variables.

Depending on the process being investigated, one could develop first-principle models (linear/non-linear (partial) differential equations) or simple linear empirical differential or difference equations built from process data only. The first type of model is always preferred; however, developing first-principle models is time consuming and consequently expensive. In this chapter, the focus will therefore be on time series models, derived from process measurements. It is assumed that outliers have already been removed from the process data set; therefore a model fit parameter as defined in chapter 19 is used to calculate how well the model fits the data.

24.1 Mechanistic Non-linear Models

Non-linear physical models tend to be relatively complex and usually require considerable development effort to arrive at an adequate form. In addition, they require intimate process knowledge and understanding. Their primary use has been in process design and simulation. It is usually the steady-state version of these models that is used to size new equipment or to simulate the process at different conditions such as feed rate, temperature and pressure, etc. These steady-state models are also useful for supervisory control where they can be used in an on-line or off-line mode to optimize the process conditions at various time intervals and readjust the setpoints of the controllers in a basic control structure.

Dynamic forms of these mechanistic models may be useful in either simulation, prediction or real time control. They can, among others, be used to develop start-up procedures for continuous processes. They could also be used for optimal control of batch processes, where a process variable has to follow a pre-defined trajectory.

Unless one is interested in operating the process over a wide range of conditions, the simple (linear) empirical models discussed in the next section are easier to develop.

One of the major drawbacks of empirical linear models is the limited range of applicability. Especially extrapolation capabilities of the models beyond the region for which they were developed can be poor owing to process non-linearities. One way to avoid this is to adapt the process model parameters to the changing process conditions.

24.2 Empirical (linear) Dynamic Models

Many empirical methods exist where an *a priori* model structure has to be chosen and subsequently the model parameters are identified by a suitable technique. The method employed could for example be a least squares method.

Consider a process with an input $u(k)$ and an output $y(k)$, where k is the discrete time value. Assuming that the signals can be related by a linear process, we can write:

$$A(z^{-1})y(k) = \frac{B(z^{-1})}{F(z^{-1})} u(k - p) + \frac{C(z^{-1})}{D(z^{-1})} e(k) \tag{24.1}$$

Process Dynamics and Control: Modeling for Control and Prediction. Brian Roffel and Ben Betlem.

where z^{-1} is the shift operator, defined by $y(k-1)=z^{-1}y(k)$ and the polynomials A, B, C, D and F are given by:

$$\begin{aligned}
A(z^{-1}) &= 1 + a_1 z^{-1} + \ldots + a_{na} z^{-na} \\
B(z^{-1}) &= b_0 + b_1 z^{-1} + \ldots + b_{nb} z^{-nb} \\
C(z^{-1}) &= c_0 + c_1 z^{-1} + \ldots + b_{nc} z^{-nc} \\
D(z^{-1}) &= d_0 + d_1 z^{-1} + \ldots + b_{nd} z^{-nd} \\
F(z^{-1}) &= f_0 + f_1 z^{-1} + \ldots + b_{nf} z^{-nf}
\end{aligned} \tag{24.2}$$

where p is the time delay (sampling intervals) or dead time between the process input and output and e the modeling error. There are two ways of looking at this error, when the polynomials $C/D = 1$, e can be considered as the modeling error. However, in all other cases when $C/D \neq 1$, e is usually assumed to be white noise and the polynomials C and D are fit such that the remaining modeling error is as small as possible. A white noise signal is an uncorrelated signal with average zero and variance equal to one. In the literature (Ljung, 1987) several special cases of model types are presented:

- ARX model structure:

  $$A(z^{-1})\,y(k) = B(z^{-1})\,u(k-p) + e(k) \tag{24.3}$$

  which implies that $nc = nd = nf = 0$.
  Example:

  $$y(k) = a_1\,y(k-1) + a_2\,y(k-2) + b_1\,u(k-3) + b_2\,u(k-4) \tag{24.4}$$

- ARMAX model structure:

  $$A(z^{-1})\,y(k) = B(z^{-1})\,u(k-p) + C(z^{-1})\,e(k) \tag{24.5}$$

  which implies that $nf = nd = 0$.
- Output error model structure:

  $$y(k) = \frac{B(z^{-1})}{F(z^{-1})}\,u(k-p) + e(k) \tag{24.6}$$

  thus $na = nc = nd = 0$.

MATLAB is a suitable environment to identify the model parameters. The identification toolbox provides an excellent user interface for these types of model.

24.3 The Least Squares Method

As mentioned, the model parameters of the linear model can be effectively identified using a least squares method. The model equation for the different types of model can be written as a difference equation, such as, for example, the one shown in Eqn. (24.4). When the mean values of y and u are subtracted from the actual values of y and u, one may write this equation as:

$$\hat{y}(k) = \theta^T x(k) \tag{24.7}$$

where $\hat{y}(k)$ is the predicted value of $y(k)$, θ is the model parameter vector and $x(k)$ is the input vector with past values of y and u, in this particular case:

$$\theta = [a_1, a_2, b_1, b_2]^T$$
$$x(k) = [y(k-1), y(k-2), u(k-4), u(k-5)]^T \tag{24.8}$$

Let us now assume that there are M data points, i.e.

$$Y = [y^1, y^2, ..., y^M]^T$$
$$X = \begin{bmatrix} (x^1)^T \\ (x^2)^T \\ ..... \\ (x^M)^T \end{bmatrix} \tag{24.9}$$

in which Y is a vector and X a matrix, in this case a vector with M rows and a matrix with M rows and four columns.

The modeling error at every time instant can be written as:

$$E = [\varepsilon_1, \varepsilon_2, ...\varepsilon_M]^T$$
$$\varepsilon_i = y^i - (x^i)^T \theta \tag{24.10}$$

The model parameter vector can now be computed by minimizing $J = (1/2)EE^T$; this can be achieved by setting $\partial J / \partial \theta = 0$, from which:

$$\theta = \left(X^T X\right)^{-1} X^T Y \tag{24.11}$$

24.4 Cross-correlation and Autocorrelation

We have discussed the model structure and model identification procedure. If data are available, we would essentially be able to identify the model parameters. There are a few pitfalls, however. The data that are or will be collected should be appropriate data for model identification. This means that the data set should contain sufficient dynamic and static information, such that the model can be identified and reflects the true process behavior.

Often normal process operating data are not fit for dynamic model identification. It just does not contain sufficient information, resulting in a poor model. The process input is therefore usually perturbed, for example by using a Pseudo-Random Binary Sequence (PRBS) or any other form of perturbation. The PRBS signal provides a sequence of upward and downward steps as shown in Fig. 24.1.

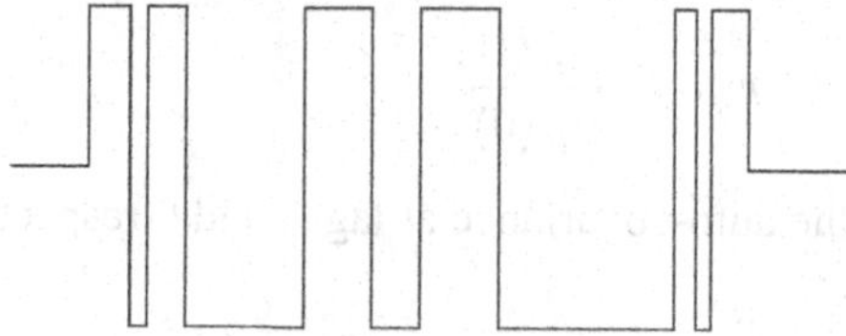

Fig. 24.1. PRBS sequence.

Before introducing any arbitrary PRBS signal, however, a step change is recommended. This gives us an idea of the gain of the process and the major time constant. The magnitude of the PRBS step should be as large as possible, with process gain information it can be calculated what process deviation can be tolerated.

Once a proper PRBS signal has been identified and process data has been collected, one should determine the cross correlation of the process input–output signals and the autocorrelation of the process output.

The cross correlation function is a tool that can be used to check whether there is sufficient impact of the process input on the process output, i.e. whether two time series are correlated. This can already be determined from a step test, but the computation of the cross correlation function is a better test and it provides additional information.

Cross covariance is defined as the product of the process output signal with the process input signal, while lagging it by $i = 1, 2, \ldots$ sampling intervals.

Given an input data series $u_1, u_2, u_3, \ldots. u_n$ and an output data series $y_1, y_2, \ldots. y_n$, the cross covariance coefficient at lag i can be computed from:

$$c_{uy}(i) = \frac{1}{N}\sum_{k=1}^{N-i}(u_k - \bar{u})(y_{k+i} - \bar{y}) \quad i = 0,1,2\ldots \tag{24.12}$$

where $c_{uy}(i)$= cross covariance at lag i, $\bar{u}$ and $\bar{y}$ = mean of the u and y series respectively. The maximum correlation between u and y does not occur at lag $i = 0$, owing to the discrete sampling of the input and output signals. The maximum correlation usually occurs at a later time i (>0), the maximum time period indicates the dead time between the input data series and output data series.

It is usually more convenient to work with the dimensionless cross correlations. The cross correlation coefficient, $r_{uy}(i)$, at lag i is given by:

$$r_{uy}(i) = \frac{c_{uy}(i)}{\sqrt{c_{uu}(0)\, c_{yy}(0)}} \tag{24.13}$$

where $c_{uy}(i)$ = the cross covariance at lag i, $c_{uu}(0)$ and $c_{yy}(0)$ are the auto-covariances at lag 0 for the input and output series respectively.

The autocorrelation is a tool that gives us an indication whether the current value of the process output $y(k)$ depends on previous values of the process output, for example $y(k-1)$, $y(k-2)$, etc. Given a series $y_1, y_2, y_3, \ldots, y_n$, the auto-covariance, $c_{yy}(i)$, at lag i can be calculated from:

$$c_{yy}(i) = \frac{1}{N}\sum_{k=1}^{N-i}(y_k - \bar{y})(y_{k+i} - \bar{y}) \quad i = 0, 1, 2\ldots\ldots \tag{24.14}$$

where N = number of observations and $\bar{y}$ = mean of the series. At lag $i = 0$ the autocovariance reduces to the variance, σ^2, of the series.

It is more convenient to work with normalized autocorrelation, $r_{yy}(i)$, which can be calculated from:

$$r_{yy}(i) = \frac{c_{yy}(i)}{c_{yy}(0)} \tag{24.15}$$

where $c_{yy}(i)$ and $c_{yy}(0)$ are the auto-covariance at lags i and 0 respectively.

Example

File p11.mat contains process data, u is the input data and y the output data. The cross-correlation between u and y can be computed as well as the autocorrelation for y. This has been done using program F2402.m. The results are shown below in Figs. 24.2 and 24.3.

As can be seen, there is a strong cross correlation between the process input and output data, which peaks at lag one and two, there is also a strong autocorrelation which means that the current value of $y(k)$ depends strongly on the previous value $y(k-1)$. The cross-correlation results indicate that the data can in principle be used for model identification and that there is a dead time between process input and output of one to two time intervals.

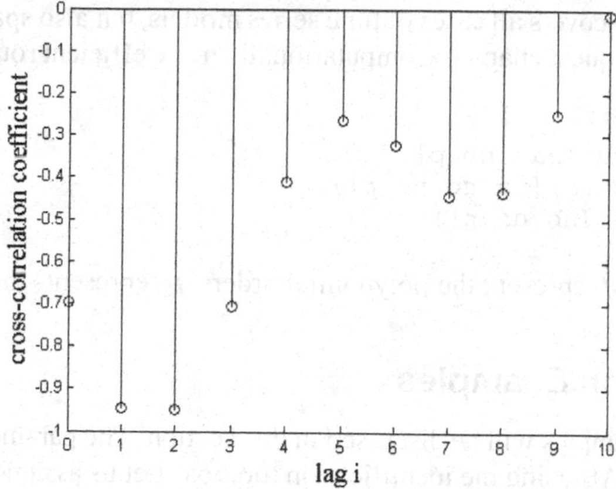

Fig. 24.2. Cross-correlation between process input u and process output y.

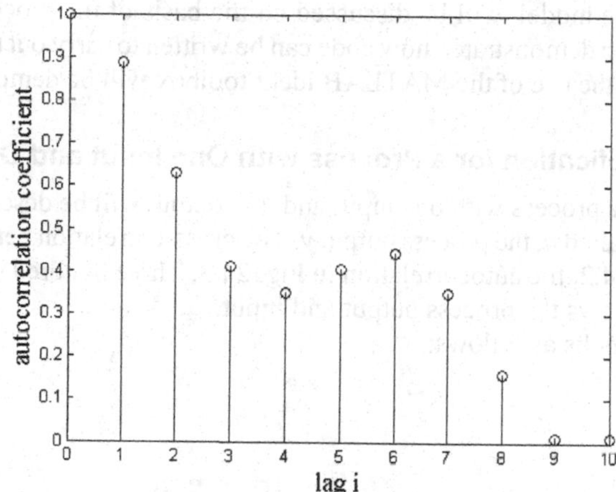

Fig. 24.3. Autocorrelation for process output y.

24.5 The Prediction Error Method

Based on the prediction error method, which minimizes the error between the prediction and the actual process output data, MATLAB (2005) can be used to construct models of basically any structure. For this general model the method

```
model = pem(data,nn)
```

is available, where data are the output-input data matrix and nn gives all the orders and delays:

```
nn = [na nb nc nd nf p]
```

for the general polynomial model shown in Eqn. (24.1).

The PEM command covers all cases of time series models, but also space models which will be discussed in a subsequent chapter. Computationally more efficient routines are available for special cases:

```
model = arx(data,[na, nb p])
model = armax(data,[na nb nc p])
model = oe(data,[nb nf p])
```

where *na*, *nb*, *nc* and *nf* represent the polynomial orders, *p* represents the dead time.

24.6 Identification Examples

Two identification examples will be discussed in this section. The parametric model identification is done in MATLAB using the identification toolbox. Let us assume that you have copied the files on your distribution disk to the directory C:\MatlabR14\work directory. Start Matlab by double clicking the Matlab icon. You are now ready to run the identification examples.

In this section the design of three different parametric models, an ARX model, an ARMAX model and an OE type model, will be discussed on the basis of the process data provided in p11.mat. First it will be demonstrated how code can be written to carry out the identification exercises, subsequently the use of the MATLAB ident toolbox will be demonstrated.

24.6.1 Model Identification for a Process with One Input and One Output

The identification of a process with one input and one output will be described in this section. The process input is called *u*, the process output *y*. The cross-correlation between *u* and *y* was already shown in Fig. 24.2, the autocorrelation in Fig. 24.3. Three hundred data points were collected. Figure 24.4 shows the process output and input.

The Matlab code looks as follows:

```
load p11
%
z = [y u];
%
% plot all measurements
%
idplot(z), pause % Press ENTER to continue
close
%
% remove any trend and make the data zero-mean:
%
z = dtrend(z);
%
model1 = arx(z,[2 2 1])
%
% plot the model and the fit
%
yhat = idsim(z(:,2),model1);
fprintf('The fit for the ARX model is %5.2f percent\n', ...
   fit(z(:,1),yhat));
plot([z(:,1) yhat]);
```

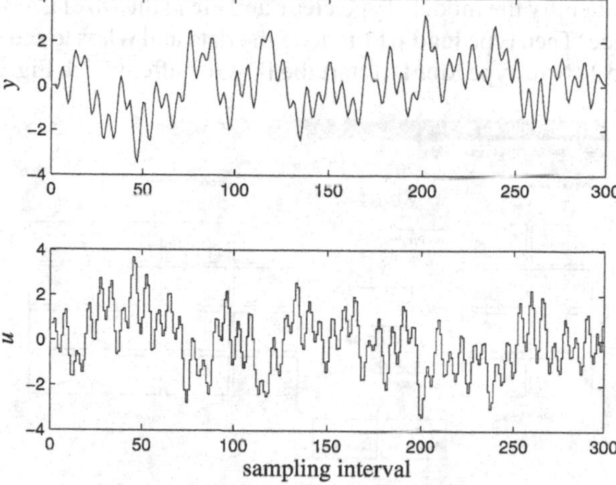

Fig. 24.4. Detrended process data.

The ARX model has the general format as shown in Eqn. (24.3).Using least squares error minimization, the following result is found:

$$\begin{aligned} A(z) &= 1 - 1.0870\, z^{-1} + 0.3135\, z^{-2} \\ B(z) &= -0.5855 + 0.3551\, z^{-1} \end{aligned} \tag{24.16}$$

The model residuals can be calculated and the autocorrelation and cross correlation of the residuals with the input are calculated and shown in Fig. 24.5.

If the model structure is satisfactory, there should be no autocorrelation of the residuals, nor should there be any cross-correlation between the residuals and the process input. The model fit is about 93%, and it is not surprising that the results for the different models are more or less the same for a process with one input and one output. This is due to the similar model structure and order. For a process with one output and more than one input, for example two inputs, the model structure is different. This will be discussed in the following section.

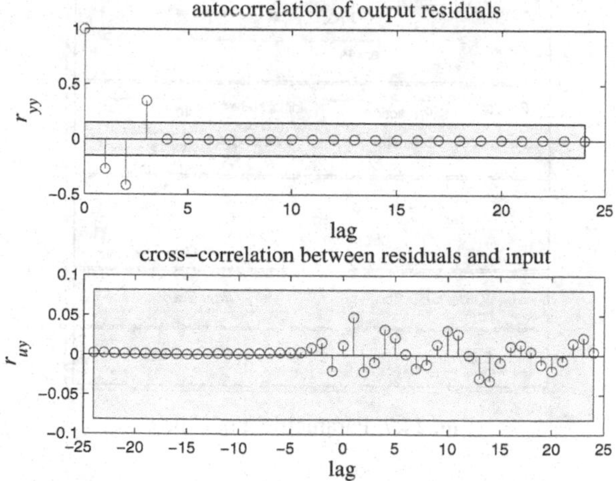

Fig. 24.5. Auto- and cross correlation of residuals.

Instead of the MATLAB command mode, one could also use the identification graphical user interface (ID-GUI) to identify the model. Type **clear** and **clc** in the MATLAB command window to clear the workspace. Then type **load p11** to load the data and **whos** to see the variable names, they should be *X* and *Y*. Now type **ident** to start the ID-GUI after which Fig. 24.6 should appear.

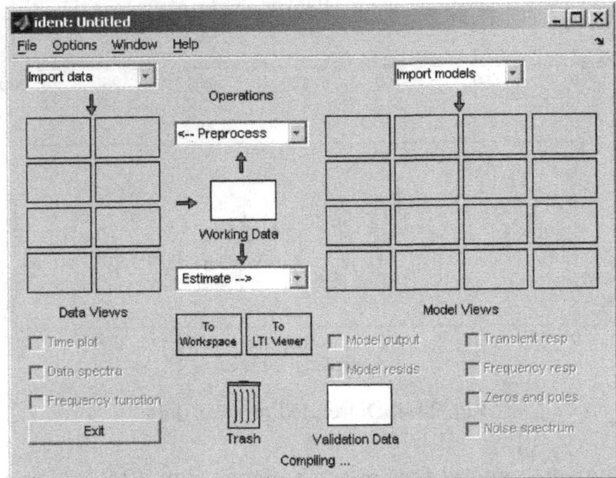

Fig. 24.6. Identification graphical user interface.

Select **Import data,** enter **Time domain data ...** and select the following workspace variables, input: *u*, output: *y*. The window *Working Data* should show that *mydata* is the working data set. Go to the preprocessor and select **Remove means**, a new data set *mydatad* is created. Drag this data set to the **Working Data** window, and perform a Remove trends operation. Now a data set *mydatadd* is created. Make this data set the Working Data set and also the Validation Data set. Now select Estimate, Parametric models, after which the parametric model box, Fig. 24.7 appears.

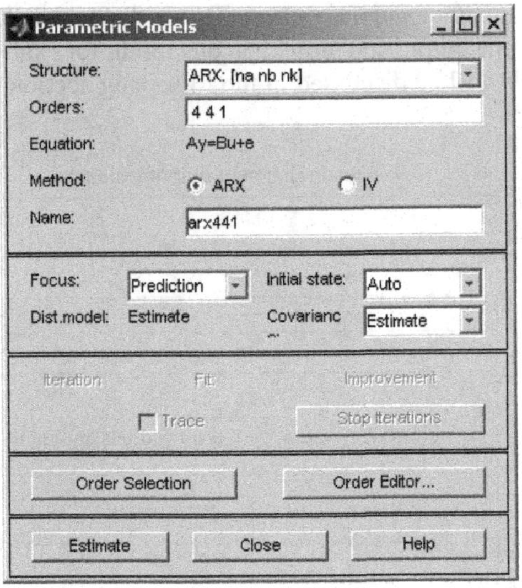

Fig. 24.7. Parametric model box.

The standard model that is chosen is a [4 4 1] model. Select [2 2 1] and **Estimate,** this will create an arx221 model. Selection of Model Output will create the model/measurement plot as shown in Fig. 24.8.

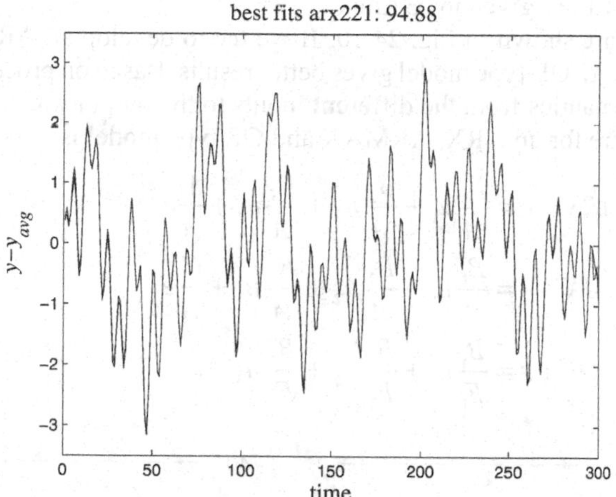

Fig. 24.8. Model output and prediction.

The "best fits" parameter indicates the percentage of output variation that is reproduced by the model.

By selecting **Estimate, Parametric models** again in the ID-GUI we could easily select another model order. Through trial and error one could try to find the best model, although for controller design a low order model is preferred.

If one is not to sure what the order of the model should be, the order selection option is a good feature. A sample for this case is shown in Fig. 24.9.

As can be seen, a model with eight parameters: $na = 4$, $nb = 4$, $nk = 1$(or p, the dead time) leaves 4% of the output variation unexplained. For models with more parameters, there is little advantage in reducing the model error, the percentage unexplained output remains around 4%, this will be primarily process noise.

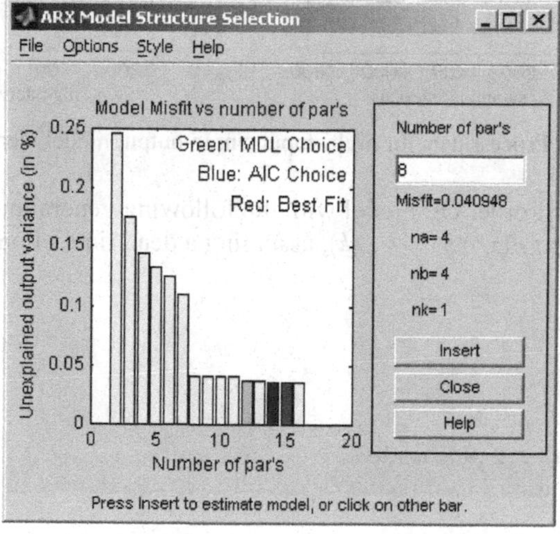

Fig. 24.9. Order selection option for ARX models.

24.6.2 Identification of Processes with Multiple Inputs

The model identification for a process with one output and multiple inputs is similar to the identification of a process with one input and one output. Data for a process with three inputs u_1, u_2 and u_3 and one output y are given in p31.mat.

The process data are shown in Fig. 24.10. If we try to develop an ARX or ARMAX type model, the fit is poor. An OE type model gives better results. Based on process knowledge this is not surprising, the dynamics from the different inputs to the output was found to be different.

The model structure for an ARX, ARMAX and OE type model is:

$$
\begin{aligned}
ARX: y &= \frac{B_1}{A}u_1 + \frac{B_2}{A}u_2 + \frac{B_3}{A}u_3 + \frac{1}{A}e \\
ARMAX: y &= \frac{B_1}{A}u_1 + \frac{B_2}{A}u_2 + \frac{B_3}{A}u_3 + \frac{C}{A}e \\
OE: y &= \frac{B_1}{F_1}u_1 + \frac{B_2}{F_2}u_2 + \frac{B_3}{F_3}u_3 + e
\end{aligned}
\tag{24.17}
$$

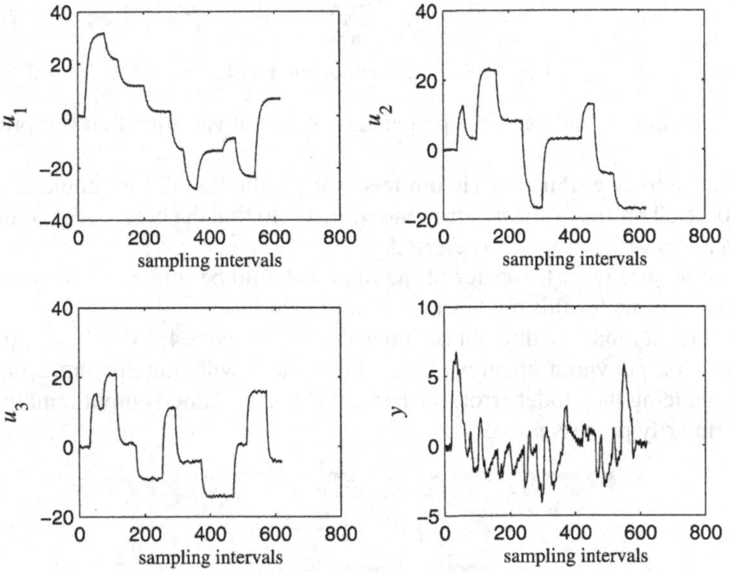

Fig. 24.10. Process data for multi-input single-output model identification.

Let us identify a third-order OE model with the following general model structure: $y(k) = B(z^{-1}) * u\,(k\text{-}p)/F(z^{-1}) + \ldots + e(k)$, assuming a dead time of one sampling period:

```
nb = [3 3 3];
nf = [3 3 3];
nk = [1 1 1];
nn = [nb nf nk];
model = oe(z,nn);
yhat = idsim(z(:,2:4),model);
```

Note: If the dead time is estimated incorrectly, the residuals will show an autocorrelation that falls outside the confidence limits. In addition, the residuals will not average around zero and the model fit will be poor.

Identification results are shown in Fig. 24.11. Agreement with the original data is good and the three identified models give good predicted values. The model fit is 81.5%.

The identification can also be done using the MATLAB graphical identification user interface. Different model orders and dead times can quickly be tried and results plotted. The models can be viewed using the LTI viewer.

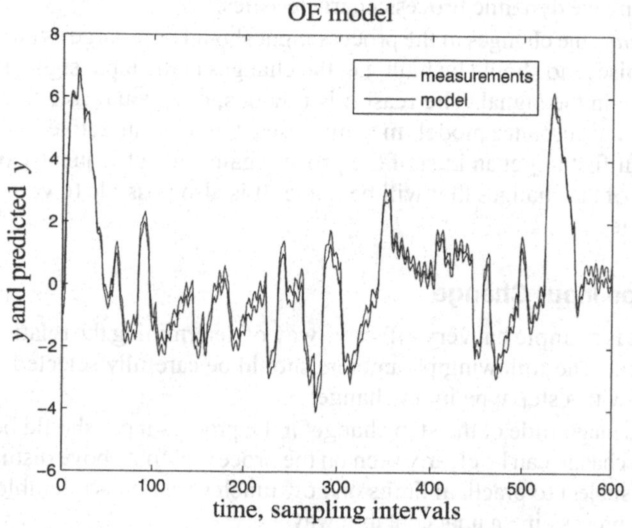

Fig. 24.11. Measured and modeled process output as a function of time.

24.7 Design of Plant Experiments

The most crucial step in process identification is the collection of meaningful data, which contains sufficient information about the dynamics and statics of the process. In some cases normal operating data may suffice, but this is not often the case. This is because this type of data often usually comes from a controlled process and will therefore only vary around a steady-state value of the process variables. In addition, when there is a process excursion, this is often due to a disturbance and the identification procedure may in that case identify a disturbance model rather than a process model. Another problem may be that data are missing owing to instrument failure, in that case more data has to be collected or new data has to be generated. Finally, there may be a high degree of correlation between the process inputs, for example, two flows may be correlated because ratio control exists between the two flows. Determining the impact of each flow individually on the process output will then be difficult.

As a result, it is in most cases necessary to design a plant experiment to ensure that there is sufficient information in a collected data set.

24.7.1 Process Input Changes

To generate an informative data set of a process, it will be necessary to vary the inputs of the process, thereby generating changes in the process output(s).

There are numerous types of input change that will provide useful information about the process; each type should have the following characteristics:

- *uncorrelated input pattern*: if one wishes to determine the relationship between each process input and each process output separately, it is necessary that the changes in process inputs are uncorrelated.
- *proper frequency content*: the frequency at which a process input is changed depends on the major time constant between process inputs(s) and output(s). At high switching frequency the process output will usually change very little, making the signal not useful for identification purposes. At low switching frequency, the process reaches steady state in all cases and the steady-state characteristics of the process can easily be determined, but it will be more difficult to determine the dynamic process characteristics.
- *proper magnitude*: the changes in the process input should be as large as possible. In addition, the signal to noise ratio should be high, i.e. the changes in the input signal should be a multiple of the noise in the signal. The reason is obvious, one would like to develop a process model and not a disturbance model. In some cases it may be attractive to give a step change in process input first to get an idea of the process gain and subsequently one may calculate the magnitude of the changes that will be made. It is also possible to vary the magnitude of the input changes.

24.7.2 Step Type Input Change

This type of input is a simple but very effective way of determining the relationship between an input and an output. The following parameters should be carefully selected when designing a plant experiment with a step type input change:

- *magnitude*: the magnitude of the step change in the process input should be such that the effect of the step change can be clearly seen on the process output above disturbances and process noise. It is subject to practical limits on how much variation is tolerable in order to avoid perturbing the process in an unacceptable way.
- *data collection sampling interval*: a rule of thumb for the data collection sampling interval T is to choose T as $0.25\tau \leq T \leq 0.5\tau$ where τ is the major time constant of the process.
- *minimum set of data*: the step response should reach at least 95% of the total expected output change, which is usually reached at a time beyond three times the major process time constant plus dead time.

Once the data have been collected, it can be analyzed by assuming a simple first-order plus dead time model (i.e. process gain, time constant and time delay) and fitting the model to the process data.

24.7.3 PRBS Type Input Change

A pseudo-random binary sequence (PRBS), shown in Fig. 24.12, is often used in parametric model identification.

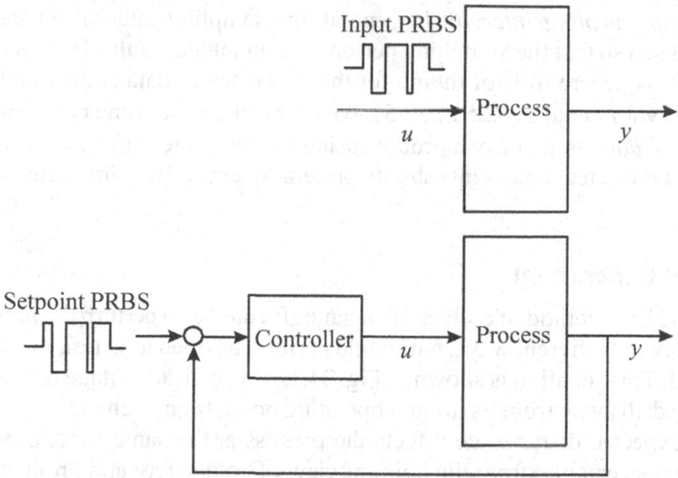

Fig. 24.12. Type of experiment (a) open loop (b) closed-loop.

The input is varied between a low and a high limit in a predetermined fashion. By using a random number generator, one can decide at every switching period whether to change to the other level or remain at the present level. The following parameters that must be chosen when designing a plant experiment that uses a PRBS type input sequence:

- *switching period*: the switching period should be chosen such that the data contains sufficient dynamic and static information. A general rule of thumb for overdamped systems is to choose the switching time T_s such that $0.33\,\tau \leq T_s \leq 3\,\tau$, where τ is the major time constant of the process.
- *magnitude*: as already mentioned, the magnitude of the input variation should be as large as possible. Saturation of the process inputs should be avoided, however. This means that it is not advisable to change a valve position in the region of its fully open or fully closed position. The signal to noise ratio should preferably be larger than five.
- *length of the input sequence*: there is no strict requirement on the length of the input sequence, however, generally one can say that more data points will lead to a more accurate model. When generating a long PRBS sequence of length L, it is not necessary that the entire sequence is unique. A unique subsequence of length L_s can be repeated as many times as required until the longer PRBS sequence is generated. The unique PRBS length (L_s) should be equal to at least five times the major time constant of the process plus dead time (i.e. $L_s \geq 5\tau + \theta$).
- *uncorrelated input sequences*: identifying a process with multiple inputs, it is necessary that the various input sequences are uncorrelated. This can be achieved by using different lengths of the unique PRBS sub-sequences (i.e. lengths of 2^n-1 switching periods, $n \geq 5$). If possible, change one process input at a time, while keeping the other inputs at a constant value.

- *data collection sampling interval*: in general, the sampling interval for data collection T should be chosen so that the switching period T_s is an integer multiple of the data collection sampling rate. A general rule of thumb for the frequency of data collection is to choose the sampling interval T equal to $0.25\tau \ldots 0.5\tau$, where τ is the major time constant of the process.
- *minimum set of data*: to identify a process using modern time series identification methods, the number of collected data points should preferably exceed ten times the number of model parameters.

24.7.4 Type of Experiment

As discussed, the best method of collecting meaningful data is to perturb the process input. This can be done in several different ways. One could perform an open loop test, in which the process is not controlled. This situation is shown in Fig. 24.12a. The disadvantage of this method is that the process may drift away from its normal operating point. If large changes are made in the test and a large unexpected disturbance affects the process at the same time, process conditions might become unacceptable, from the point of view of both safety and product quality.

If the test is performed during a production run, off-spec material may be produced; this can usually be avoided when the process is controlled. This situation is shown in Fig. 24.12b. The process will now remain in the region of its normal operating point.

There are now two options for performing a PRBS test. One could add the PRBS changes to the current setpoint, as shown in Fig. 24.12b, or one could add the PRBS changes to the valve signal (controller output). Knowledge about the controller model is not required, provided the proper signals are measured, which are the process input and process output.

Files referred to in this chapter:
p11.mat: process data for single input-single output process
p31.mat: data file for process with three inputs and one output
F2402.m: calculation of cross- and autocorrelation
F2404.m: code for SISO model identification
F2410.m: code for MISO model identification

References

Ljung, L. (1987) *System Identification – Theory for the User*, Prentice Hall, Englewood Cliffs, N.J.
MATLAB *System Identification Toolbox, User's Guide, 2005*, The Math Works, Inc., John Wiley.

25 Discrete Linear and Non-linear State Space Modeling

In this chapter, discrete linear-state space models will be discussed and their similarity to ARX models will be shown. In addition Wiener models are introduced. They are suitable for non-linear process modeling and consist of a linear time variant model and a non-linear static model. Several examples show how to develop both types of models.

25.1 Introduction

For the description of state space models two common notations are used: the state space or Output-Error form is defined as:

$$\begin{aligned} x(k+1) &= Ax(k) + Bu(k) \\ y(k) &= Cx(k) + Du(k) + v(k) \end{aligned} \tag{25.1}$$

in which $y(k)$ = process output, $u(k)$ = process input and $x(k)$ = state vector of size nx, $v(k)$ = noise vector. $A(nx,nx)$, $B(nx,nu)$, $C(ny,nx)$ and $D(ny,nu)$ are matrices, with nx = number of state variables, nu = number of process inputs and ny = number of measured variables. Figure 25.1 shows the schematic representation of the model.

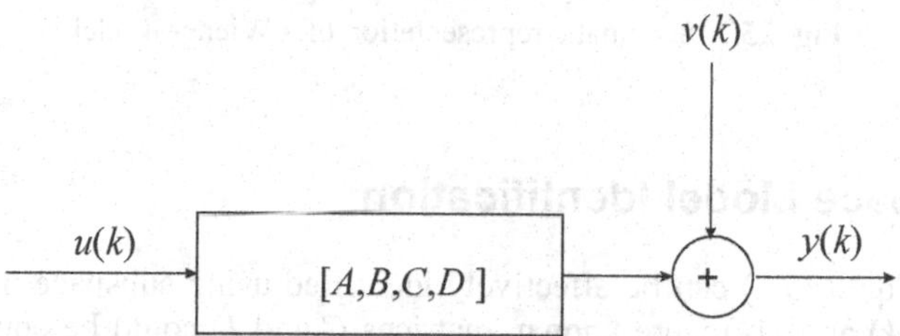

Fig. 25.1. Schematic representation of the output error model.

The other form of the state space model is the innovations form, described as:

$$\begin{aligned} x(k+1) &= Ax(k) + Bu(k) + w(k) \\ y(k) &= Cx(k) + Du(k) + v(k) \end{aligned} \tag{25.2}$$

which can also be written in a more general format as:

$$\begin{aligned} x(k+1) &= Ax(k) + Bu(k) + Ke(k) \\ y(k) &= Cx(k) + Du(k) + e(k) \end{aligned} \tag{25.3}$$

State space models are very similar to ARX models, which were discussed in the previous chapter. A process output y for an ARX model can be described by:

$$y(k) = G(z^{-1})u(k-p) + d(k) \tag{25.4}$$

with u the process input and p the dead time between input and output.

If only one delay is used in the expressions for the ARX model, then it can easily be shown that G in Eqn (25.4) can be written as:

$$G(z^{-1}) = C(z^{-1}I_{nx} - A)^{-1}B + D \tag{25.5}$$

Process Dynamics and Control: Modeling for Control and Prediction. Brian Roffel and Ben Betlem.

where I is the $nx \times nx$ identity matrix. If more than one delay period is present, this can be accommodated by increasing the size of the state vector x.

The structure of a Wiener model is similar to the structure of an Output Error model, discussed in the previous chapter. It can be described by:

$$\begin{aligned} x(k+1) &= Ax(k) + Bu(k) \\ y(k) &= Cx(k) + Du(k) \\ z(k) &= f(y(k)) + v(k) \end{aligned} \tag{25.6}$$

Figure 25.2 shows the schematic representation of the Wiener model. Note that Eqn. (25.6c) represents a static non-linear relationship. An example where such a model might be useful is in high-purity distillation columns, where in many cases the logarithm of (1-purity) is proportional to the reflux flow rate.

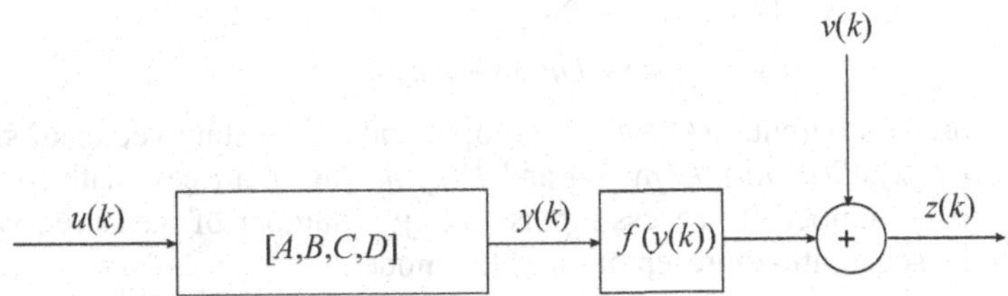

Fig. 25.2. Schematic representation of a Wiener model.

25.2 State Space Model Identification

The matrices in Eqn. (25.2) can be effectively identified using subspace methods. If the sequence of $y(k)$, $x(k)$ and $u(k)$ were known, matrices C and D could be computed from Eqn. (25.2b) using a least-squares method, with e being the residual. Subsequently, Eqn (25.2a) would form another least squares calculation yielding matrices A, B and K. In other words, when the states x are known, we can use linear least squares to compute the model matrices.

The states in Eqn. (25.2) are now being formed as linear combinations of the k-step ahead predicted outputs ($k = 1, 2, \ldots$). The literature on state space identification has shown how the states can be estimated directly from the process data by certain projections. (Verhaegen, 1994; van Overschee and de Moor, 1996; Ljung and McKelvey, 1996). The MATLAB function n4sid (Numerical Algorithms for Subspace State Space System Identification) uses subspace methods to identify state space models (Matlab 2000, van Overschee and de Moor, 1996) via singular value decomposition and estimates the state x directly from the data.

Ljung and McKelvey (1996) propose a procedure where first an ARX model is developed which is used for j-step ahead predictions while u is constant and then the states x are formed by multiplication of the predictor vector by a state space basis. Subsequently, matrices A, B, C, D and K are calculated using a least squares approach.

The Wiener model contains a non-linear relationship. There are various methods to approximate this relationship, such as Tchebychev polynomials (Kreyszig, 1999) or neural networks. The use of neural networks for the approximation of non-linear relationships will be discussed in chapter 27; therefore the first approach will be explained shortly.

Tchebychev polynomials have the following shape:

$$C_n(y) = \cos(n \arccos y) \tag{25.7}$$

From Eqn. (25.7) the following two expressions can be derived:

$$\begin{aligned} C_{n+1}(y) &= \cos[n \arccos y + \arccos y] \\ C_{n-1}(y) &= \cos[n \arccos y - \arccos y] \end{aligned} \tag{25.8}$$

Using the well-known rule:

$$\cos(\alpha + \beta) = \cos\alpha \cos\beta + \sin\alpha \sin\beta \tag{25.9}$$

this leads to:

$$\begin{aligned} C_{n+1}(y) + C_{n-1}(y) &= 2\cos[n \arccos y]\cos[\arccos y] \\ &= 2\cos[n \arccos y]\, y \end{aligned} \tag{25.10}$$

or

$$C_{n+1}(y) + C_{n-1}(y) = 2yC_n(y) \tag{25.11}$$

which can be written as the recursive equation:

$$C_{n+1}(y) = 2yC_n(y) - C_{n-1}(y) \tag{25.12}$$

This recursive equation for different values of n becomes:

$$\begin{aligned} (n=0) &\quad C_0(y) = 1 \\ (n=1) &\quad C_1(y) = y \\ (n=2) &\quad C_2(y) = 2y^2 - 1 \\ (n=3) &\quad C_3(y) = 4y^3 - 3y \\ (n=4) &\quad C_4(y) = 8y^3 - 8y^2 + 1 \end{aligned} \tag{25.13}$$

25.3 Examples of State Space Model Identification

In this section the development of linear as well as non-linear Wiener models will be illustrated.

25.3.1 Linear State Space Model

In the previous chapter a linear ARX model was developed for the Box and Jenkins (1976) gas furnace data. The model was second order and the delay was 3 minutes. Let us develop a second-order state space model and compare it with the ARX model. This is done in file F2503.m. Figure 25.3 shows the comparison between the two models.

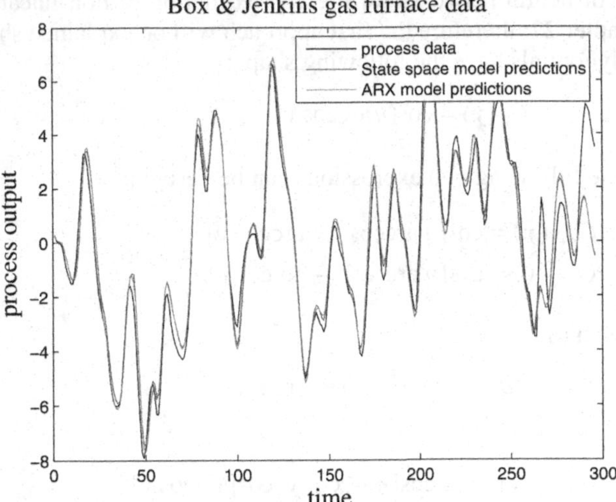

Fig. 25.3. Comparison between state space and ARX model.

The code for computing a state space model is straightforward:

```
%
% Load data from file boxj: input=MethFd, output=CO2Conc
%
load boxj;
MethFd   = dtrend(MethFd);
CO2Conc = dtrend(CO2Conc);
nrows    = length(MethFd);
%
% develop a linear state space model of order 2
%
model = n4sid([CO2Conc MethFd], 2);
ye     = idsim(MethFd,model1);
```

The data are loaded, detrended, and function n4sid is used to develop the model. The state space model explains 92.7% of the variance in the process output, whereas the ARX model explains 92.4% of the variance, hence the performance of the models is similar.

25.3.2 Non-linear State Space Model

In file nlp11.mat process data are stored for a process with one input and one output. The process is non-linear, hence we may expect that a linear model will not perform as well as a non-linear model. A non-linear Wiener model is developed, the code is stored in file wiener1.m. Some of the required functions that are used originate from the Control and Systems Library SLICOT (van den Boom *et al.*, 2002). This software is available as freeware.

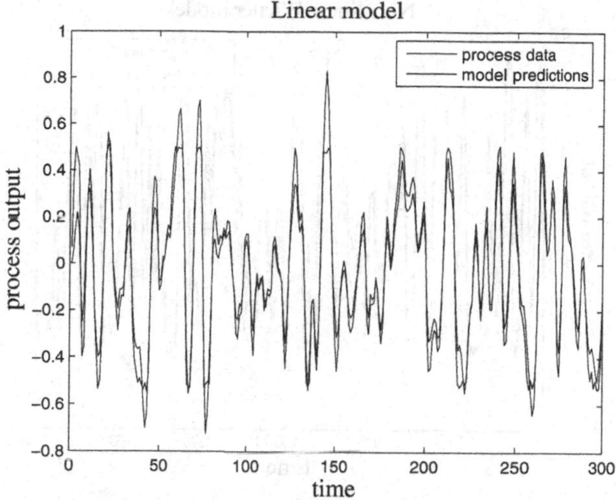

Fig. 25.4. Linear model fit for non-linear process.

Figure 25.4 shows the fit for a second order linear state space model. The percentage variance of the process output explained is 92.6%, which is still reasonable.

Figure 25.5 shows the non-linear relationship between the output of the linear state space model y and the non-linear process output z.

As can be seen, the real process output z is more or less linearly related to the output of the state space model y; however, it is limited for low and high values of y.

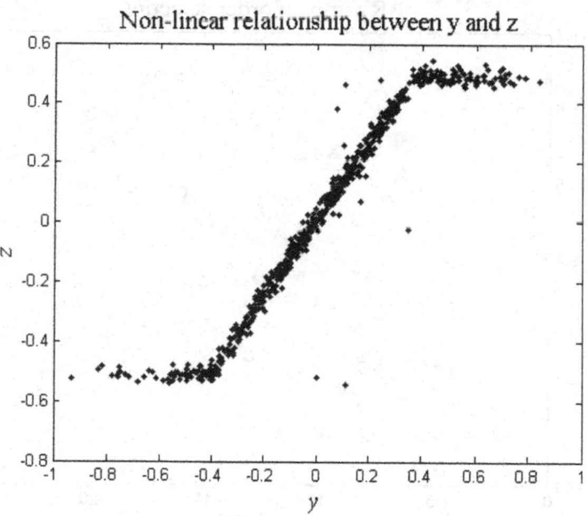

Fig. 25.5. Non-linear relationship between y and z.

Figure 25.6 shows the model predictions. The percentage variance of the process output explained is now 96.4%, which is higher than for the linear state space model.

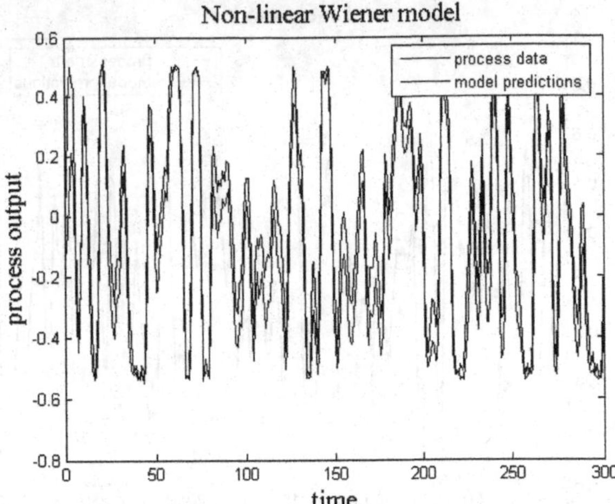

Fig. 25.6. Wiener model fit for non-linear process.

It can clearly be seen that the non-linear model performs better in predicting the peaks (minimum and maximum values) in the process output.

In file nlp22.mat data are stored for a process with two inputs and two outputs. We are not sure whether the process is severely non-linear and will therefore compare different models. First of all, an ARX model will be tested. The result for modeling the first output y_1 as a function of the two inputs u_1 and u_2 is shown in Fig. 25.7.

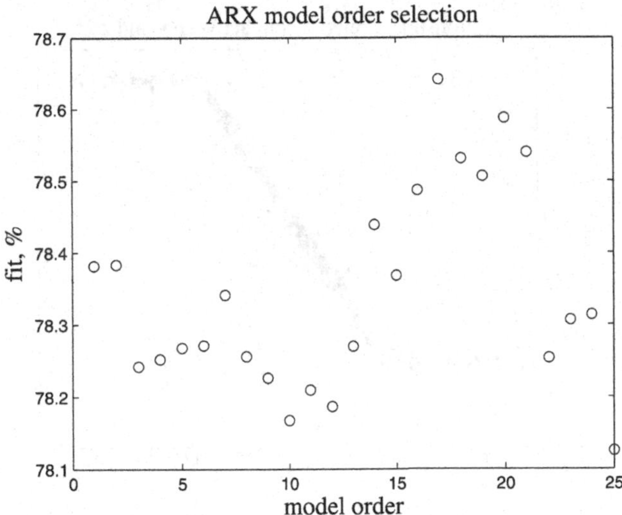

Fig. 25.7. ARX model fit as a function of model order.

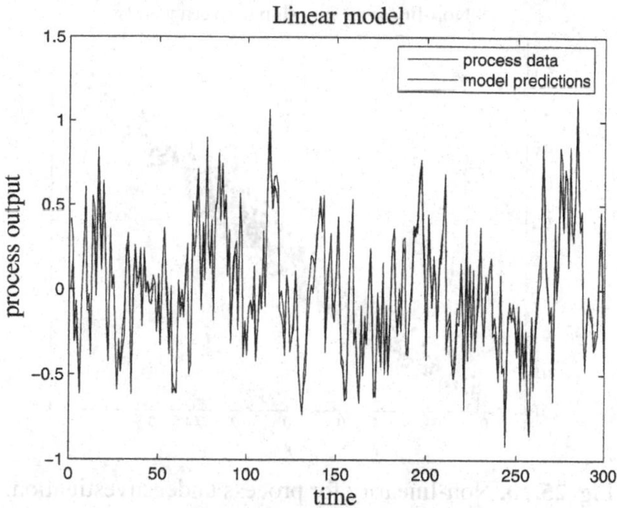

Fig. 25.8. Linear state space model fit.

As can be seen, a maximum fit of about 78.5% can be achieved. Therefore a linear state space model is tried. The result for a second order model is given in Fig. 25.8. The percentage variance explained is now 94.8%, which is considerably higher than for the ARX model. Since we are not sure whether the process contains significant non-linearities, Let us try a second-order non-linear Wiener model. The result is shown in Fig. 25.9.

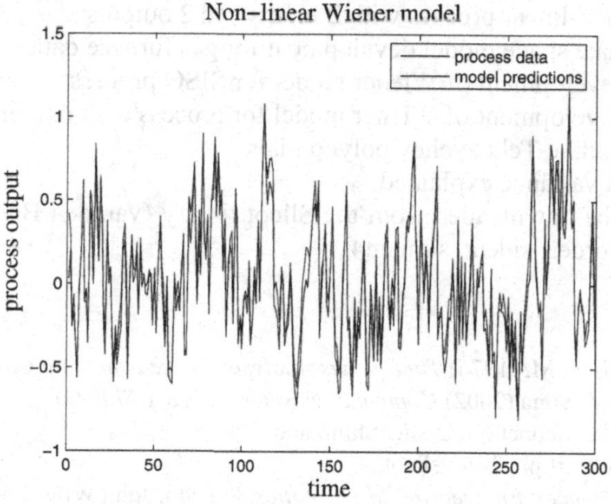

Fig. 25.9. Non-linear Wiener model fit.

The percentage variance explained is now 95.8%, which is not significantly higher than for a linear state space model. Let us investigate whether the process is indeed more or less linear. This can be shown by plotting the output of the linear state space model against the measured process output. This is done in Figure 25.10.

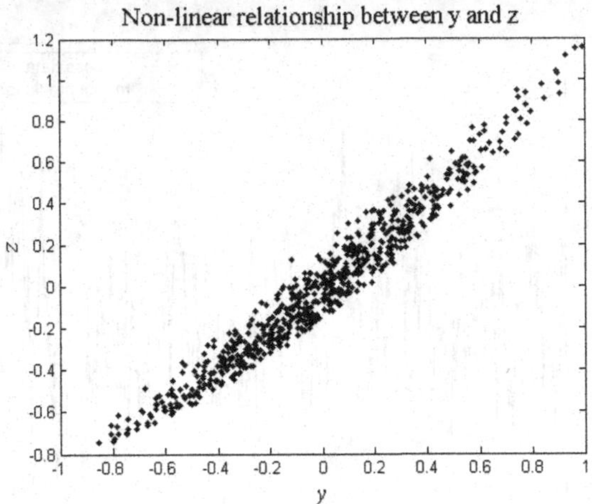

Fig. 25.10. Non-linearity for process under investigation.

As can be seen, there is a slight non-linearity, but this is not significant, hence this is the reason that the linear state space model is already performing well.

One should therefore be careful in making models unnecessarily complicated when linear models perform well.

Files referred to in this chapter:
boxj.mat: Box & Jenkins gas furnace data
nlp11.mat: data for non-linear SISO process
nlp22.mat: data for non-linear process with 2 inputs and 2 outputs
F2503.m: code for state space model development for gas furnace data
F2504.m: code for development of Wiener model for SISO process
F2507.m: code for development of Wiener model for process with one output and two inputs
tcheb.m: code to calculate Tchebychev polynomials
vae.m: calculation of variance explained

Files that should be downloaded from the Slicot library (van den Boom *et al.*, 2002) are: findbd, findr, ldsim, order, sident, slmoen4.

References

Box, G.E.P. and Jenkins, G.M. (1976) *Time Series Analysis, Forecasting and Control*, Holden-Day.

van den Boom, A. and V. Sima (2002) *Control and Systems library SLICOT*, http://www.win.tue.nl/niconet/NIC2/slicot.html and http://www.slicot.de/start.php?site=slicot.

Kreyszig, E. (1999) *Advanced Engineering Mathematics*, 8th edn, John Wiley and Sons Inc.

Ljung, L. and McKelvey, T. (1996) Subspace identification from closed loop data. *Signal Processing*, **52**, 209–15.

MATLAB *System Identification Toolbox, User's Guide, 2000*, The Math Works, Inc. John Wiley.

van Overschee, P. and de Moor, B. (1996) *Subspace Identification of Linear Systems: Theory, Implementations, Applications*, Kluwer Academic Publishers.

Verhaegen, M. (1994) Identification of the deterministic part of MIMO state space models. *Automatica*, **30**, 61–74.

26 Model Reduction

In this chapter several model reduction techniques will be discussed. The first method is based on frequency response matching, other methods make use of conversion of the model structure to a state space model and subsequently truncating the states that have a minimum impact on the input–output relationship. The main indicator used for this purpose is the so-called Hankel singular value. In addition, the model structure is converted to a balanced realization, after which the reduction techniques can be applied. Several examples are given on how to apply the different methods.

26.1 Model Reduction in the Frequency Domain

A linear transfer function of a process model in the Laplace domain can be given by:

$$F(s)=\frac{1+a_1 s+.....+a_n s^n}{1+b_1 s+...+b_m s^m} \tag{26.1}$$

Suppose it is required to find a reduced transfer function model, given by:

$$G(s)=\frac{1+a_{1r} s+...+a_{nr} s^{nr}}{1+b_{1r} s+...+b_{mr} s^{mr}} \tag{26.2}$$

where $nr<n$ and $mr<m$.

The reduced model should represent the original process as closely as possible. In order to compare the performance of the reduced model to the more complex model, one could define a quadratic performance criterion, such as:

$$J=\int_0^T (f(t)-g(t))^2\,dt \tag{26.3}$$

where $f(t)$ is the response of the process in the time domain and $g(t)$ the response of the reduced model in the time domain. There are two problems with a criterion such as Eqn. (26.3): (i) a proper value for T must be chosen and (ii) the performance index is dependent on the type of input.

Because of the latter limitation, many authors suggest to redefine Eqn (26.3) for the frequency domain. The deviations between the original system and the reduced system are then minimized over a particular frequency range.
In the frequency domain, Eqn. (26.1) can be written as:

$$F(j\omega_i)=a(j\omega_i)+jb(j\omega_i) \tag{26.4}$$

and Eqn. (26.2) is similarly:

$$G(j\omega_i)=c(j\omega_i)+jd(j\omega_i) \tag{26.5}$$

for the frequency $\omega = \omega_i$.

Over the frequency domain of interest, one could then rewrite performance index in the frequency domain:

$$J_\omega = \sum_{i=N1}^{N2} \left[\left(a(\omega_i) - c(\omega_i)\right)^2 + \left(b(\omega_i) - d(\omega_i)\right)^2 \right] \tag{26.6}$$

26.2 Transfer Functions in the Frequency Domain

This section will recapture the basics of frequency plots as discussed in chapter 9. Let us assume a first-order transfer function of the following form:

$$G(s) = \frac{K}{\tau s + 1} \tag{26.7}$$

Let the process be subjected to a sinusoidal input $f(t) = A sin\omega t$. The frequency response of G can then be obtained by substituting $s = j\omega$ into Eqn. (26.7).

The transfer function is a complex number for any given frequency with a real part and an imaginary part. According to Eqn. (9.20), the magnitude or amplitude and the phase angle ϕ of G become:

$$AR = |G(j\omega)| = \frac{\prod_{i=1}^{n} |G_i(j\omega)|}{\prod_{j=n+1}^{m} |G_j(j\omega)|} \tag{26.8}$$

$$\phi = arg\left[G(j\omega)\right] = \sum_{i=1}^{n} arg\left[G_i(j\omega)\right] - \sum_{j=n+1}^{m} arg\left[G_j(j\omega)\right]$$

where i are the numerator terms of the transfer function and j the denominator terms. Applying Eqn. (26.8) to Eqn. (26.7) results in:

$$AR = \frac{K}{\sqrt{\omega^2 \tau^2 + 1}} \tag{26.9}$$

$$\phi = tan^{-1}(-\omega\tau)$$

Diagrams where the amplitude and phase angle are plotted against the frequency are called Bode diagrams. Figure (26.1) shows an example of a Bode diagram for a simple second-order process using the MATLAB Bode function (file F2601.m):

$$G(s) = \frac{3.19 + 2.4561 s}{1 + 1.3092 s + 0.8521 s^2} \tag{26.10}$$

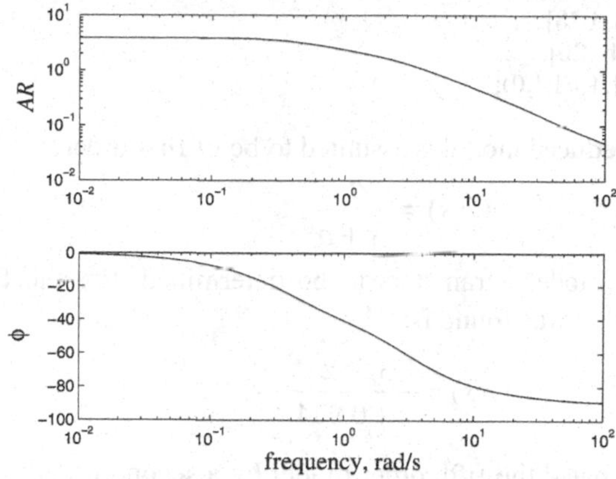

Fig. 26.1. Bode diagram for simple second-order system.

In the case of model order reduction, a note of caution should be made. In the time domain, the range of time constants which have to be considered, depends on the area of application of the model. For control applications, small time delay times and time constants are of interest. However, for dynamic optimization purposes, these small delays and lags can be ignored, since only the time constants which influence the course of the dynamics are important.

For the frequency domain, similar criteria are applicable. In control applications, the higher frequency range is of interest, owing to the sensitivity of the PID controller settings to the phase shift. It makes a large difference, whether a response is described by a first-order, a second-order, a time delay plus first-order or a lead lag (Padé) approximation. The phase shift of a first-order and lead-lag approximation remains limited, whereas the phase shift of a delay approximation will go to negative infinity.

For dynamic optimization purposes, only the lower frequency range has to be considered and a first-order approximation of process dynamics is often adequate.

26.3 Example of Basic Frequency-weighted Model Reduction

As pointed out in section 26.1, one way of performing frequency weighted model reduction is by ensuring that the real and imaginary parts of the original model and the reduced model are similar.

An example is taken from Luus (1999). The following detailed model is given:

$$F(s)=\frac{156+369s+264s^2+80s^3+10s^4}{40+148s+173s^2+84s^3+21s^4+2s^5} \tag{26.11}$$

It is desirable to simplify the fifth-order model, say, to a first-order model. Equation (26.6) is used as a minimization criterion and the code for the exercise is given in file F2602.m. The MATLAB function fmincon is used to perform the search for the optimum model parameters. Since the MATLAB function idpoly is used to construct the detailed model, Eqn. (26.11) is written in the following form:

$$F(s)=\frac{78+184.5s+132s^2+40s^3+5s^4}{20+74s+86.5s^2+42s^3+10.5s^4+s^5} \tag{26.12}$$

and subsequently the model can be defined as:

```
B1 = [0 5  40 132  184.5 78];
F1 = [1 10.5 42  86.5 74   20];
model1 = idpoly(1,B1,1,1,F1,1,0);
```

The structure of the reduced model is assumed to be of first order:

$$G(s) = \frac{a}{s+b} \tag{26.13}$$

hence there are two model parameters to be determined. For the frequency range $0.01 < \omega < 100$, the model that was found is:

$$G(s) = \frac{3.552}{s+0.934} \tag{26.14}$$

Luus (1999) approximated the fifth order model by a second-order model and found the following result:

$$G(s) = \frac{3.9+3.7766\,s}{1+2.3001s+0.7896\,s^2} \tag{26.15}$$

The Bode plot for the fifth-order model (Eqn. (26.12)), the second-order model (Eqn. (26.15)) and the first-order model (Eqn. (26.14)) are shown in Fig. (26.2).

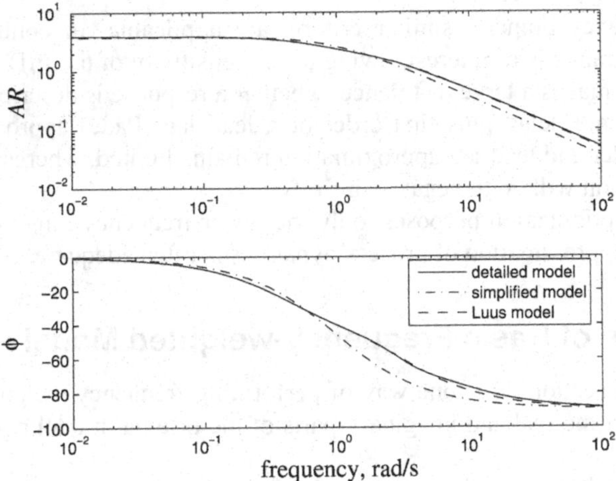

Fig. 26.2. Frequency responses of the detailed and reduced models.

There is a slight difference in phase shift between the fifth order model and first-order model, the amplitude ratio is only slightly different for high frequencies. As can be seen, the second-order model approximation is almost perfect. However, depending on the application of the model, a first-order model approximation may be adequate. The step responses of the models are shown in Fig. 26.3.

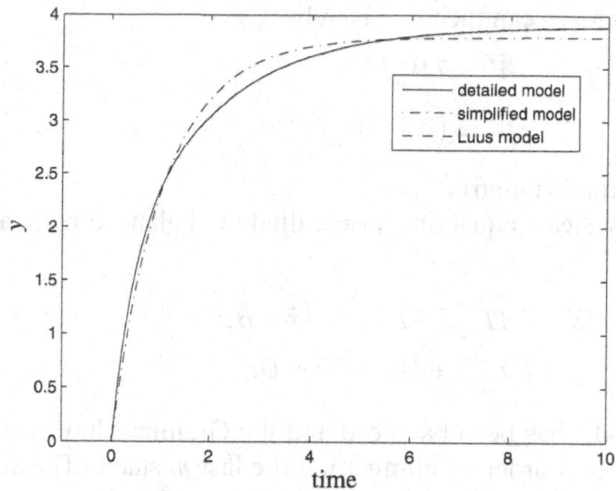

Fig. 26.3. Step responses of the detailed and reduced models.

26.4 Balancing of Gramians

High-order models are often a result of models consisting of many differential equations or partial differential equations that have been converted into ordinary differential equations. These types of model are adequate for simulations studies but are not suitable for online use. A popular technique of model reduction that does not make use of error minimization is the model balancing method. The procedure is to find observability and controllability Gramians so as to determine which states have the largest overall contribution to the model. In systems theory and linear algebra, a Gramian matrix is a real-values symmetric matrix that can be used for a test for linear independence of functions. A system is called controllable if all states x can be influenced by the control input vector u, a system is observable if all states can be determined from the measurement vector y.

Balancing of Gramians can effectively be seen as a method of model reduction while maintaining observability and controllability.

Assume a state space model has the following structure:

$$\begin{aligned} x(k+1) &= Ax(k) + Bu(k) \\ y(k) &= Cx(k) + Du(k) \end{aligned} \tag{26.16}$$

The continuous (discrete) controllability gramian W_c can be defined as:

$$W_c = \int_0^\infty e^{At} BB^T e^{A^T t} dt, \quad W_c = \sum_{k=0}^{\infty} A^k BB^T (A^T)^k \tag{26.17}$$

and the continuous (discrete) observability gramian W_o as:

$$W_o = \int_0^\infty e^{A^T t} C^T C e^{At} dt, \quad W_o = \sum_{k=0}^{\infty} (A^T)^k C^T C A^k \tag{26.18}$$

The gramians can only be computed for stable systems. A system is called frequency weighted balanced if the matrices W_c and W_o are both equal and diagonal:

$$W_c = W_o = \Sigma^2 = diag(\sigma_i),\ i = 1,2,..,n \tag{26.19}$$

where σ_i are the Hankel singular values.

The transformed Gramians can then be given by:

$$\overline{W}_c = T W_c T^T$$
$$\overline{W}_o = \left(T^{-1}\right)^T W_o T^{-1} \tag{26.20}$$

where T is a transformation matrix.

The transformed systems equations, also called the balanced realization, consequently become:

$$\dot{\overline{x}} = TAT^{-1}\overline{x} + TBu = \overline{A}\overline{x} + \overline{B}u$$
$$y = CT^{-1}\overline{x} + Du = \overline{C}\overline{x} + Du \tag{26.21}$$

If the state space model has been balanced and the Gramians have m small entries, it is possible to reduce the model order by eliminating the last m states. The states that correspond to the largest singular values have the largest impact on the input–output relationship.

26.4.1 Model Reduction by State Truncation

The first method of model reduction is to partition the state vector into important states ($\overline{x}_1$) and less important states ($\overline{x}_2$),(Hahn and Edgar, 2002):

$$\begin{pmatrix} \dot{\overline{x}}_1 \\ \dot{\overline{x}}_2 \end{pmatrix} = \begin{pmatrix} \overline{A}_{11} & \overline{A}_{12} \\ \overline{A}_{21} & \overline{A}_{22} \end{pmatrix} \begin{pmatrix} \overline{x}_1 \\ \overline{x}_2 \end{pmatrix} + \begin{pmatrix} \overline{B}_1 \\ \overline{B}_2 \end{pmatrix} u$$
$$y = \begin{pmatrix} \overline{C}_1 & \overline{C}_2 \end{pmatrix} \begin{pmatrix} \overline{x}_1 \\ \overline{x}_2 \end{pmatrix} + Du \tag{26.22}$$

State truncation is the truncation of the least important states, as a result of which Eqn. (26.22) reduces to:

$$\dot{\overline{x}}_1 = \overline{A}_{11}\overline{x}_1 + \overline{B}_1 u$$
$$y = \overline{C}_1\overline{x} + Du \tag{26.23}$$

This method has one disadvantage: it will not guarantee consistent steady-state behavior.

26.4.2 Model Reduction by Residualization

In this case the system is truncated but consistent steady-state behavior is guaranteed. This is achieved by setting $\dot{\overline{x}}_2 = 0$ in Eqn. (26.24). The model equations become then:

$$\begin{pmatrix} \dot{\overline{x}}_1 \\ 0 \end{pmatrix} = \begin{pmatrix} \overline{A}_{11} & \overline{A}_{12} \\ \overline{A}_{21} & \overline{A}_{22} \end{pmatrix} \begin{pmatrix} \overline{x}_1 \\ \overline{x}_2 \end{pmatrix} + \begin{pmatrix} \overline{B}_1 \\ \overline{B}_2 \end{pmatrix} u$$
$$y = \begin{pmatrix} \overline{C}_1 & \overline{C}_2 \end{pmatrix} \begin{pmatrix} \overline{x}_1 \\ \overline{x}_2 \end{pmatrix} + Du \tag{26.24}$$

Eliminating the states $\bar{x}_2$ from Eqn. (26.24) results in the following simplified equation:

$$\begin{aligned}\dot{\bar{x}}_1 &= \left(A_{11} - A_{12}A_{22}^{-1}A_{21}\right)\bar{x}_1 + \left(B_1 - A_{12}A_{22}^{-1}B_2\right)u \\ y &= \left(C_1 - C_2A_{22}^{-1}A_{21}\right)\bar{x}_1 - C_2A_{22}^{-1}B_2u\end{aligned} \tag{26.25}$$

26.4.3 Balancing the Model Equations of a Reactor Model

In this section an example will be given of the balanced realization of a chemical reactor model, as given by Hahn and Edgar (2002). The authors describe the system matrices by:

$$A = \begin{bmatrix} -2 & 0 & 0 \\ 1 & -1.1 & 0 \\ 0 & 0.1 & -1 \end{bmatrix},\ B = \begin{pmatrix} 2 \\ 0 \\ 0 \end{pmatrix},\ C = \begin{pmatrix} 0 & 0 & 1 \end{pmatrix},\ D = 0 \tag{26.26}$$

The following steps should now be followed in MATLAB to calculate the Gramians, transformation matrix, Hankel singular values and balanced system matrices:

Step 1: convert the model to a state space model by executing:
model = ss(A, B, C, D);

Step 2: compute the gramians from:
Wc = gram(model,'c');
Wo = gram(model,'o');

Step 3: factorize Wc, i.e. $R*R^T$=Wc, this can be done by:
R = chol(Wc);

Step 4: compute $Q = R*Wo*R^T$ and factorize it by using singular value decompsition:
[U,S,V] = svd(R*Wo*R');

Step 5: the Hankel singular values σ can now be computed from:
sigma = sqrt(S);

Step 6: the transformation matrix follows from:
T = inv(R'*U/sqrt(sigma));

Step 7: the balanced system matrices can be computed from:
Ab = T*A*inv(T);
Bb = T*B;
Cb = C*inv(T);
Db = D;

The program that executes these steps is given in Eqn2629.m. The results from the calculations are:

$$T = \begin{bmatrix} -0.0773 & -0.1845 & -2.5300 \\ 0.0887 & 0.0859 & -2.7758 \\ 0.0434 & -0.1528 & 1.3434 \end{bmatrix} \tag{26.27}$$

and

$$\Sigma = \begin{bmatrix} 0.0594 & 0 & 0 \\ 0 & 0.0153 & 0 \\ 0 & 0 & 0.0013 \end{bmatrix} \tag{26.28}$$

from which it can be seen that the third Hankel singular value is small compared with the other two. This means that the model can be reduced to a two-state model, truncating the third state. The balanced three-state model becomes:

$$\begin{bmatrix} \dot{\bar{x}}_1 \\ \dot{\bar{x}}_2 \\ \dot{\bar{x}}_3 \end{bmatrix} = \begin{bmatrix} -0.2012 & -0.6210 & 0.2212 \\ 0.6210 & -1.0310 & 1.1060 \\ 0.2212 & -1.1060 & -2.8680 \end{bmatrix} \begin{bmatrix} \bar{x}_1 \\ \bar{x}_2 \\ \bar{x}_3 \end{bmatrix} + \begin{bmatrix} -0.1546 \\ 0.1773 \\ 0.0689 \end{bmatrix} u$$

$$y = \begin{bmatrix} -0.1546 & -0.1773 & 0.0869 \end{bmatrix} \begin{bmatrix} \bar{x}_1 \\ \bar{x}_2 \\ \bar{x}_3 \end{bmatrix} \tag{26.29}$$

There is another way of reducing the reactor model, by making use of Hankel norm reductions. It will not be discussed here, the interested reader is referred to Weiland and Stoorvogel (1997) and Curtain and Sasane (2003).

26.5 Examples of Model State Reduction Techniques

In this section, MATLAB will be used to illustrate the differences between the different model reduction techniques that have been discussed in the previous section. Data is obtained for a process with two inputs and two outputs and was taken from Zhu (2001). The file is called glassdata2.mat and can be downloaded from Zhu (2001). In this exercise only output y_1 will be modeled.

First an ARX model will be developed. Using file F2604.m, Fig. 26.4 was created showing the percentage fit against the model order.

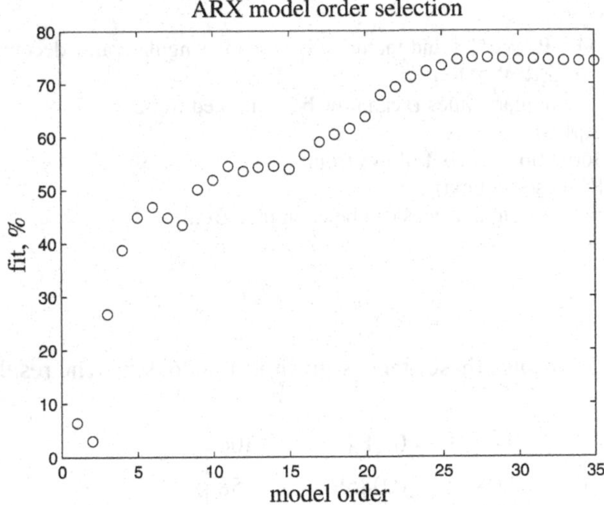

Fig. 26.4. ARX Model fit versus model order.

As can be seen, the higher the model order, the higher the percentage fit. From an order of twenty-five onward, there seems to be little improvement. Therefore a 25th order ARX model is selected as a stating point for the model reduction exercise.

File modred25.m is used to compare the various model reduction techniques and show their results. First the system is balanced and the Hankel singular values or Gramian diagonal elements are calculated for a 25th order state space model. Figure 26.5 shows the result.

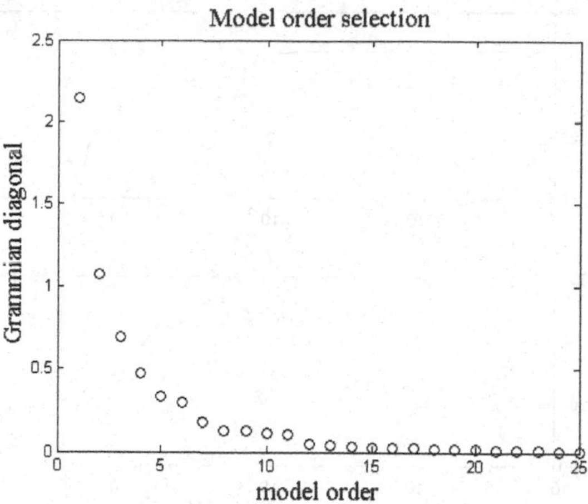

Fig. 26.5. Gramian diagonal versus state space model order.

It can be seen that the first 10 states are really significant, states 11:25 are less relevant. Therefore a 10th order state space model could be developed as a simplification of the 25th order ARX model. One should note that a 10th order state space model still contains a considerable number of model parameters.

File modred25 compares the following techniques:

- residualization
- balanced realization and truncation
- Hankel norm approximation

Figure 26.6 shows the bode diagram for (y_1, u_1), Fig. 26.7 the bode diagram for (y_1, u_2) using residualization.

It can be seen that the original model and the reduced model using residualization have the same frequency response for low to moderate frequencies, for moderate to high frequencies the two frequency response curves start to deviate somewhat, which can be expected. Figure 26.8 shows the step responses for the models, it can be seen that the two models behave almost identical and also have the same steady-state value, which is important if the model is to be used for prediction purposes.

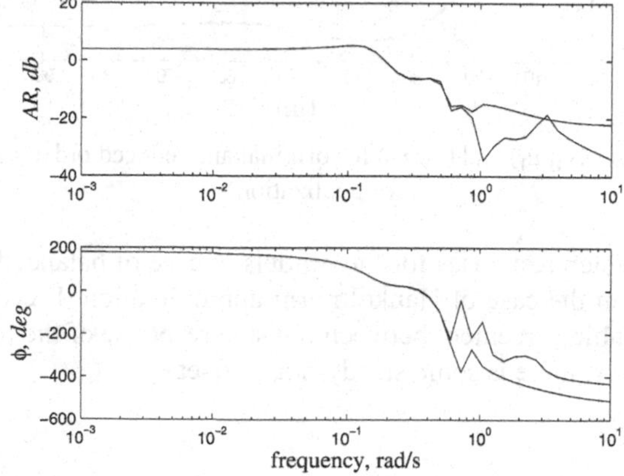

Fig. 26.6. Bode diagram for (y_1, u_1) for original and reduced 10th order state space model using residualization.

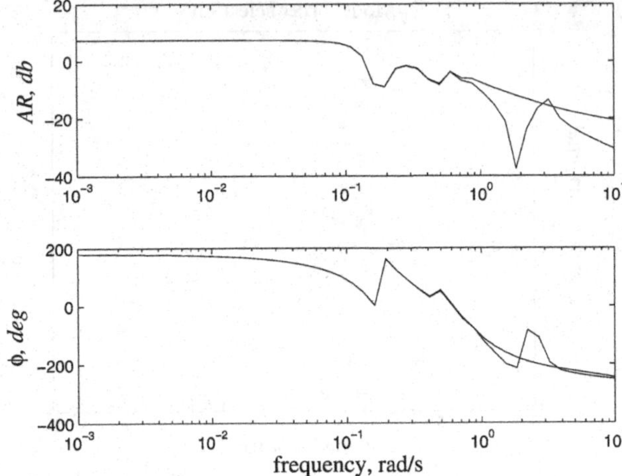

Fig. 26.7. Bode diagram for (y_1, u_2) for original and reduced 10th order state space model using residualization.

When the model is going to be used for control purposes, a steady-state model offset is usually not so important because most control strategies can cope with limited errors in dynamics and process gain.

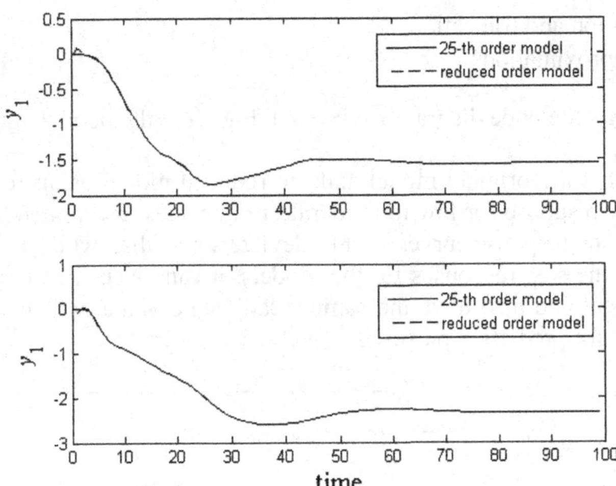

Fig. 26.8. Step responses (y_1, u_1) and (y_1, u_2) for original and reduced order state space model using residualization.

Figure 26.9 shows the step responses for the models in case of balanced realization and truncation and Fig. 26.10 in the case of Hankel norm approximation. It can be seen that in both cases there is a reasonable agreement between the step response of the original model and the reduced model, however, there is some steady-state offset.

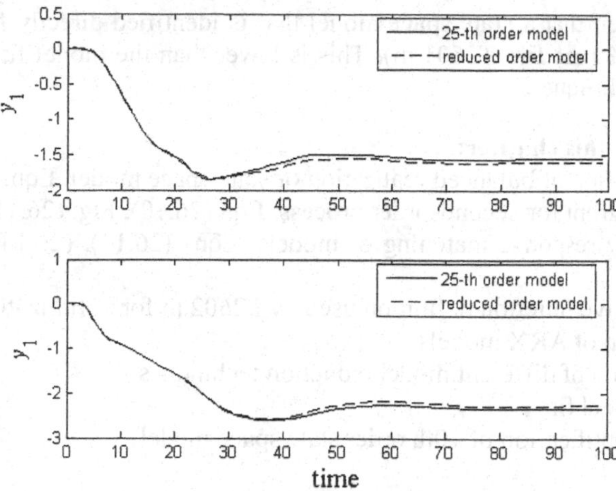

Fig. 26.9. Step responses (y_1, u_1) and (y_1, u_2) for original and reduced model in case of model truncation.

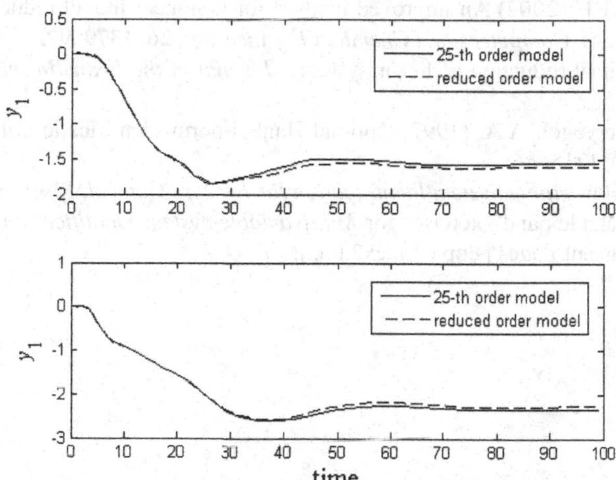

Fig. 26.10. Step responses (y_1, u_1) and (y_1, u_2) for original and reduced model in case of Hankel norm approximation.

Table 26.1 shows the model fit for the different reduced models and the original 25th order ARX model. As can be seen, the model reduction does lead to some loss in model fit. This does not mean, however, that the reduced model would be unsuitable for prediction or control purposes.

Table 26.1. Comparison of model fit percentages for different methods.

	model fit, %	**steady-state offset**
25th order ARX	69.50	—
10th order state space model using:		
residualization	63.41	no
truncation	67.73	yes
Hankel approximation	66.51	yes, some
10th order state space model directly	58.75	—

It can also be seen that a state space model that is identified directly from the data has a model fit of 58.75% (use file T2601.m). This is lower than the model fit percentage for the model reduction techniques.

Files referred to in this chapter:
Eqn2629.m: calculation of balanced realization of state space model, Eqn. (26.29)
F2601.m: Bode diagram for second-order process, Eqn (26.10), Fig. (26.1)
F2602.m: frequency response matching of models, Eqns (26.11), (26.14), (26.15) and Fig. (26.2)
F2602sub.m: objective function definition used by F2602.m for optimization
F2604.m: generation of ARX models
F2605.m: comparison of different model reduction techniques
fit.m: calculates model fit
T2601.m: direct identification of 10th order state space model

References

Curtain, R.F. and Sasane, A.J. (2003) Hankel norm approximation for well-posed linear systems. *Systems and Control Letters*, 48 (5), 407–14.

Hahn, J. and Edgar, T.F. (2002) An improved method for nonlinear model reduction using balancing of empirical gramians. *Computers and Chemical Engineering*, **26**, 1379–97.

Luus, R. (1999) Optimal reduction of linear systems. *Journal of the Franklin Institute*, **336** (3), 523–32.

Weiland, S. and Stoorvogel, A.A. (1997) Optimal Hankel-norm identification of dynamical systems. *Automatica*, **33** (7), 1235–46.

Zhu, Y.C., (2001) *Multivariable System Identification for Process Control*, Elsevier Science, Oxford.

Zhu, Y.C. (2001) Examples and exercises for *Multivariable system identification for process control*, http://www.er.ele.tue.nl/pages/people/index2.html.

27 Neural Networks

Artificial Neural Networks have gone through a rapid development and grown past the experimental stage to become implemented in a wide range of engineering applications, such as for example state estimation, pattern recognition, signal processing, process modeling, process quality control and data reconciliation.

The neural network is capable of modeling non-linear systems. On the basis of supplied training data the neural network learns (trains) the relationship between the process input and output. The data have to be examined carefully before they can be used as a training set for a neural network. The training sets consist of one or more input data and one or more output data. After the training of the network, a test-set of data should be used to verify whether the desired relationship was learned.

In practical applications a neural network can be used when the exact model is not known. It is a good example of a 'black-box' technique. By no means, however, should the neural network be seen as the ultimate solution for problems with undefined or only partially defined models. The main reason is that it gives no additional information about the physical relationships and thus it will give no physical insight into the process.

27.1 The Structure of an Artificial Neural Network

More than 50 different types of neural network exist. Certain networks are more efficient in optimization; others perform better in data modeling and so forth. According to Basheer (2000) the most popular neural networks today are the Hopfield networks, the Adaptive Resonance Theory (ART) networks, the Kohonen networks, the counter propagation networks, the Radial Basis Function (RBF) networks, the backpropagation networks and recurrent networks.

To explain the basics of the structure of a neural network it is best to choose the simplest form of a neural network, the feed-forward neural network. In Fig. 27.1 a feed-forward network is shown.

Artificial neural networks consist of different types of layers. There is the input-layer, one or more hidden layers and an output layer. All these layers can consist of one or more neurons. A neuron in a particular layer is connected to all neurons in the next layer, which is why this is called a feed-forward network. In other networks the neurons might be connected otherwise. An example of a different network is a recurrent neural network where there are also links that connect neurons to other neurons in a previous layer. A fully connected network is a network in which all the neurons from one layer are connected to all neurons in the next layer.

Process Dynamics and Control: Modeling for Control and Prediction. Brian Roffel and Ben Betlem.

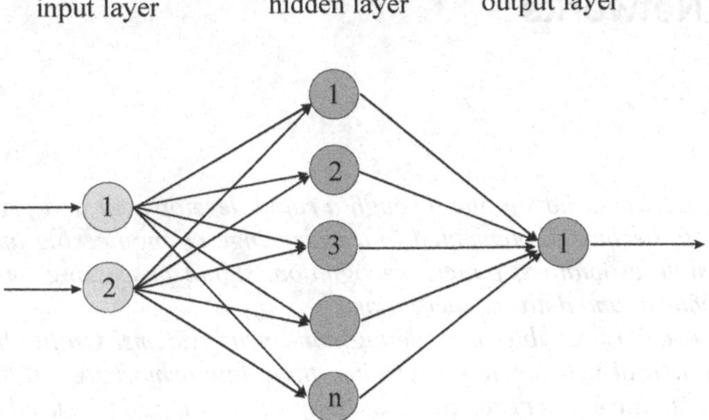

Fig. 27.1. Example of a multi-layer feed-forward neural network.

To explain how the network calculates the output, first the smallest part of the network will be observed closer: the neuron.

A neuron is a single computational point in the network. It receives one or more inputs either from a neuron in a previous layer or (when it is situated in the input layer) from the outside world. The neuron determines if the output is transmitted to neurons in the next layer. A schematic display of a neuron is given in Fig. 27.2.

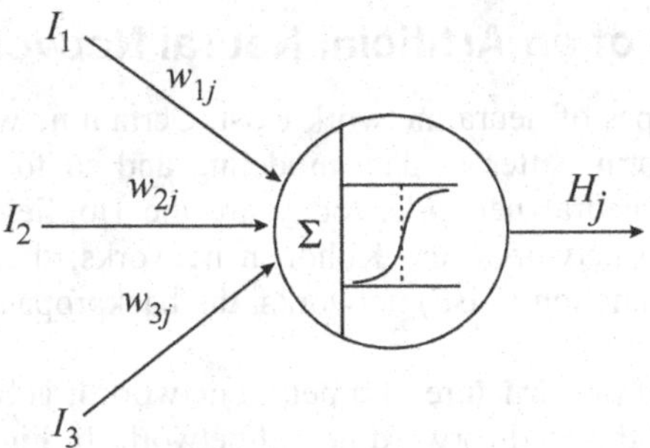

Fig. 27.2. A hidden neuron.

When the neuron receives inputs from neurons in a previous layer, I_i, they are first multiplied by a weight w_{ij}. This weight is specific for the link that connects the two neurons. No two links in a network are the same and therefore, in general, no two weights in a network are the same.

In the neuron all the weighed inputs are added and can be compared to a threshold μ. The j-th neuron in the hidden layer receives N inputs from the input layer which have connections strengths or associated weights w and fires as soon as the weighed sum together with the bias b_j exceeds the threshold:

$$h_j = \sum_{i=1}^{N} w_{ij} I_i + b_j \geq \mu_j \tag{27.1}$$

Usually, in cases of process modeling, this threshold is replaced by an activation function, making the activation of the neuron continuously valued. However this function has to be differentiable, and it has to saturate at both ends. When the neuron fires, this means that its output H has a value different than zero. The type of network that is used will determine the actual value of H. Sometimes H is simply one (a binary network), for other networks H is determined by the activation function of the neuron. This function usually has the weighed inputs and sometimes the threshold of the neuron as parameters. In these cases it can be described as a non-linear function of the sum of the weighed inputs together with the bias:

$$H_j = g(h_j) \tag{27.2}$$

where g denotes a non-linear activation function.

Typical choices for the activation function are, among others, the sigmoid function:

$$g(h) = \frac{1}{1+e^{-\beta h}} \tag{27.3}$$

the hyperbolic tangent function:

$$g(h) = tanh(\beta h) = \frac{e^{\beta h} - e^{-\beta h}}{e^{\beta h} + e^{-\beta h}} \tag{27.4}$$

or the Gaussian function:

$$g(h) = e^{-h^2} \tag{27.5}$$

The computed output of the sigmoid and Gaussian function can only fall between 0 and 1 and therefore, the output data used to train the neural network needs to be scaled to the range of 0–1 (–1 and 1 for the hyperbolic tangent function). Since the extremes of 0 to 1 occur only when the input to the sigmoid function is $-\infty$ to $+\infty$, the output data are often scaled to be between approximately 0.1 to 0.9. However, non-scaled data may be used if a linear activation function is used for the output layer.

When the neuron is situated in an input or hidden layer its output will be sent to all the neurons in the next layer. They will use it, after it is multiplied with the weight of the link, as input. When the neuron is situated in the output layer, its output is (along with other outputs from neurons in the same layer) the output of the network.

From the structure of a neural network it can be stated that the information learned is stored in the weights of the links. For different problems different structures can be used. The links do not have to point in the same direction, recurrent networks, in which delayed outputs are used as inputs, are also very common for process modeling since it can describe time dependency. Also all neurons in a network can be connected with each other. For this type of network, the input, hidden and output neurons have to be strictly defined. However, the basic principle as explained above is valid for all these networks.

Hertz *et al.* (1995) described in their work that in most cases a neural network consisting of three layers will give good results. This statement is supported by the literature where neural networks were applied in process control. In all of these cases a network consisting of an input layer, one hidden layer and an output layer was used.

The optimal structure of the network is usually determined by trial and error, however, a three-layer network in which the number of hidden neurons is twice the number of input neurons, is often a good starting point.

27.2 The Training of Artificial Neural Networks

After the specific initial structure of the neural network is determined it still needs to be trained to learn the process. Different methods of training exist, with the standard back propagation algorithm (Rumelhart *et al.*, 1986) being the most popular. This algorithm will be explained in the next paragraph.

To train a neural network, first a training-set and a test-set of sample data from the process has to be generated. The training-set is build up of pairs of input and output data, called patterns. These patterns are not necessarily unique, an output is allowed to have more than one different inputs; however, an input can only have one distinctive output.

The data in these sets have to be a good representation of average process data and therefore have to be composed carefully. This means that both these sets have to be stripped of any bad data such as errors due to malfunctions in the process or instrumentation. This does not mean, however, that the neural network cannot be trained to recognize these features. It can also be trained to recognize these errors and determine their probable causes.

After the data sets are screened, they will be presented to the network. During the training, the weights are updated in such a way that the square sum of the difference between real output value y and the output of the network $\hat{y}$, is minimized:

$$E(w) = min\frac{1}{2N}\sum_{p=1}^{N}\sum_{k=1}^{M}(\hat{y}_k^p - y_k^p)^2 \tag{27.6}$$

In this equation N is the number of patterns and M the number of output neurons, the weight vector w is the parameter vector that minimizes E.

During the training all the input/output pairs of the training-set will be presented multiple times to the network. The number of times that the entire training-set is presented to the network is called the number of epochs that is used to train the network.

There are basically four different approaches for training the neural network. The first approach is the error correction learning rule, where the error between the output of the network and the measured output is used to adjust the network weights simultaneously. A second approach is Boltzman learning, which is similar to error correction learning, however, the output of a neuron is based on a Boltzman statistical distribution.

The third approach is Hebbian learning, where learning is done locally by adjusting the weight based on the activities of the neurons.

The fourth approach is competitive learning, where neurons compete in such a way that only one neuron will be activated in a given iteration.

27.3 The Standard Back Propagation Algorithm

To explain the back propagation algorithm, a simple feed-forward neural network of three layers (input, hidden and output) is used. The network input will be denoted by x_i, the output of the hidden neurons by H_i, that of the output neurons by $\hat{y}_i$. The weights of the links between the input and the hidden layer are written as w_{ij}, where i refers to the number of the input neuron and j to the number of the hidden neuron. The weights of the links between the hidden and the output layer are denoted as w_{jk}, again j stands for the number of the hidden neuron and k for the number of the output neuron. An example network with notation is shown in Fig. 27.3. This network has three input neurons, three hidden neurons and two output neurons, in this case the input layer passes on the inputs, i.e. $I_j = x_j$.

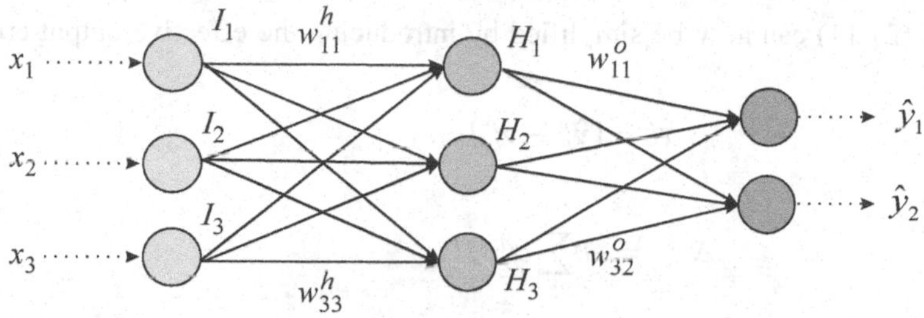

Fig. 27.3. A three layer neural network showing the notation for units and weights.

The input consists of several patterns that will be labeled by a superscript p, so input i is set to x_i^p when pattern p is presented. An input pattern is one single network input combination, it has one single, however not necessarily unique, network output combination. This means that the training set consists of a certain number of input/output pairs $\{x_i^p, y_k^p\}$. The neurons in the network all have the same activation function $g()$.

Given pattern p, hidden unit j receives a net input from the input layer as follows:

$$h_j^p = \sum_i w_{ij}^h x_i^p \tag{27.7}$$

The output of the hidden neuron j is:

$$H_j^p = g(h_j^p) \tag{27.8}$$

Therefore output neuron k receives the following input:

$$o_k^p = \sum_j w_{jk}^o H_j^p = \sum_j w_{jk}^o g\left(\sum_i w_{ij}^h x_i^p\right) \tag{27.9}$$

The output of this neuron is:

$$\hat{y}_k^p = g(o_k^p) \tag{27.10}$$

In most cases, a linear output function is used, in which case $\hat{y}_k^p$ is equal to o_k^p.

During the training, the weights are updated in order to minimize the difference between the network output and the desired output, as stated in Eqn. 27.6.

The number of presentations of the entire pattern to the network is called an *epoch*. Many epochs are generally needed before the error is acceptably small. The weights are updated every epoch; however, sometimes it becomes necessary to update them after a few patterns or even after one pattern. In the latter case the first summation (over all patterns) as well as the division by N disappears from the equation. The gradient descent rule is used to update the weights, the weights of the links between the hidden layer and the output layer are updated as follows:

$$\Delta w_{jk}^o = -\eta \frac{\partial E}{\partial w_{jk}^o} = -\eta \frac{\partial E}{\partial \hat{y}_k^p} \frac{\partial \hat{y}_k^p}{\partial o_k^p} \frac{\partial o_k^p}{\partial w_{jk}^o} = -\eta \sum_p \left(\hat{y}_k^p - y_k^p\right) H_j^p \tag{27.11}$$

In this equation η is the learning rate, it determines the step-size. A low rate means slower convergence, but if the rate is too high the minimum might not be found, as the steps are too large.

Equation (27.11) can now be simplified by introducing the effective output error α_k^p, in which case:

$$\alpha_k^p = \left(\hat{y}_k^p - y_k^p\right) \tag{27.12}$$

resulting in:

$$\Delta w_{jk}^o = -\eta \sum_p \alpha_k^p H_j^p \tag{27.13}$$

For the links between the input layer and the hidden layer, the following equation is used:

$$\Delta w_{ij}^h = -\eta \frac{\partial E}{\partial w_{ij}^h} = -\eta \frac{\partial E}{\partial H_j^p} \frac{\partial H_j^p}{\partial h_j^p} \frac{\partial h_j^p}{\partial w_{ij}^h} \tag{27.14}$$

which results in:

$$\Delta w_{ij}^h = -\eta \sum_p \sum_k \left(\hat{y}_k^p - y_k^p\right) w_{jk}^o g'\left(h_j^p\right) x_i^p \tag{27.15}$$

With the effective output error α_k^p this results in:

$$\Delta w_{ij}^h = -\eta \sum_p \sum_k \alpha_k^p w_{jk}^o g'\left(h_j^p\right) x_i^p \tag{27.16}$$

This equation can also be simplified further if the effective output error of the hidden layer α_j^p is defined:

$$\alpha_j^p = g'\left(h_j^p\right) \sum_k w_{jk}^o \alpha_k^p \tag{27.17}$$

Equation (27.13) now becomes:

$$\Delta w_{ij}^h = -\eta \sum_p \alpha_j^p x_i^p \tag{27.18}$$

Generally, with an arbitrary number of layers, the back propagation update rule has the following form:

$$\begin{aligned} &\textit{weight correction } \Delta w_{lm} = -(\textit{learning rate } \eta) * (\textit{local gradient } \alpha_m) * \\ &\qquad (\textit{input signal of node } z_l) \end{aligned} \tag{27.19}$$

where z equals $\hat{y}$ for an output node and z equals H for a hidden node. The general rule for the adaptation of weights is also known as the *generalized delta rule.*

A momentum term is often added in weight updating in order to avoid local minima and search instability. In case of an input node, Eqn. (27.20) then becomes:

$$\Delta w_{lm}^{new} = -\eta \alpha_m x_l + \mu \Delta w_{lm}^{old} \tag{27.20}$$

This rule is called the modified delta rule. The standard backpropagation algorithm has been modified in several ways to achieve a better search and accelerate and stabilize the training process (Looney, 1996; Masters, 1994).

27.4 Recurrent Neural Networks

An important factor in the popularity of feed-forward networks is that it has been shown that a continuous valued neural network with a continuous differentiable non-linear transfer function can approximate any continuous function arbitrarily well (Cybenko, 1989). The feed-forward architecture shown in Fig. 27.1 is typically used for steady-state functional approximation or one-step-ahead dynamic prediction. However, if the model is to be used to predict also more than one time step ahead, recurrent neural networks should be used, in which delayed outputs are used as neuron inputs

It was found by MacMurray and Himmelblau (1995) that externally recurrent networks (ERNs) have the best performance in predicting the process output many steps into the future. Figure 27.4 displays the ERN, it is the feed-forward network (FFN) from Fig. 27.1 that now has externally recurrent connections with one or more sampling delays (z^{-1}) between the network output and the network inputs (see also chapter 24). The network is sometimes called a tapped-delay line network.

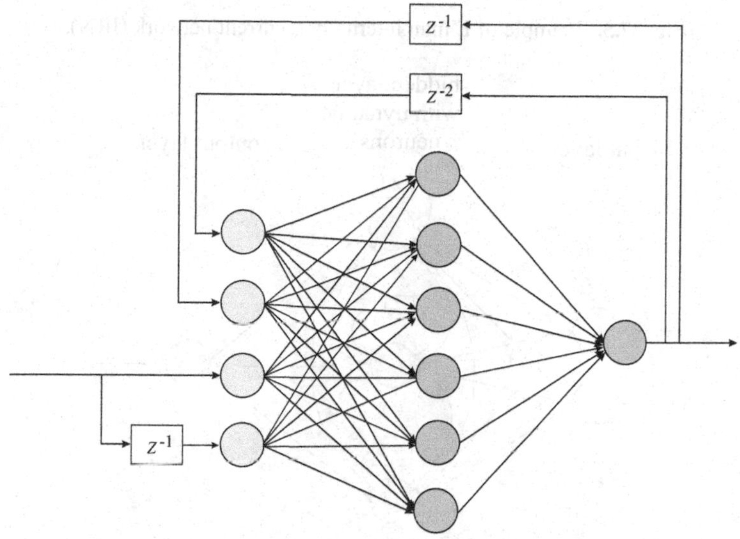

Fig. 27.4. Externally recurrent network (ERN)

Hence, the essential difference between a FFN in making predictions and an ERN is that the ERN network uses values of the process outputs predicted by the net rather than measured process values.

The adverb 'externally' is used to distinguish from internally recurrent (Elman) networks (IRN) as shown in Fig. 27.5, diagonally recurrent networks (DRN) as shown in Fig. 27.6 and combinations of ERN and IRN.

There are numerous configurations of the Elman network and the diagonal recurrent network. The weight-updating scheme depends on the structure of the network and will be different for each type of network.

The prediction made by an ERN can be expressed as a function of the lagged process inputs and predicted process outputs:

$$\hat{y}(k) = f\left(\hat{y}(k-1),\ldots\ldots,\hat{y}(k-N),u(k-1),\ldots\ldots,u(k-M)\right) \tag{27.21}$$

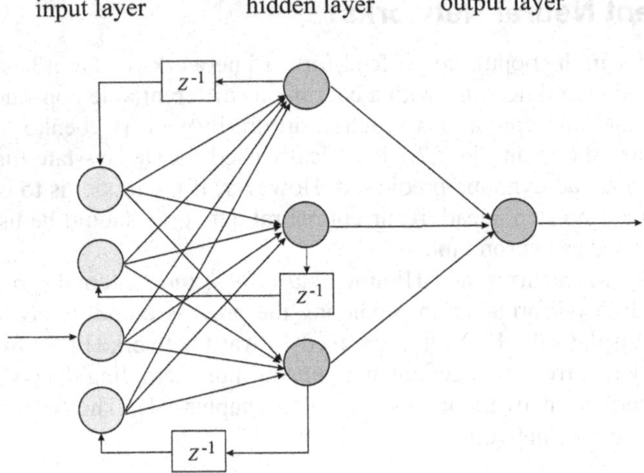

Fig. 27.5. Example of Elman internally recurrent network (IRN).

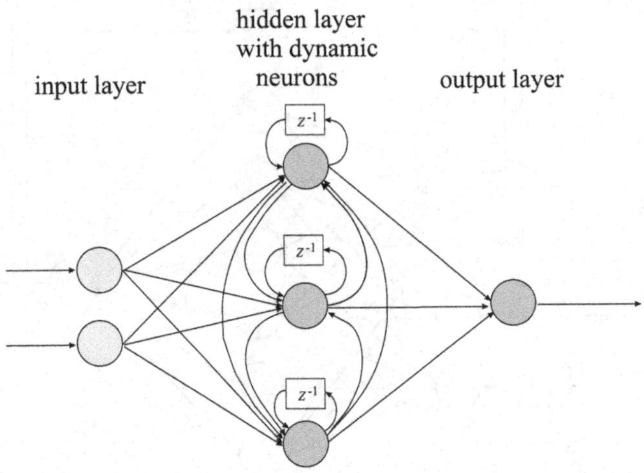

Fig. 27.6. Diagonal recurrent (dynamic) network (DRN).

in which $\hat{y}(k)$ refers to the prediction (estimate) of the process output at sampling interval k, f is a non-linear function (determined by the neural network).

N and M are integers representing the number of inputs to the neural network. If the function f is linear then the model is an ARMAX model, as discussed in chapter 24. It is good practice to start with a linear ARMA model first, if poor modeling results are obtained, the non-linear version should be investigated.

Several investigators have studied the problem of error amplification if $\hat{y}$ is replaced by the measured value y, and concluded that the use of $\hat{y}$ provided better prediction. For this reason, the ERN is used to model process behavior.

The externally recurrent network (ERN) of Eqn. (27.21) is sometimes called an N-step ahead prediction model, whereas the expression for the feed-forward network (FFN):

$$\hat{y}(k) = f\big(y(k-1),\ldots\ldots, y(k-N), u(k-1),\ldots\ldots, u(k-M)\big) \tag{27.22}$$

is sometimes called a one-step ahead prediction model (Ljung, 1987; Narendra and Parthasarathy, 1990).

Ljung gives a thorough discussion of the advantages and disadvantages of these two modeling approaches. The parallel model is usually preferable for control applications, since the model may be required to predict several steps into the future and not just one. Hence, it is then necessary to feed back values that the model has predicted in the absence of any measurements. The parallel model is also applicable on a stand-alone basis that may be useful in cases when the measurement fails.

The learning or training of a feed-forward neural network is different from a recurrent neural network. For the feed-forward neural network, the back propagation algorithm, as discussed in section 27.3, can be written as:

$$w_{i+1} = w_i - \eta \frac{\partial E}{\partial w_i} \tag{27.23}$$

where i is the iteration cycle, $\partial E / \partial w_i$ is the gradient descent of E with respect to the parameter matrix w_i and η is the learning rate. Tan and Saif (2000) derived the expression for the weight updating of the externally recurrent network; it will be closely followed here.

Assume that the dynamic model is of the form:

$$\hat{y}(k) = f(\hat{y}(k-1),\ u(k-1), w_i) \tag{27.24}$$

The derivative of $f(.)$ with respect to w is:

$$\frac{\partial \hat{y}(k)}{\partial w_i} = \frac{\partial f(.)}{\partial w_i} + \frac{\partial f(.)}{\hat{y}(k-1)} \frac{\partial \hat{y}(k-1)}{\partial w_i} \tag{27.25}$$

The backpropagation learning algorithm is usually slow in converging, one way to improve the training is through second-order-based approaches such as the Newton method and the Levenberg-Marquardt method. The former has the following updating formula:

$$w_{i+1} = w_i - \eta \left[\frac{\partial^2 E}{\partial w_i{}^2} \right]^{-1} \frac{\partial E}{\partial w_i} \tag{27.26}$$

where $\partial^2 E / \partial w_i{}^2 = [\partial \hat{y} / \partial w_i][\partial \hat{y} / \partial w_i]^T + \theta(w_i)$ is the term of the second-order derivative. Using the Gauss–Newton method, the higher order part $\theta(w_i)$ is assumed to be zero, hence the algorithm becomes:

$$w_{i+1} = w_i - \eta \left[\left(\frac{\partial \hat{y}}{\partial w_i} \right) \left(\frac{\partial \hat{y}}{\partial w_i} \right)^T \right]^{-1} \frac{\partial E}{\partial w_i} \tag{27.27}$$

The learning rate η is used to adjust the stability and convergence rate, especially when the optimum is very flat. The Levenberg-Marquardt modification of the Gauss–Newton method introduces an additional factor $\mu >= 0$, such that:

$$w_{i+1} = w_i - \eta \left[\left(\frac{\partial \hat{y}}{\partial w_i} \right) \left(\frac{\partial \hat{y}}{\partial w_i} \right)^T + \mu I \right]^{-1} \frac{\partial E}{\partial w_i} \tag{27.28}$$

The factor μ can vary between 0 and ∞. When $\mu = 0$, the Gauss–Newton algorithm is obtained. If μ approaches infinity, the algorithm becomes the gradient descent method:

$$\mu \to \infty: \quad w_{i+1} = w_i - \mu^{-1}\eta \frac{\partial E}{\partial w_i}$$

$$\mu \to 0: \quad w_{i+1} = w_i - \eta \left[\left(\frac{\partial \hat{y}}{\partial w_i} \right) \left(\frac{\partial \hat{y}}{\partial w_i} \right)^T \right]^{-1} \frac{\partial E}{\partial w_i} \tag{27.29}$$

It should be mentioned that the dynamic gradients are calculated in order to satisfy the training of the neural network with feedback inputs.

The basic steps of the Levenberg-Marquardt method are:

i. Set $w_i = w_0$ and use a large starting value of μ
ii. Calculate the dynamic gradients $\partial E / \partial w_i$ and $\partial \hat{y} / \partial w_i$
iii. Update w_i
iv. If $|E(w_{i+1})| \leq |E(w_i)|$ then decrease μ
else increase μ
v. If $|E(w_{i+1})| \leq \varepsilon$, then stop, else goto step (ii)

27.5 Neural Network Applications and Issues

Neural networks can be very efficient in solving certain types of problem. Some of the applications of neural networks will therefore be discussed in this section.

27.5.1 Neural network applications

As discussed before, there are numerous applications of neural networks, some specific ones will be discussed in this section.

- **Pattern recognition.** Pattern classification is concerned with the assignment of a new pattern to one of several pre-specified classes based on one or more properties that characterize a class. Applications in clinical medicine, microbiology and image processing are given by Penny and Frost (1996), Basheer and Hajmeer (2000), Mattone *et al.* (2000), and Egmont-Petersen *et al.* (2002).
- **Clustering.** In this case, neural networks are used to assign similar patterns to the same cluster. Often Kohonen networks are used. Typical applications can be found in the area of chemical analysis (Tokutaka *et al.*, 1999) and weather pattern recognition (Ambroise *et al.*, 2000).
- **Modeling and forecasting.** Modeling involves training the neural network on input-output data, such that an existing relationship between input and output data are represented sufficiently accurate. The relationship can be either static, in which case usually a feed-forward network is used. It could also be that the relationship is dynamic, in this case usually a recurrent neural net is used. Numerous applications can be found in the literature in different application areas, such as waste water plant modeling, data reconciliation, and so forth, see for example Miller *et al.* (1997), Meert (1998), Zhao *et al.* (1999), Basak *et al.* (2000) and Veltri *et al.* (2002).
- **Optimization.** In optimization, it is required to find a solution that optimizes an objective function subject to a set of constraints. Hopfield networks proved to be very effective in solving nonlinear optimization problems. Some examples are the application in the shortest path problem (Bousono-Carzon *et al.*, 1997) and in combinatorial optimization problems (Colorni, *et al.*, 1996).

- **Process control and identification.** In process control, a neural network can be used to calculate the process inputs to bring the process outputs to a desired target. Yonghong and van Cauwenberghe (1996) discuss the design of a neural predictive controller, Cohen *et al.* (1997) apply a neural network to the sequencing control of a batch reactor. Menhaj and Salmasi (2000) use a neural network for parameter estimation and Hafner *et al.* (2000) use neural networks for diesel engine control design.

27.5.2 Neural network issues

Neural networks cannot be successfully developed without taking a number of important issues into account. They will be discussed in this section.

Fault detection. Fault detection is concerned with the detection of abnormal events, for example the failure of an instrument in a process plant. Application examples are given, among others, by Fuente and Vega (1999) and Guglilmi *et al.* (1995).

Basheer *et al.* (2000) give a comprehensive treatment of neural network issues, such as data preprocessing, normalization, network initialization, etc. Some of the issues mentioned in this section are based on this comprehensive review.

Data base size. Neural network models will depend on the data base size. Data used for training should include the entire operating range such that the model can be used to interpolate. Neural network models are poor extrapolators, i.e. they generally predict poorly outside the operating range.

The data set is usually divided into three data sets: the training set, the data set and the validation set. The training set should include data from the entire operating range and is used to update the weights of the network. The test set is different from the training set and is used to test the response of the network for untrained data. If the response is poor, the network configuration has to be changed or more training cycles should be applied.

Although there are numerous guidelines for selecting the sizes of the different data sets, a reasonable choice is to take 65% of the entire data set for training, 25% for testing and 10% for validation.

Data preprocessing. The data should preferably be evenly distributed over the entire operating range of the process. If the process data are noisy, the noise should preferably be removed by using an appropriate smoothing or filtering technique. In addition, outliers should be removed, since they would affect the accuracy and prediction capability of the model.

Data normalization. Process values usually take arbitrary values and it can be expected that they will not all be of the same magnitude. It is therefore recommended to scale all process values between 0.1 and 0.9 to avoid saturation of the hidden nodes and to ensure that all process variables have an impact on the output.

Network initialization. In the literature there are any guidelines for initializing neuron weights. It is recommended to initialize the network weights and threshold values randomly to small values ranging from –0.30 to +0.30.

The learning rate. A high learning rate will accelerate the training process, however, may also lead to instabilities. Some authors suggest values of the learning rate between 0.1 and 10; others suggest values between 0.3 and 0.6. It is our experience that generally the value should be small and be in the range of 0–1.

Size of the hidden layer. One hidden layer is usually sufficient to approximate any continuous function. The selection of the number of hidden nodes is not a simple task. A network with too few hidden nodes will provide a linear estimate of the actual trend, whereas too many hidden nodes will cause the network to follow the noise due to over-parameterization.

Even though there are many rules of thumb it is our experience that one should start with a small number of hidden nodes and gradually increase this number in order to meet the model accuracy requirements.

27.6 Examples of Models

The first example to be considered is a static function approximation. The MATLAB neural network toolbox (Mathworks, 2006) will be used. A two-layer feed-forward neural network, consisting of a hidden layer and output layer, will be trained to approximate the sum of two sine waves. The function can be described by:

$$y = 0.5 * \sin(\pi x) + 0.5 * \sin(2\pi x), \quad x \in [0,2] \tag{27.30}$$

Data are generated at equidistant points and some noise is added:

```
randn('seed',192736547);
x = 0:0.01:2.0;
y = 0.5*sin(pi*x)+0.5*sin(2*pi*x)+0.1*randn(size(x-1));
```

The vectors x (input) and y (output) contain 201 data points each. The relationship between x and y is shown in Fig. 27.7.

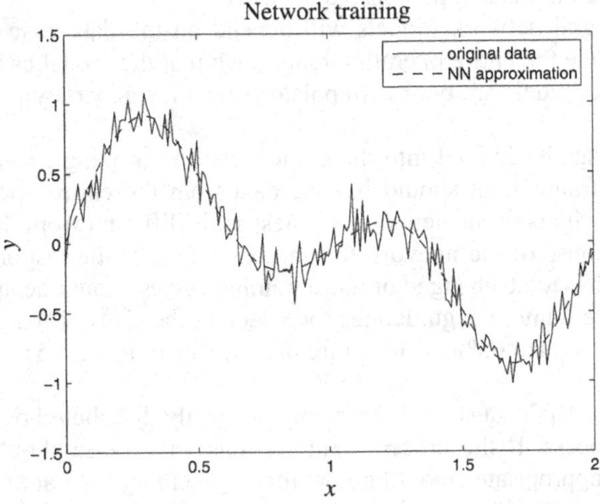

Fig. 27.7. Relationship between x and y to be approximated.

A network configuration now has to be selected. We will choose a feed-forward network with four neurons in the first layer with a sigmoid transfer function, one neuron in the output layer with a linear transfer function. The function **newff** can be used to create the network. The first argument in the call gives the ranges of the input signal, the last argument the training algorithm (one of the 14 available). TRAINBR is a network training function that updates the neuron weight and bias values according to Levenberg-Marquardt optimization. It minimizes a combination of squared errors and weights and then determines the correct combination to produce a network that generalizes well. The process is called Bayesian regularization. A number of 150 epochs is chosen as a maximum and every 10 epochs informative information is displayed.

```
net=newff(minmax(x),[4 1],{'tansig','purelin'},'trainbr');
net.trainParam.epochs  = 150;
net.trainParam.show    = 10;
net = init(net);
[net, tr] = train(net,x,y);
```

The created net is stored in the object named net. The net can now be used to test and/or validate the model. For this purpose new points are created within the region that was selected for training.

```
x1 = 1.015:0.01:1.615;
y1 = 0.5*sin(pi*x1)+0.5*sin(2*pi*x1)+0.1*randn(size(x1-1));
out1 = sim(net,x1);
plot(x1, y1, 'b',x1,out1,'r');
```

The results are shown in Fig. 27.8. Usually the model will perform well with data from within the training region.

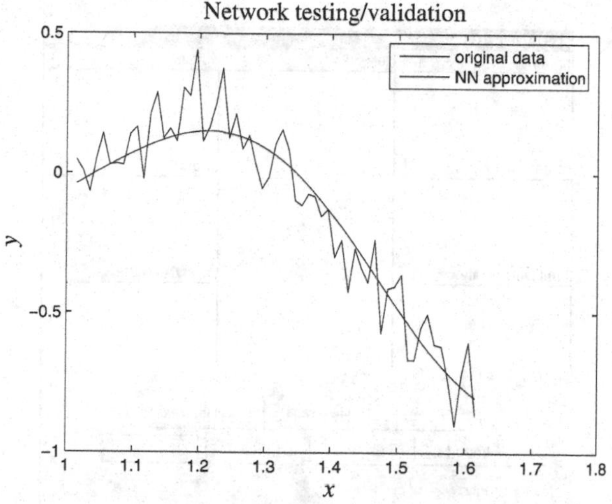

Fig. 27.8. Validation of the model with data from the training region.

We will now generate data outside the training region and thus test the extrapolation capability of the model. The result is shown in Fig. 27.9.
As can be seen, the model approximates the function not too well outside the training region. This is generally the case with black box models, unless the process is linear and a linear model was developed. The code for this example is given in file F2708.m.

The neural network can also be developed using the MATLAB graphical user interface. To start it, type **nntool**. The window shown in Fig. 27.10 will appear.

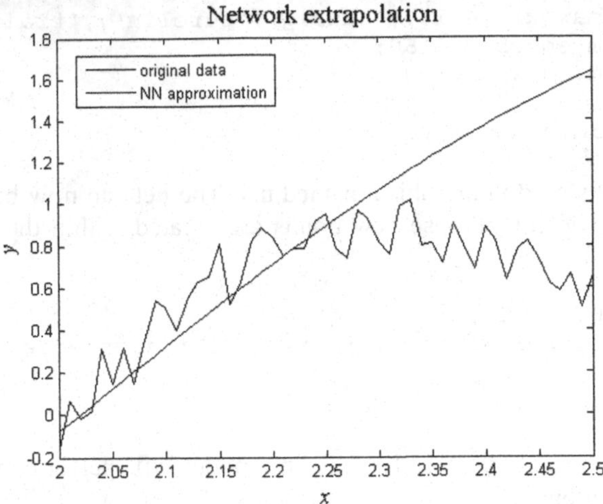

Fig. 27.9. Extrapolation capability of the model.

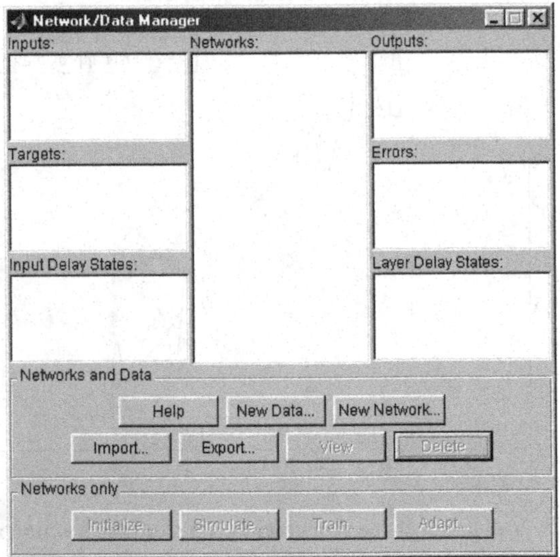

Fig. 27.10. NNTOOL graphical user interface.

First generate the input (x) and output (y) data. On the interface, select **Import** and import x as inputs and y as targets. Select **New Network**, on the next window that appears select x for **Input ranges,** the range [0 2] should be shown as input range.

For the **training function** select TRAINBR. Select **layer 1**: Number of neurons 4, transfer function TANSIG, select **layer 2**: Number of neurons 1, transfer function PURELIN. Select **Create**, this will create a network called network1.

Select **network1** and select **Train**. Select as inputs x and as outputs y, and select **Train network**.

The network1_ouputs and network1_errors are now generated. By exporting (select Export in the Network/Data Manager) them to the workspace they can be plotted. When additional data in the training region or outside the training region is generated, the model can be used to simulate the output. In addition, the network weights can easily be viewed and the network for this example is shown in Fig. 27.11.

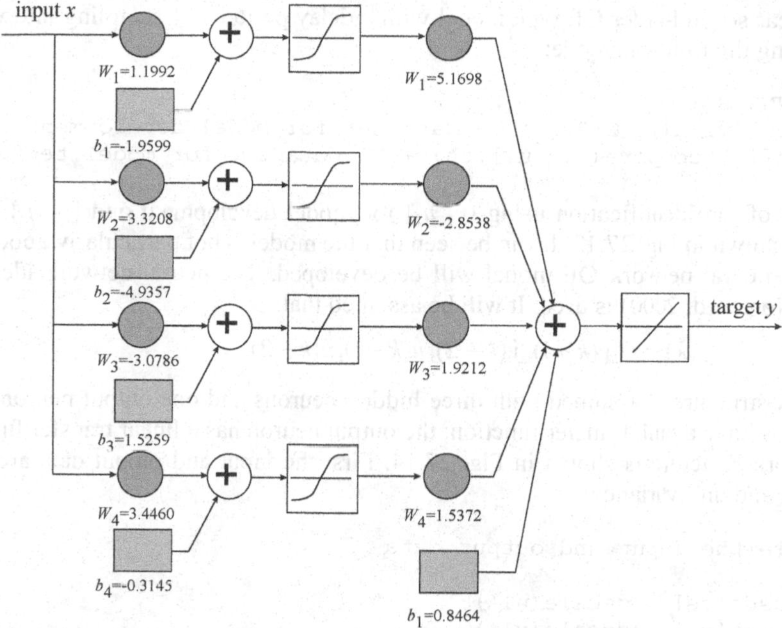

Fig. 27.11. Network for sine wave approximation.

Example 2. The second example is a dynamic process described by the following equation:

$$y(k) = y(k-1)/(2.5 + 1.5y^2(k-1)) - 0.3y(k-2)u(k-2) + 0.6u(k-1) \quad (27.31)$$

where $u(k) = -1 \ldots +1$, $k = 1 \ldots 499$.

Data generated with this model are stored in file NNsiso.mat. The data are shown in Fig. 27.12.

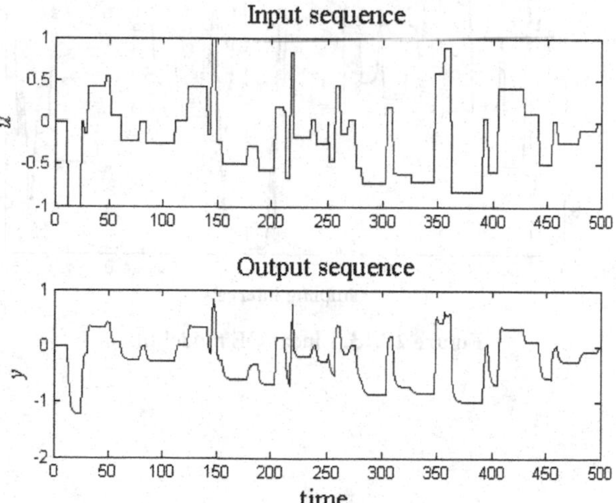

Fig. 27.12. Process data used for identification.

First a linear second-order OE type model with a delay period of 1 sampling interval is generated, using the following code:

```
load NNSISO;
th = oe([y1 u1],[2 2 1]); % data set for model development
figure(1), compare([y2 u2],th,1); % data set for model test
```

The result of the identification using $[y_1\ u_1]$ for model development and $[y_2\ u_2]$ for model testing, is shown in Fig. 27.13. It can be seen that the model is not particularly good.

Next, a neural network OE model will be developed. The neural network identification toolbox (Nørgaard, 2000) is used. It will be assumed that:

$$y(k) = f(\hat{y}(k-1), \hat{y}(k-2), u(k-1), u(k-2) \tag{27.32}$$

A network structure is assumed with three hidden neurons and one output neuron. The hidden neurons have a tanh transfer function, the output neuron has a linear transfer function. The network structure is shown in Fig. 27.14. First the input and output data are scaled to zero mean and unit variance.

```
% Scale the input and output data
%
[u1s,uscales] = dscale(u1');
[y1s,yscales] = dscale(y1');
u2s = dscale(u2',uscales);
y2s = dscale(y2',yscales);
```

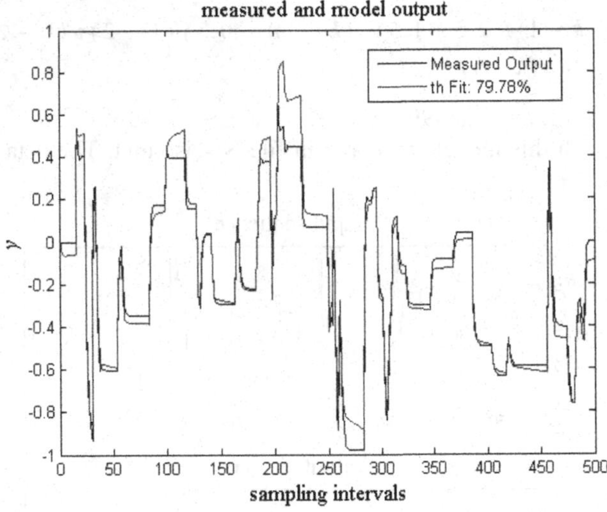

Figure 27.13. Linear OE model fit.

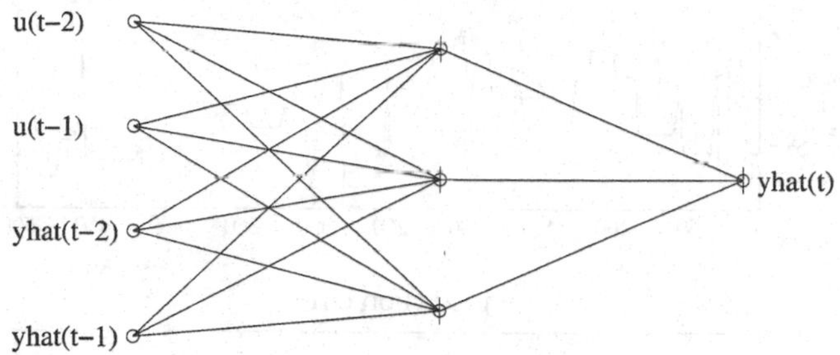

Fig. 27.14. Neural network OE model structure.

Next, the network is defined and trained:

```
NetDef = ['HHH';'L--'];
NN = [2 2 1];
trparms = settrain;
%
% Set the maximum number of iterations = 200
% The weight decay parameter D = 1.E-3
%
trparms = settrain(trparms,'maxiter',200,'D',1E-3);
[W1,W2,NNSEvec] = nnoe(NetDef,NN,[],[],trparms,y1s,u1s);
```

Now that the network has been developed, using the $[y_1\ u_1]$ data, it can be tested using the $[y_2\ u_2]$ data:

```
%
% Rescale the weights so that the unscaled data can be used
% for validation using the test data
%
[w1,w2] = wrescale('nnoe',W1,W2,uscales,yscales,NN);
[yhat,NSSE] = nnvalid('nnoe',NetDef,NN,w1,w2,y2',u2');
```

Figure 27.15 shows the result of the neural network OE model predictions and the actual output data. As can be seen, the predictions using the neural network model are far better than from the linear OE model. The fit for the test data is 98.4%. The program to execute this example is given in F2712.m.

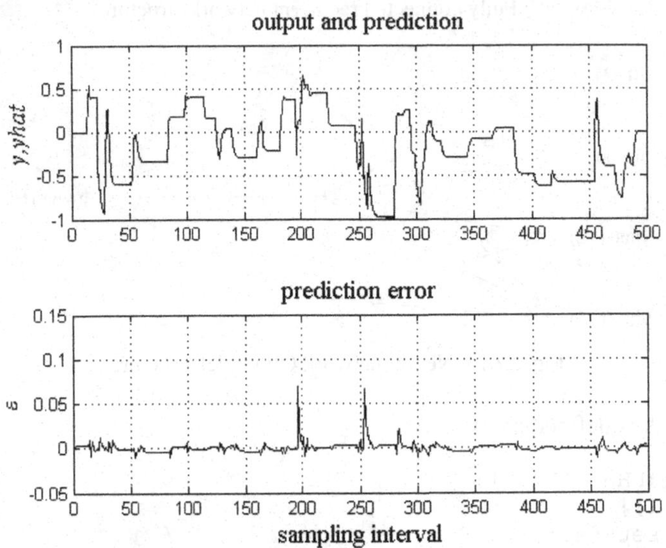

Fig. 27.15. Neural network OE model fit.

In this example, a fully connected network was used, i.e. every input had connections to every hidden neuron and every hidden neuron had a connection to the output neuron. It is important to realize that the number of model parameters of the neural network with three hidden neurons and one output neuron is a multiple of the number of parameters of the OE linear model, however, increasing the number of model parameters in the linear OE model does not improve the model fit.

The neural network identification package also has an option to prune the network. This feature is very time consuming; however, it takes out the links that are not useful and thus creates a reduced network with better generalization properties. It is beyond the scope of this chapter to explain the mechanism of pruning in detail. The procedure is to eliminate one weight and evaluate its impact on the prediction error. The impact of all weights are evaluated, the superfluous weights are eliminated according to the so-called *Optimal Brain Surgeon* strategy (Nørgaard *et al.*, 2000). Using F2716.m yields the pruned network shown in Fig. 27.16. Dotted lines indicate negative network weights, solid lines positive weights.

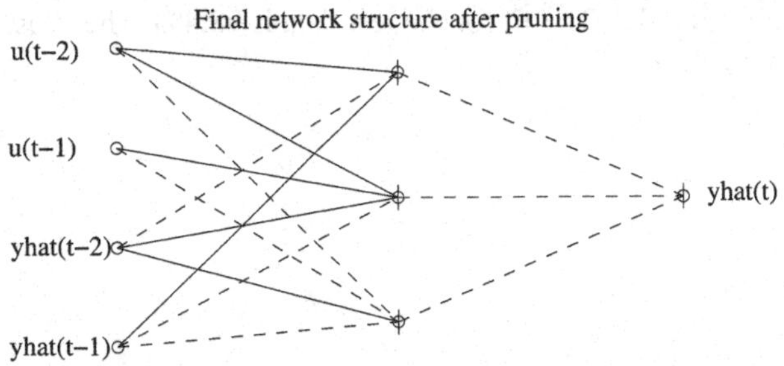

Fig. 27.16. Network structure after pruning.

The original fit for the test set of 98.4% has now been increased to 99.4%. The approximation of the response is even better than shown in Fig. 27.15.

The weight matrix w_1 for the unscaled data to the hidden layer is (numbered from bottom (1) to top (3) in Fig. 27.16). It can be seen that there is one link (from input 3 to HN3 at the top) that have been pruned (weight = 0), since they apparently contributed little to a reduction of the prediction error.

```
 input1   input2   input3    input4    bias
-0.0605   0.0780  -0.0887   -0.0539   0.1961   HN1 (bottom)
-0.0327   0.3302   0.0182    0.2353  -1.3918   HN2
 0.0606   0.1957   0.0000    0.0181   0.2916   HN3 (top)
```

Note that there are four inputs to the network ($\hat{y}$ (*k*–1), $\hat{y}$ (*k*–2), *u*(*k*–1) and *u*(*k*–2)) and a bias term which is equal to one.

The weight vector w_2 from the hidden layer to the output is:

```
-7.2572   -2.0468   -4.5373   0.9610
```

The fourth element of the weight vector w_2 is the bias term.

If the neural network is simplified to two hidden neurons and one output neuron the model fit is still 96.9%, hence it is still better than the linear model.

Files referred to in this chapter:
F2708.m: sine wave approximation
F2712.m: code for generation of neural net model
F2716.m: code for neural network pruning
Fit.m: used to calculate model fit
nnsiso.mat: data for non-linear SISO process

References

Ambroise, C., Seze, G., Badran, F. and Thiria, S. (2000) Hierarchical clustering of self-organizing maps for cloud classification. *Neurocomputing*, **30**, 47–52.

Basak, S. C., Grunwald, G.D., Gute, B.D., Balasubramanian, K. and Opitz, D. (2000) Use of statistical and neural net approaches in predicting toxicity of chemicals. *Journal of Chemical Information and Computer Sciences*, **40** (4), 885–90.

Basheer, I.A. and Hajmeer, M. (2000) Artificial neural networks: fundamentals, computing, design and application. *Journal of Microbiological Methods*, 43, 3–31.

Buscono-Calzon, C. and Figueiras-Vidal, A.R. (1997) A bank of Hopfield neural networks for the shortest path problem. *Journal of Computational and Applied Mathematics*, **82** (1–2), 117–128.

Cohen, A., Janssen, G., Brewster, S.D., Seeley, R., Boogert, A.A., Graham, A.A., Mardani, M.R., Clarke, N. and Kasabov, N.K. (1997) Application of computational intelligence for on-line control of a sequencing batch reactor at Morrisville sewage treatment plant. *Water Science and Technology*, **35** (10), 63–71.

Colorni, A., Dorigo, M., Maffioli, F., Maniezzo, V., Righini, G. and Trubian, M. (1996) Heuristics from nature for hard combinatorial optimization problems. *International Transactions in Operational Research*, **3** (1), 1–21.

Cybenko, G. (1989) Approximations by superpositions of a sigmoidal function. *Mathematics of Control Signal and Systems*, **2**, 202–314.

Egmont-Petersen, M., de Ridder, D. and Handels, H. (2002) Image processing with neural networks, a review. *Pattern Recognition*, **35**, 2279–2301.

Fuente, M.J. and Vega, P. (1999) Neural networks applied to fault detection of a biotechnological process. *Engineering Applications of Artificial Intelligence*, **12** (5), October, 569–84.

Guglilmi, G., Parisini, T. and Rossi, G. (1995) Fault diagnosis and neural networks: a power plant application. *Control Engineering Practice*, **3** (5), 601–20.

Hafner, M., Schuler, M., Nelles, O. and Isermann, R. (2000) Fast neural networks for diesel engine control design. *Control Engineering Practice*, **8** (11), 1211–21.

Hecht-Nielsen, R. (1987) Counterpropagation networks. Proceedings of the 1987 IEEE First International Conference on Neural Networks, vol. II, pp. 19–32.

Hertz, J., Krogh, A. and Palmer, R.G. (1995) *Introduction to the Theory of Neural Computation*, Addison-Wesley Publishing Company, Redwood City.

Ljung, L. (1987) *System identification: Theory for the User*, Prentice Hall, Englewood Cliffs, NJ, USA.

Looney, C.G. (1996) Advances in feedforward neural networks: demystifying knowledge acquiring black boxes. *IEEE Transactions on Knowledge and Data Engineering*, **8** (2), 211–26.

MacMurray, J.C. and Himmelblau, D.M. (1995). Modeling and control of a packed distillation column using artificial neural networks. *Comp. Chem. Eng.*, **19**, 1077–1088.

Masters, T. (1994) *Practical Neural Network Recipes in C^{++}*, Academic Press, Boston, MA, USA.

The Mathworks, Description of the neural network toolbox: http://www.mathworks.com/products/neuralnet/.

Mattone, R., Campagiorni, G. and Galati, F. (2000) Sorting of items on a moving conveyer belt, part 1: a technique for detecting and classifying objects. *Robotics and Computer Integrated Manufacturing*, **16** (2–3), 73–80.

Meert, K. (1998) A real-time recurrent learning network structure for data reconciliation. *Artificial Intelligence in Engineering*, **12** (3), 213–8.

Menhaj, M.B. and Salmasi, F.R. (2000) A novel neuro estimator and it's application to parameter estimation in a remotely piloted vehicle. *Engineering Applications of Artificial Intelligence*, **13** (4), 459–64.

Miller, R.M., Itoyama, K., Uda, A., Takada, H. and Bhat, N. (1997) Modeling and control of a chemical waste water treatment plant. *Computers and Chemical Engineering*, **21**, Supplement 1, S947–52.

Narendra, K.S. and Parthasarathy, K. (1990) Identification and control of dynamical systems using neural networks. *IEEE Transactions on Neural Networks*, **1**, 4–27.

Nørgaard, M., Ravn, O., Poulsen, N.K. and Hansen, L.K. (2000) *Neural Networks for Modeling and Control of Dynamic Systems: A Practitioner's Handbook*, Springer Verlag, http://www.iau.dtu.dk/research/control/nnsysid.html.

Penny, W. and Frost, D. (1996) Neural networks in clinical medicine. *Medical Decision Making*, **16** (4), 386–98.

Rumelhart, D.E., Hinton, G.E. and Williams, R.J. (1986) Learning internal representations by error propagation. *Parallel Distributed Processing*, **1**, 318–62.

Tan, Y. and Saif, M. (2000) Neural-networks-based nonlinear dynamics modeling of automotive engines. *Neurocomputing*, **30**, 129–42.

Tokutaka, H., Yoshihara, K., Fujimara, K., Obu-Cann, K. and Iwamoto, K. (1999) Application of self-organizing maps to chemical analysis. *Applied Surface Science*, **144**, 59–63.

Veltri, R.W., Chaudhari, M., Miller, M.C., Poole, E.C., O'Dowd, G.J. and Partin, A.W. (2002) Comparison of logistic regression and neural net modeling for prediction of prostate cancer pathologic stage. *Clinical Chemistry*, **48** (10), 1828–34.

Yonghong, T. and van Cauwenberghe, A.R. (1996) Optimization techniques for the design of a neural predictive controller. *Neurocomputing*, **10** (1), 83–96.

Zhao, H., Hao, O.J. and McAvoy, T.J. (1999) Approaches to modeling nutrient dynamics: ASM2, simplified model and neural nets. *Water Science and Technology*, **39** (1), 227–34.

28 Fuzzy Modeling

Fuzzy modeling is a useful technique for the description of non-linear systems. In fuzzy modeling, non-linear process behavior is approximated by multiple linear models with fuzzy transitions. These fuzzy models employs fuzzy sets to describe the continuous domains of input and output variables by dividing these domains into a small number of overlapping regions which constitute so-called linguistic values (for example High, Medium, Low) of the input and output variables. In each region a simple model is formulated, thus establishing a link between the model input domain and the output domain. The most frequently used fuzzy models basically fall into two categories, which differ in their ability to represent information. The first type is the so-called Mamdani models. These models are based on collections of IF–THEN rules with vague predicates and fuzzy reasoning. The second type is the so-called Takagi-Sugeno (TS) models. These models are formed by logical IF–THEN rules that have a simple linear relationship as consequent part.

28.1 Mamdani Fuzzy Models

One can distinguish different types of fuzzy models, each having their own structure and area of application. The Mamdani models are a category of models of the linguistic type. They could, for example, consist of the following fuzzy rules:

$$\begin{aligned} &R^1\!: \quad IF\ x\ is\ A^1\ THEN\ y\ is\ B^1 \\ &R^2\!: \quad IF\ x\ is\ A^2\ THEN\ y\ is\ B^2 \\ &R^3\!: \quad IF\ x\ is\ A^3\ THEN\ y\ is\ B^3 \end{aligned} \tag{28.1}$$

The ‘If’ part of the model is often called the premise or antecedent part of the model; the ‘Then’ part is called the consequent part of the model.

The variable x is the input variable, it assumes crisp values with corresponding linguistic fuzzy values $\{A^1, A^2, A^3\}$, represented by the membership functions $\mu_A(x)$. The process of converting crisp values into membership functions is called fuzzification. The variable y is an output variable, it assumes linguistic fuzzy values $\{B^1, B^2, B^3\}$ with corresponding membership functions $\mu_B(y)$. Each rule in the above model describes a fuzzy region in the input-output space. The set of rules together partitions the input-output space into a set of overlapping fuzzy regions as shown in Fig. 28.1. A typical example of a rule in a Mamdani fuzzy model is:

$$IF\ x_1\ is\ High\ AND\ x_2\ is\ Small\ THEN\ y\ is\ Low \tag{28.2}$$

Process Dynamics and Control: Modeling for Control and Prediction. Brian Roffel and Ben Betlem.

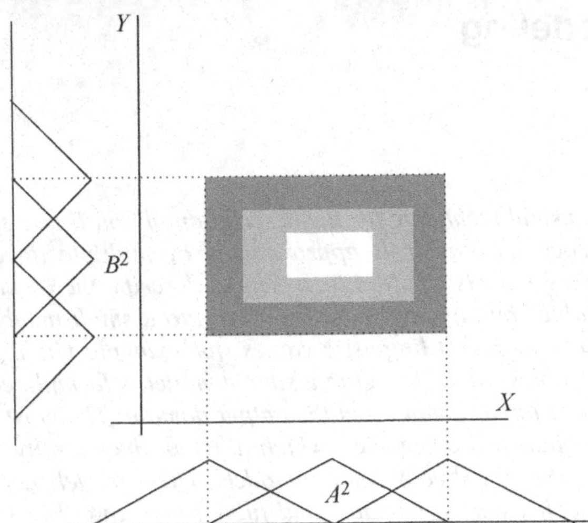

Fig. 28.1. Division of the input space into fuzzy regions, showing the rule IF x is A^2 Then y is B^2.

The rule base, together with the formulation of the fuzzy sets, forms the knowledge base of the fuzzy system.

By weighting, fine-tuning of the consequents of the rules is possible, without changing the reference fuzzy sets. This process is called defuzzification.

If the consequent fuzzy sets B^i are single numbers b_i, the resulting model is called singleton type fuzzy model.

28.2 Takagi-Sugeno Fuzzy Models

Takagi and Sugeno (1985) proposed models where the consequent part of the rule is described by a linear regression model. These models are easier to identify because each rule describes a fuzzy region in which the output depends on the inputs in a linear manner. An example of such a model is shown in Eqn. (28.3):

$$\begin{aligned} R^1:&\quad IF\ x\ is\ A^1\ THEN\ y = a_1x + b_1 \\ R^2:&\quad IF\ x\ is\ A^2\ THEN\ y = a_2x + b_2 \\ R^3:&\quad IF\ x\ is\ A_3\ THEN\ y = a_3x + b_3 \end{aligned} \tag{28.3}$$

or in general format:

$$R^i:\quad IF\ x\ is\ A^i\ THEN\ y_i = a_i^T x + b_i,\quad i=1,2,\ldots,N \tag{28.4}$$

The model as shown in Eqn. (28.4) can represent multi-input, multi-output static and dynamic systems. The global output of the system y can be calculated using the outputs of the individual consequents y_i, using the weighted mean formula:

$$y = \frac{\sum_{i=1}^{N} \mu_{A^i}(x) y_i}{\sum_{i=1}^{N} \mu_{A^i}(x)} \tag{28.5}$$

N is the number of rules in the rule base. μ_{A^i} is the degree of fulfillment (or degree of membership) of the antecedent of rule i. If the antecedent is a multidimensional proposition, such as shown in Eqn. (28.6):

$$IF\ x_1\ is\ A^{i,1}\ AND\ x_2\ is\ A^{i,2}\ AND\ ....\ THEN\ y_i = a_i^T x + b_i \tag{28.6}$$

then the degree of fulfillment is calculated as:

$$\mu_{A^i}(x) = \mu_{A^i}(x_1) \wedge \mu_{A^i}(x_2) \wedge ..... \tag{28.7}$$

where $\wedge$ is the minimum operator.

In Fig. 28.2 an example with three TS rules is shown constituting a local linear description of the functional relationship between y and x.

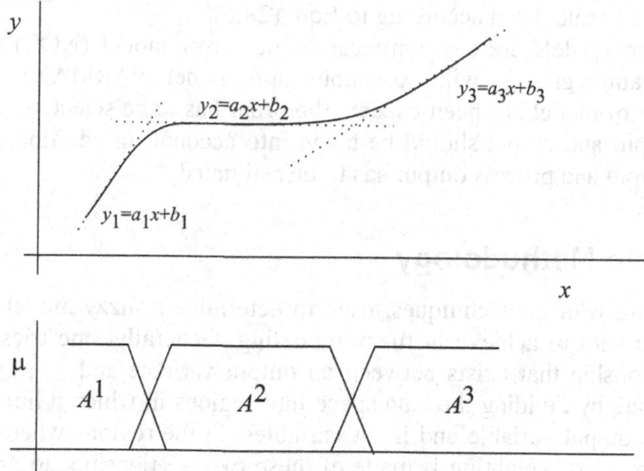

Fig. 28.2. Local linear description using a TS model.

Rather than assuming three rules, one could also decide to make a more accurate approximation by dividing the x-region into five sections and create a linear model in each of these sections. In addition, one could also assume different membership functions; this will be illustrated in section 28.4.

Modeling of dynamic systems

Takagi-Sugeno models are particularly suited to model dynamic systems (de Bruin and Roffel, 1996). The most common structure is the NARX (Non-linear autoregressive with exogenous input) model, which can represent a large class of discrete time nonlinear systems.

The model relates a predicted output to a collection of past input-output data:

$$\hat{y}(k+1) = F(y(k),....., y(k-n_y-1), u(k),...., u(k-n_u-1)) \tag{28.8}$$

where k is the discrete time, n_y and n_u are integers related to the order of the system. The process is approximated by a MISO fuzzy model with rules of the following structure:

$$R^i : If\ U^i = A^i\ And\ y^i = b^i\ Then \tag{28.9}$$

$$y(k) = p_0 + \sum_{j=1}^{m} p_j{}^u u_1(k-j) + ... + \sum_{l=1}^{n} p_l{}^y y(k-l)$$

$y(k)$ is the model output, $u(k-j)$ are past inputs and $y(k-l)$ are past outputs. A^i is a matrix with fuzzy sets, b^i is a vector with fuzzy sets and:

$$U = \begin{bmatrix} u_1(k-1) & u_1(k-2) \ldots & u_1(k-m) \\ u_2(k-1) & u_2(k-2) \ldots & u_2(k-m) \\ \vdots & & \\ u_N(k-1) & u_N(k-2) \ldots & u_N(k-m) \end{bmatrix} \tag{28.10}$$

is a matrix with fuzzified past inputs. The overall degree of membership of the premise of rule R^i can be calculated as:

$$\mu_i = \min(A^i, b^i) \tag{28.11}$$

The model output is calculated according to Eqn. (28.5).

Other common models are the nonlinear output error model (NOE) and the nonlinear moving average autoregressive with exogenous input model (NARMAX).

Once the type of model has been chosen, the order has to be selected, i.e. how many past values of the input and output should be taken into account. In addition, the dead time between process input and process output has to be estimated.

28.3 Modeling Methodology

Before proceeding with the techniques, used to determine a fuzzy model, it is necessary to realize what one tries to achieve in fuzzy modeling. Generally, one tries to approximate a non-linear relationship that exists between an output variable and one or more input variables. This is done by dividing the data space into regions in which a linear relationship exists between the output variable and input variables. In the regions where two relationships overlap, an appropriate weighting is made of these two relationships to produce an accurate approximation of the measured data. A fuzzy model consists therefore of multiple linear relationships between variables.

What should be done then is to determine regions or operating ranges in which linear relationships can be developed. This is called data clustering, sometimes called fuzzy clustering, it will be discussed in a following section. The region for which the model will be valid corresponds to the premise part of the model.

As with all modeling exercises, one should realize how many degrees of freedom a fuzzy model contains. Suppose an output variable depends on two input variables as shown in Eqn. (28.12).

$$IF\ x_1\ is\ A_1\ AND\ x_2\ is\ A_2\ \ THEN\ y = a_1 x_1 + a_2 x_2 + b_1 \tag{28.12}$$

In the antecedent part of the model, each rule now contains two membership functions A_1 and A_2 which contain, as will be demonstrated later, two degrees of freedom each. The consequent part of the model contains three degrees of freedom, leaving us with 7 degrees of freedom. Therefore, a fuzzy model with 10 rules may easily contain 70 degrees of freedom. Not all degrees of freedom will affect the output in the same manner, i.e. the sensitivity of the output to the degrees of freedom is not the same. The sensitivity is usually the highest for the consequent parameters.

28.4 Example of fuzzy modeling

Let us assume that we like to approximate the function $z = x^3$, as shown in Fig. 28.3, in the region $x = [-2, 2]$ by a Takagi-Sugeno fuzzy model.

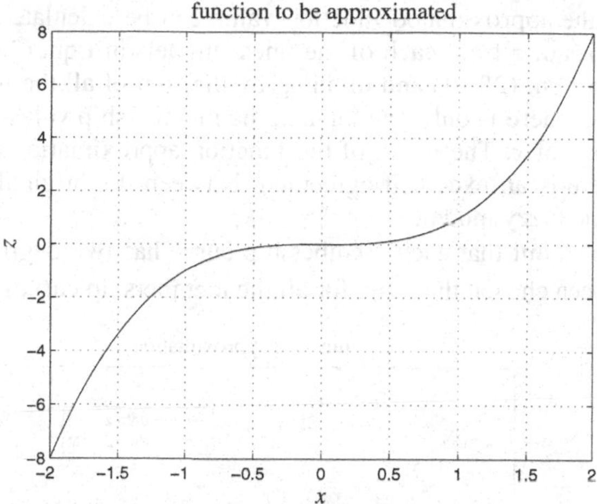

Fig. 28.3. Example of function to be approximated.

Approximating this relationship by three linear functions might be a somewhat crude approximation, we therefore choose to develop five linear functions for the following values of x: $\bar{x}$ = [–1.75, –1.1, 0, 1.1, 1.75]. From the slope to the curve in any of these points the following five linear models can be derived:

$$
\begin{aligned}
z_1 &= 11.87 + 10.05x, && \textit{for large negative } x \\
z_2 &= 2.07 + 3.738x, && \textit{for small negative } x \\
z_3 &= 0.9739x, && \textit{for } x \textit{ around zero} \\
z_4 &= -2.07 + 3.738x, && \textit{for small positive } x \\
z_5 &= -11.87 + 10.05x, && \textit{for large positive } x
\end{aligned}
\tag{28.13}
$$

For this illustration Gaussian membership functions of the following form are chosen:

$$\mu(x) = e^{-(x-\bar{x})^2/2\sigma^2} \tag{28.14}$$

in which $\bar{x}$ is the average and σ the variance around the average. Since $\bar{x}$ was already chosen, σ still has to be selected. Through trial and error it was found that 0.424 was a suitable value. The five membership functions are shown in Fig. 28.4.

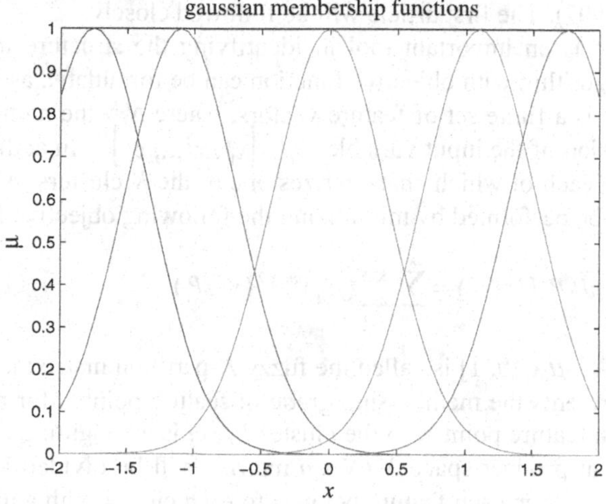

Fig. 28.4. Gaussian membership curves.

Using Eqn. (28.5), the approximated function value can be calculated (using F2803.m), by multiplying the values of z from each of the linear models in Eqn. (28.13), by the membership value of x from Eqn. (28.14) and dividing by the sum of all the membership values for each value of x. Since there is only one input x, the membership value of y becomes equal to the membership value of x. The result of the function approximation is shown in Fig. 28.5. As can be seen, there is an excellent agreement between the original function and the approximation using the fuzzy model.

It should be pointed out that each membership curve has two degrees of freedom, $\overline{x}$ and σ, although σ has been chosen the same for all the membership curves.

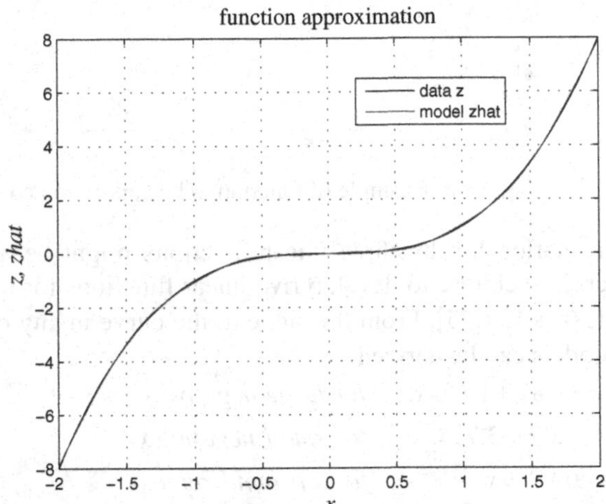

Fig. 28.5. Function approximation by fuzzy models.

28.5 Data Clustering

Several techniques have been proposed in the literature for extracting rules and memberships of the fuzzy model. The most popular ones are fuzzy clustering (Yoshinari *et al.*, 1993; Babuska and Verbruggen, 1994a,b; de Bruin and Roffel, 1996) and neural learning methods (Jang, 1992; de Oliveira, 1993). In this section fuzzy clustering will be explained. A quantitative analysis of fuzzy clustering can be found in (de Bruin and Roffel, 1996; Babuska and Verbruggen, 1997). The first article will be followed closely.

Fuzzy clustering is an important tool in identifying the structure in data. In general, a fuzzy clustering algorithm with objective function can be formulated as follows: suppose $X = \{x^j \mid j=1, 2, \ldots, N\}$ is a finite set of feature vectors, where N = the number of measurements and n = the dimension of the input variables $x_j = \left[x_1^j, x_2^j \ldots, x_n^j\right]^T$. In addition, $P = (P_1, P_2, \ldots, P_K)$ are prototypes, each of which characterizes one of the K clusters. A partition of X into K fuzzy clusters will be performed by minimizing the following objective function:

$$J(P,U^*;X) = \sum_{i=1}^{K}\sum_{j=1}^{N}(\mu_{ij})^m d^2(x_j,P_i) \tag{28.13}$$

where $U^* = [\mu_{ij}]_{K\times N}$, $\mu_{ij} \in [0, 1]$ is called the fuzzy K-partition matrix; it has to satisfy certain conditions. μ_{ij} represents the membership grade of feature point x_j for cluster P_i, $d(x_j, P_i)$ is the distance from a feature point x_j to the cluster P_i, m is a weighting exponent. This means that the input-output product space X ($N \times n$ matrix) will be divided into K clusters. If the fuzzy exponent $m > 1$ then each feature belongs to each cluster with a membership value $\mu =$

[0, 1]. The total of all membership values of a feature is equal to 1. A cluster can have different shapes, depending on the selection of prototypes. The calculation of the membership values depends on the definition of the distance measure. If a feature is closer to a cluster, its membership value will be higher.

According to the choice of prototypes and the definition of the distance measure, different fuzzy clustering algorithms are obtained. If the prototype of a cluster is a point – the cluster center – it will give spherical clusters, if the prototype is a line it will give tubular clusters, and so on. The Gustafson-Kessel (GK) algorithm is one of the clustering algorithms which are the most appropriate for linear or planar clusters.

28.5.1 The Gustafson-Kessel (GK) Clustering Algorithm

The GK algorithm (Gustafson and Kessel, 1979), searches for ellipsoïdal clusters. It can be used for linear or planar clusters because this type of cluster can be viewed as a special case of ellipsoids for which one or more radii are zero.

In the GK algorithm, the distance from a point x_j to a cluster P_i is:

$$d^2(x_j, P_i) = (x_j - v_i)^T M_i (x_j - v_i), \tag{28.14}$$

where v_i and $M_i = |F_i|^{1/n} F_i^{-1}$ are the cluster center and a positive-definite symmetric matrix related to the covariance matrix F_i of the i-th prototype, and n is the dimension of the input-output product space. The shapes of the i-th cluster can be described, by the scatter matrix (Windham 1983).

$$S_i = \sum_{j=1}^{N} (u_{ij})^m \left(M_i^{1/2}(x_j - v_i)\right)\left(M_i^{1/2}(x_j - v_i)\right)^T. \tag{28.15}$$

If the set of data concentrated around the center forms an ellipsoidal shaped cluster, then the principal axes of the ellipsoid will be given approximately by the eigenvectors of S_i , and the relative length of the axes by the corresponding eigenvalues.

The GK algorithm can be formulated as follows: given a set of data $\{x_j \mid j = 1, 2, \ldots, N\}$ and initial guesses for cluster center v_i , covariance matrix F_i , and the fuzzy partition matrix $U^* = [\mu_{ij}]$, the following steps should be performed iteratively:

1. Compute the distances:

$$d^2(x_j, P_i) = (x_j - v_i)^T M_i (x_j - v_i), \tag{28.16}$$

2. Compute the membership values:

$$\mu_{ij} = \frac{[d^2(x_j - v_i)]^{-1/(m-1)}}{\sum_{l=1}^{K} [d^2(x_j - v_l)]^{-1/(m-1)}} \tag{28.17}$$

if $d^2(x_j, P_i) = 0$ for some i=k, set μ_{kj}=1, and $\forall i \neq k$, μ_{ij}=0.

3. Compute new cluster centers:

$$v_i = \frac{\sum_{j=1}^{N} \mu_{ij}^m x_j}{\sum_{j=1}^{N} \mu_{ij}^m} \tag{28.18}$$

4. Compute new covariance matrices:

$$F_i = \frac{\sum_{j=1}^{N} \mu_{ij}^m (x_j - v_i)(x_j - v_i)^T}{\sum_{j=1}^{N} \mu_{ij}^m} \tag{28.19}$$

until a specified convergence criterion is reached. The cluster center matrix is randomly initialized whereas the covariance matrix s is initialized with the identity matrix. The fuzzy GK clustering algorithm contains two parameters, a weighting or fuzzy exponent m (for a crisp model $m = 1$, fuzzy model $m > 1$ but usually $m = 2$) and the number of clusters K.

28.5.2 Modified Compatible Cluster Merging (MCCM) Algorithm

Developing a fuzzy model of a dynamic nonlinear process requires the tuning of many parameters. Doing this heuristically is time consuming. A clustering technique provides a much easier way of doing this. However, the number of clusters and the number of rules should be determined *a priori*. To limit the number of rules, it is recommended to merge clusters which show a certain degree of conformity. Kaymak and Babuska (1995) developed the modified compatible cluster merging (MCCM) algorithm for this purpose. The key elements of this algorithm are criteria which measure the degree of compatibility between clusters. These criteria are based on the cluster distances and eigenvectors of the covariance matrices of the clusters which the GK algorithm finds. Suppose that the centers of two clusters are v_i and v_j and that the eigenvectors are $\{\phi_{i,1}, \ldots, \phi_{i,n}\}$ and $\{\phi_{j,1}, \ldots, \phi_{j,n}\}$. The vectors are arranged in descending order. The following criteria can be defined for merging clusters:

$$\left|\phi_{i,n} \cdot \phi_{j,n}\right| \geq c1_{ij},\ c1_{ij} \text{ close to } 1 \tag{28.20}$$

$$\left\|v_i - v_j\right\| \leq c2_{ij},\ c2_{ij} \text{ close to } 0 \tag{28.21}$$

The first criterion states that the clusters should be merged if the hyperplanes are almost parallel, the second criterion states that the clusters should be sufficiently close to be merged. By evaluating these criteria for all pairs of clusters, matrices $C1[c1_{ij}]$ and $C2[c2_{ij}]$ are obtained, whose elements indicate the degree of similarity between the i-th and j-th clusters measured according to the corresponding criterion. The aforementioned two criteria are by themselves not sufficient for successfully establishing which clusters should be merged. Therefore a decision-making algorithm is used. The algorithm takes as its inputs the matrices C1 and C2 and maps every element onto a two-dimensional space using two membership functions, resulting in matrices $\widetilde{C}1$ and $\widetilde{C}2$. The membership functions indicate the degree of compatibility between two clusters, based on the evidence from the corresponding criterion. Subsequently, a similarity matrix can be calculated and one can decide whether to merge clusters (de Bruin and Roffel, 1996).

Qualitative explanation of clustering

The foregoing explanation of fuzzy clustering may be difficult to understand if one is not familiar with fuzzy techniques, therefore a more qualitative analysis, will also be given, it is taken from Babuska and Verbruggen (1996).

The idea of fuzzy clustering is shown in Fig. 28.6. In this case, the data are clustered into two groups with prototypes p_1 and p_2, using the Euclidean distance measure of Eqn. (28.14).

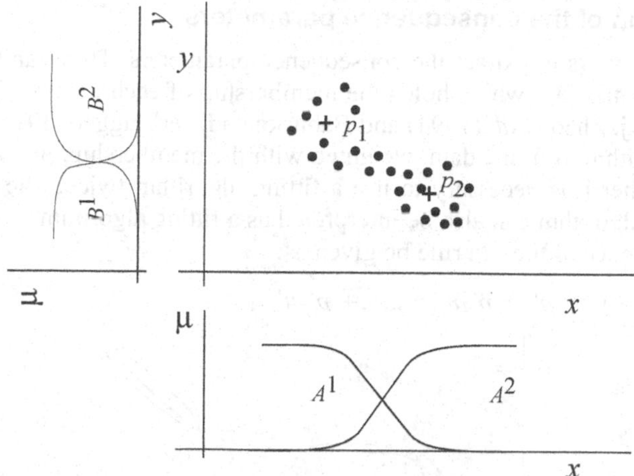

Fig. 28.6. Rule-based interpretation of fuzzy clustering.

The partitioning of data can be expressed in the so-called fuzzy partition matrix whose elements μ_{ij} are the degrees of membership of the data points $[x_i,y_i]$ in a fuzzy cluster with prototype p_j.

The fuzzy partition matrix follows directly from the clustering calculations. Rules can subsequently be extracted by projecting the clusters onto the model variables. Figure 28.6 shows a data set with two clusters and two rules: If x is A^1 then y is B^2; If x is A^2 then y is B^1.

The form of the membership functions obtained by cluster projection will depend on the data distribution. The membership functions for fuzzy sets A^1, A^2, B^1 and B^2 are thus generated by projecting the clusters onto the variables x and y respectively.

A cluster can have different shapes, depending on the choice of prototypes. The calculation of the membership values is dependent on the definition of the distance measure. If a feature is closer to a cluster, its membership value will be higher.

According to the choice of prototypes and the definition of the distance measure, different fuzzy clustering algorithms are obtained. If the prototype of a cluster is a point – the cluster center – it will give spherical clusters, if the prototype is a line it will give tubular clusters.

In view of the linear form of the consequence part use in fuzzy models, an obvious choice for fuzzy clustering is the fuzzy C-varieties or Gustafson-Kessel algorithm, in which linear or planar clusters are allowed as prototypes to be sought.

The number of clusters and thus the number of rules in the fuzzy model has to be specified before clustering. When the process is slightly non-linear, a few rules may provide a good enough approximation of reality. However, if the process is very non-linear, the number of rules has to be increased. The higher the number of rules, the higher the number of model parameters. One should be careful to avoid over-fitting. One method to reduce the number of rules that look similar is compatible cluster merging.

Cluster merging starts with a high number of clusters and gradually merges similar clusters. The initial number of clusters should be set sufficiently high to capture the non-linearity of the regression hyper-surface sufficiently.

The number of clusters is iteratively reduced by merging clusters that are sufficiently close and parallel as shown in Fig. 28.7.

The degree of similarity and a degree of compatibility between two clusters should be calculated in order to judge whether the clusters should be merged.

Cluster merging has the advantage that the number of rules in the model is reduced. It is possible to merge more than two clusters at the time.

Direct calculation of the consequence parameters

There are different ways to extract the consequence parameters. They can be extracted from the membership matrix U – which holds the membership of each feature xj to each cluster k – and the data set xj. Zhao *et al.* (1994) and Babuska and Verbruggen (1995) used a weighted least squares algorithm to fit the data, weighted with the membership, into a hyperplane. It is questionable whether it is necessary to use a fitting algorithm twice. The fuzzy Gustafson-Kessel clustering algorithm can also be interpreted as a fitting algorithm.

Let the consequence of the *i*-th rule be given as:

$$y^i = p^i{}_1 + p^i{}_2 u^i{}_1 + \ldots\ldots + p^i{}_n u^i{}_{n-1} \tag{28.22}$$

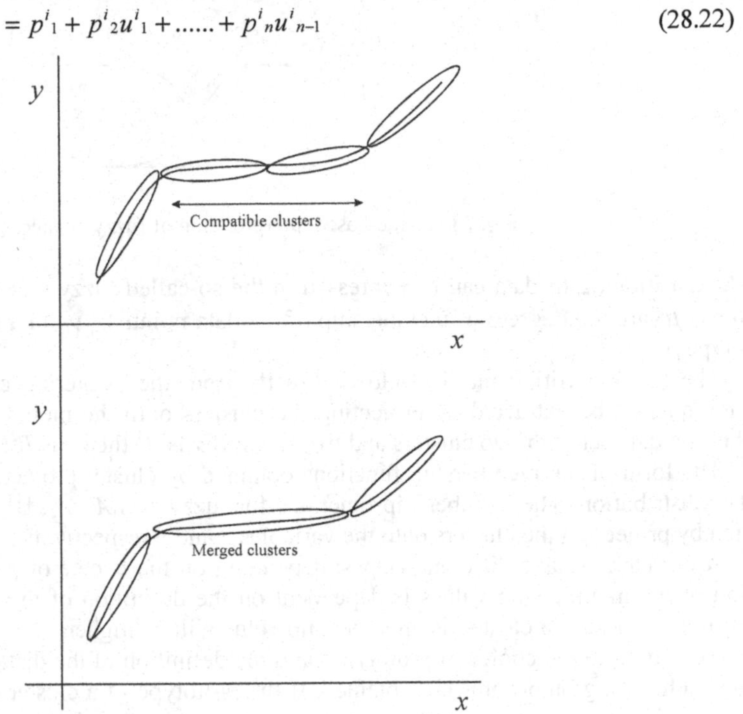

Fig. 28.7. Merging of compatible clusters.

which can be rewritten for *n* equations as:

$$y^i = U^i p^i \tag{28.23}$$

in which $p^i = [p_1, \ldots , p_n]$ are the consequence parameters, y^i is the output of the *i*-th rule, and U is the input matrix, defined as:

$$y^i = \begin{bmatrix} y_1 \\ y_2 \\ \cdots\cdot \\ y_n \end{bmatrix}; \quad U^i = \begin{bmatrix} 1 & u_{1,1} & \cdots & u_{n-1,1} \\ 1 & u_{1,2} & \cdots & u_{n-1,2} \\ \cdot\cdot & \cdots & & \\ 1 & u_{1,n} & \cdots & u_{n-1,n} \end{bmatrix} \tag{28.24}$$

It can easily be seen that the equation for solving p^i is obtained by inverting the input matrix U^i:

$$p^i = [U^i]^{-1} y^i \tag{28.25}$$

Optimization of model parameters

Many papers about fuzzy modeling use optimization of model parameters. Often a least squares method is used. Optimization of consequence parameters may give a perfect global fit, but can also lead to bad local representation of the system. In this case a minor change in the premise parameters may give a major change in the consequence parameters.

Optimization of premise parameters can lead to crisp boundaries of rules instead of fuzzy boundaries. The fuzzy clustering algorithm is already an optimization algorithm, for these reasons it is suggested not to perform an additional optimization step.

28.6. Non-linear Process Modeling

In this section three examples fuzzy models will be discussed.

28.6.1 Non-linear Process 1

The first model that is used for illustration of the proposed identification technique is a nonlinear dynamic process described by Zhao *et al.* (1994):

$$y(k) = y(k-1)^3 - 0.2|y(k-1)|u(k-1) + 0.08u(k-1)^2 \quad (28.26)$$

Let the input signal u for identification be a uniformly distributed noise in the interval [–2, 2], the number of sample data $N = 1001$, and assume the number of fuzzy rules is not known *a priori*.

The hyperplane that shows the relationship between $y(k)$ and $u(k–1)$ and $y(k–1)$ is shown in Fig. 28.8.

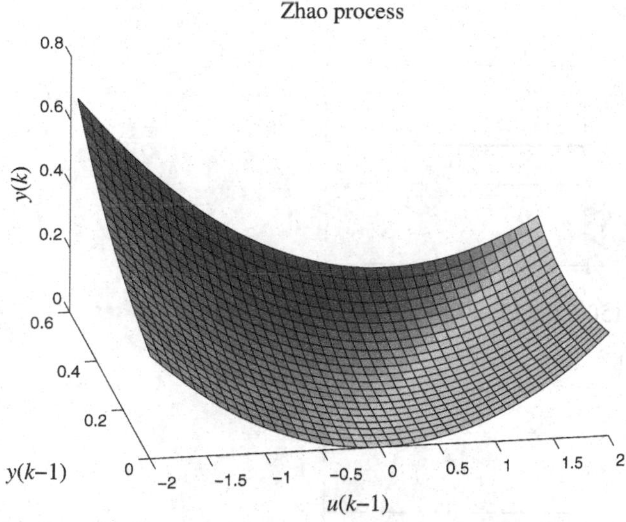

Fig. 28.8. The surface of the model Eqn. (28.26).

The modified cluster merging technique is used with a threshold value of 0.3. Kaymak (1995) recommends to choose the threshold between 0.75 and 0.55, depending on the non-linearity of the system. A value of 0.3 will merge more clusters and this threshold value may later have to be changed.

The clustering software computes the fuzzy K-partition matrix U (a matrix with membership values of each feature to each cluster), and the consequence parameters p_i for each rule i.

By projecting the fuzzy K-partition matrix on the input variables $u(k-1)$ and $y(k-1)$, an initial fuzzy model with four rules, which determine four hyperplanes, is obtained:

RULE 1:

If $u(k-1)$ is

AND

$y(k-1)$ is

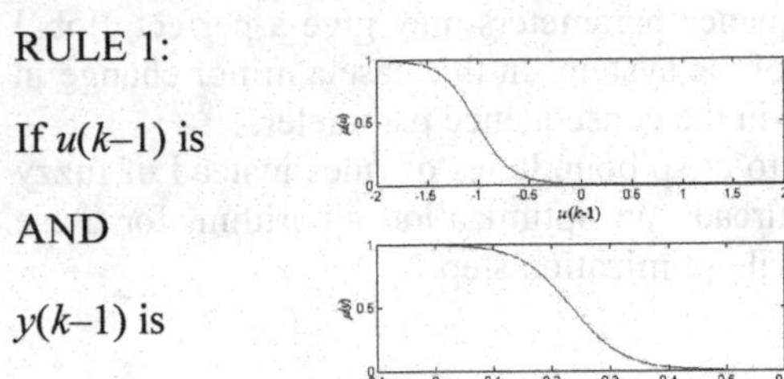

THEN $y(k) = -0.228 + -0.271*u(k-1) + 0.404*y(k-1)$

RULE 2:

If $u(k-1)$ is

AND

$y(k-1)$ is

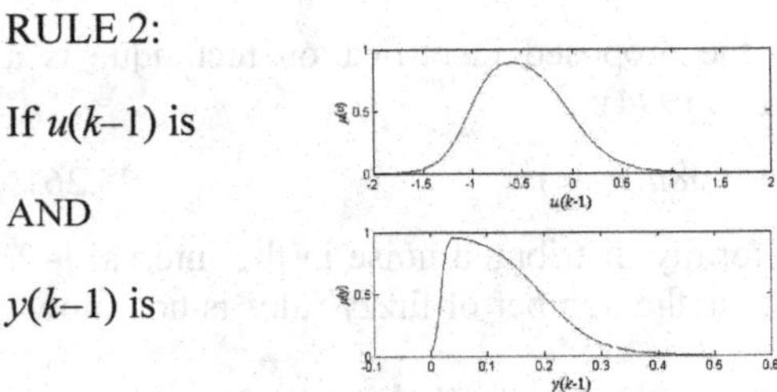

THEN $y(k) = -0.034 + -0.108*u(k-1) + 0.196*y(k-1)$

RULE 3:

If $u(k-1)$ is

AND

$y(k-1)$ is

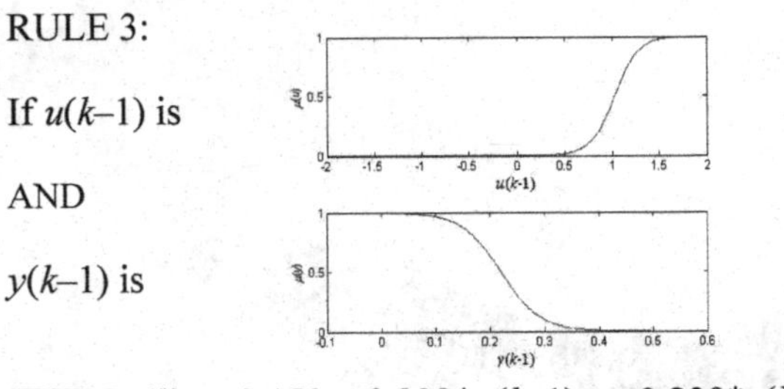

THEN $y(k) = -0.150 + 0.222*u(k-1) + -0.230*y(k-1)$

RULE 4:

If $u(k-1)$ is

AND

$y(k-1)$ is

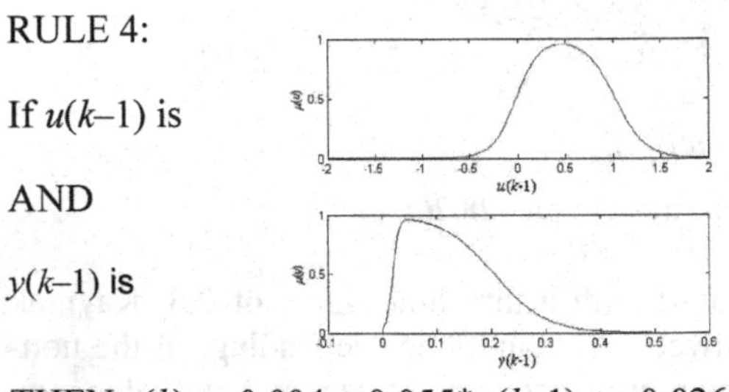

THEN $y(k) = -0.004 + 0.055*u(k-1) + -0.026*y(k-1)$

Figure 28.9 shows the process values $y(k)$ and predicted values. As can be seen, the fit is acceptable.

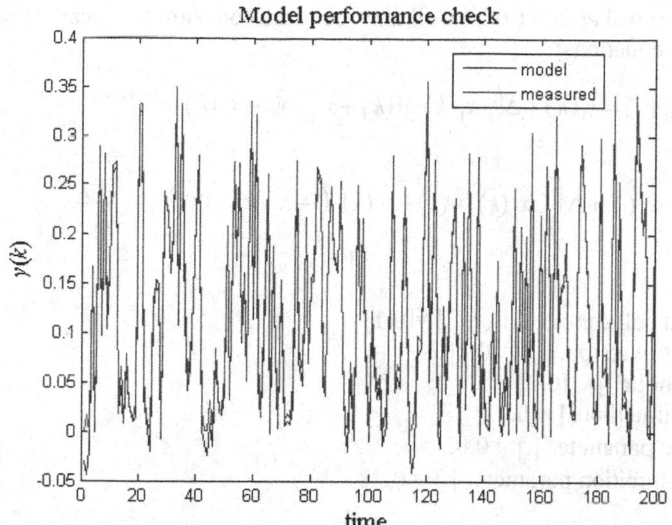

Fig. 28.9. Model prediction and process values for model with 4 rules.

The program to generate these results is the file F2808.m.

The orientation of the clusters is shown in Fig. 28.10, the graph plots $\mu(y(k-1))$ against $u(k)$ and $y(k)$. This information can be obtained from the fuzzy K-partition matrix.

The membership functions can easily be fitted into a sigmoidal shape, this would lead to a slightly different model, and the consequence parameters might have to be optimized in order to assure optimal model performance.

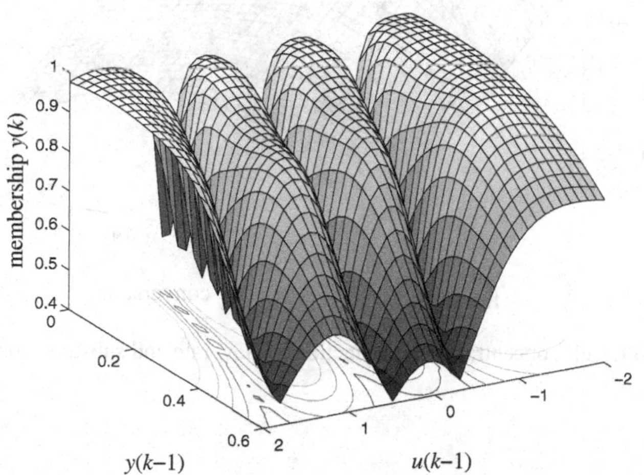

Fig. 28.10. The fuzzy K-partition matrix indicating the cluster orientation[1].

[1] Reprinted from de Bruin, H.A.E. and Roffel, B. (1996) A new identification method for fuzzy linear models of non-linear dynamic systems. *Journal of Process Control*, **6** (5), 277–93, with permission from Elsevier.

28.6.2 Non-linear Process 2

Another interesting problem is shown in de Bruin and Roffel (1996). It is a bioreactor model described by Agrawal *et al.* (1982). Cell and substrate concentrations can be described by the following set of equations:

$$x_1(k+1) = x_1(k) + \Delta\left(-x_1(k)\cdot w(k) + x_1(k)(1-x_2(k))\,e^{x_2(k)/\gamma}\right) \tag{28.27}$$

$$x_2(k+1) = x_2(k) + \Delta\left(-x_2(k)\cdot w(k) + x_1(k)\cdot(1-x_2(k))\cdot e^{x_2(k)/\gamma}\frac{1+\beta}{1+\beta-x_2}\right) \tag{28.28}$$

in which:

x_1 normalized cell concentration [.] =[0,1]
x_2 substrate conversion, [.] = [0,1]
w flow rate [hour^{-1}] = [0,2]
Δ sampling time [hour] = 0.01
β growth rate parameter [.] = 0.02
γ substrate inhibition parameter [.] = 0.48

Figure 28.11 shows the cell concentration x_1 as a function of the flow rate w and substrate conversion x_2.

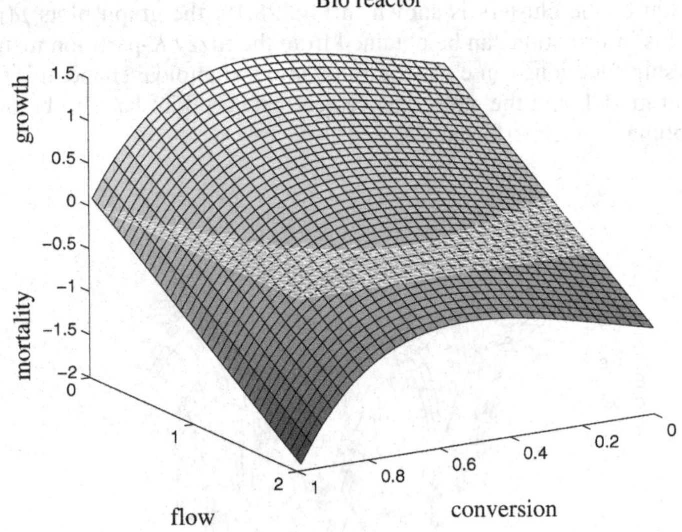

Fig. 28.11. Cell concentration as a function of flow rate and substrate conversion.

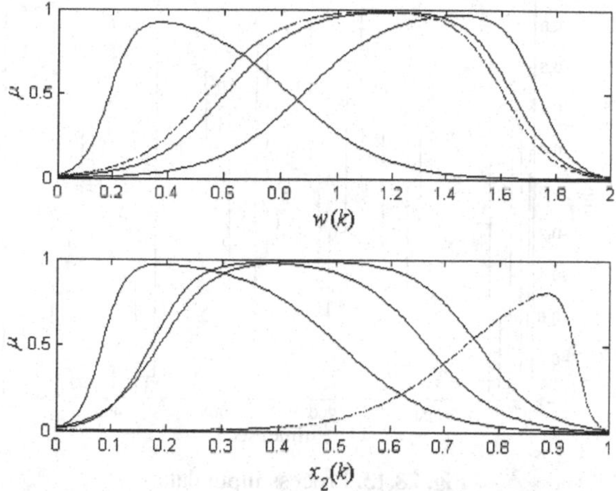

Fig. 28.12. Membership values for the flow rate and the substrate conversion.

The fuzzy model generation can be done using file F2811A.m. The membership functions for the flow rate and substrate conversion are shown in Fig. 28.12.

Each graph has ten membership functions; one function belonging to one particular rule. The membership functions for the cell concentration have a similar pattern.

The fuzzy rule base that is found for a cluster merging threshold value of 0.2 are:

rule 1: $\Delta x_1(k) = 0.136 - 0.510*w(k) + 0.114*x_2(k) + 0.939*x_1(k)$
rule 2: $\Delta x_1(k) = 1.067 - 0.762*w(k) + 0.511*x_2(k) - 0.334*x_1(k)$
rule 3: $\Delta x_1(k) = 0.160 - 0.133*w(k) - 0.051*x_2(k) + 0.168*x_1(k)$
rule 4: $\Delta x_1(k) = 3.689 - 0.685*w(k) - 3.260*x_2(k) - 0.524*x_1(k)$

The percentage fit of the model is 74.2%, if more rules are used the percentage fit obviously increases, as do the number of model parameters.

28.6.3 Non-linear Process 3

Another example to illustrate the capabilities of a fuzzy model is described in Babuska and Verbruggen (1994). The authors pose the following challenging process model:

$$y(k) = y(k-1) + u(k-1) * e^{-3|y(k-1)|} \tag{28.29}$$

For a given pattern of process input data, the model is used to generate output data, it is stored in file babver.mat. The data are shown in Figs. 28.13 and 28.14.

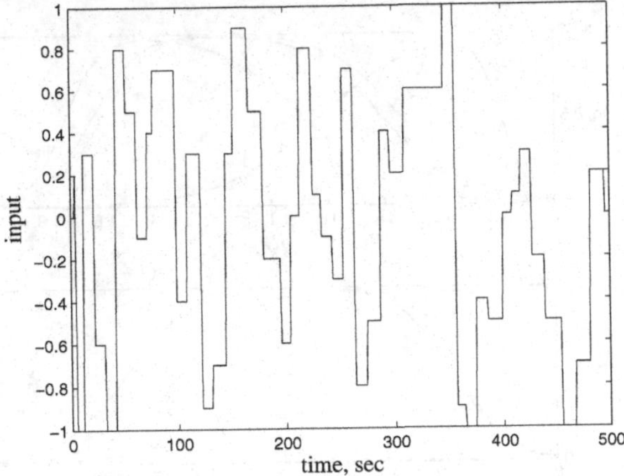

Fig. 28.13. Process input data.

According to the authors, the performance of a linear model is poor and even the application of neural network techniques cannot produce a good model.

Using the function genfis2 of the MATLAB fuzzy control toolbox (2005), the desired model (F2813.m) can easily be found. The model prediction is also shown in Fig. 28.14. As can be seen, this prediction is nearly perfect, with only a rule base of three rules. Babuska and Verbruggen report seven rules, but their rule format is slightly different:

$$IF\ (y(k{-}1) = mf1)\ AND\ (u(k{-}1) = \text{mf2})\ THEN\ y(k) = a{*}y(k{-}1) + b{*}u(k{-}1) \qquad (28.30)$$

A larger number of rules provide only a marginally better result. It should be noted that the clustering algorithm used here is slightly different from the previous two, in this case spherical clusters are used since it is assumed that the points are bounded by a hypercube (Chiu, 1994), whereas in the first two cases ellipsoidal clusters were used.

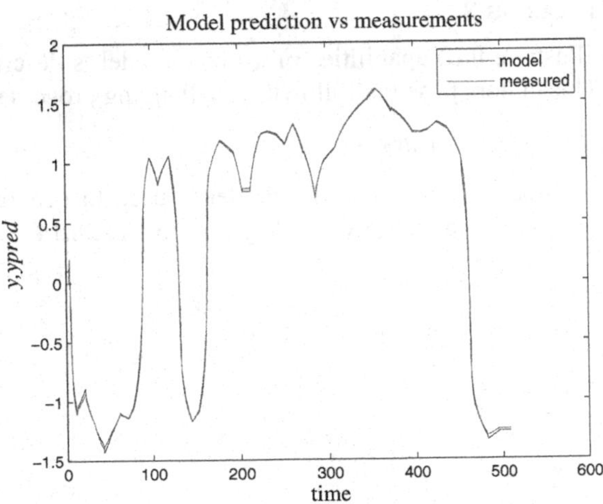

Fig. 28.14. Fuzzy model prediction and actual process data.

Files referred to in this chapter:
F2803.m: illustration of calculation of fuzzy approximation
F2808.m: model generation for non-linear process 1
F2811A.m: model generation (GK algorithm) for non-linear process 2
F2811B.m: model generation (MATLAB toolbox) for non-linear process 2
F2813.m: model generation for non-linear process 3
zhao.dat: data for non-linear process 1
bioreac.dat: data for non-linear process 2
babver.mat: data for non-linear process 3
bvfuzmod.fis: fuzzy inference system generated by F2810.m
fuzmod.mat: fuzzy model generated by F2805.m
Pgenfuz.m: TSK-model generation
Pgenmem.m: membership generation
Pgenrule.m: fuzzy rule generation
Pgenvar.m: input-output variable generation
Pmcma.m: TSK model identification documentation
Pmcma.dll: TSK model identification
Pplotfuz.m: plot TSK model
Ppsigmfguess: makes initial guess for parameters of sigmoid membership function

References

Agrawal P., Lee, C., Lim, H.C. and Ramkrishna, D. (1982) Theoretical investigations of dynamic behavior of isothermal continuous strirred tank biological reactors. *Chemical Engineering Science*, **37** (3), 453–62.

Babuska, R. and Verbruggen, H.B. (1994) Applied fuzzy modeling. Proceedngs of the IFAC Symposium on Artificial Intelligence in Real Time Control, Valencia, Spain, pp. 61–6.

Babuska, R. and Verbruggen, H.B. (1994) Fuzzy modeling and its application to non-linear control. IFAC Congress SICICA '94, June 8–10, Budapest, Hungary.

Babuska, R. and Verbruggen, H.B. (1996) An overview of fuzzy modeling for control. *Control Engineering Pratice*, **4** (11), 1593–1606.

Babuska, R. and Verbruggen, H.B. (1997) Constructing fuzzy models by product space clustering, in *Fuzzy Model Identification* (eds H. Hellendoorn and D. Driankov), Springer Verlag.

de Bruin, H.A.E. and Roffel, B. (1996) A new identification method for fuzzy linear models of non-linear dynamic systems. *Journal of Process Control*, **6** (5), 277–93.

Chiu, S.L. (1994) Fuzzy model identification based on cluster estimation. *Journal of Intelligent and Fuzzy Systems*, **2**, 267–78.

Gustafson, D.E. and Kessel, W.C. (1979) Fuzzy clustering with a fuzzy covariance matrix. Proceedings of the IEEE CDC, San Diego, CA (USA), pp. 761–66.

Jang, J.S.R. (1992) Self learning fuzzy controller based on temporal back propagation. *IEEE Transactions on Neural Networks*, **3** (5) 714–23.

Kaymak U. and Babuska, R. (1995) Compatible cluster merging for fuzzy modeling, FUZZ-IEEE/IFES '95, Yokohama, Japan, March 95.

MATLAB (2005) *Fuzzy Logic Toolbox, User's Guide, Version 2*, 4th edn, The Mathworks.

de Oliveira, J.V. (1993) Neuron inspired rules for fuzzy relational structures. *Fuzzy Sets and Systems*, **57** (1), 41–55.

Takagi T. and Sugeno, M. (1985) Fuzzy identification of systems and its application to modelling and control. *IEEE Transactions on Systems, Man, and Cybernetics* **15**, 116–132.

Windham M.P. (1983) Geometrical fuzzy clustering algorithms. *Fuzzy Sets and Systems*, **10**, 271–9.

Yoshinari, Y., Pedrycz, W. and Hirota, K. (1993) Construction of fuzzy models through clustering techniques. *Fuzzy Sets and Systems*, 54, 157–65.

Yang, M.S. (1993) A survey of fuzzy clustering. *Mathematical and Computer Modelling*, **18** (11), 1–16.
Zhao, J., Wertz, V. and Gorez, R. (1994) A fuzzy clustering method for the identification of fuzzy models for dynamic systems. Proceedings of the 9th IEEE International Symposium on Int. Control, Columbus, Ohio, USA, pp. 172–7.

29 Neuro Fuzzy Modeling

Neuro fuzzy modeling is a useful technique that combines the advantages of neural networks and fuzzy inference systems. In this approach, the fuzzy model is architecturally the same as a neural network. In this case one could use, for example error back-propagation to train the network to find the parameters of the fuzzy model. The most well-known method is the so-called ANFIS method: the Adaptive-Network based Fuzzy Inference System. The method will be explained in this chapter and several examples will be developed as an illustration.

29.1 Introduction

In the previous chapters, neural networks (chapter 27) and fuzzy systems (chapter 28) were discussed. As was pointed out, neural networks have two significant benefits: they are capable of learning non-linear mappings of numerical data and they perform parallel computation. The weakness of neural systems is that the knowledge is distributed over the entire network in the network weights. The value of each of the individual weights has little meaning and it is therefore hard to interpret the meaning of a neural network structure and its accompanying weights.

In fuzzy systems we saw that human-understandable linguistic terms can be used to capture and express the knowledge about the system.

The aim of neuro-fuzzy systems is to combine the advantages of both approaches. The knowledge about the system is expressed as linguistic fuzzy relationships and the learning schemes of neural networks are used to train the system. In addition, neuro-fuzzy systems allow incorporation of numerical and linguistic data into the system. The neuro-fuzzy system is capable of extracting fuzzy knowledge from numerical data. In this chapter only the fuzzy reasoning rules of the Sugeno type function approximator will be considered:

$$IF\ x_1=A_1\ AND\ x_2=A_2\ THEN\ y=ax_1+bx_2+c \tag{29.1}$$

The neuro-fuzzy approach can easily be extended to include other types of fuzzy reasoning rules.

29.2 Network Architecture

The fuzzy logic inference system for identification of the Sugeno type model can be implemented as a five layer network (Ojala, 1994; Sfetsos, 2000) and is shown in Fig. 29.1.

Consider a system with two inputs x_1 and x_2 and one output y. Further assume that the rule base contains two rules which are:

$$\begin{aligned} &IF\ x_1=A_1\ AND\ x_2=B_1\ THEN\ y=z_1 \\ &IF\ x_1=A_2\ AND\ x_2=B_2\ THEN\ y=z_2 \end{aligned} \tag{29.2}$$

z_i represents the linear relationship as shown in Eqn. (29.1).

Process Dynamics and Control: Modeling for Control and Prediction. Brian Roffel and Ben Betlem.

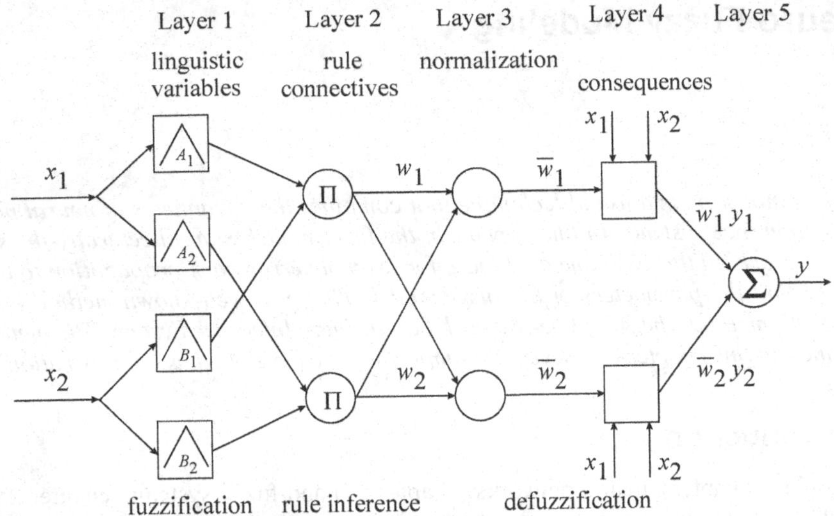

Fig. 29.1. Neural-fuzzy inference system.

The operation of the neuro-fuzzy approach can be described by the following steps:

Layer 1: Fuzzification
Each neuron in this layer corresponds to a linguistic label. The crisp inputs x_1 and x_2 are fuzzified by using membership functions of the linguistic variables A_i and B_i. Usually, triangular, trapezoid or Gaussian membership curves are used. For example, the Gaussian membership function is defined by:

$$\mu_i(x) = exp\left(-\frac{1}{2}\frac{(x-\bar{x})^2}{\sigma^2}\right) \tag{29.3}$$

where $\bar{x}$ is the center and σ the width of the membership function.

Layer 2: Rule nodes
The second layer contains one node for each one fuzzy IF … THEN rule. Each rule node performs a connective operation between rule antecedents (the premise or IF part). Usually the minimum or dot product is used as AND intersection. The union or OR is usually performed by using the maximum operator. The firing strength of the fuzzy rules can be computed according to:

$$w_i = \min\{\mu_{A_i}(x_1), \mu_{B_j}(x_2)\}, \quad i = 1,2,\ j = 1,2 \tag{29.4}$$

or

$$w_i = \mu_{A_i}(x_1) \times \mu_{B_j}(x_2), \quad i = 1,2,\ j = 1,2 \tag{29.5}$$

Layer 3: Normalization
In this layer the firing strengths w_i of the fuzzy rules are normalized according to:

$$\overline{w}_i = \frac{w_i}{\sum_{i=1}^{2} w_i} \tag{29.6}$$

Layer 4: Consequence layer
This label is related to the consequence fuzzy labels z_i. The values of z_i are multiplied by the normalized firing strengths $\overline{w}_i$ according to:

$$y_i = \overline{w}_i z_i = \overline{w}_i (a_i x_1 + b_i x_2 + c_i) \tag{29.7}$$

Layer 5: Summation
This layer computes the overall output as a summation of the incoming signals:

$$y = \sum_{i=1}^{2} y_i \tag{29.8}$$

The last layer defuzzifies the computed values of y_i using an average weighting procedure. A backpropagation training method can be employed to find the optimal values for the parameters of the membership functions and a least squares procedure for the linear parameters of the fuzzy rules, in such a way as to minimize the error between the calculated output and the measured output.

Other learning algorithms include Adaptive vector quantization (Kong and Kosko, 1992), hybrid learning algorithms (Nie and Linkens, 1993; Jang, 1993; Jang and Sun 1995) and the orthogonal least squares method (Wang and Mendel, 1992).

29.3 Calculation of Model Parameters

Jang (1993) and Jang and Sun (1995) uses a combination of the least squares method and gradient descent for tuning a so-called ANFIS structure. This structure stands for Adaptive Network based Fuzzy Inference System and it unifies both neural network and fuzzy models. The system is initialized with several membership functions and a fixed rule base. The learning scheme consists of two separate passes: a forward pass in which the consequence parameters (layer 4 and 5) are identified by the least squares method and a backward pass in which the antecedent parameters (layer 1) are updated by using a gradient descent algorithm. Layer 2 and 3 involve straightforward computations based on the results from layer 1.

It can easily be seen that the consequence parameters can be computed from a least-squares approach. The output y in Fig. 29.1 and Eqns. (27.6) and (27.5) can be written as:

$$\begin{aligned} y &= \overline{w}_1 y_1 + \overline{w}_2 y_2 = \\ &= \overline{w}_1 (a_1 x_1 + b_1 x_2 + c_1) + \overline{w}_2 (a_2 x_1 + b_2 x_2 + c_2) = \\ &= (\overline{w}_1 x_1) a_1 + (\overline{w}_1 x_2) b_2 + (\overline{w}_1) c_1 + (\overline{w}_2 x_1) a_2 + (\overline{w}_2 x_2) b_2 + (\overline{w}_2) c_2 \end{aligned} \tag{29.9}$$

which is linear in the consequent parameters (a_1, b_1, c_1, a_2, b_2, c_2), hence a least squares algorithm is suitable for model identification.

29.3.1 Batch Least Squares

In this section the batch least squares method will be briefly discussed as the solution to the linear identification problem. Suppose the physical system should be identified and an experimental input–output data set for the process is available. In linear system identification a model with the following structure is often identified:

$$y(k) = \sum_{i=1}^{n} a_i y(k-i) + \sum_{i=1}^{m} b_i u(k-i) \tag{29.10}$$

in which $y(k)$ and $u(k)$ are the process output and input at time k respectively. The procedure to calculate the model parameters was already outlined in section 24.3, it will be briefly summarized. Equation (29.10) can also be written as:

$$\hat{y}(k) = \theta^T x(k) \tag{29.11}$$

with

$$x(k) = [y(k-1),...y(k-n),u(k),...,u(k-m)]^T \tag{29.12}$$

and

$$\theta = [a_1,...a_n,b_1,...,b_m]^T \tag{29.13}$$

The system dimension is $N = n+m+1$, thus $x(k)$ and θ are $N \times 1$ vectors, $x(k)$ is called the regression vector.

In the batch least squares approach a vector Y and X are defined:

$$Y = [y^1, y^2,..., y^M]^T \tag{29.14}$$

as a vector of $M \times 1$ output data points coming from the experimental data set, and:

$$X = \begin{bmatrix} (x^1)^T \\ (x^2)^T \\ ... \\ (x^M)^T \end{bmatrix} \tag{29.15}$$

which is a $M \times N$ matrix of x^i vectors also coming from the experimental data set. The parameter vector can then be obtained from:

$$\hat{\theta} = (X^T X)^{-1} X^T Y \tag{29.16}$$

If the process data show sufficient variation, it guarantees that the matrix $X^T X$ is invertible and for a sufficient number of data points M and a linear process, a good model can be found.

29.3.2 Recursive Least Squares

In the batch approach, all data points $\{x_i,y_i\}$ have to be available in advance and all data points are used simultaneously in the identification process. This is good for the identification of models where the output depends on known inputs. If the model output at time k, however, also depends on model outputs in the past, say, k–1 and k–2, these model outputs first have to be calculated by using the model that we have to identify. Also, if the model is changing, we may continue to collect data that should be processed. For these cases the recursive least squares algorithm has been developed. The algorithm for recursive estimation of parameters can be given by (Roffel *et al.*, 1989; Passino and Yurkovich, 1998):

$$\begin{aligned} P(k) &= \frac{1}{\lambda}\left(I - P(k-1)x^k(\lambda I + (x^k)^T P(k-1)x^k)^{-1}(x^k)^T\right)P(k-1) \\ \hat{\theta}(k) &= \hat{\theta}(k-1) + P(k)x^k(y^k - (x^k)^T \hat{\theta}(k-1)) \end{aligned} \tag{29.17}$$

in which λ is the so-called forgetting factor, usually of the order of 0.99 … 0.999, P is the covariance matrix of x, I is the identity matrix and θ the parameter vector of model coefficients.

29.3.3 Identification of Parameters in T-S Models

The algorithms summarized in the previous two sections can easily be applied to identify the consequence part of a fuzzy Takagi-Sugeno model.

The model output for a fuzzy T-S model can be written as:

$$y = \frac{\sum_{i=1}^{R} g_i(x)\mu_i(x)}{\sum_{i=1}^{R} \mu_i(x)} \tag{29.18}$$

where R is the number of rules required to achieve a good fit. The function $g_i(x)$ is defined as:

$$g_i(x) = a_{i,0} + a_{i,1}x_1 + \ldots + a_{i,n}x_n \tag{29.19}$$

Combining Eqns (29.18) and (29.19) results in

$$y = \frac{\sum_{i=1}^{R} a_{i,0}\mu_i(x)}{\sum_{i=1}^{R} \mu_i(x)} + \frac{\sum_{i=1}^{R} a_{i,0}x_1\mu_i(x)}{\sum_{i=1}^{R} \mu_i(x)} + \ldots + \frac{\sum_{i=1}^{R} a_{i,0}x_n\mu_i(x)}{\sum_{i=1}^{R} \mu_i(x)} \tag{29.20}$$

Now define:

$$\xi_i(x) = \frac{\mu_i(x)}{\sum_{i=1}^{R} \mu_i(x)} \tag{29.21}$$

and

$$\begin{aligned} \xi(x) = [&\xi_1(x), \xi_2(x), \ldots, \xi_R(x), x_1\xi_1(x), x_1\xi_2(x), \ldots x_1\xi_R(x), \\ &x_n\xi_1(x), x_n\xi_2(x), \ldots, x_n\xi_n(x)]^T \end{aligned} \tag{29.22}$$

and

$$\theta = [a_{1,0}, a_{2,0}, \ldots a_{R,0}, a_{1,1}, a_{2,1}, \ldots, a_{R,1}, a_{1,n}, a_{2,n}, \ldots, a_{R,n}]^T \tag{29.23}$$

then the prediction of y can be written as:

$$\hat{y}(x) = \theta^T \xi(x) \tag{29.24}$$

which represents a Takagi-Sugeno system which is linear in the parameters. The batch or recursive least-squares algorithm can be applied to identify the model parameters.

29.4 Identification Examples

In this section several examples will be shown that illustrate the approaches discussed in the previous section.

29.4.1 Approximation of a Sinusoidal Function

A function $f(x)$ is given as a function of x and has a sinusoidal shape with decreasing amplitude. The interval range of x is [0.0 … 3.0]. Five bell-shaped membership functions are defined over the interval [0 ... 3]. ANFIS is used to update the membership functions such that the error between the data points and the model will be minimized over the entire interval.

The initial membership functions are shown in Fig. 29.2, the final membership functions in Fig. 29.3.

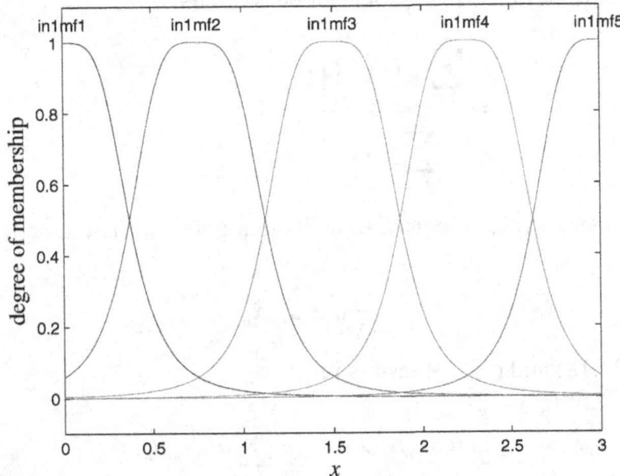

Fig. 29.2. Initial membership functions for sinusoidal function.

As can be seen, the width of some of the membership functions has changed. Figure 29.4 shows the model prediction error as a function of the number of training cycles with the input–output data. Figure 29.5 shows the input–output data and the model predictions; as can be seen, there is an excellent fit. The example is coded in file F2902.m.

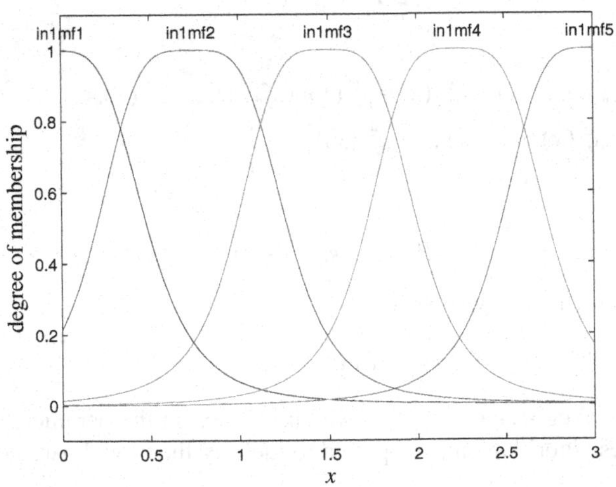

Fig. 29.3. Final membership functions for sinusoidal function.

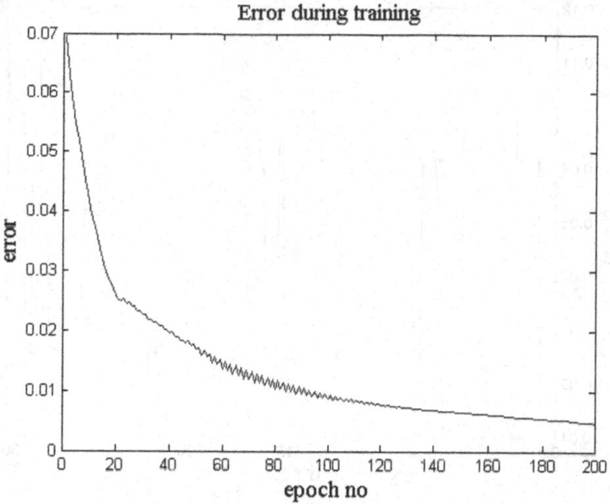

Fig. 29.4. Training error as a function of the number of epochs.

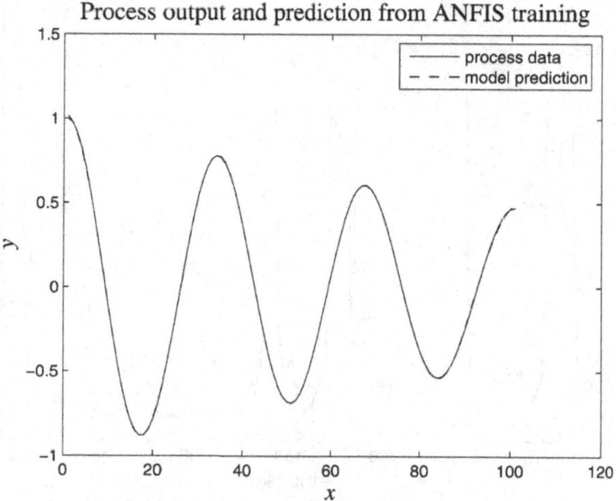

Fig. 29.5. Output data and model predictions.

29.4.2 Non-recurrent pH Neutralization Model

The next example of the use of ANFIS is the development of a model for the neutralization of a caustic solution with an acid flow. The process is described in detail in (Zhang and Morris, 1999). Training data can be generated by running pHprocess.mdl. The process input (acid flow) is shown in Fig. 29.6; the process output (pH) is shown in Fig. 29.7.

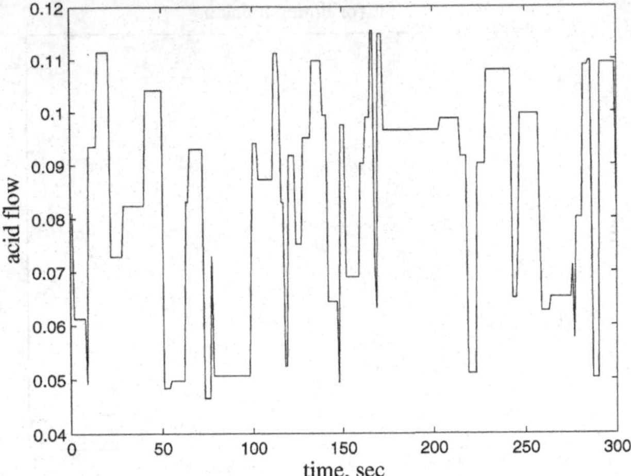

Fig. 29.6. Acid flow as a function of time.

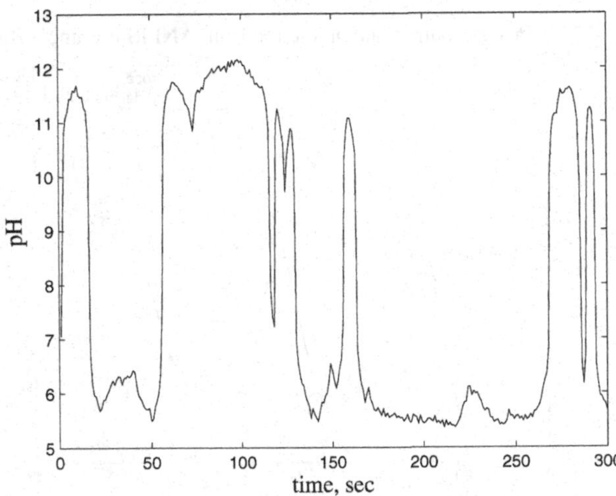

Fig. 29.7. pH as a function of time.

A dynamic process model for this process can, according to Zhang and Morris, be developed in two different ways:

$$\begin{aligned}\hat{y}(k+1) &= ay(k-1) + by(k-2) + cF_a(k-1) + dF_a(k-2) + e \\ \hat{y}(k+1) &= a\hat{y}(k-1) + b\hat{y}(k-2) + cF_a(k-1) + dF_a(k-2) + e\end{aligned} \tag{29.25}$$

in which y is the measured pH, $\hat{y}$ the predicted pH and F_c the acid flow. The first model in Eqn. (29.25) is called a non-recurrent model or one step-ahead prediction model. This is obvious: if the measured values of the pH and acid flow are given, the predicted pH can be generated from this information. This can either be done by using ANFIS, which develops the model and updates the membership functions, or membership functions could be assumed and the least squares method used to find the consequence part of the local models.

The second type of model is called recurrent model or N-step-ahead prediction model; the recursive least squares algorithm can be used to identify this type of model. Only after the pH has been predicted at a particular time can the next predicted value be calculated. In addition to the update of the model coefficients, also the membership functions could be updated using a gradient search method.

Let us first look at a model as formulated in Eqn. (29.25a). In this case the regression vector is defined as x = [pH(k–1) pH(k–2) F_a(k–1) F_a(k–2)]. However, first data clustering is used to see how many membership functions would be needed for the pH and how many for the acid flow. The data clustering is performed by using F2908.m. The results are given in Fig. 29.8. It can be seen that there are three distinct clusters for the pH, one for low pH values and one for high pH values and one for intermediate pH values.

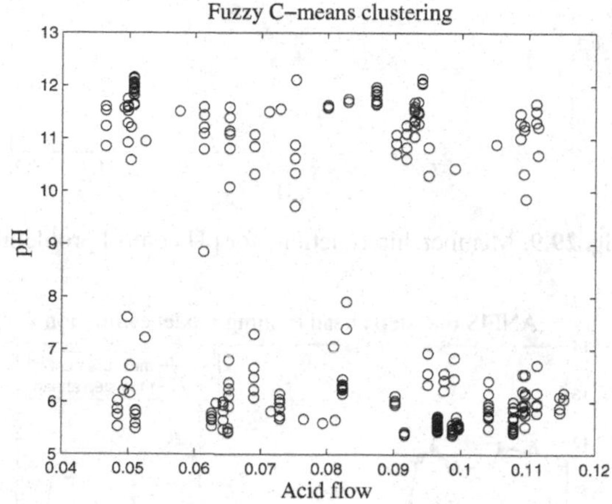

Fig. 29.8. Data clustering for the pH neutralization problem.

However, there are not really any distinct clusters for different values of the acid flow. Therefore a model will be developed with a premise part that only depends on the pH; three linguistic terms will be used: pH = Low, Medium, High. ANFIS will be used to develop the model.

The initial ANFIS model is described in pHmodel0.fis; it has three rules with sigmoid membership functions, a premise part as just described and a model structure as shown in Eqn. (29.25a). Since the model coefficients are unknown, the initial values of a … d are chosen equal to zero and a value of e = 7 is set for all local models. The program is coded in F2909.m.

Using the initial model, ANFIS updates the membership functions and computes the model coefficients. The final membership functions are shown in Fig. 29.9.

The model structure is shown in Eqn. (29.30) and the prediction of the pH in Fig. 29.10. The final model is stored in pHmodel1.fis. The average deviation in the predicted pH value is 0.05 pH units.

$$\begin{aligned}
&IF\ \ pH = Low\ THEN \\
&\quad p\hat{H}_{k+1} = 0.509\, pH_k + 0.331\, pH_{k-1} - 3.791\, Fc_k - 0.665\, Fc_{k-1} + 1.314 \\
&IF\ \ pH = Medium\ THEN \\
&\quad p\hat{H}_{k+1} = 0.303\, pH_k + 0.195\, pH_{k-1} - 99.120\, Fc_k - 29.602\, Fc_{k-1} + 14.841 \\
&IF\ \ pH = High\ THEN \\
&\quad p\hat{H}_{k+1} = 0.582\, pH_k + 0.288\, pH_{k-1} + 1.069\, Fc_k - 3.789\, Fc_{k-1} + 1.706
\end{aligned} \qquad (29.26)$$

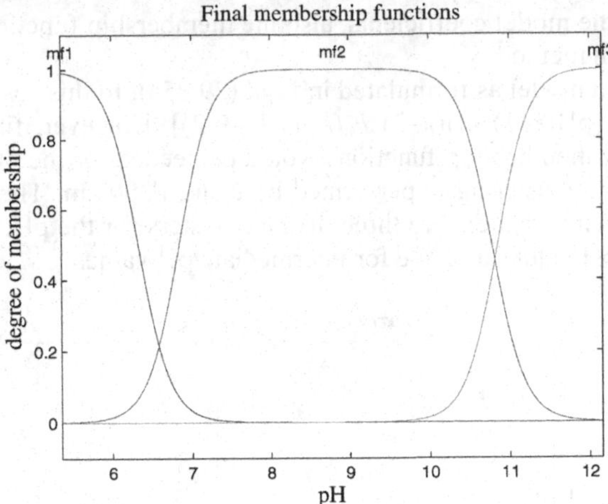

Fig. 29.9. Membership functions for pH control problem.

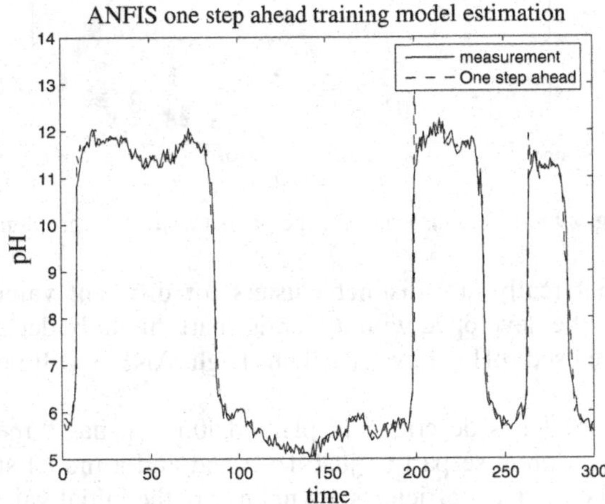

Fig. 29.10. Final prediction using ANFIS.

Assuming certain membership functions that look similar to the ones shown in Fig. 29.9, ordinary least squares can also be used to compute the consequence part of each rule of the non-recurrent model. The program that performs this is coded in F2911.m, the result of the model identification is shown in Fig. 29.11, the model structure is similar to Eqn. (29.26), although the model coefficients are different as shown in Eqn. (29.27):

$$\begin{aligned}
&IF \;\; pH = Low \;\; THEN \\
&\quad p\hat{H}_{k+1} = -0.275\, pH_k + 0.209\, pH_{k-1} + 0.589\, Fc_k - 4.163\, Fc_{k-1} + 6.328 \\
&IF \;\; pH = Medium \;\; THEN \\
&\quad p\hat{H}_{k+1} = 0.653\, pH_k - 0.401\, pH_{k-1} - 4.885\, Fc_k - 1.795\, Fc_{k-1} + 6.754 \\
&IF \;\; pH = High \;\; THEN \\
&\quad p\hat{H}_{k+1} = -0.032\, pH_k + 0.160\, pH_{k-1} - 4.902\, Fc_k - 1.947\, Fc_{k-1} + 10.669
\end{aligned} \tag{29.27}$$

The model performance is in this case also somewhat less. The average deviation in the predicted pH value is 0.20 pH units.

The fact that the model coefficients are different might indicate that it is not absolutely necessary to include second-order terms in the model difference equation, but that a model based on only the process values one time step ago already gives a good fit. In addition, there seem to be multiple models that do give an acceptable prediction.

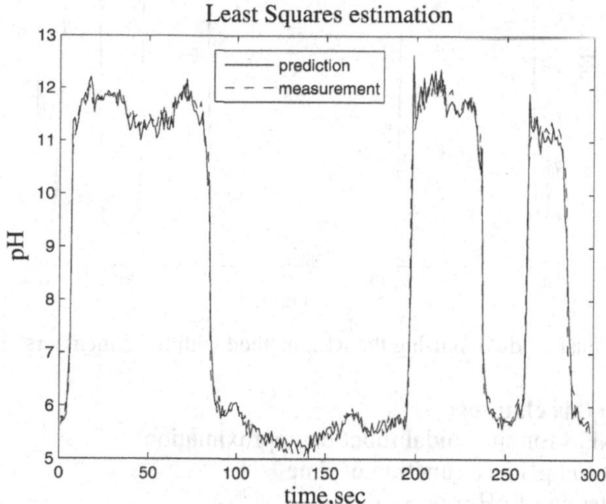

Fig. 29.11. Final prediction using the LS method with fixed membership functions.

29.4.3 Recurrent pH Neutralization Model

The most simple model identification procedure for recurrent model identification is to assume fixed membership functions and use recursive least squares to identify the consequence part of the model. The model structure in this case is the model as shown in Eqn. (29.25b).

Each time a new predicted value of the pH is calculated it is used in the next step of the identification procedure. The program that performs the identification is coded in F2912.m.

The model that is determined is given in Eqn. (29.28):

$$\begin{aligned}
&IF\ \ pH = Low\ THEN \\
&\quad p\hat{H}_{k+1} = -0.610\, p\hat{H}_k + 0.297\, p\hat{H}_{k-1} + 2.363\, Fc_k - 5.992\, Fc_{k-1} + 7.709 \\
&IF\ \ pH = Medium\ \ THEN \\
&\quad p\hat{H}_{k+1} = 0.802\, p\hat{H}_k - 0.552\, p\hat{H}_{k-1} + 8.025\, Fc_k + 3.767\, Fc_{k-1} + 5.579 \\
&IF\ \ pH = High\ THEN \\
&\quad p\hat{H}_{k+1} = -0.187\, p\hat{H}_k + 0.215\, p\hat{H}_{k-1} - 5.922\, Fc_k - 1.053\, Fc_{k-1} + 11.754
\end{aligned} \tag{29.28}$$

and the predicted pH is shown in Fig. 29.12.

Note that there is not much difference between Figs. (29.11) and (29.12). In this case, the average deviation in the predicted pH value is 0.21 pH units, even though the model structure is substantially different from those in Eqns. (29.27) and (29.28). In some cases, however, a recurrent model may be preferred over a non-recurrent model, depending on the application of the model.

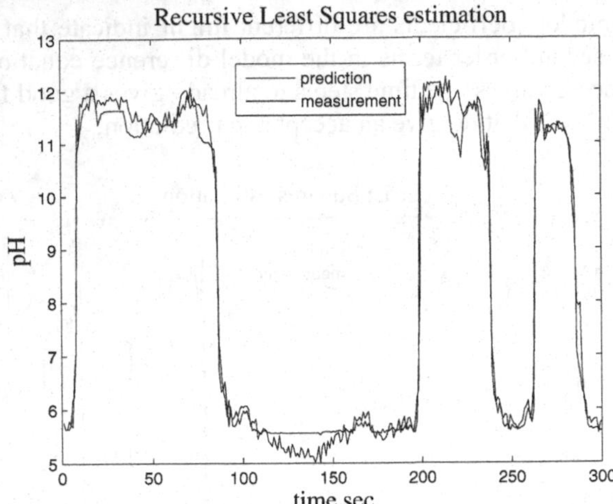

Fig. 29.12. Final prediction using the RLS method with fixed membership functions.

Files referred to in this chapter:
F2902.m: using ANFIS for sinusoidal function approximation
F2906.m: acid flow and pH as a function of time
F2908.m: data clustering of pH process data
F2909.m: using ANFIS for pH non-recurrent model development
F2911.m: using ordinary least squares for pH non-recurrent model development
F2912.m: using recursive least squares for pH recurrent model development
F1.mat: values of the acid flow used by pHprocess.mdl
members.m: illustration of creating sigmoid membership functions
membership.m: function used by members.m
phcalc.m: S-function used by pHprocess.mdl
phmodel0.fis: initial model for pH problem
phmodel1.fis: final model for pH problem
pHprocess.mdl: model to generate training and validation data for the neutralization process
trndata.mat: training data for pH problem
tstdata.mat: test data for pH problem
valdata.mat: pH process validation data

References

Jang, J.S.R. (1993) ANFIS: adaptive-network-based fuzzy inference system. *IEEE Transactions on Systems, Man and Cybernetics*, **23** (3), 665–85.

Jang, J.S.R. and Sun, C.T. (1995) Neuro-fuzzy modeling and control. *Proceedings of the IEEE*, **83** (3), 378–405.

Kong, S.G. and Kosko, B. (1992) Adaptive fuzzy systems for backing up a truck and trailer. *IEEE Transactions on Neural Networks*, **3** (2), 211–23.

Nie, J. and Linkens, D.A. (1993) Learning control using fuzzified self-organizing radial basis function networks. *IEEE Transactions on Fuzzy Systems*, **1** (4), 280–7.

Ojala, T. (1994) Neuro fuzzy systems in control. Tampere University of Technology, Finland. Ph.D. thesis.

Passino, K.M. and Yurkovich, S. (1998) *Fuzzy Control*, Addison-Wesley.

Roffel, B., Vermeer, P.J. and Chin, P.A. (1989) *Simulation and Implementation of Self-tuning Controllers*, Prentice Hall.

Sfetsos, A. (2000) A comparison of various forecasting techniques applied to mean hourly wind speed time series. *Renewable Energy*, **21**, 23–35.

Wang, L.X. and Mendel, J.M. (1992) Fuzzy basis functions, universal approximation and orthogonal least squares learning. *IEEE Transactions on Neural Networks*, **3** (5), 807–14.

Zhang, J and Morris, A.J. (1999), Recurrent neuro-fuzzy networks for nonlinear process modeling. *IEEE Transactions on Neural Networks*, **10** (2), 313–26.

30 Hybrid Models

In this chapter hybrid models will be discussed. They consist of a dynamic physical framework augmented with black-box static sub-models, which are derived from process data. This way, it is possible to combine prior knowledge and information from measurements in an optimal manner. These models have good dynamic properties and are suitable for optimization purposes in an operation range determined by the range of the measurements. The dynamic framework consists of the mass, component and energy balances. Through this framework, hybrid models have a dynamic behavior that corresponds well to the original process, and their performance exceeds time-series or tapped-delay line models. Their fit to process data assures good static behavior and the dynamic behavior is corrected for assumptions and simplifications in the framework. To illustrate the influence of prior knowledge and framework simplifications, three different hybrid models will be built by means of fuzzy relationships.

30.1 Introduction

Dynamic modeling for optimization and control requires models that describe the essential dynamic characteristics of the process under study. For complex nonlinear processes, a simple and effective model may be difficult to derive. The success of data-driven approaches may be limited in an industrial environment where the process is subject to control, and measurements may be difficult to obtain, while development of a first principles model may be time consuming and expensive. In such cases hybrid models may be attractive. These hybrid models consist of a framework of dynamic mass and energy balances, supplemented with sub-models describing additional equations, such as mass transformation and transfer rates. In this chapter, a structured approach for designing this type of model will be presented. Some other approaches have been presented in this area. Most notable are the publications by Psichogios and Ungar (1992), and Thompson and Kramer (1994), in which artificial neural networks have been used to augment the first principles information and by Roubos *et al*. (1999). The last reference focuses mainly on the parameter identification step.

Four main sources of information are available when constructing models, ranging from general to process specific information (Fig. 30.1). Physical understanding forms the basis of the model. This understanding is based on abstract physical and chemical concepts, such as balances, equilibria and driving forces and on more process unit specific principles, such as separation stages and reactor types. Process measurements are the most important source of information of a specific process. While first principles provide general information about the behavior of the process, process measurements are required to identify a suitable process model. In addition to process measurements, human experience is an important source of information because it can be used to learn more about dependencies of relevant phenomena of a specific process. A human can, based on his or her experience, establish whether certain effects are important or negligible. Hybrid models enable us to construct models for a specific process, for a required operating range, using as much information as is available.

Process Dynamics and Control: Modeling for Control and Prediction. Brian Roffel and Ben Betlem.

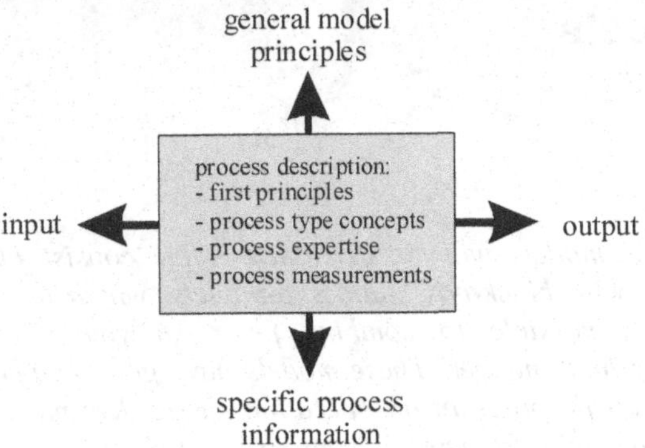

Fig. 30.1. Modeling approach.

30.2 Methodology

30.2.1 Structure

For the development of hybrid models, a structured modeling approach will be used (van Lith *et al.*, 2002, and van Lith, 2002). This approach consists of four main phases as indicated in Fig. 30.2. Compared with the general methodology for deterministic model design, the difference concerns mainly the distinction into two separate design steps (Fig. 1.2). The modeling problem is divided into smaller sub-problems. In the analysis phase, the model objective and model quality requirements are formulated. This phase will produce an environmental model defining the inputs and outputs of the model to be designed. In the second phase, the framework consisting of physical dynamic balances is designed by distinguishing the characteristic phenomena and key components. The framework identifies the balances, known relationships and sub-models to be identified using data. This can be considered as the global design of the hybrid model. In this phase first principles knowledge, process conceptions and expertise is used. In the next phase, the sub-models are designed. This is done in two steps: modeling of all sub-models independently and subsequently combining them. The different sub-models are integrated within the framework to form the hybrid model, which describes the earlier defined environmental model as appropriately as possible. In this phase, process measurements are used. The hybrid model is evaluated in a fourth phase. In the next sections, the four phases are clarified by means of the bioreactor example (van Lith, 2002).

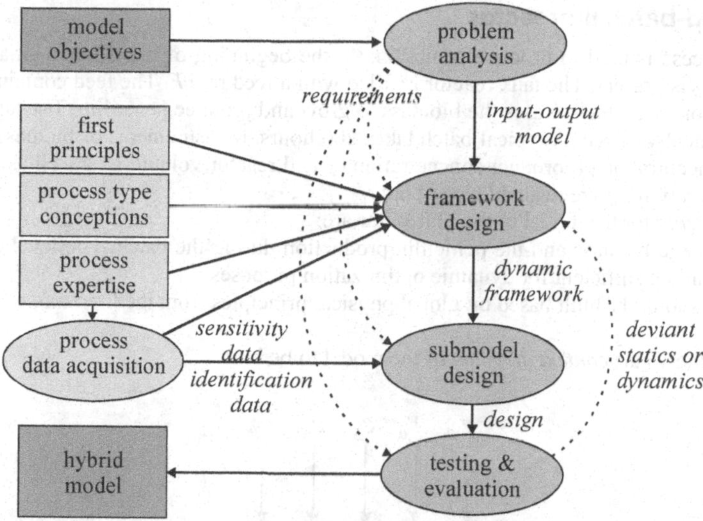

Fig. 30.2. Phases for the development of a hybrid model.

Throughout this chapter, the steps will be illustrated using a fed-batch bioreactor for penicillin production (van Lith, 2002). The process possesses the properties for which hybrid modeling is useful: a large operating range and uncertainty about the phenomena that play a role, such as biomass growth. In addition, the fuzzy sub-models that will be developed are double-input single-output systems. This allows simple visualization of the model and therefore provides a better understanding of the results. Because of the simplicity, it provides a good basis for illustrating and evaluating the various tools that are used during hybrid model design.

30.2.2 Problem Analysis

The goal of this phase is to determine the context of the problem and to gain initial information about the process. First, the description and purpose of the process is formulated and the available process information is listed. Subsequently, the process model objective, quality requirements, level of detail and the domain range are defined. It has practical and economic advantages if a model can be built on the readily available process information (that is using information that is available without performing specific experiments), be it experimental data or human expertise. This will not always be possible, but the available process information should be analyzed before the model structure is designed. The result of the analysis phase is a set of requirements for the environmental model of the process to be built.

Example: Fed-batch bioreactor

A fed-batch process is used to produce penicillin. At the beginning of the batch, a small culture with concentration c_{X0} is present. The tank reactor is filled with a feed rate F. The feed contains the substrate with concentration c_{SF}, which causes the biomass to grow and produce penicillin. The tank reactor may be assumed as ideally mixed. A typical batch takes 200 hours. Measurements of biomass concentration c_X, substrate concentration c_S, product concentration c_P and reactor volume V as well as the feed rate F and feed concentration c_{SF} are available every hour.

The requirements for the model of the bioreactor are:

1. Describe of the biomass and the penicillin production during the batch. The level of detail of the model should be sufficient for dynamic optimization purposes.
2. The model should be built based on global physical principles from the literature.

Figure 30.3 shows the *context diagram* of the model to be built.

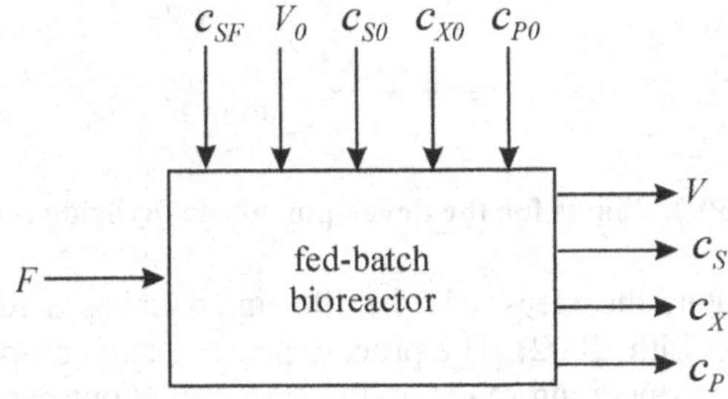

Fig. 30.3. Environmental model of a fed-batch bioreactor.

30.2.3 Framework Design

During the framework design phase (Fig. 30.4), first principles, process type conceptions and process expertise are combined to form a physical framework. In addition, the parameters that will be modeled by sub-models have to be identified. The output of each sub-model is the parameter concerned. However, independent variables are not always clear. It may be necessary that those input variables are determined by means of a sensitivity analysis. The conceivable candidates for these independent variables are all state and control variables used in the framework. In this example fuzzy logic will be used to build these sub-models, but other black-box modeling techniques, which are discussed in the previous chapters, such as, for example neural networks, may also be appropriate. The result of the framework design phase is the framework, which can be represented by a behavioral model. This is a data flow diagram of the model structure.

The state variables are the output variables in the environmental model (Fig. 30.3): V, c_S, c_X and c_P. The model is fairly similar to the one discussed in chapter 17. The volume balance is straightforward; the accumulation is given by the feed rate F.

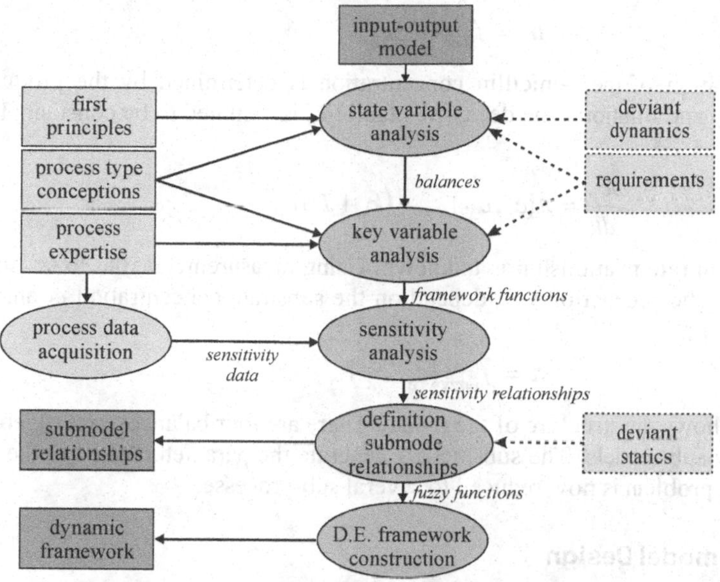

Fig.30.4. Steps in framework design phase.

$$\frac{dV}{dt} = F \tag{30.1}$$

The dilution is the result of the changing volume:

$$D = \frac{F}{V} \tag{30.2}$$

The substrate is added by the feed stream and is consumed by the biomass with consumption rate σ. In addition, the dilution influences the substrate concentration. The substrate balance can be given by:

$$\frac{dc_S}{dt} = -\sigma(c_S, c_X)\, c_X + D\,(c_{SF} - c_S) \tag{30.3}$$

The consumption is used for biomass growth, penicillin production and biomass maintenance and can be described by a known function:

$$\sigma = \frac{\mu^*(c_S, c_x)}{Y_{XS}} + \frac{\pi(c_S, c_x)}{Y_{PS}} + \frac{m_{xm} c_x}{c_x + 10} \tag{30.4}$$

where Y_{XS}, Y_{PS} and m_{xm} are constants.

The biomass accumulation is determined by biomass growth, biomass decay and a dilution factor:

$$\frac{dc_X}{dt} = \left(\mu^*(c_S, c_X) - D\right) c_X \tag{30.5}$$

The biomass growth and decay relationships are unknown. It is assumed that the difference between growth and decay can be lumped into a biomass net growth rate μ^*. The biomass growth will depend on the substrate concentration c_S and the biomass concentration c_X. Using available measurements, a fuzzy sub-model can be built.

$$\mu^* = f_{\text{fuzzy}}(c_S, c_X) \tag{30.6}$$

The accumulation of the penicillin concentration is determined by the production rate π, product decay and dilution. The decay rate factor k_h is assumed to be constant. The penicillin balance becomes:

$$\frac{dc_P}{dt} = \pi(c_S, c_X)\, c_X - (k_h + D)\, c_P \tag{30.7}$$

The production rate relationship is unknown. Using measurements, the fuzzy sub-model can also be built. The production will depend on the substrate concentration c_S and the biomass concentration c_X.

$$\pi = f_{\text{fuzzy}}(c_S, c_X) \tag{30.8}$$

Figure 30.5 shows the structure of the model. There are four balances, one algebraic equation and the fuzzy sub-models. The sub-models generate the parameters used in the balances and the modeling problem is now reduced to several sub-processes.

30.2.4 Sub-model Design

The sub-model design phase consists of the following steps (Fig. 30.6):

- **Data acquisition.** To be able to perform the identification of the various model parameters, sufficient process data needs to be available. This data can be gathered by doing experiments specifically designed for obtaining process behavior information if this information is not readily available. In two of the three examples below, this data can be obtained by performing several simulation "experiments" under different operating conditions. The "measurements" of the state profiles can then be used as identification data. Parameter or state estimation techniques can also be used to obtain the correct information.

  When designing experiments, care has to be taken that the data will best reveal the behavior of the sub-process. Experiments can be designed more effectively when knowledge that is available is used. In addition, exploratory experiments can be useful. Additional experiments are then designed on the basis of these initial results.

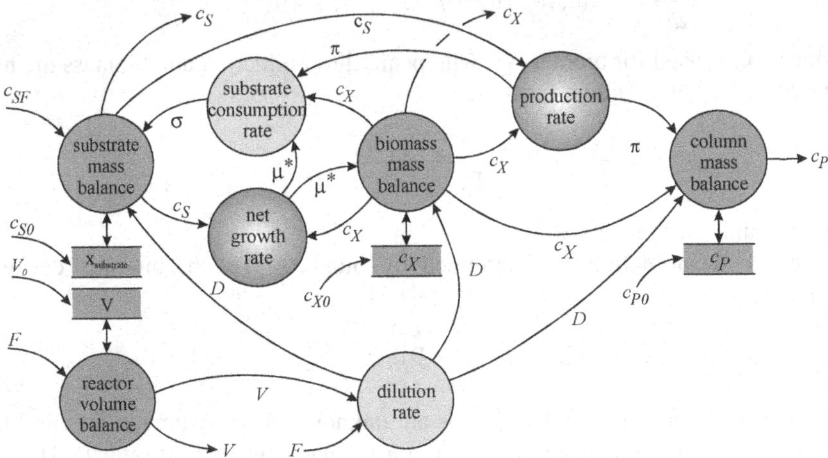

Fig. 30.5. Behavior model of a fed-batch bioreactor.

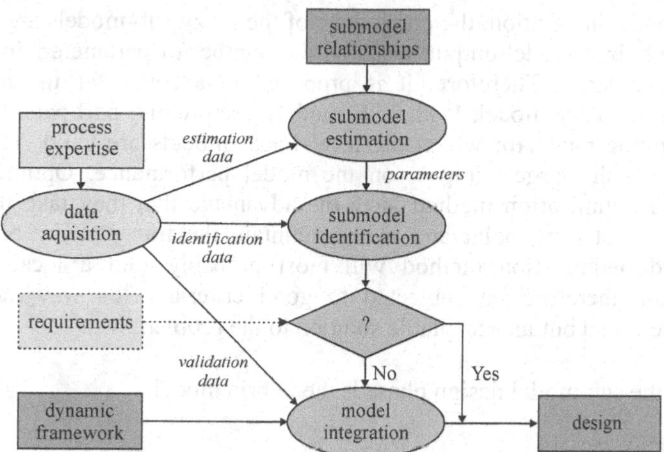

Fig. 30.6. Steps in the sub-model design phase.

The following steps should be performed:

- **Sub-process behavior estimation.** Since the sub-process outputs cannot be measured directly, they should be estimated. Several techniques are available for this estimation, such as PI-estimation or Kalman filtering. In this chapter, PI-estimation will be applied (van Lith et al., 2001). The PI-estimator is a simple and useful alternative to more elaborate estimation approaches. Its simple nature makes it possible to set it up quickly. Care should be taken when there is large coupling between two or more parameters that are estimated simultaneously. For many applications, however, the approach can be applied without many problems. Since the model is divided into several smaller sub-processes, there will be several small estimation problems.
- **Sub-model identification.** Fuzzy clustering provides a good way to identify the fuzzy sub-models. It is an unsupervised learning algorithm and requires little *a priori* model structure information. In addition, because of the structure optimization algorithm, it is insensitive to initialization. It also derives a fuzzy model with independent rules directly from the data, which results in models that are not likely to show anomalous extrapolation behavior.

  The fuzzy sub-models are identified using the data obtained in the previous step. The clustering method used is the Gustafson-Kessel clustering algorithm (chapter 28), which is very flexible in describing complex systems and insensitive to scaling of the data or initialization. It is iteratively used together with Modified Compatible Cluster Merging, to merge compatible clusters that show a certain degree of conformity in order to reduce the number of clusters. The iteration is necessary because the number of clusters in the Gustafson-Kessel clustering algorithm needs to be determined beforehand, which is not always possible.

  Takagi Sugeno Kang (TSK) type fuzzy models are used. This type of fuzzy models can be interpreted as a collection of local linear sub-models and is extremely suitable to describe highly nonlinear relationships based on process data. In addition, the model structure is simple and transparent. Many good algorithms for the identification of TSK models from data are available, including algorithms with structure optimization.
- **Sub-model integration.** The sub-models are integrated to form the hybrid model, which involves connecting the sub-models and optimization of the overall hybrid model performance. Since there may be interactions caused by this integration, an optimization of the hybrid model may need to be done. Also errors from sub-process estimation or sub-model identification can manifest themselves in the hybrid model.

During sub-model integration, the parameters of the fuzzy sub-models are optimized with respect to the hybrid model output. Usually, the number of parameters in the fuzzy sub-models is quite large. Therefore, it is proposed to account for the meaning of the parameters of the fuzzy model. With TSK models, the premise part parameters determine only the operating range for which the local linear models are valid. The consequence parameters have the largest impact on the model performance. Optimizations with a gradient-based optimization method have the advantage that they take the initial fuzzy sub-models as a starting point and hence maintain the transparency of the model. A gradient-based optimization method will most probably find a local optimum. The solution should therefore be subjected to good criteria. This way, not the optimal parameters are found but an acceptable solution to the problem.

The result of the sub-model design phase is the hybrid model.

Example: Fed-batch bioreactor

- **Data Acquisition**

For the identification of the two fuzzy relationships (Eqns. (30.6) and (30.8)), sufficient measurements are needed to characterize these relationships appropriately. This means that the data need to be distributed over a sufficiently large domain in the input space, so that all operating conditions are covered. The input space is the same for both relationships; the input space is formed by the concentrations c_S and c_X. Since the fuzzy relationships are static, no dynamic data are required. This means that a limited number of experiments will be sufficient; only steady-state measurements of the process operated under different conditions need to be obtained.

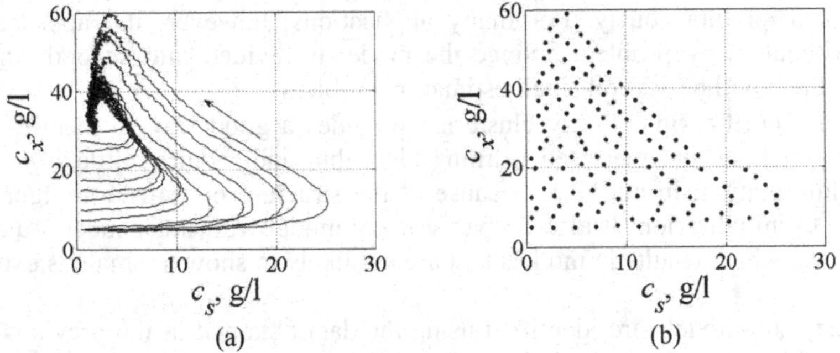

Fig. 30.7. Input space distribution of experiments (a) and reduced input space (b).

Figure 30.7a shows the trajectories of several typical batch runs in the input space. Each run can be divided into three phases. At start-up, the biomass concentration c_X is low and the substrate concentration c_S is zero. During the initial phase, c_S increases fast by the substrate feed and, consequently, during the second phase the biomass will rise. During the final phase, exhaustion of the substrate will follow. In order to obtain a good distribution, 16 different experiments were simulated with varying initial conditions for the initial biomass concentration c_x and feedrate F.

The data set can be presented directly to the clustering algorithm. However, owing to the nature of the batch runs, the data are not evenly distributed in the input space. During the latter hours of a batch, a kind of "pseudo-steady-state" is obtained in which c_X and c_S do not vary as much as during the first hours. This causes data features to lie closer together in the area where c_X is high and c_S is low than data features in the area where c_X is low and c_S is high. Since the clustering approach uses least squares calculation, results may be biased because of this. This problem is solved by reduction of the data set:

all features, for which the distance in the input space to another feature is smaller than a predetermined threshold, are removed. This threshold is set manually by finding a balance between the number of data features in the reduced data set and the average distance between these features. The threshold was set to 2 and the distribution of the resulting data set is shown in Fig. 30.7b.

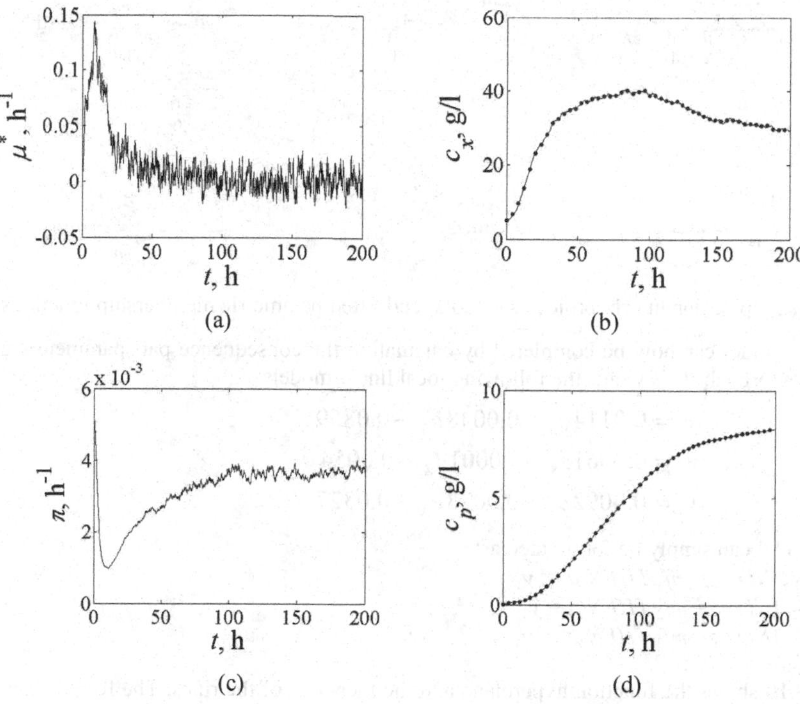

Fig. 30.8. PI-estimates of μ (a), c_X (b), π (c) and c_P (d). Dots represent the measurements and a line the estimates.

- **Estimation of the sub-model**

The parameters μ^* and π of the bioreactor can be estimated by two PI-estimators (van Lith, 2002). The estimates (Fig. 30.8) are good; the trajectories of c_X and c_P are followed closely.

- **Sub-model identification**

Fuzzy clustering will be used to build a model of the net growth rate as a function of the substrate concentration c_S and biomass concentration c_X for the hybrid model of the bioreactor. Assume that an input–output data set has been created using all identification experiments and that the PI-estimator data are available in the form of measurements. This means that input–output data are available of c_S and c_X and estimates of μ^* and π. The data set can be presented directly to the clustering algorithm. The initial number of clusters was ten. This number was reduced to three by the merging algorithm. The next step is to project the fuzzy partition matrix onto the c_S and c_X axis, so that parametric membership functions can be determined. This is shown in Fig. 30.9.

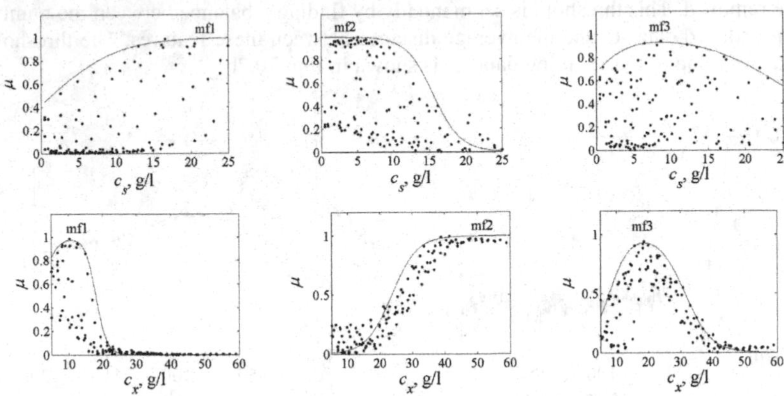

Fig. 30.9. Fuzzy partition matrix projections (dots) and fitted parametric membership functions (lines).

The fuzzy model can now be completed by calculating the consequence part parameters using the least squares approach. This yields the following local linear models:

$$\begin{aligned} y_1 &= 0.0114c_S + 0.0048c_X - 0.0859 \\ y_2 &= 0.0061c_S - 0.0001c_X - 0.0054 \\ y_3 &= 0.0092c_S - 0.0028c_X - 0.0327 \end{aligned} \tag{30.9}$$

The fuzzy model can simply be formulated as:

If $c_S = mf_1$ *AND* $c_X = mf_1$ *THEN* $\mu^* = y_1$
If $c_S = mf_2$ *AND* $c_X = mf_2$ *THEN* $\mu^* = y_2$
If $c_S = mf_3$ *AND* $c_X = mf_3$ *THEN* $\mu^* = y_3$

Figure 30.10 shows the function hyperplane and the locations of the rules. The Root Mean Squared Error with respect to the identification data is 0.0248. This error is mainly the result of a poor fit in the region where c_X is low; the error with respect to data features in the region where c_X is high is 0.0091.

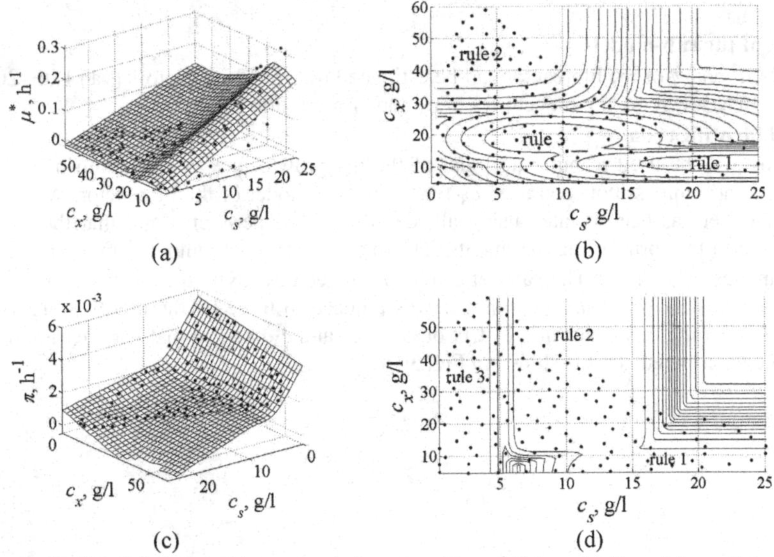

Fig. 30.10. Fuzzy relationships and rule locations for μ^* ((a) and (b)) and for π ((c) and (d)). Dots indicate identification data.

A more complex fuzzy relationship can provide a better description. However, the observed behavior of the net growth rate is not so complex that a more complex fuzzy relationship is justified from a transparency point of view. A final evaluation of the fuzzy model should be done after integration in the hybrid model structure. Although fuzzy logic is a black box technique, a posteriori analysis of the model shows that the three rules represent three phases during a batch.

- **Model integration**

The initial performance of the hybrid model for one of the validation sets, before optimization, is shown in Fig. 30.11. It can be seen that for this run, the concentration c_P starts to decrease after about 160 hours, which is not possible. This is caused by anomalous behavior of π. The production rate decreases after 160 hours, which results in a decrease in production of P. Because the amount of P in the reactor is diluted, the net result is a decrease in the concentration of P. The increase in S is caused by a decrease in the substrate consumption rate, which is a result of the decrease in π.

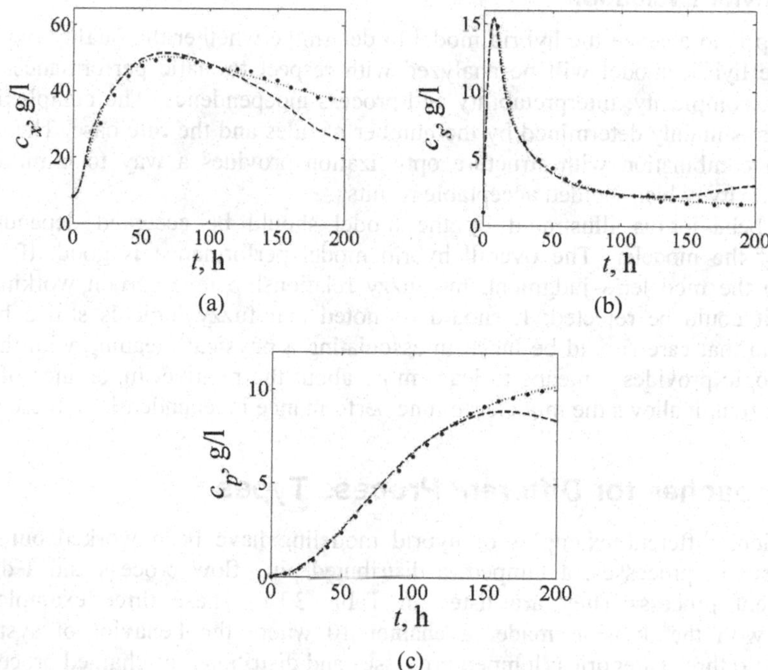

Fig. 30.11. Validation results for and after optimization for c_X (a), c_S (b) and c_P (c). Dots indicate measurements, dashed line hybrid model before optimization and dotted line hybrid model after optimization.

The goal is to improve hybrid model performance with respect to the errors in the biomass concentration c_X, the substrate concentration c_S and the product concentration c_P. This will be done by using the measurements from the identification batch runs as a reference. The objective function is formulated according to:

$$J = \frac{1}{2} \sum \left(e_S^2 + e_X^2 + e_P^2 \right) \tag{30.10}$$

in which e_S, e_X and e_P are the normalized error signals in c_S, c_X and c_P.

The number of parameters in the fuzzy sub-models is quite large. One rule of the fuzzy model for the net growth rate, for example, contains about 10 parameters, depending on the type of membership function that is used. Owing to the "curse of dimensionality" the number of parameters increases exponentially for systems with higher dimensions. Therefore, only the consequence parameters have been optimized.

Although it is possible to optimize the fuzzy models for μ^* and π sequentially, it is interesting to see how the large-scale algorithm deals with optimization of a large set of parameters in a hybrid model. Therefore, all the parameters of the two fuzzy models will be optimized simultaneously. The result is that a set of 66 parameters will be optimized; 48 premise part parameters and 18 consequence part parameters. The premise part parameters are constrained; the bounds are set at the initial values ±10%. No constraints were placed on the consequence part parameters. The results of the optimization are shown in Fig. 30.11. (for clarity, only some of the measurements are shown). The anomalous behavior has been removed and model offset has been reduced to acceptable levels.

30.2.5 Behavior Evaluation

The next step is to analyze the hybrid model to determine whether the quality requirements are met. The hybrid model will be analyzed with respect to static performance, dynamic performance, complexity, interpretability and process independence. The complexity of the fuzzy models is mainly determined by the number of rules and the rule base. The clustering algorithm in combination with structure optimization provides a way to minimize fuzzy model complexity, which yielded acceptable results.

Whether behavior as illustrated by the model should be accepted depends on the objectives of the modeler. The overall hybrid model performance is good. If, however, according to the modeler's judgment, the fuzzy relationship in a certain working area is unrealistic, it could be rejected. It should be noted that fuzzy logic is still a black box technique and that care should be taken in associating a physical meaning with the results. Yet, fuzzy logic provides a means to learn more about the relative importance of working areas. In addition, it allows the modeler to tune performance independently in these areas.

30.3 Approaches for Different Process Types

In this section, different examples of hybrid modeling have been worked out for three different types of processes: a lumped, a distributed plug flow process and a distributed counter-current process. They are listed in Table 30.1. These three examples are in accordance with the division made in chapter 10 where the behavior of systems was partitioned into three categories: lumped processes and distributed or chained processes with and without feedback. For each example, the main advantages and disadvantages of hybrid modeling are briefly discussed. The hybrid models of a polymer reactor and pulp digester were derived from existing detailed simulation models. The hybrid model design and the evaluation of the results are based on measurements generated by the original simulations.

For the polymer reactor the required inputs of the fuzzy sub-models are obtained through a sensitivity analysis. The effect of lumping and data reduction on the accuracy of the model will be examined.

For the digester the different sections of the process are assumed to be lumped. It appeared that the simulation results improved owing to optimization of the entire integrated model. The extrapolation capabilities of the model are also tested.

In the case of the batch distillation process another strategy is followed. As a starting point a simple model is assumed. This model is adapted until the essential process dynamics are accounted for.

Table 30.1. Examples for hybrid modeling.

Process type	Example	Dynamics	Statics	Approach
lumped	polymer CSTR	based on balances ↑	based on relationships ↑	all balances are considered, sensitivity analysis to design static relationships
distributed	pulp digester	↕	↕	order reduction by lumping of phases and sections by physical insight
counter current flow	batch distillation column	based on global dynamics	based on data	covering the essential dynamics by approximating of overall behavior

30.3.1 Lumped Process: Polymer CSTR

In the case of lumped processes, hybrid model design focuses on reducing a complex set of static relationships with many fit parameters by overall fuzzy sub-models. These sub-models can be derived from measurements. A polymer reactor has been used as an example. In the liquid phase reactor, which is ideally stirred, monomer reacts to polymer by means of a single site metallocene catalyst. The inputs of the reactor are listed in Table 30.2: the load F_{in}, the hydrogen fraction of the feed x_{H2}, the activated catalyst fraction $x_{c,a}$, the non-activated catalyst fraction $x_{c,na}$, the monomer fraction x_m and the jacket temperature T_{jacket}. The most important outputs are the molecular weight distribution *MWD* (or q, the chain termination probability which is inversely proportional to *MWD*) and the conversion ζ or ζ^* (feed mass fraction conversion). The model is described in detail by Roffel and Betlem (2003). The requirements for the model of this reactor are:

1. The hybrid model should at least describe the molecular weight distribution MWD and the conversion ζ^*.
   The *MWD* can be characterized by a chain termination probability q. The *MWD* can be calculated from q by the Schultz-Flory relationship.
2. Dynamic and static behavior of the hybrid model should match the physical model.
   This requires the modeling of the known states of the system in the hybrid model to be similar to that in the physical model.
3. The input–output structure of the hybrid model needs to be compatible with the physical model.
4. The hybrid model needs to be less complex than the physical model.
   Less complexity is a prerequisite for dynamic optimization calculations of the project concerned.

For this example the procedures for the hybrid model design and the sub-model design, as described above, are followed. As a consequence of the requirements, for the hybrid model, all known balances involved are incorporated: density balance, energy balance and component balances for monomer, hydrogen, non-activated catalyst, and activated catalyst. The volume of the reactor is constant. The reaction rates for the chain propagation, chain termination, catalyst activation and catalyst deactivation, determine the shifts between the component balances. For these rates, fuzzy relationships are built.

From the sensitivity analysis (Table 30.2), it appeared that the jacket temperature T_{jacket} and hydrogen fraction x_{H2} are the dominant influences.

Table 30.2. Sensitivities of the model outputs as a function of the model inputs.

independent variables	first-order effects	
	ζ average: 0.1803	Q average: 3.728×10^{-3}
F_{in}	–0.0147	-0.023×10^{-3}
T_{jacket}	0.1691	-0.145×10^{-3}
T_{in}	0.0180	-0.007×10^{-3}
$x_{c,na,in}$	0.0396	0.005×10^{-3}
$x_{c,a,in}$	0.0557	0.004×10^{-3}
$x_{H2,in}$	0.1052	0.804×10^{-3}
$x_{m,in}$	–0.0453	-0.065×10^{-3}

Therefore, in the fuzzy relationships the reaction rates are made dependent on T_{jacket} and x_{H2} and when necessary on the related component fraction. Hereby, an overall description of the reaction rate is derived and it is not necessary to describe explicitly the adsorption rate from the monomer into the amorphous polymer phase. Figure 30.12 shows the hybrid model consisting of the mentioned balances and fuzzy sub-models. Several data sets with different number of data points were used to fit the model. For the model identification, no state or parameter estimation was necessary.

Data collection for a polymer reactor is not simple. It is interesting to examine the minimum amount of data required to design a sufficiently accurate model. The required number of identification data depends mainly on the number of model parameters. The number of parameters is listed in Table 30.3.

Table 30.3. Fuzzy sub-model characteristics for polymer reactor, *R* are reaction rates for activation, propagation and deactivation.

dependent variables	independent variables	number of membership functions	number of parameters premise part	number of parameters consequence part	number of weighting parameters
q	$[H_2]$, $T_{reactor}$	3	$2 \times 3 \times 4 = 24$	$3 \times (2 + 1) = 9$	—
R_a	$x_{non\text{-}act.cat.}$, $[H_2]$, $T_{reactor}$	2	$3 \times 2 \times 4 = 24$	$2 \times (3 + 1) = 8$	—
R_p	$x_{act.cat.}$, $[H_2]$, $T_{reactor}$	3	$3 \times 3 \times 4 = 36$	$3 \times (3 + 1) = 12$	—
R_d	$x_{act.cat.}$, $[H_2]$, $T_{reactor}$	3	$3 \times 3 \times 4 = 36$	$3 \times (3 + 1) = 12$	—

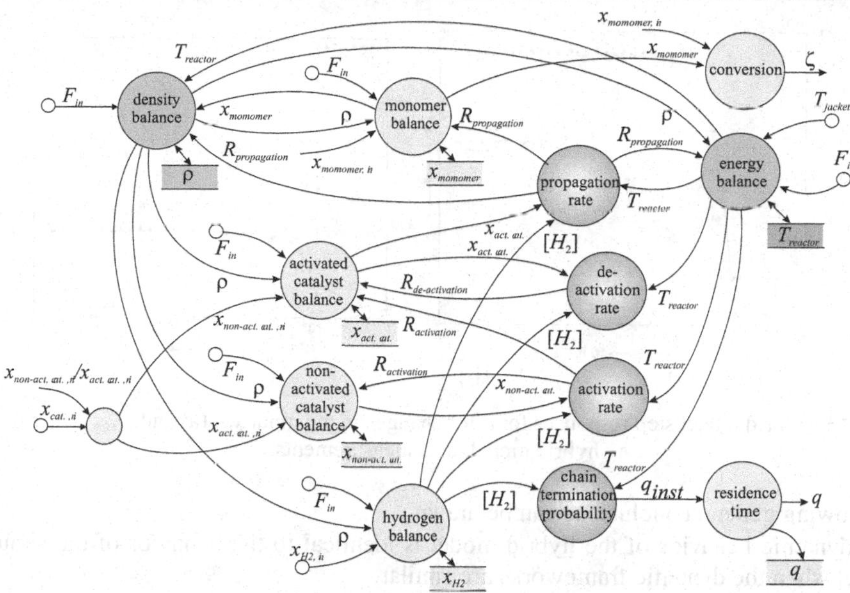

Fig. 30.12. Hybrid model of polymer reactor.

For the sub-model of q, which depends on two input variables, T and the H_2 composition, the minimum number of data points required for accurate fuzzy sub-model development was investigated. The data was obtained by adding random signals to the input variables in the reactor simulation. There are two requirements: the data should be evenly distributed over the entire set and the data density should be an independent variable. To obtain an evenly distributed data set, the number of data points was reduced by sequentially removing data points within a certain distance of a randomly chosen data point. By varying this distance it is possible to produce different reduced data sets. The base case counted 200 data points.

From the base case it appeared that the premise part should contain three membership functions. Therefore 10–15 evenly distributed data points were sufficient. The number of premise parameters is also shown in Table 30.3, but these parameters are not independent of each other. Identification of the consequence part of the model requires many observations. This should be of the order of two to three times the number of consequence part parameters, which is nine. This is supported by the simulations. With a data set size of 15, no accurate model could be realized, but 24 evenly distributed data points appeared to be sufficient.

Figures 30.13 and 30.14 show the dynamic results for the conversion ζ^* and the termination probability q after step changes in the independent variables T_{jacket} and x_{H2}.

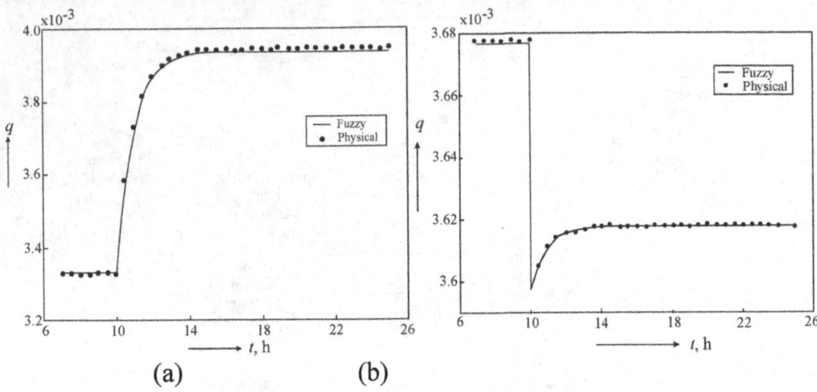

Fig. 30.13. Hybrid model step responses for q for changes in the input x_{H2} (a) and T_{jacket} (b). Solid line hybrid model, dots measurements.

The following general conclusions can be drawn:

- The dynamic behavior of the hybrid model is identical to the behavior of the simulation model when the dynamic frameworks are similar.
- The static behavior shows differences in variables for which the sensitivity is large. In this case it concerns variations in T_{jacket}.
- Static relationships can be combined. The transfer from the monomer liquid phase to the amorphous polymer phase can be ignored and incorporated in the fuzzy relationships for the reaction rates.
- From reduced data sets it appeared that still an accurate hybrid model could be designed, although it was more difficult to find a good set of clusters. However, model extrapolation capabilities became less reliable.

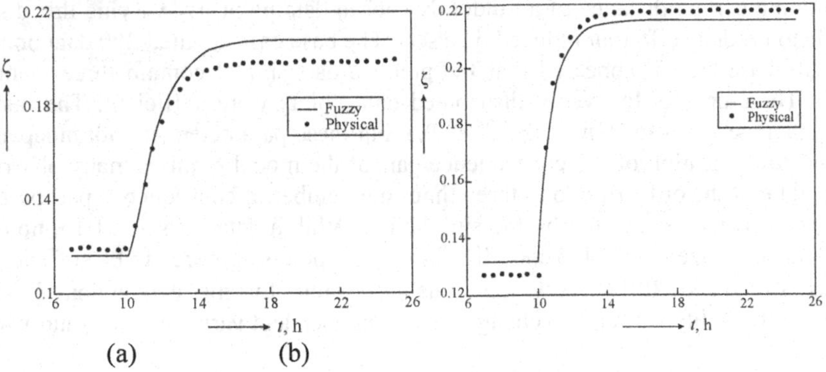

Fig. 30.14. Hybrid model step responses for ζ^* for changes in the input x_{H2} (a) and T_{jacket} (b). Solid line hybrid model, dots measurements.

30.3.2 Plug Flow Process: Digester

In the case of distributed plug flow processes, hybrid model design focuses on reducing the number of sections by overall fuzzy relationships. A pulp digester has been used as an example. The original model is the so-called Purdue Model (Smith and Williams, 1974), which is a detailed and industrially accepted digester model.

The continuous pulp digester is a unit operation designed to convert wood chips into a cellulose fiber pulp by delignification. The pulping of the wood chips is accomplished by cooking the wood chips in a hot solution of sodium hydroxide and sodium sulfide, referred to as "white liquor". The digester is essentially a tubular reactor, where wood chips travel from the inlet at the top to the outlet at the bottom of the digester. A schematic overview of the digester is given in Fig. 30.15a. The digester is divided into five functional zones. In the impregnation zone, the wood chips are brought into contact with a co-current flow of white liquor. The digester contains a heating zone for gradually heating the white liquor to the desired reaction temperatures. The actual delignification takes place in the cooking zone. In the washing zone, a counter current flow washes the degradation products from the pulp. A process flow referred to as "filtrate" is used as washing liquor. In the cooling zone, part of the injected filtrate flow goes down and thus travels co-current with respect to the wood chips to provide cooling.

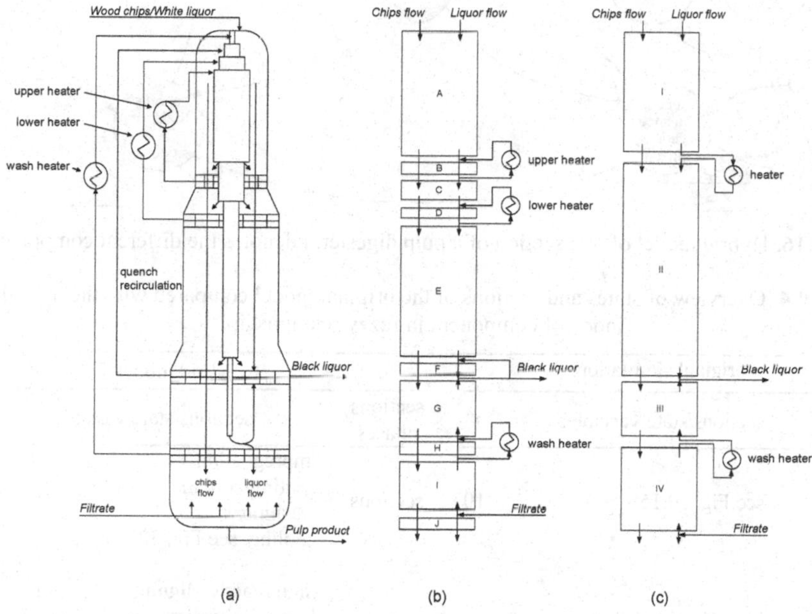

Fig. 30.15. Pulp digester (a), simulation model (b) hybrid model (c).

At the bottom of this zone, the outlet device is located. The most digester input variables are the wood chips flow $\phi_{wood,in}$, the liquor flow $\phi_{liqour,in}$ and the heat supply $Q_{heating}$ and the most important output variables are the Kappa number κ, which is a measure for the amount of lignin in the wood chips and the yield γ, which is a measure of the amount of wood recovered in comparison to the amount fed to the digester.

The original simulation model, on which the hybrid model will be based, counts ten sections, as shown in Fig. 30.15b, which are described by partial differential equations. The requirements for the model of this reactor are:

1. The hybrid model should at least describe the key variables: the kappa number and the yield.
2. Dynamic and static behavior of the hybrid model should match the physical model.
3. The input–output structure of the hybrid model needs to be compatible with the physical model.
4. The hybrid model needs to be less complex than the original model.

Less complexity is a prerequisite for dynamic optimization calculations of the project concerned.

A hybrid model has been built according the procedure discussed above. The number of sections is reduced from ten to four.

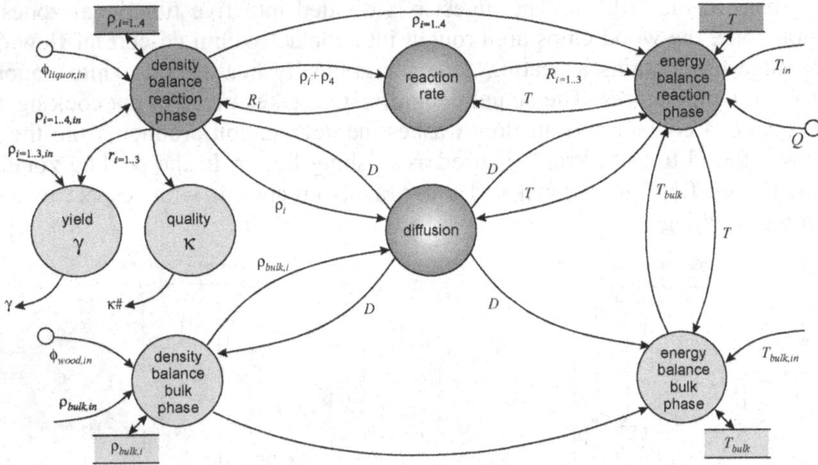

Fig. 30.16. Hybrid model of one section of a pulp digester. *i* denotes the different components.

Table 30.4. Overview of states and sections of the original model compared with the hybrid mode. [a]index of component in fuzzy relationship.

original simulation			hybrid model			
sections phases	sections/state variables	#	sections phases	sections/state variables		#
section	see Fig. 30.15b	10	sections	impregnation heating/cooking washing cooling, see Fig. 30.15c		4
chip phase	high reactive lignin low reactive lignin cellulose galactoglucomannan araboxylan temperature	6	reaction phase states	high reactive lignin low reactive lignin carbohydrates active liquor passive liquor dissolved lignin dissolved carbohydrates temperature	1[a)] 2 3 4	8
entrapped phase	active alkali passive alkali active hydrosulfide passive hydrosylfide dissolved lignin dissolved carbohydrates	6				
liquor phase	active alkali passive alkali active hydrosulfide passive hydrosylfide dissolved lignin dissolved carbohydrates temperature	7	liquor phase states	active liquor passive liquor dissolved lignin dissolved carbohydrates temperature		5

The chosen structure of one section is shown in Fig. 30.16 and the reduction of the number of states is listed in Table 30.4.

The apparent choice is to derive the hybrid model structure from the original simulation by model reduction. By formulating interpretability requirements, the model reduction is performed from a physical point of view instead of a mathematical point of view. This way, requirement four is met. Since the hybrid model describes the internal behavior of the digester, the data needs to be available in the form of "state profiles", that represent the states as a function of the location in the reactor. The data were obtained from nine identification sets and for the validation eight sets were available.

The fuzzy sub-model characteristics for the pulp digester are shown in Table 30.5.

Table 30.5. Fuzzy sub-model characteristics for pulp digester; *R*, reaction rate; *ρ*, density; *T*, temperature.

dependent variables	independent variables	number of membership functions	number of parameters premise part	number of parameters consequence part	number of weighting parameters
R_1	$\rho_1, \rho_4, T_{react}$	4	$3 \times 4 \times 4 = 48$	$4 \times (3+1) = 16$	0
R_2	$\rho_2, \rho_4, T_{react}$	6	$3 \times 6 \times 4 = 72$	$6 \times (3+1) = 24$	6
R_3	$\rho_3, \rho_4, T_{react}$	6	$3 \times 6 \times 4 = 72$	$6 \times (3+1) = 24$	6
D	T_{react}	2	$1 \times 2 \times 4 = 8$	$2 \times (1+1) = 4$	0

From the lignin and carbohydrate densities, the kappa number κ and yield γ can be derived, according to:

$$\kappa = \frac{\rho_1 + \rho_2}{0.00153 \sum_{i=1}^{3} \rho_i} \tag{30.11}$$

and

$$\gamma = \frac{\sum_{i=1}^{3} \rho_{i,exit}}{\sum_{i=1}^{3} \rho_{i,entry}} \tag{30.12}$$

To identify the fuzzy models for reaction rates and diffusion coefficient, input–output data are required. The inputs can be obtained from the simulation model. However, the outputs ($R_1(\rho_1,\rho_4,T)$, $R_2(\rho_2,\rho_4,T)$, $R_3(\rho_3,\rho_4,T)$ and $D(T)$) of the fuzzy models cannot be readily obtained. The outputs are therefore derived from observed behavior by PI-estimation (van Lith *et al.*, 2001). The reaction rates and diffusion coefficient can be estimated with the help of the mass balances of species of the hybrid model and the steady state profiles from the identification experiments. The estimators were tuned manually by comparing the concentration profiles with the reference values from the simulation model.

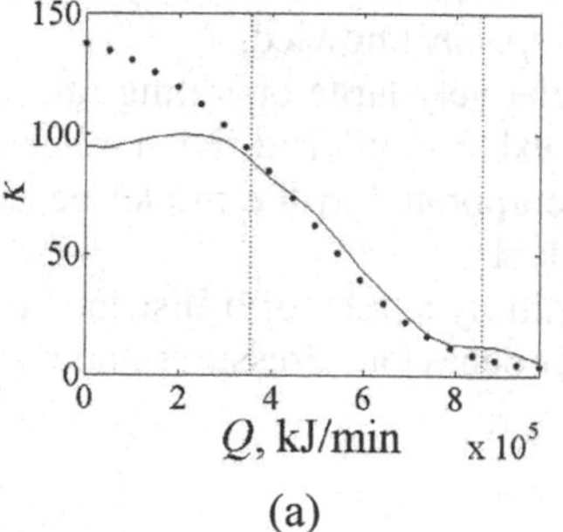

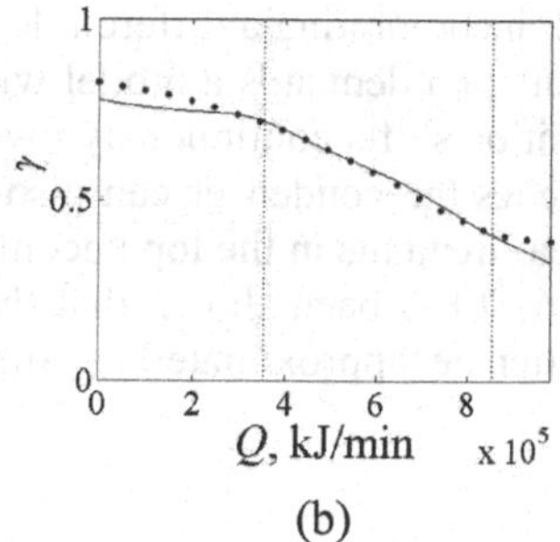

(a) (b)

Fig. 30.17. Static extrapolation results for κ(a) and γ(b) as a function of the load change in $Q_{heating}$. Solid line hybrid model, dots measurements. Vertical lines indicate the identification domain.

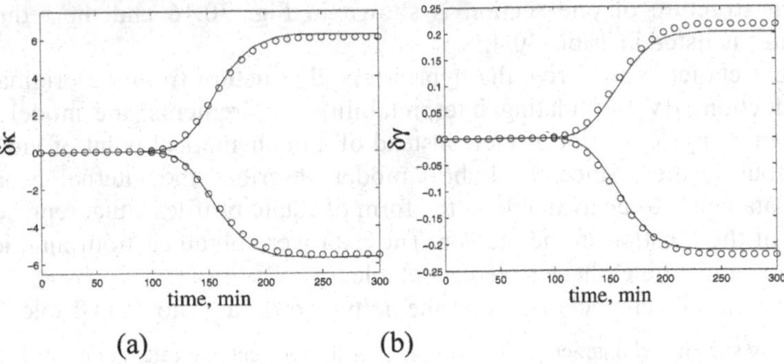

Fig. 30.18. Hybrid model step response for changes in κ(a) and γ(b) for a positive and negative load change in $Q_{heating}$. Solid line hybrid model, dots measurements.

The fuzzy sub-models for reaction rates and the diffusion rate were combined. The hybrid model showed errors mainly caused by the model reduction. Therefore, the fuzzy sub-models were adjusted to improve the performance by optimization. Figures 30.17 and 30.18 show the static and the dynamic results of the hybrid model for the kappa number κ and the yield γ due to step changes in the heat supply. The following general conclusions can be drawn:

- Analysis of the static behavior shows that with respect to interpolation, the hybrid model performs well and with respect to extrapolation, performance is determined by the quality of the data driven fuzzy models (Fig. 30.17).
- The dynamic behavior of the hybrid model is identical to the behavior of the simulation model (Fig. 30.18). Hybrid fuzzy-first principles models can match desired dynamic behavior if the model structure represents the essential dynamic characteristics of the process. This was achieved by determining the essential characteristics of the process from a physical point of view, for given operating conditions.
- Despite a substantial model reduction, an accurate hybrid model can be obtained. The number of states has been reduced from 19 to 13, while the number of model sections has been reduced from 10 to 4.

30.3.3 Counter-current Process: Batch Distillation Column

In the case of a chained process, hybrid model design focuses on exploring the essential process dynamics. The objective is to develop a hybrid model that can simulate a batch run, including start-up. Such a model can be used for simulation studies or the optimization of operating conditions, such as the determination of the optimal batch time. However, the main purpose of this work is to investigate the influence of additional knowledge for this process in the partly data driven modeling approach. Therefore, three different hybrid models will be developed, each incorporating a different level of *a priori* knowledge.

A batch distillation demands a model with a relatively large operating range, because the bottom composition shifts continuously owing to exhaustion. The start-up behavior after the vapor flow reaches the condenser can also be incorporated in the model because from that moment the measurements in the top become available.

In chapter 16, it has been shown that the overall dynamics of a distributed process with feedback can often be approximated by first-order behavior. Pressures and compositions on adjoining trays mutually affect each other, a stationary situation is only achieved when all trays are at steady state. Therefore, the dynamics can be lumped. However, for the static relationships of the deterministic model we still rely on tray-to-tray calculations. These calculations contain fit parameters such as the tray vapor efficiency. Therefore, an overall black-box model, which describes the static relationship between top composition as a function of the controls such as energy supply and reflux flow, and state variables of the bottom, could simplify the model substantially.

Based on physical considerations, a model structure can be derived which correlates the two dependent variables (vapor flow or production and top composition or quality) to the state variables (bottom composition and bottom contents) and the control variables (energy supply and the reflux). Fuzzy logic can be used to derive the description from the observations, without the need for rigorous tray-to-tray modeling.

The basic framework can be described by three balances: the mass balance, the component balance and the heating-up of the column at start-up. These balances are minimally needed to describe the long-term changes of the column behavior. The mass balance is:

$$\frac{dM_{col}}{dt} = -V(1 - R^*) \tag{30.13}$$

where R^* is the reflux fraction (see distillation chapter 16), V is the vapor flow and M_{col} the column mass holdup. The average column composition x_{col} is:

$$M_{col}\frac{dx_{col}}{dt} = -V(1 - R^*)(x_{n+1} - x_{col}) \tag{30.14}$$

The heating-up of the column can be described by the increase of the vapor flow:

$$\tau_V \frac{dV}{dt} = V_{ss} - V \tag{30.15}$$

In this equation V_{ss} is the steady state value of the vapour flow when the column is heated up en τ_V is a fit parameter. The most important input variable is the reflux (R^*). V_{ss} is constant since the heat supply is constant. The output variable is the top fraction (x_{n+1}).

Model I

The top composition x_{n+1} is described by a fuzzy relationship without dynamics:

$$x_{n+1} = f_{\text{fuzzy}}(R^*, x_{col}, V) \tag{30.16}$$

Model II

The top composition x_{n+1} is described by a fuzzy relationship:

$$x_{n+1,k+1} = f_{\text{fuzzy}}\left(x_{n+1,k}, R_k^*, x_{col,k}, V_k\right) \tag{30.17}$$

Dynamic behavior has been incorporated by a first-order AutoRegressive with exogenous input (ARX) structure in which the subscript k denotes the time step.

Model III

The first-order overall column dynamics can be described by:

$$\tau_{x,\,column}\frac{dx_{n+1}}{dt} = x_{n+1}^* - x_{n+1} \tag{30.18}$$

where x^*_{n+1} is the pseudo static top composition, that is described by a fuzzy relationship:

$$x^*_{n+1} = f_{\text{fuzzy}}\left(R^*, x_{col}, V\right) \tag{30.19}$$

The dynamic behavior is explicitly modeled by a first-order time constant. Figure 30.19 shows the complete model III.

The three hybrids models were built according to the procedure discussed in this chapter. Model characteristics are shown in Table 30.6.

For the identification three data sets and two validation sets were available, which all have different initial bottom holdup and compositions. The fuzzy model will again be identified using fuzzy clustering, for which input–output data are required.

Table 30.6. Fuzzy sub-model characteristics for distillation column.

dependent variables	independent variables	number of membership functions	number of parameters premise part	number of parameters consequence part	number of weighting parameters
$x_{top,static}$	R^*, V, x_{col}	3	$3 \times 3 \times 4 = 36$	$3 \times (3+1) = 12$	—

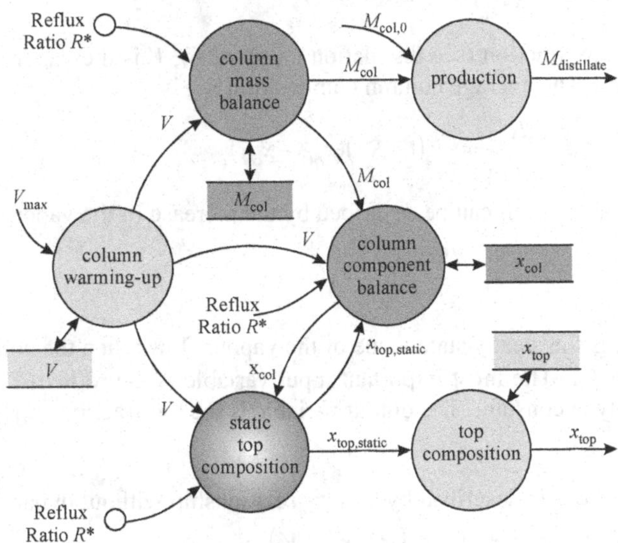

Fig. 30.19. Hybrid model of batch distillation column.

The inputs are available, the output is not available. Therefore, the output x^*_{n+1} is estimated using Eqn. 30.19 and a PI-estimator. The parameters of the estimator were determined manually. For each of the identification runs, different settings were obtained. Results of the estimates of x^*_{n+1} for one of the identification sets are given in Fig. 30.20; other results are similar. The corresponding estimates of x_{n+1} are also given. The estimates are acceptable, although the noise level is higher than in the measurements. This is the result of the filtering effect, which is represented by Eqn. 30.18.

During the batch the top composition is controlled by the reflux fraction. To test whether the column models behave like the process itself, the derived hybrid models are used to simulate batch runs using constant quality control.

For this purpose, a PI-controller is designed which controls the simulated product quality by manipulating the reflux fraction R^*. Similar to the experimental set up, the controller was switched on if the deviation was less than 0.002 from setpoint.

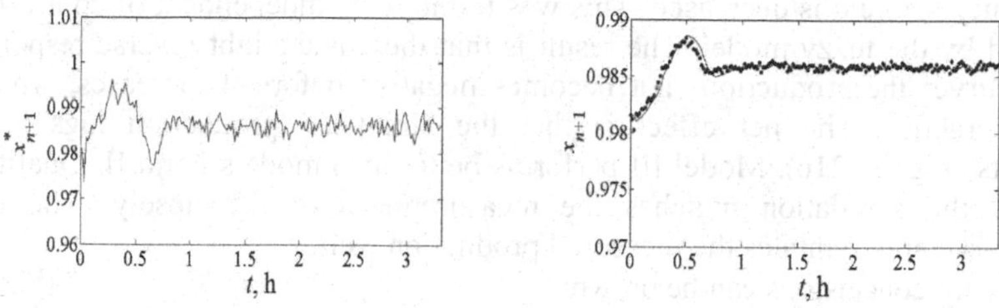

Fig. 30.20. PI estimation results of x^*_{n+1} and corresponding estimates of x_{n+1}, measurements (dots) and estimates (lines).

The setpoint was identical to the setpoint for the experimental batch runs and the controller was tuned manually. Figure 30.21 shows the results for a simulated batch run with initial conditions taken from one of the three batch runs. The results for the two validation sets were similar.

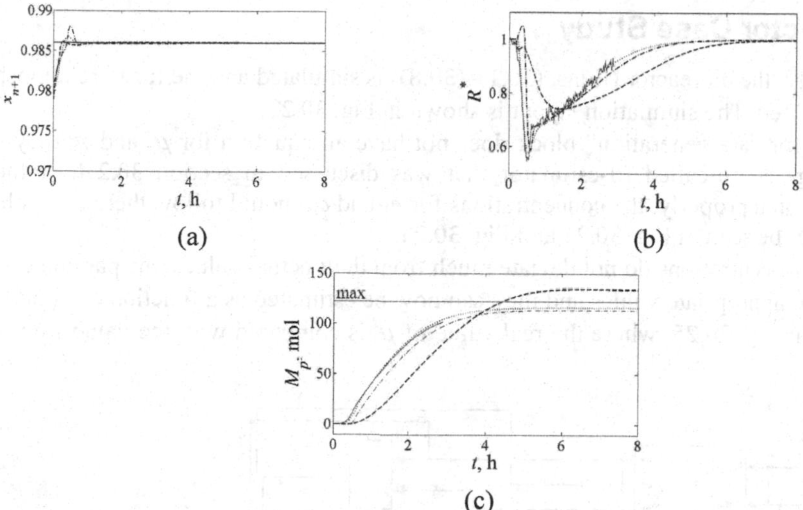

Fig. 30.21. Batch distillation simulation with constant quality control: quality control results (a), reflux fraction R^* (b), production (c). Measurements(solid line), model I (dashed), model II (dash-dot) and model III (dot).

The controller was able to achieve good setpoint control quickly for all models. However, the controller gain is negative for model I. This means that in order to increase product quality, the reflux fraction R^* is decreased. On average, the simulated reflux is lower than the measurements, which results in a production that is higher than the theoretical maximum (Fig. 30.21c). This behavior is caused by the fact that model I neglects information about the dynamics of the relationship between R^* and x_{n+1}, which is characterized by the dominant time constant (chapter 16). Although the model performs acceptably if measurements of the manipulated variable R^* are supplied, it performs unacceptable during simulation. To improve performance, additional information about the dynamics is required.

Compared with model I, model II performs considerably better during simulation (Fig. 30.21). The manipulated variable R^* follows the measurements closely, which indicates that the closed loop dynamics of the simulation approximates the actual experimental setup. However, when the controller is switched on, the reflux fraction is increased and becomes larger than one, before it is decreased. This was found to be independent of controller tuning and is caused by the fuzzy model. The result is that there is a slight inverse response in the production curve: the production first becomes negative before it increases, which is not possible in practice. The net effect is that the simulated production lags behind the measurements (Fig. 30.21c). Model III performs better than models I and II. Quality control is good and the simulation matches the measurements of R^* closely. The simulated production curve approximates the measured production well.

The following conclusions can be drawn:

- Information about the dynamic behavior is incorporated in the form of prior knowledge. The model, which incorporates the most dynamic prior knowledge, performs best.
- By separating the dynamic and static properties during the design of the data driven part of the hybrid model, the measurements are used more effectively.
- The comparison of the different model structures shows that with respect to product quality, the measurements of the controlled process only provide sufficient information to derive static characteristics.

30.4 Bioreactor Case Study

In this case study, the bioreactor (Eqns. (30.1)–(30.8)) is simulated and the fuzzy relationship for μ^* is determined. The simulation layout is shown in Fig. 30.22.

The "model for data generation" block does not have an equation for μ^* and π, they are estimated using the so-called PI-estimator that was discussed in section 30.2.4. If these values are estimated properly, the concentrations for c_X and c_P should follow there real values closely. This can be seen in Fig. 30.23 and Fig. 30.24.

Since these concentrations do not deviate much from their actual values, the parameters μ^* and π must have appropriate values and they can now be estimated as a function of c_X and c_S. This is shown in Fig. 30.25, where the real value of μ^* is compared with the value from the fuzzy model.

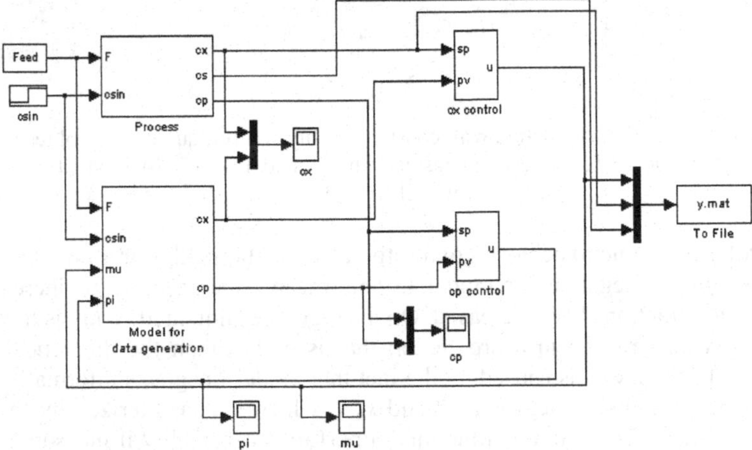

Fig. 30.22. Simulation block diagram for bioreactor.

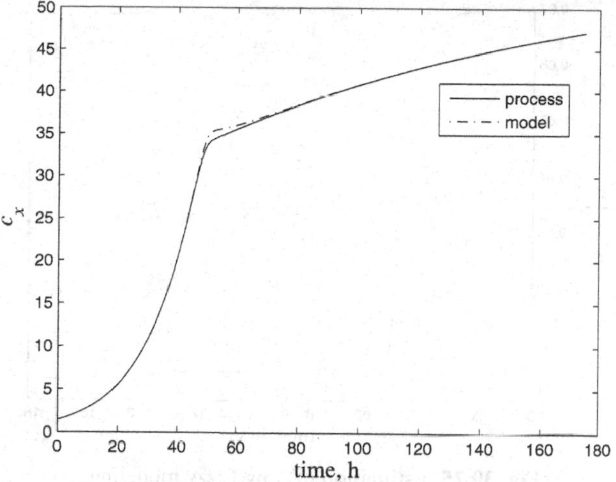

Fig. 30.23. Real and estimated concentration c_X.

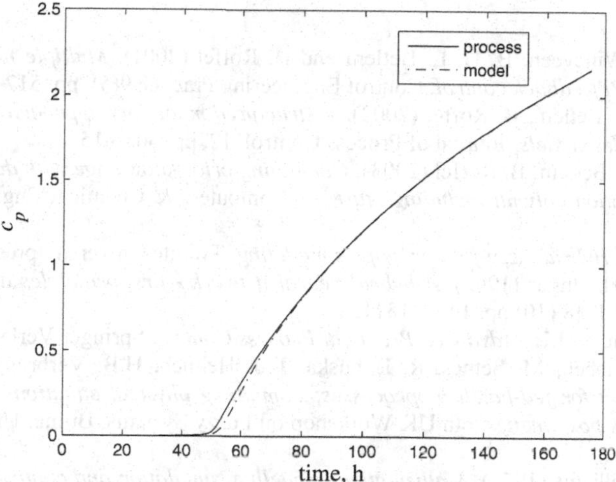

Fig. 30.24. Real and estimated concentration c_P.

If necessary, a more accurate model could be developed when multiple runs are made under different conditions, i.e. when more data points are available.

Files used in this chapter:
F3024.m: development of fuzzy model for μ
penilFB.mdl: penicillin process
penprocess.m: penicillin process
penmodel.m: model for penicillin process with μ and π estimated

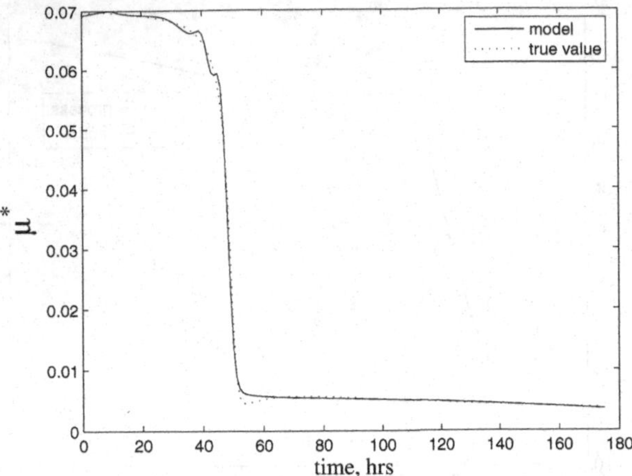

Fig. 30.25. Estimation of μ by fuzzy modeling.

Literature

Lith, P. F. van, H. Witteveen, B. H. L. Betlem and B. Roffel (2001). *Multiple nonlinear parameter estimation using PI feedback control.* Control Engineering Practice 9(5), pp. 517–531.

Lith, P.J. van, B.H.L. Betlem, B. Roffel (2002). *A structured modelling approach for dynamic hybrid fuzzy-first principles models*, Journal of Process Control, 12, pp. 605-615.

Lith, P.J. van, B.H.L. Betlem, B. Roffel (2003). *Combining prior knowledge with data driven modeling of a batch distillation column including start-up*, Computers & Chemical Engineering, 27(7), pp. 1021–30.

Lith, P.J. van (2002). *Hybrid fuzzy-first principles modeling.* Twente University press, Enschede.

Psichogios, D.C., L.H. Ungar (1992). *A hybrid neural network - first principles approach to process modelling*, AIChE J. 38 (10) pp. 1499–1511.

Roffel, B. and Betlem, B.H.L., *Advanced Practical Process Control*, Springer Verlag, 2003.

Roubos, J.A., P. Krabben, M. Setnes, R. Babuska, J. J. Heijnen, H.B. Verbruggen (1999). *Hybrid model development for fed-batch bioprocesses; combining physical equations with the metabolic network and black-box kinetics*, 6th UK Workshop on Fuzzy Systems, Brunel University, Uxbridge, UK, pp. 231–9.

Smith, C.C., T. J. Williams (1974). *Mathematical modeling, simulation and control of the operation of kamyr continuous digester for kraft process*, Tech. Report 64, Purdue University, PLAIC.

Thompson, M.L., M.A. Kramer (1994), Modeling chemical processes using prior knowledge and neural networks, AIChE J. 40 (8) pp. 1328–40.

31 Introduction to Process Control and Instrumentation

In this chapter the goal for controlling the process will be discussed first, followed by a review of different elements of the control loop. Subsequently, different types of measuring device and their associated dynamics, as well as the characteristics of the correcting device, will be briefly explained. The discrete and continuous controller equation and a simulation of a control loop will be discussed.

31.1 Introduction

A standard feedback control loop consists of several elements, as shown in Fig. 31.1.

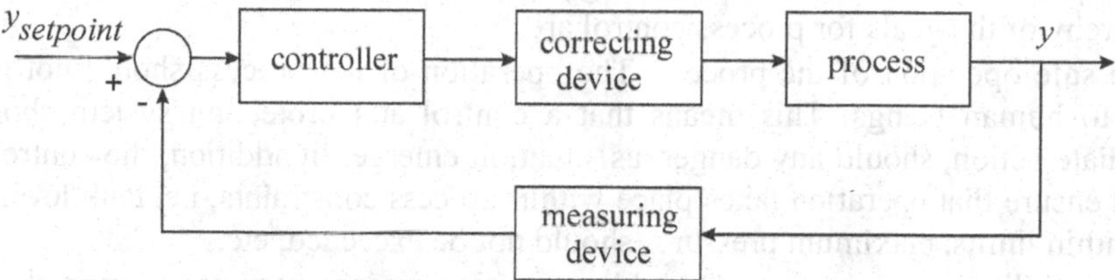

Fig. 31.1. Standard feedback control loop.

The process output can be a flow, level, concentration or another type of measurement. The sensor or measuring device to measure the value of the process output variable can therefore be very different. Whether the dynamics of the measuring device should be taken into account when analyzing the control loop depends on major time constant of the sensor compared with the major time constant of the process. In a subsequent section, different types of measuring devices will be briefly reviewed.

In a feedback control loop, the value of the process measurement is usually compared with a target value or setpoint and subtracted from it. The difference or error serves as input to a controller. This is often a proportional-integral or proportional-integral-derivative controller. The controller calculates a change of the signal to the control valve or in a more general sense the signal to the correcting device. The correcting device adjusts in most cases a process flow. In a single-loop controller one measured variable is controlled by one manipulated variable. The performance of the loop depends on the gain, linearity and dynamics of all elements.

Suppose the process is a reactor with a cooling jacket, as shown in Fig. 31.2.

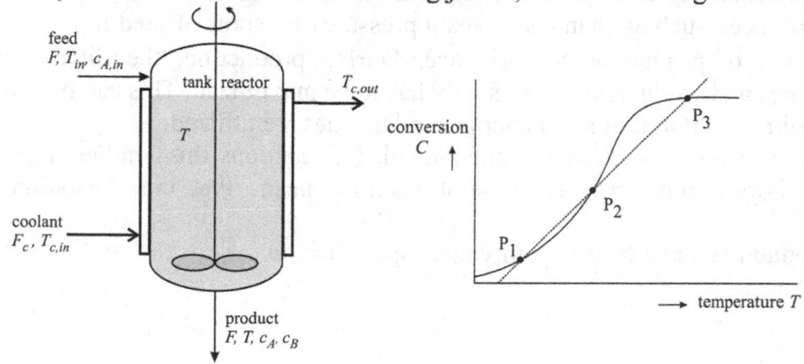

Fig. 31.2. Stability of reactor with cooling jacket.

Process Dynamics and Control: Modeling for Control and Prediction. Brian Roffel and Ben Betlem.

One possibility of maintaining the reactor temperature is by adjusting the coolant flow through the jacket. In this case the measuring device could be a thermocouple, protected by a thermowell. As explained in chapter 12, there are three possible operating points of which only two are stable. Operating point P_2 is an unstable operating point, the reactor can only be kept in this point by a temperature controller.

There are several types of control valve; the type to be used depends on the characteristics of the process and characteristics of the measuring device. This will be discussed in a subsequent section.

Before discussing the separate elements of the control loop in more detail, first the goals of process control will be discussed.

31.2 Process Control Goals

The development of a control scheme strongly depends on the goals as set by the designer. The hierarchy of the goals for process control are:

- Ensure safe operation of the process. The operation of the process should not pose any threat to human beings. This means that a control and protection system should take immediate action, should any dangerous situation emerge. In addition, the control system should ensure that operation takes place within process constraints, i.e. tank levels should stay within limits, maximum pressures should not be exceeded, etc.
- Ensure stability of the process. Unstable processes are not very common in the process industries. An example is an exothermal reactor in which the reactor temperature increases as a result of an increase in cooling water temperature. The heat production, which depends exponentially on the temperature, increases more than can be removed by the heat exchanger. The heat removal depends linearly on the temperature. Consequently, the reactor temperature will continue to increase. Without temperature control, stable operation cannot be realized (Fig 31.2). There are two stable operating points, P_1 and P_3. Without control, the reactor cannot be kept at operating point P_2 (see also chapter 12).
- Suppress disturbances. A process does not usually stay in the same operating point because disturbances have an impact on the operation. The disturbances can be external disturbances, such as changes in process throughput, changes in feed composition, cooling temperature or even weather conditions. Internal disturbances often have a long-term effect, such as catalyst decay or equipment fouling. Control has to ensure that process conditions are constant by compensating the disturbances. Fig. 31.3 shows a reactor temperature control scheme that not only ensures that the reactor operation is stable, in addition it ensures constant product specifications when the quality is dependent on the temperature (in many cases it is), it keeps reactor operation within a small operating range and disturbances, such as changes in steam pressure) are compensated for.
- Optimization of production performance. During production, the difference between product sales and production costs usually has to be maximized. This can be achieved if:
  - the minimum amount of raw materials and utilities are utilized;
  - optimal process conditions are maintained, for reactions this implies that conversion and selectivity is maximal, for separations this means that the separation energy is minimal;
  - the production quantity and quality meet specification.

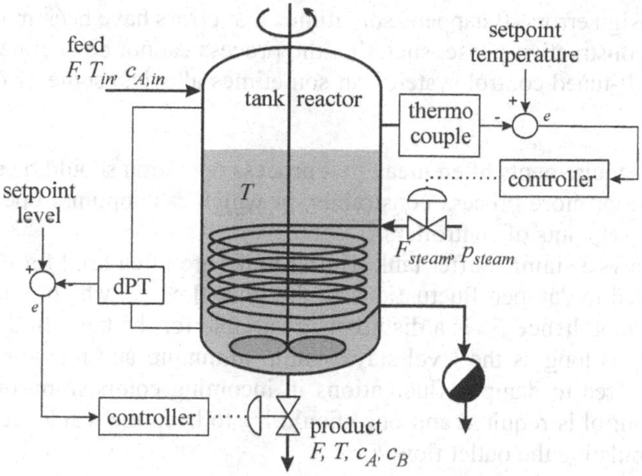

Fig. 31.3. Reactor operation to ensure stability and suppress disturbances.

Suppose a reaction A → B → C takes place in the reactor, the concentration profiles for different values of the product of rate constant and residence time could look like the ones shown in Fig.31.4.

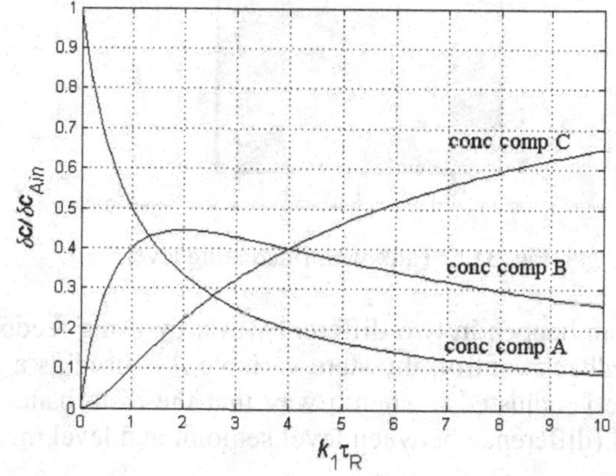

Fig. 31.4. Goal of control demonstrating performance optimization.

C is an undesired byproduct and B is the main product. Optimization of production performance in this case means controlling the reactor such that the concentration of B is maximized; however, there could be a limitation on the amount of undesirable byproduct and/or energy requirement. Optimization of production performance would mean in this case the definition of the optimal setpoint for the concentration of B in a higher-level control strategy that would download a setpoint to a composition controller, which uses the residence time of the reactor.

- Account for design errors. It happens sometimes that errors have been made in the process design and/or construction phase, such that the process cannot be operated the way it was intended. A well-tuned control system can sometimes alleviate some of the consequences of this situation.

The foregoing requirements often mean that process operation should be maintained at the intersection of one or more process constraints, in which case optimal operating conditions are translated into setpoints of controllers.

Figure 31.5 shows a simple buffer tank. Buffer tanks are often used for different purposes It could be required to dampen fluctuations in the inlet flow F_{in} which is determined by an upstream process unit, hence F_{in} is a disturbance variable for the tank. In this case not much control is needed, as long as the level stays within minimum and maximum boundaries. It could also be required to dampen fluctuations in incoming composition or temperature. In this case, level control is required and one should try to keep the residence time in the tank constant by manipulating the outlet flow F_{out}.

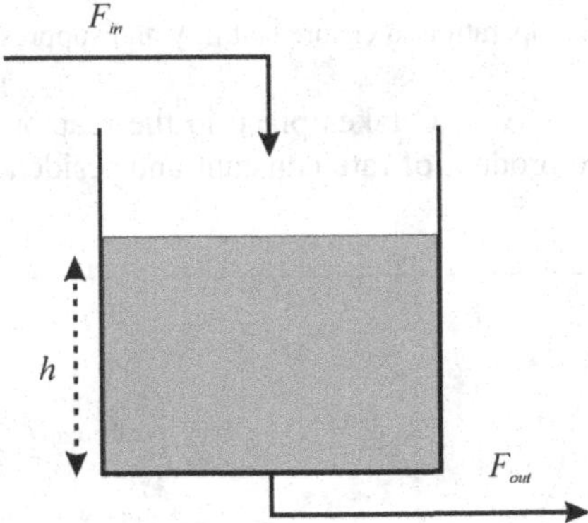

Fig. 31.5. Tank with fluctuating level.

Control of the level can happen in two different ways: by using feedback or feed-forward control. When using feedback control, the state variable h is used as a feedback signal and the manipulated variable is adjusted in such a way that the disturbance is compensated for and the error in the level (difference between level setpoint and level measurement) becomes zero (Fig. 31.6).

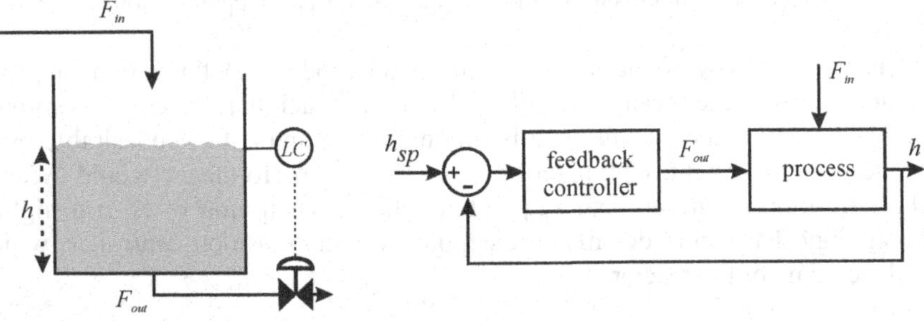

Fig. 31.6. Feedback control of level.

The disadvantage of this type of control is that action is taken only after the level deviates from its setpoint; however, this is not a problem for level control. In the case of feed-forward control the disturbance is measured and immediate action is taken to compensate the disturbance (Fig. 31.7).

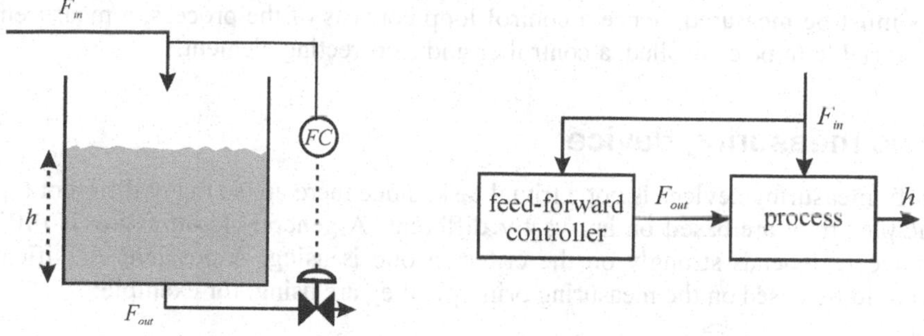

Fig. 31.7. Unfeasible feed-forward control of level.

This type of control is fast, but because there is no feedback of the state variable (level) there may be considerable difference between the measured level and its setpoint (offset). The controller model should, however:

- take all possible disturbances into account
- be very accurate
- be able to measure the disturbances accurately
- and adjust the manipulated variable exactly

Suppose a small error occurs in adjusting the flow F_{out}. After some time the tank will go empty or overflow, since the inlet flow and outlet flow do not match and there is no feedback of the accumulating variable, the level. This is the reason that feed-forward control is always used in combination with feedback control (Fig. 31.8).

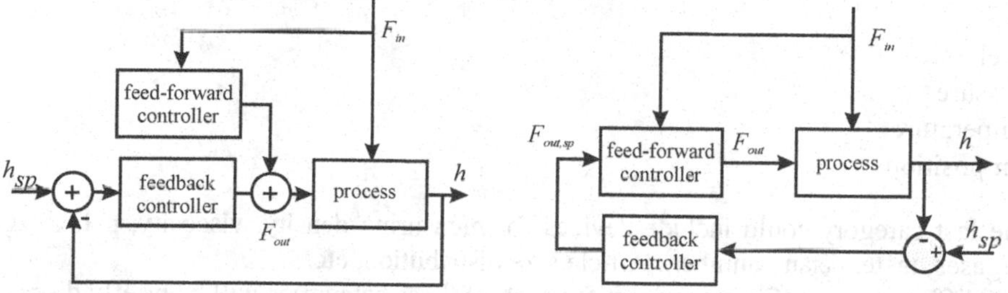

Fig. 31.8. A. Combination of feed-forward and feedback control, B. disturbance correction control.

Feed-forward control takes quick action on any disturbance while feedback ensures that any persisting or increasing offset in the level is taken care of. Feed-forward control is often used in case of slow changing processes where the state variable reacts slowly on a disturbance. This happens, for example, in distillation columns with many trays, which are subject to large disturbances (for example feed changes from furnaces). Most process units, however, can be well controlled by using feedback control only.

An important goal of process control is therefore, to ensure that state variables that are not self-regulating (such as the level in the tank) stay within certain limits. Most process conditions that are characterized by temperatures and pressures are also state variables. In the latter case, however, these variables are often self-regulating, i.e. when a disturbance acts upon the process, the state variables assume new steady-state values. State variables describe the dynamic behavior of the process. The values the state variables assume are a result of the adjustable variables (controls) and the disturbance variables. Control in a narrow sense is compensation of the influences of disturbances. In order to use process controllers, the state variables must be measured, hence a control loop consists of the process, a measurement of the state variable to be controlled, a controller and a correcting element.

31.3 The measuring device

To classify measuring devices is not a trivial task, since there are so many different types and the principles they are based on are vastly different. A general classification is difficult to make, since it depends strongly on the criterion one is using. A possible classification of sensors could be based on the measuring principle they are using, for example:
- thermal
- mechanical
- chemical
- electrical
- magnetic
- optical
- radiation

Even this classification is hard to complete, since there are always sensors that do not fall in one of the above categories.

Another classification one could make is by looking at the medium the sensors are used for, for example gases, liquids, slurries, powders, etc.

Another way of looking at sensors or measuring devices is from an industrial point of view: which sensors are very common and used mostly in the process industries. The five types that are most widely used are measurements for:
- flow
- level
- pressure
- temperature
- composition

The last category could include devices for measuring density, viscosity, pH, oxygen in stack gases, taste, octane number, particle size distribution, etc.

The different types of measurement for each of these categories will be briefly discussed.

31.3.1 Flow measurement

Volumetric flow can be measured in a variety of ways, amongst others, by using:
- orifice plates
- a venturi tube
- target flow meter
- turbine flow meter

An example of a mass flow measuring device is the Coriolis flow meter.

The most common type of measurement makes use of orifice plates, as shown in Fig. 31.9.

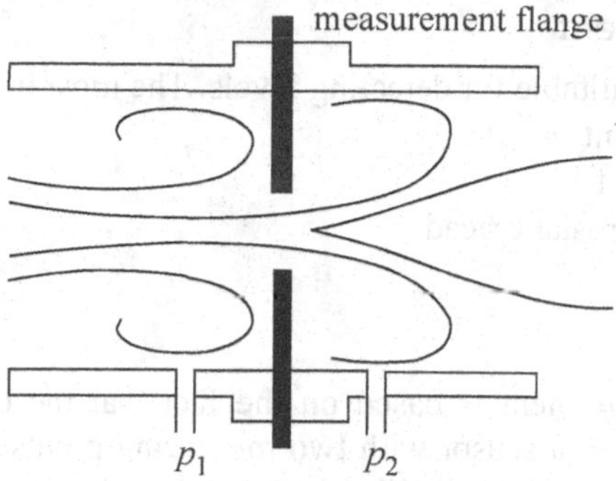

Fig. 31.9. Flow measurement by using orifice plates.

The equation for measuring the flow is:

$$\Delta p = p_1 - p_2 = \alpha \frac{1}{2} \rho v^2 \qquad (31.1)$$
$$F = \alpha^* \sqrt{\Delta p}$$

in which α^* is a constant, Δp the pressure drop and F the flow. The pressure difference is measured by a pressure difference sensor, which is discussed in section 31.3.3. The measurement is usually fast, hence it is not required to take dynamics for the device into account. It can also be seen that the flow does not vary linearly with the measured pressure drop, a square root function is required to create a linear flow measurement. The measurement is sensitive to composition and temperature, which change the density and viscosity of the fluid.

Another type of flow measurement that is sometimes used is a venturi tube, shown in Fig. 31.10. The measuring principle is the same as in the case of an orifice measurement, a restriction creates a pressure drop which is a measure for the flow. The advantage of a venturi tube over an orifice measurement is that the pressure drop is usually lower.

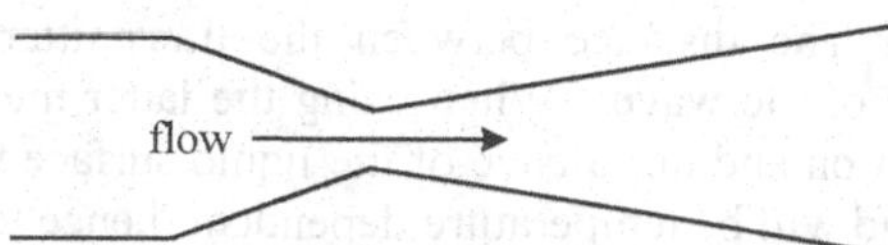

Fig. 31.10. Venturi type flow measurement.

Flow can also be measured using target flow meters. A small obstacle is inserted in the flow direction in a pipe and the force on this obstacle is proportional to the flow through the pipe.

Another common device for flow measurement is the magnetic flow meter. The meter generates a voltage as a result of flow through a magnetic field, this signal is proportional to the flow.

Turbine flow meters are used, especially in clean, non-corrosive environments. The meter produces pulses proportional to the fluid velocity.

In many applications, such as monomer addition to polymerization reactors, it can be desirable to measure the mass flow, rather than the volumetric flow. The so-called Coriolis flow meter uses the Coriolis effect to measure the amount of mass moving through the element. The flow runs through a U-shaped tube that vibrates in a perpendicular direction to the flow. Fluid forces running through the tube interact with the vibration, causing it to twist. The greater the angle of the twist, the greater the flow.

31.3.2 Level measurement

There are several devices suitable for detecting levels. The most important ones are:

- conductivity measurement
- capacitance measurement
- measurement of the hydrostatic head
- radar measurement
- ultrasonic measurement

The conductivity measurement is based on the fact that the conductivity of a liquid is different to that of air. When a sensor with two measuring points measures the conductivity of air, a low value will be measured. When the two measuring points are submerged in a liquid, the conductivity will be high. This type of measurement is therefore particularly suited for low and high level indication and to a lesser extent for a continuous level measurement.

The capacitance measurement is usually performed in such a way that the wall of the vessel acts as one conductor and a measuring probe as a second conductor. The capacitance between the two conductors is:

$$C = P_a KA / d \tag{31.2}$$

where C is the capacitance, P_a the absolute permittivity, K the dielectric constant of the medium, A the area between the conductors and d the distance. As the tank is filled, the area between the two conductors changes and hence the value of the capacitance changes. The dielectric constant for liquids varies considerably: for water it is about 80.0, for oil about 2.0 and for alcohol about 20.0.

Another type of level measurement makes use of radar or microwave signals. In this case the sensor is located at the top of the vessel in which the liquid height should be measured. A sensor outputs a frequency-modulated signal from 0–200 Hz. The signal that is reflected by the liquid surface is delayed in proportion to the distance between the level and the surface.

A similar measurement is the ultrasonic measurement; in this case the signal varies in the range from 20–200 kHz. The distance between the transmitter and the liquid surface determines the travel time of the waves. When using the latter measurement one should be aware of the fact that foam on and turbulence of the liquid surface will affect the reading. In addition, the speed of sound will be temperature dependent; hence temperature compensation may be required for an accurate measurement.

31.3.4 Pressure measurement

Pressure measurements, flow measurement through pressure difference measurement and level measurements by electrostatic head are closely related. They all use a pressure difference measurement. An example is the level measurement in which a diaphragm is used. A strain gauge could be used to detect the deflection of the diaphragm due to the pressure differential at its two sides.

Owing to the pressure difference p_1–p_2, the diaphragm is being displaced and the position x of the diaphragm is related to the pressure difference according to:

$$\frac{Al\rho}{g}\frac{d^2x}{dt_2} + c\frac{dx}{dt} + Kx = A\Delta p \tag{31.3}$$

in which A = capillary area, c = damping coefficient of the liquid, g = gravity constant, K = Hooke constant for the diaphragm, l = length of the capillary tube (from the measuring point p_1 to the diaphragm) and ρ = liquid density. As can be seen, the dynamics are second order.

The first term with the second-order derivative is, however, usually small in practice, thereby reducing the dynamics to first order. If this type of measurement is combined with the flow measurement using an orifice, this dynamic behavior has to be added to Eqn. (31.1). As Fig. 31.11 shows, the deflection of the diaphragm causes a change in capacitance that is detected by a bridge circuit. In this case a high-frequency, high-voltage oscillator is used to charge sensing electrode elements.

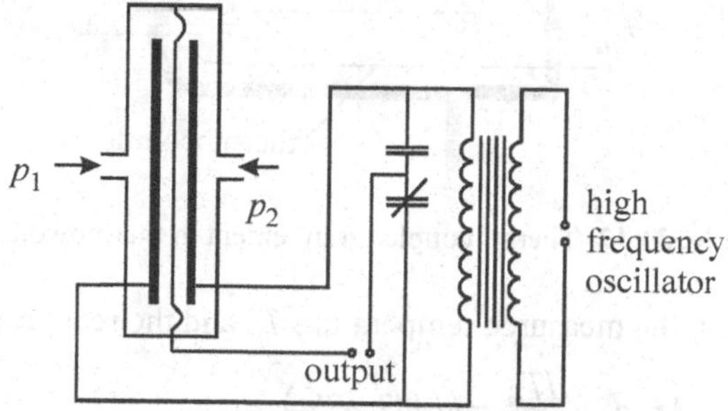

Fig. 31.11. Capacitance based pressure sensor.

A potentiometric pressure sensor is another simple but effective pressure sensing device. In this case, the wiper arm of the potentiometer is linked to the diaphragm element.

A newer type of pressure sensor is the resonant-wire pressure transducer. The wire is fixed at one end, the other end of the wire is linked to the diaphragm. An oscillator circuit ensures that the wire oscillates at its resonant frequency. When the diaphragm changes position, the tension in the wire changes, which in turn causes a change in resonance frequency. This can then be translated into a pressure change.

A pressure sensor which is well suited to measure dynamic pressure changes is a quartz crystal (piezoelectric measurement). A charge across the crystal is proportional to the force on the crystal. This force results in a deformation which causes a subsequent short lasting change in flow of electric charge. These sensors are therefore not suited to measure static pressures.

Some pressure transducers make use of magnetic phenomena. In these transducers, a pressure change produces a movement, as in the case of a diaphragm, which is translated into the change in inductance of an electric circuit.

The dynamics of pressure measurements is similar to the dynamics of level measurements.

31.3.3 Temperature measurement

The three most important types of temperature measurement are the thermocouple measurement, resistive temperature and infrared measurement. In a thermocouple, two wires of different material are joined together at both ends (junctions). A temperature leads to a change in electromotive force between the two wires. In control systems the reference junction is usually located at the electromotive force measuring device. The temperature change is often fairly linearly related to the electromotive force.

Another type of temperature measurement is the so-called resistive temperature device. Measurement is based on the fact that the electrical resistance of a material changes with temperature. If a metallic device is used, the change in resistance varies linearly with temperature; if a thermistor or ceramic semiconductor is used the resistance varies non-linearly with temperature. A common example is a PT-100 element.

Infrared temperature measuring devices rely on the fact that a varying amount of thermal radiation is emitted by a material at varying temperatures.

In the process industries often a thermocouple or resistive temperature device is used. However, submerging it directly into the gas or liquid could damage the device. Therefore often a thermowell, as shown in Fig. 31.12, is used for protection.

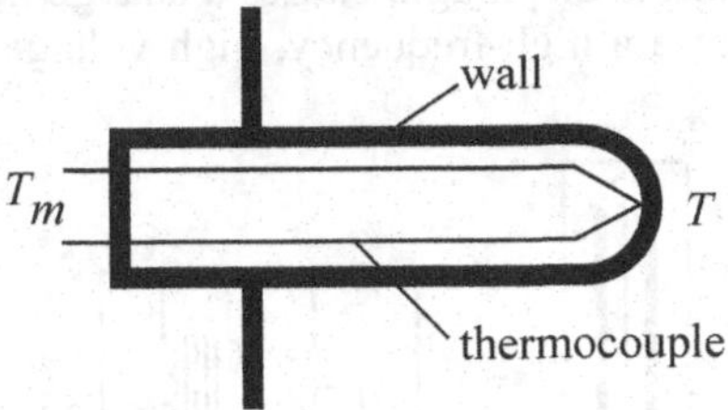

Fig. 31.12. Thermocouple arrangement in thermowell.

The relationship between the measured temperature T_m and the real process temperature T is:

$$M_w c_{p,w} \frac{dT_m}{dt} = UA\left(T - T_m\right) \tag{31.4}$$

where M_w = mass of the wall material, $c_{p,w}$ = specific heat wall material, UA = product of heat transfer coefficient and heat transfer area. The ratio $M_w c_{p,w} / UA$ defines the time constant of the system. Usually the thermowell is filled with liquid, in this case it is assumed that the mass of the liquid with which the thermowell is filled is small compared with the mass of the thermowell. If this is not the case the dynamics will become second order. In many industrial applications the time constant can be considerable, of the order of 10 … 30 seconds.

31.3.4 Quality measurement

Quality measurement is a totally different issue than flow, level, temperature or pressure measurement. The primary reason is that there are so many different quality characteristics that can be measured and they are vastly different with respect to the type of measurement and measuring principles. Just to name a few examples of quality measurements: viscosity, odor, color, taste, pH, octane number, particle size distribution, etc. In traditional chemical engineering one type of measurement that is often encountered is composition analysis by means of a gas chromatograph. Because these devices are expensive they are usually used to analyze multiple streams. To give an example, it may be required to analyze the feed flow composition, top and bottom composition of a distillation tower. If one analysis takes a few minutes (residence retention time), and three streams are analyzed it is obvious that composition measurement cannot be realized without a delay or dead time. This is characteristic for numerous quality measurements: they often have a measurement delay associated with them, which can vary from a few minutes to tens of minutes. In control this will pose a problem and a dead time compensation technique may be required to achieve adequate control loop performance.

Many measurements exist that cannot be measured online so samples have to be analyzed in the laboratory offline. It is obvious that this also introduces a dead time in the measurement, which is detrimental for control.

A delay is not present in all quality measurements. Consider, for example, a pH measurement. The response of a pH electrode to a change in pH can usually be characterized by a second-order response with a small time constant of the order of a few seconds and a larger time constant of the order of 5–30 seconds, depending on the type of electrode. However, the most important characteristic of the electrode is that it gives a non-linear response: the time constants for a positive change in pH and negative change in pH can be considerably different.

For control loop performance it is essential to have some understanding of whether the measurement has linear characteristics and whether it introduces additional first- or second-order dynamics or even a dead time in the control loop. When testing process dynamics it is hard, however, to distinguish between pure process dynamics and measurement dynamics, because we observe the measured system variable in which the measurement dynamics are already accounted for.

Table 31.1 summarizes the most frequently used types of measurements in the process industries.

31.4 The control device

In most cases the correcting device to which the controller outputs will be a control valve. Small valves can be pneumatic or electric, larger valves are usually operated by air pressure. This may introduce additional dynamics for the correcting element, which could be approximated by a first-order time constant.

There are essentially two types of control valve: a fail open and fail closed type. They are shown in Fig. 31.13.

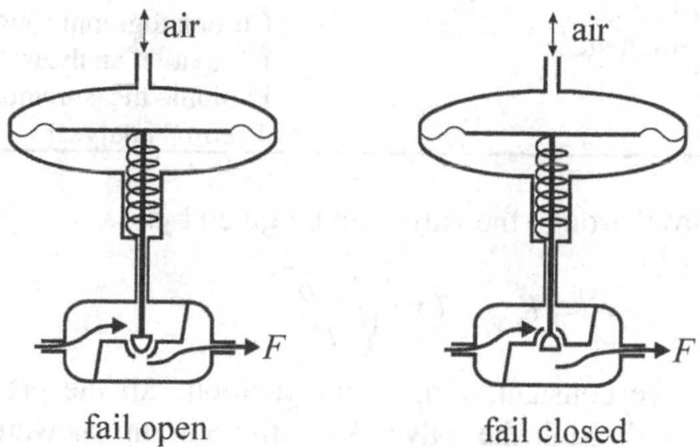

Fig. 31.13. Different types of control valve.

Which type of valve one chooses is a safety issue, if upon a major failure the valve should go to the open position (for example cooling water) or closed position (for example reactants) should be judged by the engineer. If, for example, the valve is used at the outlet of a tank, we may want to close the valve to prevent all the liquid in the tank from being drained.

Table 31.1. Frequently used measurements in the process industry.

Flow measurement	**Level measurement**
Differential pressure measurement: Orifice plates Flow nozzle Rotameters Target meters Venturi tubes *Mechanical flow measurement:* Turbine flow meter *Electronic flow measurement:* Vortex flow meter Magnetic flow meter Ultrasound *Mass flow measurement* Coriolis flow meter Hot-wire anemometer Thermal mass flow meter	*Differential pressure measurement:* Hydrostatic head *Electrical/other level measurement:* Conductance Radar Ultrasonics di-electric **Pressure measurement** *Differential pressure measurement:* Hydrostatic head *Electrical/magnetic measurement:* Capacitance measurement Piezo-electric element Magnetic transducer
Temperature measurement Thermocouples Bimetal thermometer Resistance thermometer Radiation pyrometer Infrared measurement	**Quality measurement** Conductivity pH Viscosity Infrared analyzer Mass spectrometer Radiation analyzer Chromatographic analyzer Ultraviolet analyzer Photonic measurement Thermal analyzer

The equation for flow through the valve can be given by:

$$F = K_v\, f(x) \sqrt{\frac{\Delta p}{\rho}} \tag{31.5}$$

in which K_v is the valve constant, x the stem position, Δp the pressure drop, ρ the fluid density and F the flow through the valve. K_v is the amount of water that flows through a totally opened valve at a pressure difference of 1 bar and 20 °C. The geometry of the combination of the valve seat and plug determines the valve characteristics.

The signal to the air valve (or stem position) does not necessarily vary linearly with the flow through the valve. Valves can have different characteristics, as shown in Fig. 31.14 left.

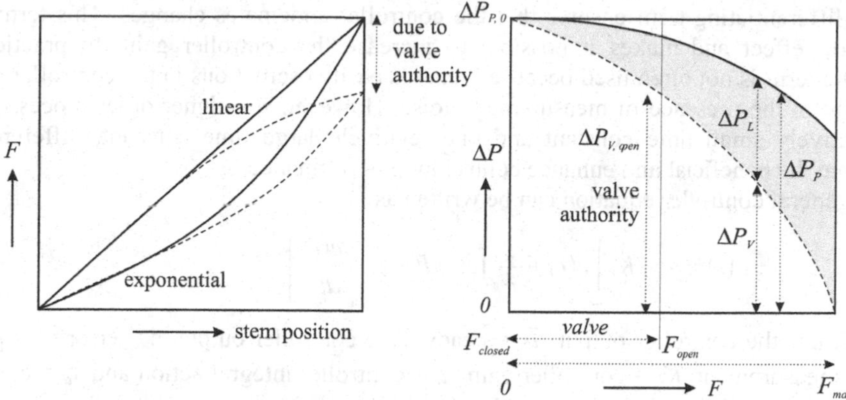

Fig. 31.14 Valve characteristics.

There essentially two major valve types: one with a linear characteristic and one with an exponential characteristic. One should keep in mind, however, that a linear valve may not behave linearly when installed in a pipeline circuit. The flow rate is not determined solely by the valve characteristics, but also depends on the resistance of the pipeline and the pump characteristics. It is therefore essential to evaluate the entire circuitry in which the valve is installed. Figure 31.14 shows how the pump pressure varies with the flow through the system. At $F = 0$ the pump head is $\Delta P_{P,0}$. The maximum flow without valve is F_{max}. The range of the valve covers only a part of the total possible range and varies from F_{closed} to F_{open}.

When a valve opening increases, three accumulative effects will occur:

- the pressure difference over the valve ΔP_V will decrease, according to Eqn. (31.5),
- the pressure difference over the pipe line ΔP_L will increase according to $\Delta P_L = \rho C_L F^2$, with C_L the resistance constant of the pipe line,
- when the flow loop contains a pump, the pump head ΔP_P will drop according to the pump characteristic which can usually be described by $\Delta P_P = \Delta P_0 - C_P F^2$; this has the same effect as an extra pipe line resistance.

The ratio between the pressure drop through the open valve and the total pressure drop can be expressed by the valve authority:

$$valve\ \ authority = \frac{\Delta P_{v,open}}{\Delta P_{P,0}} = 1 - \frac{\Delta P_{P,0} - \Delta P_{v,open}}{\Delta P_{P,0}} = 1 - \left(\frac{F_{open}}{F_{\max}}\right)^2 \tag{31.6}$$

When F_{open} is small compared with F_{max}, then a linear valve will offer linear flow operation. However, when F_{open} is relatively large, an exponential valve is required to compensate the pressure drop.

31.5 The Controller

The most commonly used controller in the process industry is a PI(D) controller, a proportional, integral and differentiating controller. The proportional term calculates a correction proportional to the error between setpoint or desired process output value and measurement. The integrating term ensures that the long-term error approaches zero. This term guarantees that there will be no offset, i.e. the process output will be on setpoint. Without this term the error can never be reduced to zero, except when coincidentally $u(t) = \overline{u}$.

The differentiating term ensures that the controller anticipates changes. This term has a stabilizing effect and makes it possible to increase the controller gain. In practice, the derivative term is not often used because it causes easily oscillations in the controller output, especially in the presence of measurement noise. However, for higher-order processes with one relatively small time constant and one relatively large time constant, differentiating action may be beneficial and enhance control loop performance.

The general controller equation can be written as:

$$u(t) = \bar{u} + K_C \left[e(t) + \frac{1}{\tau_i} \int e(t)dt + \tau_d \frac{de(t)}{dt} \right] \tag{31.6}$$

in which u = the controller output, $\bar{u}$ = steady-state controller output, e = error = setpoint – process measurement, K_c = controller gain, τ_i = controller integral action and τ_d = controller derivative action. It can easily be seen that for the proportional controller τ_i should approach infinity and τ_d approach zero. For a proportional-differential controller τ_i should be infinity.

In steady state, when $e(t)$ = 0, the control action of a proportional integral controller becomes:

$$u(t) = \bar{u} + \frac{K_C}{\tau_i} \int e(t)dt \tag{31.7}$$

31.6 Simulating the controlled process

In this section an example will be given of a simulated control loop. Let us assume that the process can be described by the following continuous transfer function:

$$\frac{\delta y}{\delta u_p} = \frac{1.2e^{-3s}}{1+20s} \tag{31.8}$$

where y is the process output and u_p the process input.

The measurement has first-order dynamics and can be given by:

$$\frac{\delta y_m}{\delta y} = \frac{1}{1+3s} \tag{31.9}$$

All time constants are given in minutes. The controller is a PID controller, which executes every minute. To simulate a discrete PID controller, Eqn. (31.6) should be differentiated and the derivatives be replaced by a backward difference, which yields:

$$u_k = u_{k-1} + K_c \left(e_k - e_{k-1} + \frac{\Delta t}{\tau_i} e_k + \frac{\tau_d}{\Delta t}(e_k - 2e_{k-1} + e_{k-2}) \right) \tag{31.10}$$

where Δt = controller execution interval. In discrete z-notation, the controller can be written as:

$$\frac{u_k}{e_k} = K_c \frac{\left(1 + \frac{\Delta t}{\tau_i} + \frac{\tau_d}{\Delta t}\right) - \left(1 + 2\frac{\tau_d}{\Delta t}\right) z^{-1} + \frac{\tau_d}{\Delta t} z^{-2}}{1 - z^{-1}} \tag{31.11}$$

It is further assumed that the correcting element has linear characteristics and that the control valve equation can be given by:

$$u_p = 0.2u \tag{31.12}$$

Figure 31.15 shows the block diagram of the controlled process.

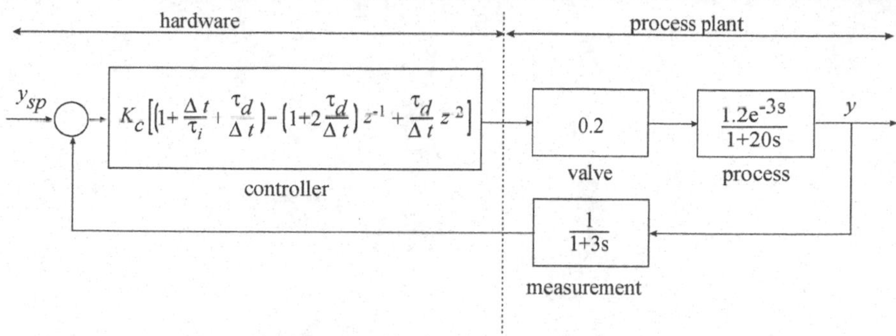

Fig. 31.15. Block diagram of simulated control loop.

This control loop is simulated and stored in F3116.mdl. The response to a setpoint change from zero to one is shown in Fig. 31.16.

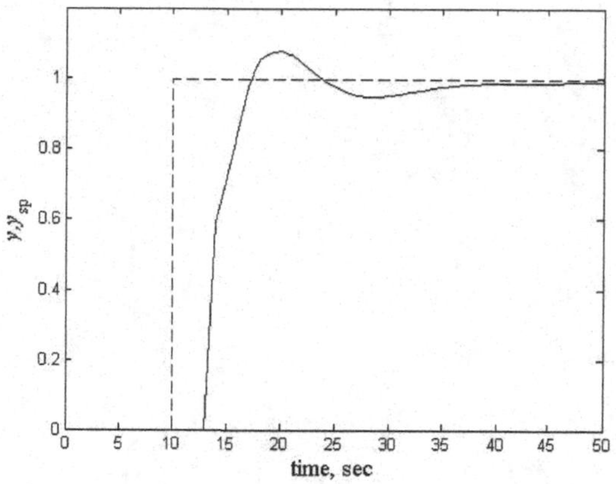

Fig. 31.16. Response of process output to setpoint change.

Files referred to in this chapter:
F3116.mdl: control structure to generate Fig. 31.16
F3116plot: plot file for Fig. 31.16

References

Childs, P.R.N., (2001) *Practical Temperature Measurement*, Butterworth-Heinemann.

Considine, D.M., 1993 *Process/Industrial Instruments and Controls Handbook*, 4th edn, McGraw Hill, Singapore.

Doebelin E.O. (1992) *Measurement Systems: Application and Design*, 4th edition , McGraw Hill, New York.

Gillum, D. (1995) *Industrial Pressure, Level and Density Measurement*, ISA Press, NewYork.

Kerlin, T.W. (1999) *Practical Thermocouple Thermometry*, ISA Press, NewYork.

Liptak B. G. (1995) *Process Measurement and Analysis*, 3rd edition, Chilton Book Company, Radnor, Pennsylvania.

Liptak, B. (2003) *Instrument Engineers' Handbook*, CRC Press.

Noltingk B.E. (1996) *Instrumentation Reference Book*, 2nd edn, Butterworth Heinemann, Oxford.

Patranabis, D. (1997) *Principles of Industrial Instrumentation*, 2nd edition, Tata McGraw Hill, New Delhi.

Smith, E. (1984) *Principles of Industrial Measurement for Control Applications*, ISA Press, NewYork.

Sohlberg, B. (1998). *Supervision and Control for Industrial processes*. Springer Verlag, Berlin.

Spitzer, D.W., 2001, Flow measurement: practical guides for measurement and control, ISA.

32 Behaviour of Controlled Processes

The purpose of a controller is to keep the controlled variable at its setpoint or bring it to setpoint. The determination of the behaviour of the controlled system is essentially not different from the determination of the behaviour of the uncontrolled system. The controlled behaviour depends strongly on the controller parameters and the type of controller that is used. Tuning of the controller parameters should lead to a matching of the behaviour of the controlled process to desired and/or specified behaviour for changes in setpoint or disturbances that act on the process.

32.1 Purpose of Control

In this chapter it will be assumed that the transfer function of the control valve is ideal, i.e. the gain is equal to one and the dynamics can be ignored. In addition, it is assumed that the transfer function of the measuring device is incorporated in the process transfer function. In that case, as can be seen from Fig. 32.1, the response of the output variable of a process for a change in the disturbance variable or control variable can be given by:

$$\delta y = G_P(s) \cdot \delta u + G_D(s) \cdot \delta w \tag{32.1}$$

in which δ is a variation around the steady state.

The gain of the transfer function can be constant or be a function of the Laplace operator s; in the latter case it is time dependent. As a result of the use of feedback, the control algorithm will try to keep the controlled variable at its desired value or setpoint.

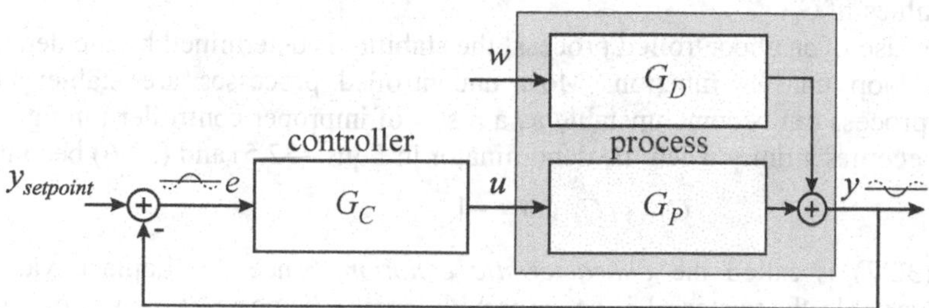

Fig. 32.1 Feedback control loop with disturbance.

In Fig. 32.1 it is assumed that measurement and control are ideal, i.e. they do not have any dynamics and the gain is equal to one. The controller has a transfer function G_C, the process transfer function is G_P and the disturbance transfer function G_D. The process output can now be written as the sum of all dependencies:

$$\delta y = G_D \cdot \delta w + G_P G_C \cdot \delta y_{setpoint} - G_P G_C \cdot \delta y \tag{32.2}$$

which means that the dependent variable y can be written as:

$$\left(1 + G_P G_C\right) \delta y = G_D \cdot \delta v + G_P G_C \cdot \delta y_{setpoint} \tag{32.3}$$

or:

Process Dynamics and Control: Modeling for Control and Prediction. Brian Roffel and Ben Betlem.

$$\delta y = \frac{G_D}{(1+G_P G_C)} \cdot \delta w + \frac{G_P G_C}{(1+G_P G_C)} \cdot \delta y_{setpoint} \qquad (32.4)$$
$$= G_{load} \cdot \delta w + G_{servo} \cdot \delta y_{setpoint}$$

G_{load} is the transfer function for the disturbance. Control now has to be such that the disturbance has no impact on the process output y; this can be achieved by proper manipulation of the process input u. This means that for ideal disturbance suppression it holds that:

$$G_{load} = \frac{G_D(s)}{(1+G_P(s)G_C(s))} \xrightarrow{for\ all\ s} 0 \qquad (32.5)$$

There is one other condition that should be satisfied. The setpoint should be tracked well; this is called servo behaviour. For ideal servo control it holds that:

$$G_{servo} = \frac{G_P(s)G_C(s)}{(1+G_P(s)G_C(s))} \xrightarrow{for\ all\ s} 1 \qquad (32.6)$$

The above-mentioned conditions can be met when $G_C G_P$ approaches infinity for all values of s.

The transfer function of the process G_P can, to a certain extent, be affected by the choice of the control scheme. It is important to choose a high gain and fast response. This will result in maximum power and speed of control, as will be explained further in section 33.1.8. However, the possibilities to achieve this are limited.

The transfer function of the controller G_C can be chosen freely, and depends on the controller tuning. Unfortunately, an infinite value of G_C cannot be realized. There are two limitations. The control input u has a limited range and can generally not be negative *(*δu *can be positive as well as negative!)*. In addition, the transfer function can become unstable for too large values of G_C.

Like the case of an uncontrolled process, the stability is determined by the denominator of the closed loop transfer function. Most uncontrolled processes are stable; however, a controlled process can become unstable as a result of improper controller tuning. The closed loop gain becomes infinity when the denominator in Eqns. (32.5) and (32.6) becomes zero:

$$G_P(s)\, G_C(s) = -1 \qquad (32.7)$$

Equation (32.7) is called the *characteristic equation.* Since the Laplace variable is an imaginary variable, the vector –1 is a vector with length (the norm of the vector) equal to one and a phase angle of –180°. This is shown in Fig. 32.2.

In this figure, a frequency is shown which is amplified by –1 by the transfer function $G_C G_P$; in other words, the gain of the transfer function is 1, while the frequency lags behind by 180°. In that case the control error is "recycled" with an opposite phase.

If negative feedback is used, the frequency will never dampen out anymore; it will continue to oscillate. The frequency at which this takes place is called crossover frequency ω_{CO}. The period with which the control is oscillating is in that case $P_{CO} = 2\pi/\omega_{CO}$. The gain

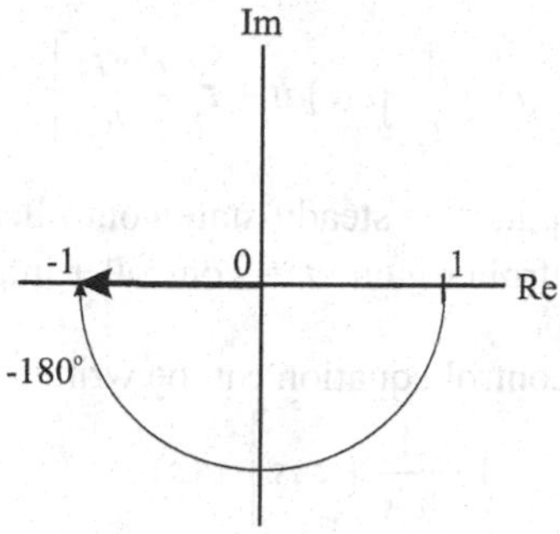

Fig. 32.2. Vector −1 in the imaginary plane.

of the controller at which the closed loop gain becomes one is called K_{CO}. If the gains of the measuring device and control valve are also one, then $K_{CO} = 1/K_P$. In this respect the Bode criterion should be mentioned which states that if at the crossover frequency ω_{CO} the corresponding amplitude ratio is less than one, the system is stable.

Generally, one will try to make the feedback action as large as possible. A proportional-differential-integral (PID) controller is well suited for this purpose. This controller is capable to make the gain large for low values of s (stationary behaviour) and high values of s (rapid changes), while in the range where the gain has to be limited in order to avoid stability problems, the gain can be kept at low values. It will be shown that improvement of control action for one frequency range always has a negative impact on control performance for another frequency range.

As can be seen from Eqn. (32.6), in steady state the servo behaviour ($s = 0$) becomes:

$$G_{servo}(s=0) = \frac{G_P(0)G_C(0)}{1+G_P(0)G_C(0)} \tag{32.8}$$

The remaining offset (lasting deviation between setpoint and measurement) becomes:

$$offset = 1 - G_{servo}(s=0) = 1 - \frac{G_P(0)G_C(0)}{1+G_P(0)G_C(0)} = \frac{1}{1+G_P(0)G_C(0)} \tag{32.9}$$

In the case of a controller with integral action, $G_C(0) = \infty$, i.e. the offset is not present.

32.2 Controller Equations

A PID controller is the controller that is used most frequently in the process industries. The controller has three terms or actions:

- A proportional term, which makes a correction proportional to the error between setpoint and measurement.
- An integrating term, which ensures that the long-term error approaches zero. This term guarantees that there will be no offset or long-term deviation from setpoint.
- A differentiating term, which ensures that the controller anticipates changes. It controls the high frequency behaviour. This term has a stabilizing effect and makes it possible to increase the controller gain.

The general controller equation can be written as:

$$u(t) = \overline{u} + K_C\left[e(t) + \frac{1}{\tau_i}\int e(t)dt + \tau_d\frac{de(t)}{dt}\right] \tag{32.10}$$

in which u = the controller output, $\overline{u}$ = steady-state controller output, e = error = setpoint – process measurement, K_c = controller gain, τ_i = controller integral action and τ_d = controller derivative action

In the Laplace domain, this control equation can be written as:

$$\delta u(s) = K_C\left[1 + \frac{1}{\tau_i s} + \tau_d s\right]\cdot e(s) \tag{32.11}$$

It can easily be seen that for the proportional controller τ_i should approach infinity and τ_d approach zero. For a proportional-differential controller τ_i should be infinity.

32.3 Frequency Response Analysis of the Process

Determination of the frequency response of the controller is relevant since it can be used for controller tuning. To calculate the frequency response, the following transformation is used (see chapter 9):

$$s = j\omega \tag{32.12}$$

Using this substitution, the amplitude ratio and phase shift of a transfer function can be determined for different values of the frequency ω.

The gain (amplitude ratio AR) is the ratio between the amplitude of the input signal and output signal of a transfer function $G(\omega)$. This is equal to the norm of the transfer function:

$$AR(G) = |G(j\omega)| = \frac{\prod_{i=1}^{n} AR(numerator\ term(i))}{\prod_{i=1}^{p} AR(denominator\ term(i))} \tag{32.13}$$

The phase shift (ϕ) is the phase lag that develops between the input signal and output signal of a transfer function $G(\omega)$. It is equal to the angle of the transfer function:

$$\begin{aligned}\phi(G) &= arg[G(j\omega)] \\ &= \sum_{i=1}^{n}\phi(numerator\ term(i)) - \sum_{i=1}^{p}\phi(denominator\ term(i))\end{aligned} \tag{32.14}$$

Example

Consider a second-order transfer function with time delay in the Laplace domain:

$$\frac{\delta y(s)}{\delta u(s)} = \frac{K_P\cdot e^{-\theta s}}{\tau^2 s^2 + 2\varsigma\tau s + 1} \tag{32.15}$$

in which y = process output, u = process input, δ = variation around the steady state and for which the following values hold: static gain K_P = 0.1, the natural time constant τ = 10 min., the damping coefficient ζ = 0.2 and the dead time θ = 3 min. As can be seen, the damping of the system is low. In the frequency domain, this transfer function becomes:

$$\frac{\delta y(\omega)}{\delta u(\omega)} = \frac{K_P \cdot e^{-j\omega\theta}}{-\tau^2\omega^2 + 2j\varsigma\omega\tau + 1} \tag{32.16}$$

The amplitude ratio can then be calculated as:

$$\begin{aligned} AR\left\{\frac{\delta y}{\delta u}\right\} &= \frac{AR(K_P)\cdot AR(e^{-j\omega\theta})}{AR(-\tau^2\omega^2 + 2j\varsigma\omega\tau + 1)} = \\ &= \frac{\sqrt{(K_P)^2}\cdot\sqrt{(\cos(\omega\theta))^2 + (-\sin(\omega\theta))^2}}{\sqrt{\left(1-\tau^2\omega^2\right)^2 + (2\zeta\tau\omega)^2}} \\ &= \frac{K_P}{\sqrt{\left(1-\tau^2\omega^2\right)^2 + (2\zeta\tau\omega)^2}} \end{aligned} \tag{32.17}$$

and the phase shift as:

$$\begin{aligned} \phi\left\{\frac{\delta y}{\delta u}\right\} &= \phi(K_P) + \phi(e^{-j\omega\theta}) - \phi(-\tau^2\omega^2 + 2j\varsigma\omega\tau + 1) \\ &= tan^{-1}\left(\frac{0}{K_P}\right) + tan^{-1}\left(\frac{\cos(\omega\theta)}{-\sin(\omega\theta)}\right) - tan^{-1}\left(\frac{2\zeta\tau\omega}{\left(1-\tau^2\omega^2\right)}\right) \\ &= 0 - \omega\theta - tan^{-1}\left(\frac{2\zeta\tau\omega}{\left(1-\tau^2\omega^2\right)}\right) \end{aligned} \tag{32.18}$$

As was derived in chapter 9, the amplitude ratio for a dead-time process is 1.0 and the phase shift $-\omega\theta$. The amplitude ratio for the process becomes then AR (second-order process) × AR(dead-time process). The phase shift of the process becomes then ϕ (second-order process) + ϕ (dead-time process). Figure 32.3 shows the Bode diagram in which the logarithm of the amplitude ratio and the phase shift ϕ are plotted against the frequency ω. For the amplitude ratio two asymptotes emerge, one for low frequencies $\omega\to 0$ (static behaviour) en one for high frequencies $\omega\to\infty$ (high-frequency behaviour). The values can easily be calculated from:

$$\begin{aligned} log\left(AR\left\{\frac{\delta y}{\delta u}\right\}\right) &= log\left(\frac{K_P}{\sqrt{\left(1-\tau^2\omega^2\right)^2 + (2\zeta\tau\omega)^2}}\right) \\ &= log(K_P) - \tfrac{1}{2}log\left(\left(1-\tau^2\omega^2\right)^2 + (2\zeta\tau\omega)^2\right) \end{aligned} \tag{32.19}$$

If $\omega << 1/\tau$, it follows that:

$$log\left(AR\left\{\frac{\delta y}{\delta u}\right\}\right) = log(K_P) \tag{32.20}$$

and if $\omega >> 1/\tau$, it follows that:

$$\begin{aligned} log\left(AR\left\{\frac{\delta y}{\delta u}\right\}\right) &= log(K_P) - \tfrac{1}{2}log\left(\left(\tau^2\omega^2\right)^2\right) \\ &= log(K_P) - 2log(\tau\omega) \end{aligned} \tag{32.21}$$

In a graph where log(AR) is plotted against log(ω) this is an asymptote with a slope of –2.

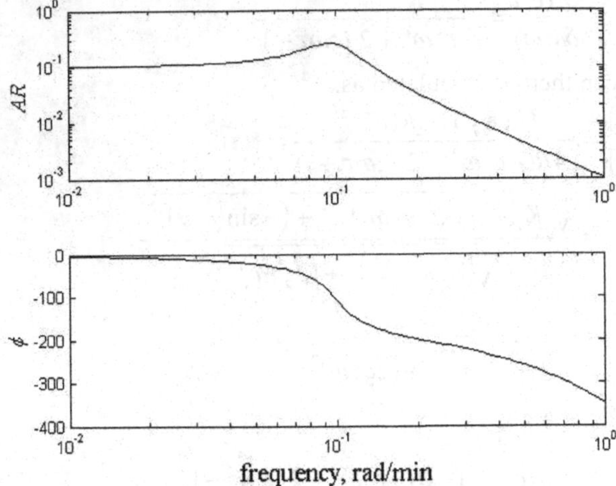

Fig. 32.3. Bode diagram for second-order process with dead time.

32.4 Frequency Response of Controllers

In this section the frequency response of a PID controller will be discussed. The transfer function of a proportional integral derivative controller is:

$$G_C(s) = K_C\left(1 + \frac{1}{\tau_i s} + \tau_d s\right) \tag{32.22}$$

In the frequency domain this can be written as:

$$\delta u(\omega) = K_C\left[1 + \frac{1}{j\omega\tau_i} + j\omega\tau_d\right]\cdot e(\omega) \tag{32.23}$$

which can also be written as:

$$\delta u(\omega) = K_C\left[1 + j\left(-\frac{1}{\tau_i\omega} + \tau_d\omega\right)\right]\cdot e(\omega) \tag{32.24}$$

The amplitude ratio and phase angle for the controller becomes then:

$$\begin{aligned} AR(G_C) &= K_C\sqrt{\left(-\frac{1}{\tau_i\omega} + \tau_d\omega\right)^2 + 1} \\ \phi(G_C) &= tan^{-1}\left(-\frac{1}{\tau_i\omega} + \tau_d\omega\right) \end{aligned} \tag{32.25}$$

For stationary behaviour, when $\omega << 1/\tau_i$, we may write:

$$AR(G_C) = K_C \frac{1}{\tau_i \omega}$$
$$\phi(G_C) = tan^{-1}\left(-\frac{1}{\tau_i \omega}\right) = -90^o + tan^{-1}(\tau_i \omega) \tag{32.26}$$

For the limit case, when ω approaches zero, the amplitude ratio will approach infinity and the phase shift ϕ will approach -90°. For low frequency the integral action dominates, leading to a high gain. Consequently, the offset will approach zero.

For high frequency behaviour, when $\omega >> 1/\tau_d$, one can write:

$$AR(G_C) = K_C \cdot \tau_d \omega$$
$$\phi(G_C) = tan^{-1}(\tau_d \omega) \tag{32.27}$$

This means that in the limit case, when ω approaches infinity, the AR will approach infinity and the phase shift ϕ will approach 90°. Hence for high frequencies the derivative action dominates. This causes a positive phase shift; consequently, the derivative action has a stabilizing effect on control loop stability. However, an ideal derivative controller cannot be realized in practice. A pure differentiating element cannot be built, because not any system is capable of amplifying high frequencies. In addition, a signal contaminated with some noise, will generate extremely large controller action.

When discussing the goal of the controller, it was mentioned that it should be tried to increase the controller gain as much as possible; however, the control loop may not become unstable. The gain may not be such that the phase shift is more than –180°. It is therefore possible to formulate requirements for a closed-loop system, based on the open-loop system. Good measures for the adjustment of the controller are the phase margin and gain margin. These are the remaining margins that are still available at a gain of one and shift of –180° respectively. This will be discussed in more detail in the next section.

Figure 32.4 shows the Bode diagram of a PI controller with integral action $\tau_i = 10$. Rather than plotting the amplitude ratio, the ratio between amplitude ratio and controller gain is plotted on the vertical axis. As can be seen from Eqn. (32.25), plotting log(AR) versus log(ω) at low frequencies yields a slope of –1. The corner frequency, where the asymptotes intersect is at $1/\tau_i$. The phase shift ranges from –90° to 0°.

Figure 32.5 shows the Bode diagram for the open loop transfer function $G_C\,G_P$. The graph is created by adding the graphs of the second-order plus dead-time process and the graph for the PI controller. At –180° the system already has a moderate gain in the order of 0.28. This is a result of the poor damping in combination with the dead time. At the crossover frequency of $1/\tau$ the second-order process already has a phase shift of –90°, the dead time adds another –17°, and the controller adds even more, bringing the total phase shift at approximately –150°.

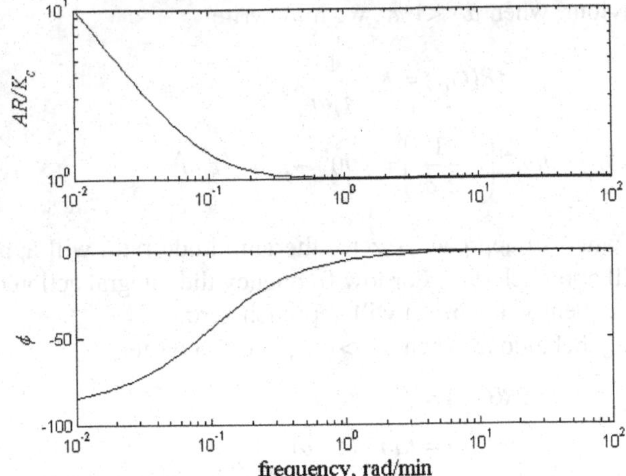

Fig. 32.4. Bode diagram for controller with integral time $\tau_i = 10$.

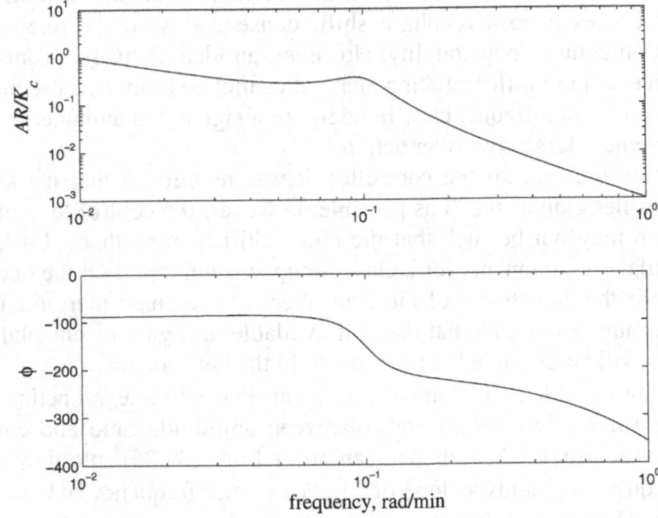

Fig. 32.5. PI control of second-order plus dead-time process with $\tau_i = 10$ min.

The controller can be chosen freely as long as $AR/K_c < 0.6 \ldots 0.8$. Integral action only is not very useful. The integral action should be added carefully since the phase shift is already large. A large derivative controller action ensures a positive phase shift, thereby making a larger integral controller action still possible. However, the application of derivative action is often limited, especially in the presence of process noise and dead time. One way to create better control would be to apply dead-time compensation, since this essentially eliminates the negative impact of dead time.

32.5 Controller Tuning Guidelines

When the phase shift is $-180°$ at $\omega = \omega_{CO}$, the crossover frequency, and the amplitude ratio is M, then, according to the Bode criterion, the control system is stable when $M < 1$ and the system is unstable when M>1. The crossover frequency and amplitude ratio M have been used in various controller-tuning guidelines. Figure 32.6 illustrates the concepts of gain and phase margins.

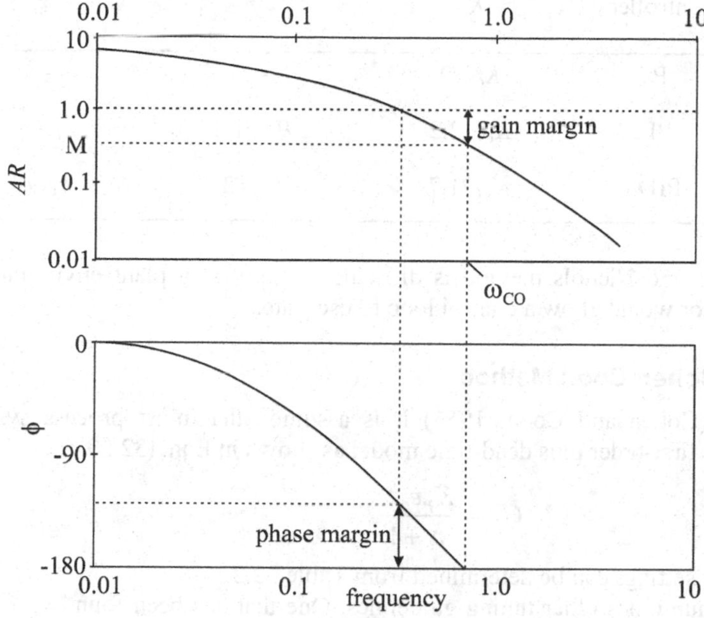

Fig. 32.6. Gain and phase margins.

The most well-known tuning guidelines that make use of these concepts are the Ziegler Nichols method and the Cohen-Coon method. There are also other controller tuning guidelines, such as dead-beat tuning and Internal Model Control tuning. It is beyond the scope of this chapter to discuss the last two methods.

32.5.1 The Ziegler-Nichols Method

The procedure for tuning a process according to this rule proceeds as follows (Ziegler and Nichols, 1942):

- use proportional control only and increase the controller gain until the system starts to oscillate. The frequency of oscillation is the crossover frequency ω_{CO}.
- determine the amplitude ratio at his frequency, let it be M.
- compute the ultimate controller gain and ultimate period of oscillation from:

$$\begin{aligned} &\textit{ultimate gain } K_{CO} = 1/M \\ &\textit{ultimate period of oscillation } P_{CO} = 2\pi/\omega_{CO} \end{aligned} \tag{32.28}$$

The controller settings for various controllers can now be calculated from Table 32.1.

Table 32.1. PID Controller settings according Ziegler and Nichols.

controller	K_C	τ_i	τ_d
P	$K_{CO}/2$	—	—
PI	$K_{CO}/2.2$	$P_{CO}/1.2$	—
PID	$K_{CO}/1.7$	$P_{CO}/2$	$P_{CO}/8$

The Ziegler and Nichols method is difficult to apply in a plant environment, since no process operator would allow a control loop to oscillate.

32.5.2 The Cohen-Coon Method

In this case (Cohen and Coon, 1953) it is assumed that most process systems can be modelled by a first-order plus dead-time model as shown in Eqn. (32.29):

$$G_P = \frac{K_P e^{-\theta s}}{\tau s + 1} \tag{32.29}$$

The controller settings can be determined from Table 32.2.

There are numerous other tuning guidelines. One that has been found very useful, based on the author's experience, are the following settings for a PI controller, also assuming a model as in Eqn. (32.29). $K_C K_P = 1 - 0.35\theta/\tau$, $\tau_i = \tau$.

Table 32.2. PID Controller settings according Cohen-Coon.

controller	K_C	τ_i	τ_d
P	$\frac{1}{K_P}\frac{\tau}{\theta}\left(1+\frac{\theta}{3\tau}\right)$	—	—
PI	$\frac{0.9}{K_P}\frac{\tau}{\theta}\left(1+\frac{\theta}{11\tau}\right)$	$\theta\frac{30+3\theta/\tau}{9+20\theta/\tau}$	—
PID	$\frac{4}{3K_P}\frac{\tau}{\theta}\left(1+\frac{3\theta}{16\tau}\right)$	$\theta\frac{32+6\theta/\tau}{13+8\theta/\tau}$	$\theta\frac{4}{11+2\theta/\tau}$

References

Cohen, G.H. and Coon, G.A. (1953) Theoretical considerations of retarded control. *Transactions of the American Society of Mechanical Engineers*, **75**, 827–34.

Ziegler, J.G. and Nichols, N.B. (1942) Optimum settings for automatic controllers. *Transactions of the American Society of Mechanical Engineers*, **62**, 759–68.

33 Design of Control Schemes

In this chapter, the design of a control scheme for an entire plant will be discussed. On the basis of the relationship between process outputs and inputs, the control scheme will be developed. The first part of the procedure is similar to the procedure for the development of an environmental model, which is identifying the inputs and outputs of the process. Measurement problems and costs of the correcting devices, however, should now also be taken into consideration. The result of this procedure is a table with interactions, in which the relationships between the manipulated and controlled variables is shown. The static relationship determines the power of control; the dynamic relationship determines the speed of control. The design procedure is illustrated by an example. Subsequently, methods for optimization and extension of the control scheme are discussed.

33.1 Procedure

It is not a trivial task to develop a systematic procedure for the design of control schemes for process units and entire plants. It is even not without risk, since the designer may think that by following the procedure the task will be completely finished. But it is of extreme importance to verify an intermediate result in different ways and if necessary to start the design procedure all over again.

It is not very meaningful to search for the ultimate control scheme for a particular process. The magnitude and nature of disturbances, the frequency of changes and the way the process is operated (for example at maximum load or maximum efficiency), the flexibility of the plant and the knowledge of operating and maintenance personnel all play a crucial role in the control scheme selection process. It is very well possible, that a particular process is heavily instrumented and automated in one plant and is hardly automated in another plant. It is therefore recommended to review the control scheme after the plant has come online and make changes in the control scheme if required (this can easily be done with modern instrumentation systems).

Sometimes, the expansion of a plant can have consequences for control of the older part of the plant. For example, after an expansion of the number of boilers, it is advantageous to operate a new boiler at full load, since it has a higher efficiency than the older boilers. This also has consequences for the control scheme for the older boilers, which should cope with the changes in steam demand.

This example illustrates the problem of system boundaries. Can a process be viewed separate from the environment? Increased integration of process units, meant to decrease intermediate material storage and increase recycling of energy and material, makes this increasingly difficult to realize. In particular, the development of separate steam and power generation units merges increasingly towards a distributed system of waste heat boilers in different process units and local turbines for the recovery of mechanical energy. The design of the control scheme for a new process leads in that case often to the redesign of the control scheme for the entire plant.

Process Dynamics and Control: Modeling for Control and Prediction. Brian Roffel and Ben Betlem.

In the next sections we will not focus on this issue, but rather focus on the processes. The sequence of the subjects might suggest a design procedure, from which one might deviate. It is recommended, however, to use the list of subjects as a "checklist", after a basic control scheme has been developed.

The next sequence of steps (see Table 33.1) can be used as a design procedure, or as a checklist afterwards if one prefers a more intuitive approach to control system design. The checklist is more or less similar to the checklist for the development of an environmental model (Table 3.1).

Table 33.1. Checklist, i.e. basic control scheme design procedure.

	Basic control scheme
i.	Operation of the process
ii.	Goals of the operation
iii.	Process boundaries and external disturbances
iv.	Controlled process conditions, accumulation variables and qualities
v.	Throughput/load and recycle flows
vi.	Correcting variables
vii.	Interaction table and count of degrees of freedom
viii.	Power and speed of control
ix.	Basic control scheme

Each subject has been worked out in more detail in the sequel.

(i) *Operation of the process.* Study the operation of the process. It is essential to understand why the process equipment is there, what its function is. The same holds for every process flow. Close contact with process developers and process designers is very useful at this point.

(ii) *Goals of the operation.* The goals of the operation are dictated by a hierarchically higher operational layer. For a plant, the goal is usually to produce a certain or maximum quantity (throughput) with a pre-defined quality. In addition, requirements may exist for side streams, waste streams and energy supply (temperature level). The goal is in that case to produce a desired quality and quality at minimum costs, in other words, minimum use of feed and energy. Different control goals result consequently in different control schemes. A simple example is shown in Fig. 33.1. A heat exchanger heats a liquid by means of condensing steam. Examples of different schemes are:

  a. a varying liquid flow (from a previous process) or maximum liquid flow is heated to a required temperature.
  b. a desired liquid flow is heated to a required temperature.
  c. all steam supply is used to heat the liquid flow to a required temperature.
  d. a maximum flow of steam is condensed by a maximum liquid flow.

  In case of a maximum or minimal requirement, usually one of the flows has no control restriction.

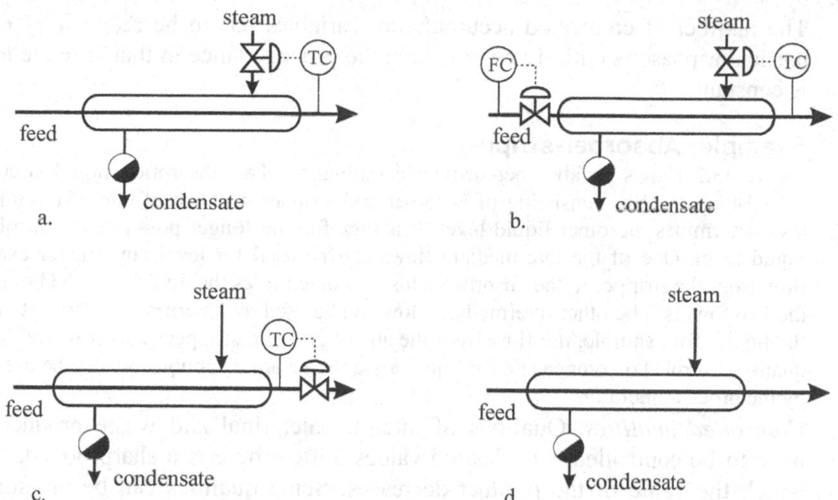

Fig. 33.1. Heat exchanger with different control schemes.

The goals of the operation are reflected in the selection of the controlled variables. The optimization will be discussed in the following sub-section.

(iii) *Process boundaries and external disturbances.* Investigate whether the boundaries of the process are also correct for the design of the control scheme. Pay special attention to utilities (steam, electricity, flue gas) and exchange of energy and material, which can lead to a widening of the process boundaries.

It would be an advantage if the process could be sub-divided into independent sub-processes for which a control scheme could then be developed separately.

(iv) *Controlled variables.* The controlled variables can be classified into three categories: process conditions, material contents and qualities.

- *Controlled process conditions.* The conditions of different process units have to be maintained at their design specification. This concerns in particular pressures (of reactors, columns, furnaces, etc), temperatures (of reactors), and residence times (of reactors).
- *Controlled material contents.* Every vessel or piece of equipment has a material content, divided over one or more phases. Select controlled variables, which ensure that the individual inventories are kept constant. Table 33.2 shows several possibilities.

  The number of controlled variables can be reduced if one can make use of self-regulation. For example, the liquid contents of the trays of distillation towers are usually not controlled, but kept within reasonable bounds by proper selection of overflow weirs.

Table 33.2. Controlled accumulation variables.

phase	controlled variable
Gas or vapor	Pressure
Liquid (entirely filled)	Pressure or pressure drop
Free liquid level	Level
Level between two liquids	Intermediate level
Fluidized bed	Pressure drop
Granular material on conveyor belt	Layer height or speed

The number of controlled accumulation variables has to be reduced by one if the particular phase is entirely contained in the process, since in that case the total mass is constant.

Example: Absorber-stripper

Figure 33.2 shows an absorber-stripper combination. The absorption liquid circulates in a closed-loop system, consisting of absorber and stripper. Control of one of the liquid levels also determines the other liquid level. It is therefore no longer possible to control the latter liquid level. One of the intermediate flows can be used for level control, for example, the flow from the stripper to the absorber. This also determines the division of the liquid between the two towers. The other intermediate flow can be used to determine the circulation speed of the liquid, for example, the flow from the absorber to the stripper. This flow can be used for quality control. To compensate for liquid losses, a separate supply flow can be used, handled by the process operator.

- *Controlled qualities.* Qualities of intermediate, final and waste products usually have to be controlled near desired values. Often there is a sharp boundary beyond which the value of the product decreases. Some qualities can be measured using robust and affordable measuring devices, which can be installed in the process. (Table 33.3).

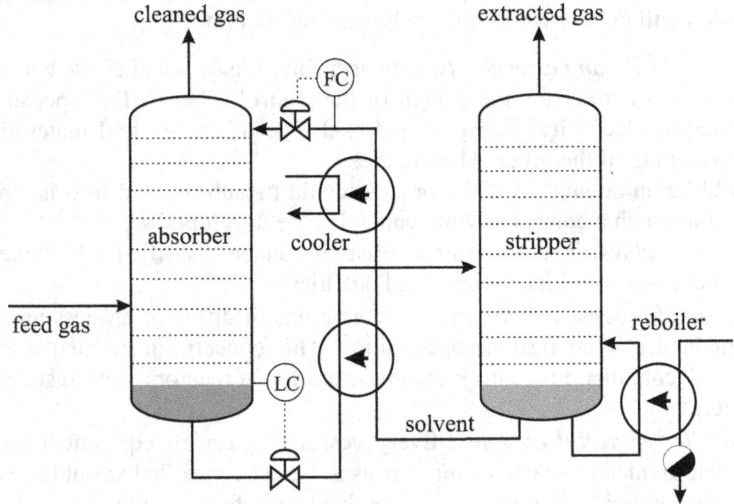

Fig. 33.2. Absorber-stipper combination.

Table 33.3. On-line measuring devices.

Not expensive and relatively reliable (provided sample is "clean")
pH, electrical conductivity (boiler feed water), viscosity, density, oxygen and carbon dioxide (flue gas)
Not very reliable and expensive, especially in maintenance
Infrared and ultraviolet spectrometers, chromatographs, boiling trajectory, flash point, etc.

The reliability depends strongly on the presence of solid particles (rust, catalyst particles, etc.) in the sample. Sample conditioning is therefore a critical step in the entire measuring system.

One could also try to estimate the quality ("inferential control") by using a suitable algorithm that uses conventional measured values of for example temperature, pressure, etc. A simple example is the pressure compensation of temperature measurements on the tray of a distillation column. For binary mixtures this yields a measure of the composition, for a multi-component system it is an approximate measure of the composition.

Often the process operators are responsible for the final control of product quality, based on "off-line" analysis of quality via the laboratory.

(v) *Throughput/load and recycle flows*. Choose a process variable (or several process variables) which determine the load of the process. It is not required to use the feed as the variable, which determines the throughput. Especially, in case of high load, can the production be maximized by keeping the variable, which is the limiting factor, at its maximum. In case of a distillation column this could be the reboiler, which should then be operated with the steam supply fully open. If one process acts as a supply to other processes, for example a central steam boiler, then the throughput is usually determined by the consumer(s).

Recycle flows can best be kept at constant flow or flow ratio, this reduces the propagation of disturbances. Disturbance propagation is more difficult to control if the recycle flow includes multiple process units.

(vi) *Correcting variables*. For total control, all process flows should be adjustable, hence in principle every process flow is a candidate for a correcting variable. This only holds under the condition that the accumulations are automatically controlled: every form of self-regulation decreases the number of correcting variables.

Investigate whether it is economically feasible to adjust the correcting variables by control valves. Control valves for pipelines with large diameter are very expensive and consume large amounts of mechanical energy. They can be omitted unless the controllability of the process is negatively affected.

An alternate for control valves are speed controllers of pumps, compressors and turbines. This is especially interesting because of the high cost of energy. Already interesting options exist for asynchronous motors.

A simple rule of thumb for the selection of adjustable process variables is:

Every incoming flow that has not been determined externally, should be controlled, except when the flow is established as a result of self-regulation of the adjustable mass- or energy accumulation.

The positioning of a valve, which creates an extra accumulation that has to be controlled, is not allowed. Similarly, two valves in a pipeline (without creating a meaningful accumulation) is not allowed either. If a flow is separated into two other flows, only two of the three flows may contain a control valve.

Examples of flows that do not have to be controlled or should not be controlled are:

- no valve should be positioned in the discharge line of a thermo-syphon reboiler, since free circulation is required.
- no valve should be positioned in a multiphase flow of vapor and liquid between a condenser and accumulator, since the liquid flow into the accumulator as a result of gravity should be unobstructed.
- no valve should be positioned in a supply or discharge line of an evaporator or condensor, since the amount of liquid that evaporates or vapor that condenses, is directly determined by the energy supply or removal.
- no valve should be positioned between pressure vessels if the pressure drop is relatively small, since otherwise an undesired pressure drop is created, one of the pressures will follow the other one which is controlled and is therefore more or less self-regulating.
- valves should not be present in fluidized flows, the gas and solids flows have to be adjusted separately.

(vii) *Interaction table and count of degrees of freedom.* From this point on, the procedure for the development of the control scheme deviates from the procedure for the development of an environmental model. The controlled variables now have to be arranged in a table versus the correcting variables. The table is used to represent power and speed of control for each combination of variables.

Compare the number of correcting variables with the number of controlled variables. If there are fewer correcting variables than controlled variables, the previous steps have to be reviewed. There is obviously no guarantee that no errors have been made. Sometimes the review process is equivalent to determining a relationship between controlled variables.

(viii) *Power of control and speed of control.* For each combination of the controlled and correcting variables in the Table, the power and speed of control have to be established. An example is shown in Table 33.5, which will be discussed in the sequel. The quality of control that can be expected for an automatic control loop, is characterized by the power of control and speed of control. The power of control is a measure for the static impact of the correcting variable on the controlled variable. If this impact is small or negligible, then the corresponding control loop will not function properly. The speed of control is a measure for the dynamic impact. This can be expressed by the oscillation frequency for proportional control close to the limit of stability. The higher the frequency, the better disturbances can be eliminated. The speed of control is limited by the apparent dead time in the control loop, which is the cumulative dead time of correcting device, process and measurement device together. The overall system response has usually one of the following shapes:

- "dead-time like" (see Fig. 33.3a)
- "first-order like" (see Fig. 33.3b)
- "second-order like" (see Fig. 33.3c)

"Dead-time like" responses are dominated by an apparent or real propagation time T_d. The most important control action that should be used is integral action.

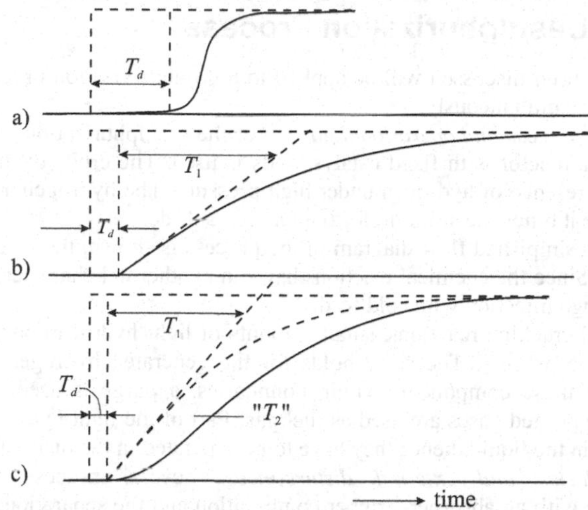

Fig. 33.3. Step responses of a closed loop control system.

Sometimes a weighting between power of control and speed of control is necessary. An example of such a situation is shown in Fig. 33.4.

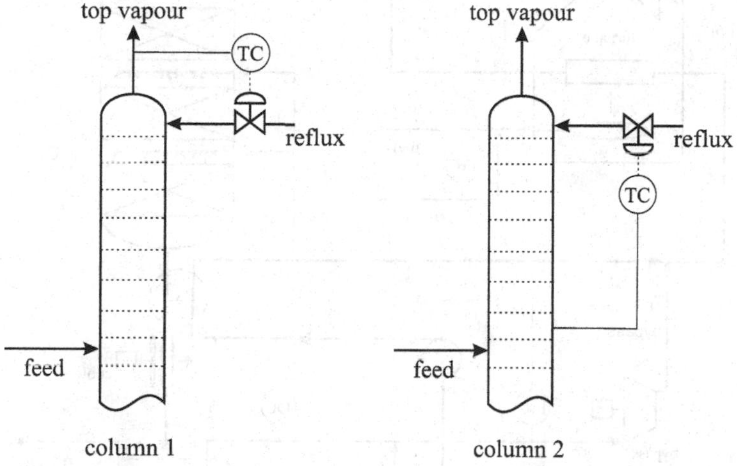

Fig. 33.4. Difference in process dynamics leading to difference in power and speed of control.

When the temperature sensor is located in the top of the column, the speed of control for the reflux is high. However, owing to small temperature fluctuations, often within 0.1 °C, the power of control is limited. If the temperature sensor is located in the middle of the column, the power of control is at its maximum and the speed of control is lower.

(ix) *Selection of basic control schemes*. Find the basic control schemes from the table by a combination of fields in the table in such a way that each controlled variable is only selected once and every correcting variable not more than once. Avoid fields that have an indication of the power of control "none" or "small". At this level, a third criterion in addition to power and speed of control should be considered, namely, control loop interaction. When control loop interaction cannot be avoided, one might have to resort to decoupling or multivariable control If more than one control scheme has been found, a proper selection can only be made after the extension of the control schemes have been reviewed.

33.2 Example: Desulphurization Process

The procedure that has been discussed will be applied to a desulphurization process (Fig. 33.5). Some points will be discussed simultaneously.

(i, ii) *Operation of the process and operational goals.* For the desulphurization of different fractions of crude oil, often a reactor with fixed catalyst beds is used. The crude oil fractions flow through the beds in the presence of hydrogen under high pressure. The hydrogen acts as a protection for the catalyst since it binds the sulfur to form hydrogen sulfide.

In Fig. 33.5 a simplified flow diagram of the process is shown, the reactor is the central part in this diagram. Since the chemical reactions have an exothermal character, intermediate cooling takes place through injection with cold feed.

As a result of cracking reactions, small amounts of light hydrocarbons are generated, which accumulate in the process. The same holds for the generated hydrogen sulfide. To keep the concentration of these components within boundaries, a purge is used. After cleaning (H_2S-elimination), the purged gases are used as fuel gas. Part of the light hydrocarbons and hydrogen sulfide dissolve in the liquid, hence they have to be separated in the distillation section.

(iii) *Process boundaries and external disturbances.* Several process units, such as the desulphurization with an absorber-stripper combination and the separation section, consisting of multiple distillation columns, have been omitted, since they operate more or less independently.

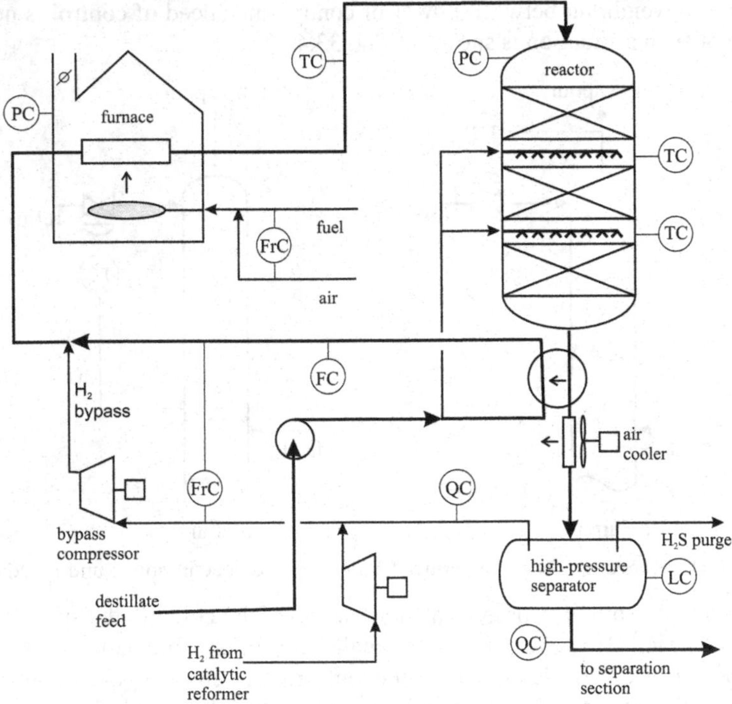

Fig. 33.5. Desulphurization of crude oil fractions.

(iv) *Controlled process conditions, material contents and qualities.*

Process conditions

The input conditions of the first catalyst bed are fixed as much as possible by control of the following variables:

- inlet temperature (TC)
- inlet pressure (PC)
- hydrogen flow in ratio with the crude oil flow (FrC)
- temperature in the separator, it has to be minimal to achieve maximum separation

The inlet conditions of the second and third bed depend strongly on the reactions in the first bed, it is therefore more difficult to keep them constant. It is, however, possible to keep the inlet temperatures constant through injection with cold feed.

Material Contents

In Fig. 33.5, the level in thc separator and the pressure in the gas circulation system are the most important indicators of material contents. In addition, the pressure in the furnace is indicative of the material contents of the furnace.

Only one pressure is controlled (at the entrance of the reactor) since it is very expensive and unnecessary to control the pressure in every process unit. This is a decision that is already made at the dcsign stage.

Qualities

The qualities that are important for control are:

- hydrogen or hydrogen sulfide fraction (QC) in the hydrogen recycle
- sulfide fraction in the crude oil

In a refinery, the sulfide fraction is a degree of freedom for optimization, since final products are made by mixing intermediate products. The value of an intermediate product is therefore a continuous function of the sulfur contents, in other words there is no unique quality requirement.

(v) *Throughput/load and recycle flows.* In Fig. 33.5 the flow of the crude oil fraction to the furnace is used for throughput control. For the circulating hydrogen flow, it was already mentioned that a ratio control is desirable, in order to ensure that there is always an excess of hydrogen.

(vi) *Correcting variables.* It cannot be economically justified, to install a control valve in the circulating gas flow, owing to the size of the control valve and thc associated pressure drop. To control this flow it is better to control the speed of the recirculation compressor, which then becomes the correcting variable.

A control valve should neither be installed in the gas-liquid mixture from the reactor, since gravity is used to transport the mixture to the separator, which also acts as a condenser.

In the case of an air cooler, sometimes the speed of rotation or the position of the rotor blades can be adjusted. Since maximum cooling is used, neither one of them will be used.

(vii) *Degrees of freedom.* The count of the controlled variables and correcting variables is shown in Table 33.4.

As can be seen, there is one more controlled variable than correcting variables. The product quality, however, depends on the reactor conditions.

The quality can only be affected by changing the reaction conditions, such as temperatures, the pressure, the hydrogen/crude oil fraction ratio and the hydrogen sulfide concentration.

Table 33.4. Controlled and correcting variables.

controlled variables	correcting variables
Crude oil fraction flow	feed
Hydrogen flow	speed bypass compressor
Hydrogen pressure	purge
Furnace pressure	flue gas to stack
Oxygen in flue gas	air
Outlet temperature furnace	fuel
Intermediate temperature (reactor)	injection cold feed
Pressure in reactor	hydrogen supply
Level in separator	liquid from separator
Sulfur contents in product	

Quality control can best be achieved by a master controller, which uses control of the most pronounced process condition as a slave controller. The control scheme is shown in Fig. 33.6.

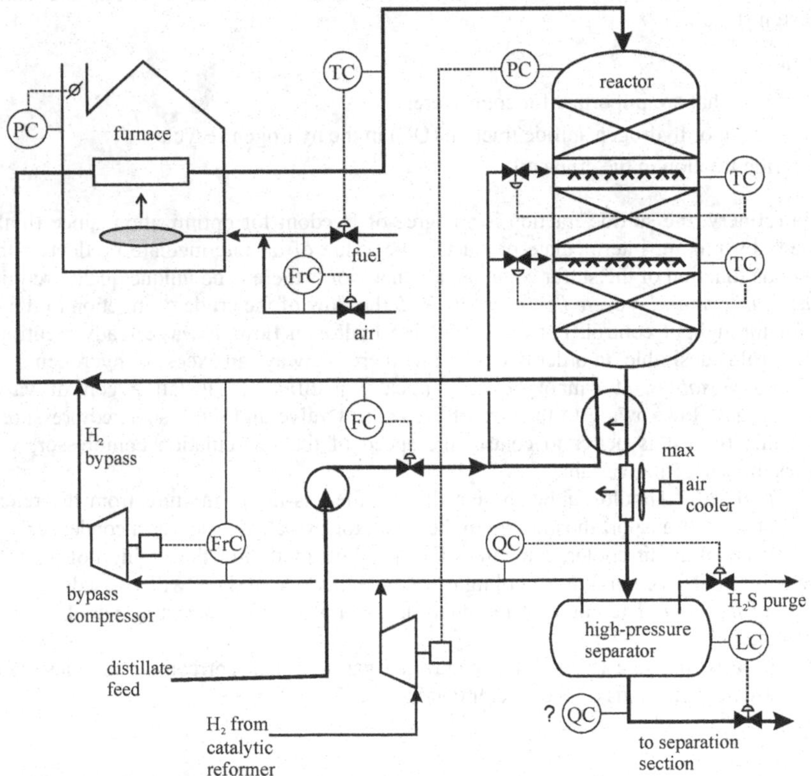

Fig. 33.6. Basic control scheme for desulphurization.

(viii) *Power and speed of control.* Table 33.5 yields only one basic control scheme on the main diagonal with two options for control of the furnace pressure and the oxygen in the flue gas. Note that for pressure control the gas supply or gas purge can be used as it influences the accumulation directly. The recycle gas flow only influences the recycle speed but not the accumulation.

Table 33.5. Interaction table showing power and speed of control for the desulphurization process.

controlled variables → correcting variables	feed flow	H_2 flow	H_2 concentration	furnace pressure	O_2 in flue gas	exit temp furnace	interm. temp reactor	reactor pressure	separator level
Feed	large fast	nil	large slow	nil	nil	large fair	large slow	moderate fair	large slow
Bypass compressor	nil	large fast	small	nil	nil	large fair	large fair	small	small
Purge	nil	nil	small fast	nil	nil	small	small	large fast	small
flue gas	nil	nil	nil	large fast	large fast	small	small	small	small
Air	nil	nil	nil	large fast	large fast	small	small	small	small
Fuel	nil	nil	small	moderate fast	large fast	large fair	large slow	large fair	large slow
Feed injection	nil	nil	small	nil	nil	nil	large fast	moderate fair	moderate slow
Supply compressor	nil	small	large high	nil	nil	small	small	large fast	small
Liquid from separator	nil	nil	nil	nil	nil	nil	nil	nil	large fast

33.3 Optimal Control

So far, we have only discussed control loops that keep process conditions constant or reject disturbances, thereby avoiding undesirable process conditions.

One could go one step further and try to keep the process in an optimal operating point. A checklist for items to be addressed is shown in Table 33.6.

Table 33.6. Items to be addressed for optimal control.

Optimal control	
i.	Degrees of freedom for optimization
ii.	Constraints
iii.	Model
iv.	Optimal operating point
v.	Active constraints
vi.	Degrees of freedom for consideration

Since the optimal operating point depends on the "mode of operation", high demands are imposed on the optimal control structure. The control scheme must be able to switch to different constraints if they become active.

(i, ii) *Degrees of freedom for optimization and constraints*. One should investigate which setpoints of the basic control scheme are candidates for an optimal setting. The same holds for the remaining correcting variables if any.

For the desulphurization process (Fig. 33.5), this investigation yields the following list of variables:

- inlet temperature of the reactor
- intermediate temperatures of the reactor
- pressure reactor
- hydrogen-crude oil ratio
- hydrogen or hydrogen sulfide fraction

The separator level does not appear in this list, since the level does not have any impact on the operation of the process (there is no optimal value of the level). The oxygen controller takes care of a local optimization by maintaining a small excess air to fuel ratio.

For every process unit and process flow the limits should be determined between normal and abnormal operation. Table 33.7 might be helpful in that respect.

(iii) *Model.* For continuous process operation a static model will suffice, supplemented with the constraints inequalities. It is not easy to develop a good model, since process knowledge is usually limited and inaccurate.

(iv) *Optimal operating point.* By using an optimization algorithm for the static model with its constraints, the optimal operating point is determined as a function of the independent variables. In most cases the optimum is located on a constraint or the intersection of several constraints. Hardly ever is the optimum a free optimum.

(v) *Active constraints*. The optimal operating point could be located on different constraints, hence, those constraints are active. For different operating conditions identify groups of constraints that are active. Investigate whether these groups of

Table 33.7. Constraints for desulphurization process.

flow or process unit	boundary	consequences of constraint passing
Process flow	Minimum quality Undesired two-phase flow Reversal of flow direction	Decrease in value Pipe burst
Pressure vessel	Maximum pressure Atmospheric pressure	Explosion Entrance of air from the environment
Liquid vessel	Maximum level Minimum level	
Liquid phase separator	Maximum intermediate level Minimum intermediate level	Wrong phase leaving
Separation column	Maximum load Minimum load	"Flooding" "Weeping"
Heat exchanger	Maximum ΔT at evaporation	Blockage of heat transfer by gas film
Filter	Maximum ΔP	Fouling
Compressor	Maximum speed of rotation Minimum speed of rotation Minimum flow	Explosion Vibration "Suction"
Furnace	Minimum under pressure Minimum excess air Flame extinguished Maximum temperature	Exit flue gasses Smoke and CO formation Explosion Pipe burst
Oxydation reaction	Maximum O_2 concentration	Explosion
Reactor	Maximum temperature	Downgrading of catalyst
Neutralization	Maximum pH Minimum pH	Corrosion and/or environmental damage

constraints can be controlled by their own correcting variable. If so, a selecting control system can be developed.

Example: Capacities distillation column

In a distillation column, the condenser capacity, the tray capacity (maximum tray pressure drop) or the reboiler capacity are limiting factors for the operation. This will be explained in more detail in the next chapter.

(vi) *Degrees of freedom for consideration.* If the optimal operating point is not located on active constraints, the optimum is determined by "degrees of freedom for consideration". Try to find an approximate model, representing the values of these degrees of freedom as a function of the independent variables or certain dependent variables.

Example: Reboiler heat distillation column

Sometimes the heat flow in the reboiler of a distillation column is a degree of freedom for consideration. This heat flow has to be approximately proportional to the feed flow. This points at using a ratio control between the feed and the heat flow to the reboiler. This ratio can be a function of the pressure in the column (affecting the separation) and specific costs.

33.4 Extension of the Control Scheme

In this section several examples will be given of extensions of the control scheme. Issues that will be discussed are:

i. *Cascade control.* One control loop sets the setpoint of another control loop. Goals for control that are hierarchical, are reflected in the control structure.
ii. *Feed-forward control.* This provides compensation for measurable disturbances. In this case disturbance information is used in the control scheme.
iii. *Compensators.* Often process behavior exhibits a dead time or inverse response in which case model-based control may be advantageous.
iv. *Selectors.* This implies switching between controllers and/or sensors.
v. *Split-range control.* This involves selection of controllers based on the range.

In the following sections, these issues will be elaborated on in more detail.

(i) *Cascade control.* A cascade control structure, also called master–slave controller, consists of two or more control loops (Fig. 33.7). The correcting value of the outer loop (the master) is the setpoint of the inner loop (the slave).

The cascade control structure can be used to increase the speed of the master loop or to reduce disturbances.

Improving the speed of the master loop

This is in particular the case if the process contains two large time constants, of which one is contained in the slave loop. This is schematically shown in Fig. 33.7. The slave control loop is shown with an ideal measurement (no measurement dynamics).

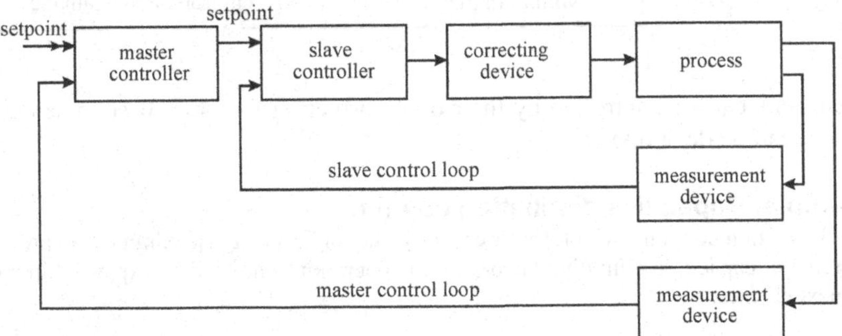

Fig. 33.7. Information flow diagram of a cascade control structure.

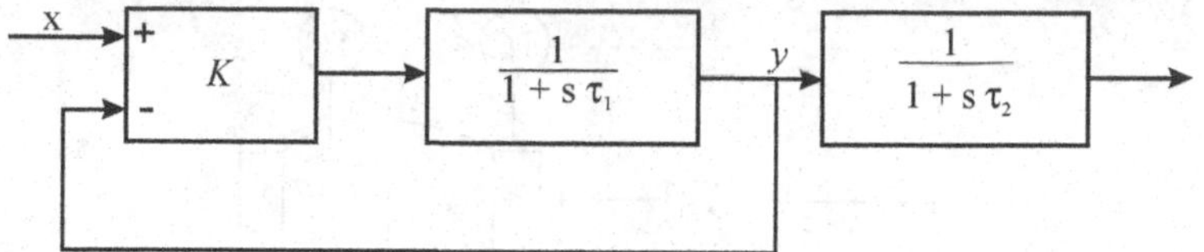

Fig. 33.8. Slave control loop.

The transfer function from x to y can be written as:

$$\frac{x}{y} = \frac{K}{1+K} \frac{1}{1+\tau_1^* s} \tag{33.1}$$

with

$$\tau_1^* = \frac{\tau_1}{1+K} \tag{33.2}$$

When $\tau_1 = 10$ minutes and $K = 9$, the slave control loop can be replaced by a first-order transfer function with a time constant of 1 minute. The value of the gain K is usually limited by secondary effects, such as non-linearities and smaller time constants, which also play a role).

Example: Reboiler level distillation column

Suppose in an evaporator as shown in Fig.33.9, the evaporation is controlled by a level controller, thus affecting the heat transfer area. However, the dynamic response is quite slow and can be represented by a first-order response with a large time constant. In order to reduce the effect of the large time constant in the condensate level, a cascade controller should be used. The steam supply is controlled by a flow controller, which determines the setpoint of the level controller. The flow controller will keep the steam supply constant rcgardless of disturbances in the steam supply pressure.

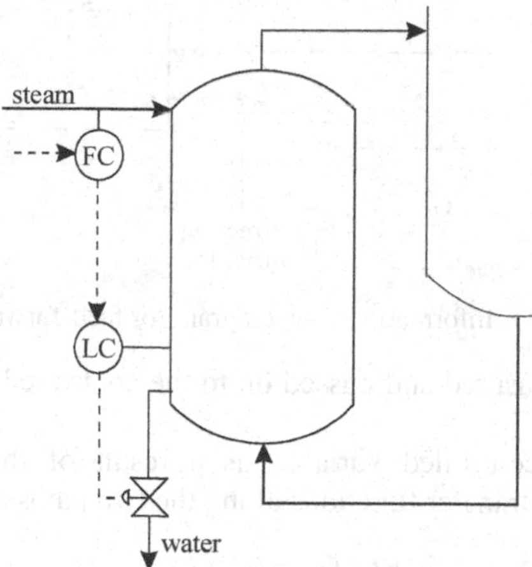

Fig. 33.9. Cascade control for an evaporator.

Reduction of disturbances that can be measured inside the slave controller

This is only relevant if the slave controller is faster than the master controller. An example of the use in reactor temperature control is shown in Fig. 33.10.

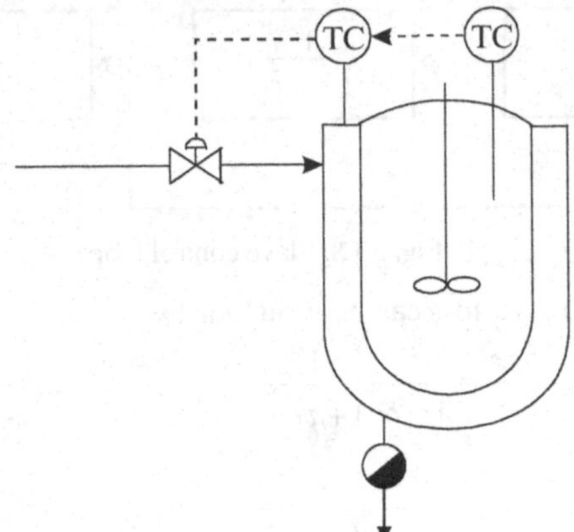

Fig. 33.10. Cascade controller structure for reactor.

Example: Reactor temperature control

Disturbances in the steam supply can be suppressed at an early stage by the slave controller. This controller is trimmed by the master controller, which controls the temperature of the reactor. The reactor temperature controller has considerably slower dynamics. Through the use of this cascade control structure disturbances are eliminated. Without this structure, the reactor temperature would first have to change before control action is taken.

(ii) *Feed-forward control.* If a measurable disturbance shows large variations and the process is slow, feedback control may not suffice. In that case it is possible to use anticipating action or feed-forward control. Figure 33.11 shows the general information flow diagram.

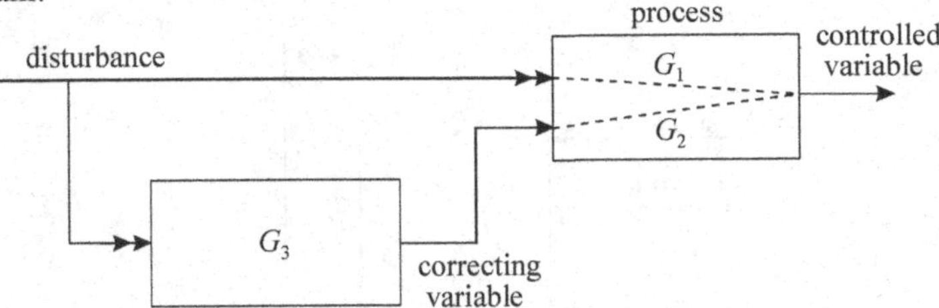

Fig. 33.11. Information flow diagram for feed-forward control.

A disturbance is measured and passed on to the controlled variable via the correcting variable.

Variations in the controlled variable as a result of the disturbance are exactly compensated for if the transfer functions along the two paths are identical:

$$G_1 + G_2G_3 = 0 \tag{33.3}$$

The transfer function of the controller should therefore, under all operating conditions, be represented by the equation:

$$G_3 = \frac{G_1}{G_2} \tag{33.4}$$

which requires sometimes extensive process testing and which is, in some cases, impossible to realize. Example: if G_1 contains no dead time and G_2 does, G_3 should contain a negative dead time.

Example: Feed-forward

Suppose for a process the following transfer functions were found:

$$G_1 = \frac{1.0e^{-5s}}{1+10s}$$
$$G_2 = \frac{0.5e^{-3s}}{1+8s} \tag{33.5}$$

then the feed-forward controller would be:

$$G_{FF} = 2e^{-2s}\frac{1+8s}{1+10s} \tag{33.6}$$

which shows a gain of 2.0, a dead time of two sampling times and a lead-lag element.

A much applied form of feed-forward control is the ratio controller, which is often used to compensate for feed flow changes. Figure 33.12 shows a typical application. The ratio controller realizes a constant ratio between fuel and air supply. The temperature controller can adjust the ratio setting. As already mentioned, feed-forward control only gives a noticeable improvement if the transfer functions along both path are more or less identical. If, for example, the transfer function from feed flow to temperature is much faster than the transfer function from fuel to temperature, the "anticipating" action will come too late and the temperature controller still has to eliminate temperature changes and in addition compensate the incorrect action of the feed-forward controller. Also, when the anticipating action is too early, the temperature controller will start to act. It can, however, be expected that in case of the furnace, both paths will show similar dynamics: a change in feed flow as well as a change in heat input, will affect the furnace tubes immediately over the full lengths of the tubes.

Differences originate from irregular heat distribution and the dynamics of the tube wall.

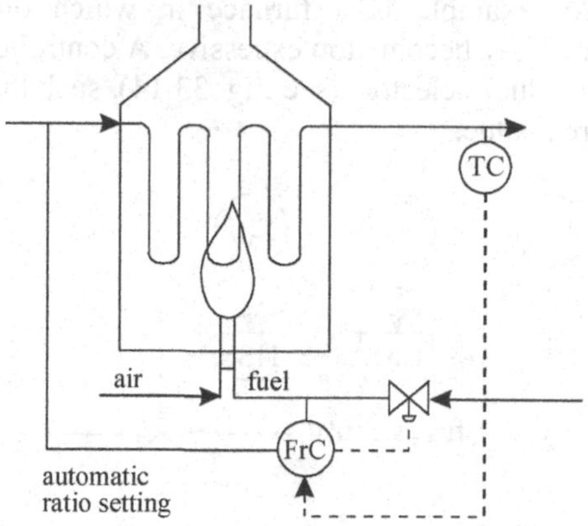

Fig. 33.12. Feed-forward control for a furnace.

(iii) *Compensators*. Many processes can exhibit dynamics that are detrimental to good control. Examples are the presence of a process dead time or inverse dynamics. Assume that we are able to factor the process dynamics into two parts, one that we shall call reversible process dynamics and another part that will be called irreversible dynamics. An example is shown in Fig. 33.13, part a shows how the controller will be configured and part b shows the equivalent block diagram in case of an ideal process model. The denominator of the transfer function from y_{sp} to y becomes now $G_C G_{P,rev}$ instead of $G_C G_P$, as a result of which the dynamics of the closed loop response have become considerably faster.

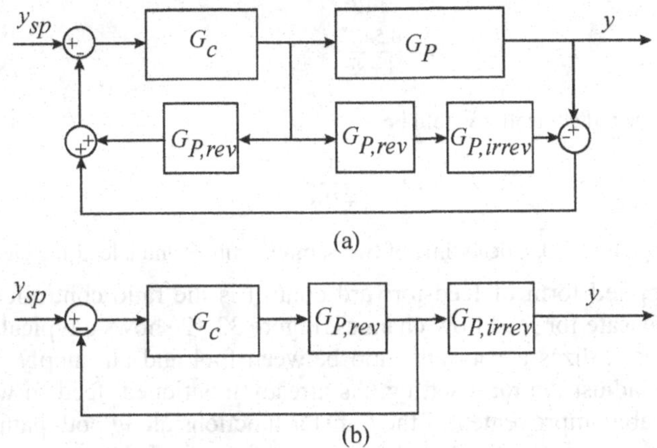

Fig. 33.13. Process control with compensation for irreversible dynamics.

Example: Dead time compensator

In the case where the process includes a dead time and the transfer function would be G_2 of Eqn. (33.5), the reversible and irreversible parts of the model would become:

$$G_{P,rev} = \frac{0.5}{1+8s}$$
$$G_{P,\,irrev} = e^{-3s} \tag{33.7}$$

(iv) *Use of selectors.* A "high" or "low-value" selector can be very useful in case of constraint control. An example is a furnace in which different temperatures are measured, none of them may become too excessive. A controller can be connected to the fuel supply, via "high-value" selectors (see Fig. 33.14), such that the controller only acts on the highest measured value.

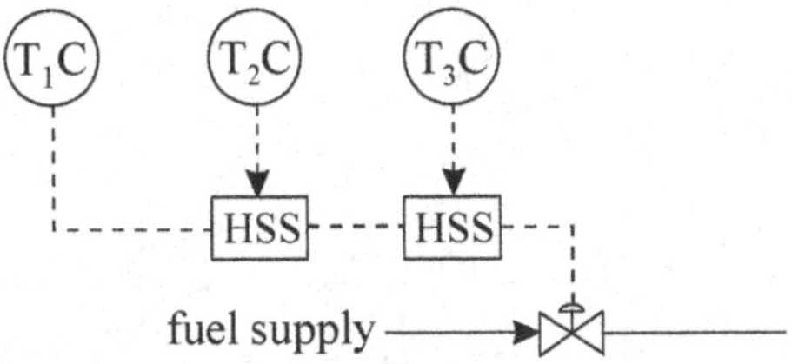

Fig. 33.14. Selector control for a furnace.

Figure 33.15 shows a selecting measurement system. The tray loading in a distillation column is monitored at three locations, the highest value is passed on to a pressure difference controller manipulating the cooling water flow to the condensor.

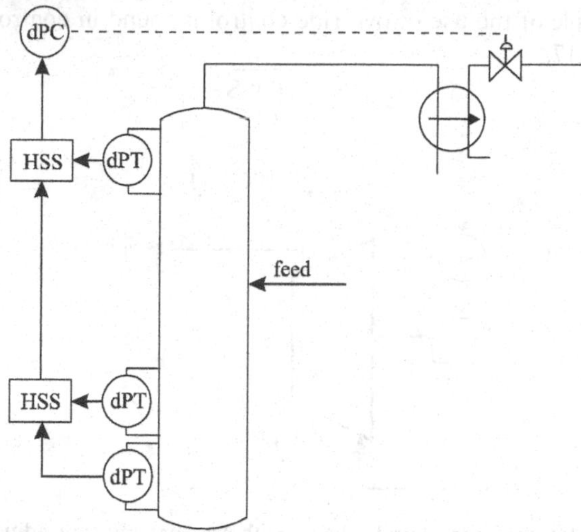

Fig. 33.15. Selecting measurement system for a distillation column.

(v) *Override control.* Complicated control loops, in which cascade control, ratio control and selectors are applied, can be found in furnace control. This is shown in Fig. 33.16.

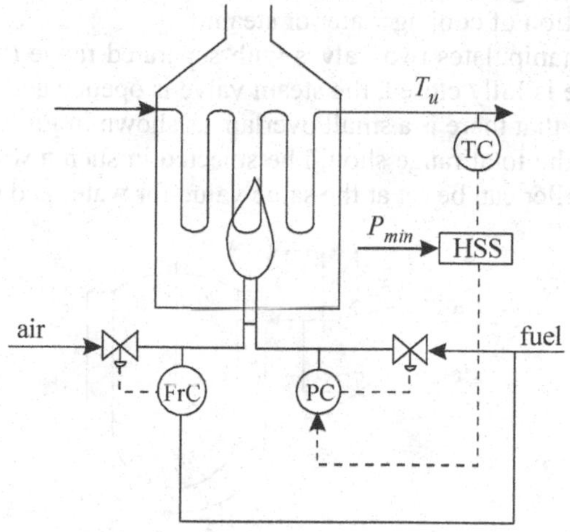

Fig. 33.16. Control scheme for a furnace.

Usually, the outlet temperature is controlled automatically. This is most likely also the highest temperature that the feed will experience. For safety reasons, the control structure is realized by adjusting the fuel supply, often via a slave controller. Figure 33.16 shows a pressure slave controller that adjusts the fuel supply to the burners. The advantages of this slave controller are:

- Disturbances in the fuel supply pressure are eliminated by the fast slave controller
- In case of liquid fuel, the burner pressure can be maintained above a minimum value, which is required for good atomization of the liquid. This can be achieved by a "high value" selector H: an instrument that passes on the highest measurement only.

Another example of the use of override control is found in control of compressors, as shown in Fig. 33.17.

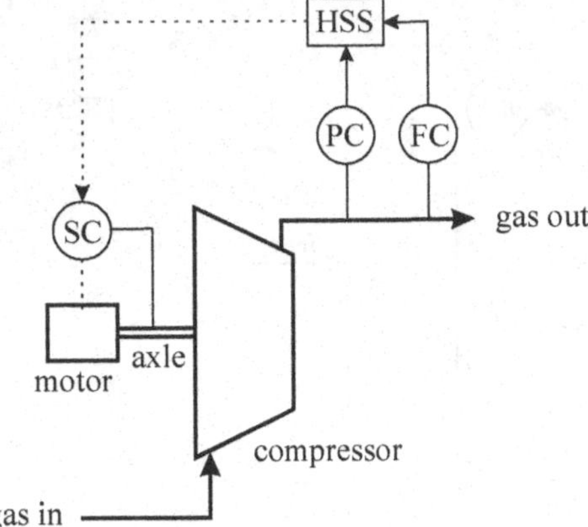

Fig. 33.17. Compressor control scheme with pressure and flow adjusting the speed.

The most efficient operation of the compressor is at maximum capacity, therefore the compressor is on flow control (cascaded to speed of rotation); however, there is a protection against high pressures.

(vi) *Split-range control.* Figure 33.18 shows temperature control in the jacket of a chemical reactor by introduction of cooling water or steam.

The controller manipulates two valves with separated range ("split range"): after the cooling water valve is fully closed, the steam valve is opened and vice versa. For smooth control it is desired that there is a small overlap, as shown in Fig. 33.19.

The division of the total range should be selected in such a way that the proportional action of the controller can be set at the same value for water and steam.

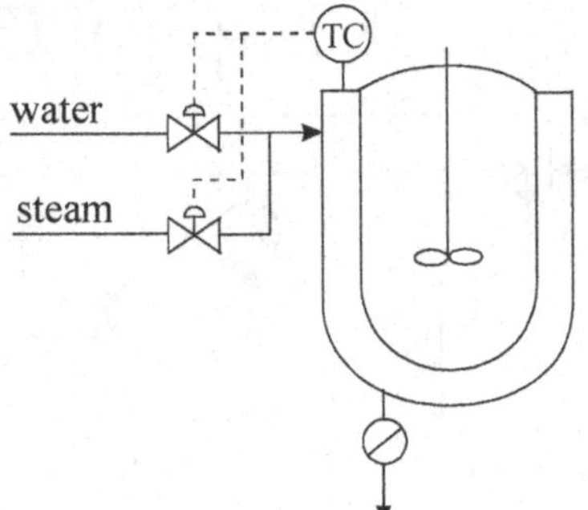

Fig. 33.18. Jacket temperature control.

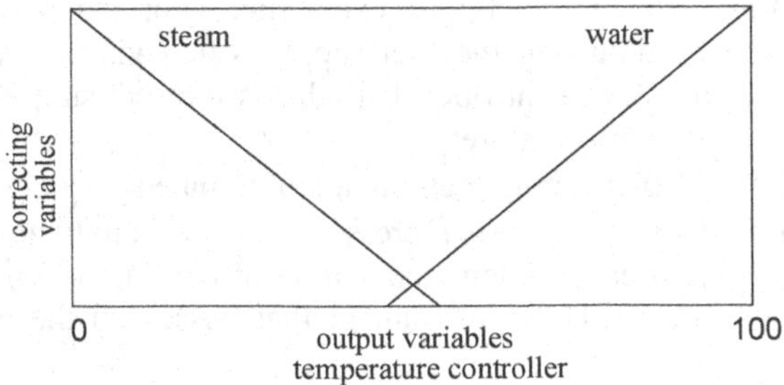

Fig. 33.19. Overlapping ranges.

33.5 Final Considerations

In this chapter, we have tried to indicate a systematic procedure for the design of the basic control scheme. In addition, some attention was paid to protection and optimal control. This did not lead to a cookbook recipe, which produces ready to install control schemes. On the contrary, the final selection of the process control system is still a matter of different aspects, affected by local circumstances. In addition, close cooperation between people from different disciplines is necessary. It is probably a good idea to highlight a few of these circumstances, although this list will not be complete.

- *Process disturbances and process noise.* The selection of the control system, its completeness and degree of complexity (degree of automation and intelligence) depends strongly on the size and frequency of disturbances and process noise. One should also take into account disturbances that are caused by humans (maintenance, switching to another mode of operation). Experience with the operation of the process will yield the relevant information and data.
- *Process instrumentation.* Extension and development of a more complicated control scheme can only succeed if the process instrumentation is up to date and functioning and if regular maintenance takes place. Integrated control systems depend strongly on the availability of measurement and correcting devices: the failure of one device might reduce the functioning of the control system from optimal to conventional.
- *Process models.* Optimal control requires good process models. The development of these models requires much effort and time, in cooperation with process experts. If this is not possible, one should not even consider starting the job.
- *Motivation of the process operators.* There is a difference between conventional and optimal process control. A conventional control scheme will try to keep process conditions constant and will keep considerable margins between the operating points and their constraints. This makes the job of the process operators easier, since they do not have to spend so much time monitoring the process and they have more time to take action when things go wrong. Optimal process control, on the contrary, will try to move the process as close as possible to the constraint(s), since this is optimal. Process operation will continuously be adapted to changes in the circumstances.
- *Motivation of the management.* Often managers and supervisors have a critical look towards more advanced control systems and advanced control strategies: it should be clear from the operational results that improvements have been achieved. But in practice this is sometimes hard to prove, since savings are usually small, perhaps 5% in energy consumption, smoother process operation, etc. These small improvements are often hard to monitor owing to process noise and measurement inaccuracies.

34 Control of Distillation Columns

In the previous chapter the procedure for the design of control schemes was discussed. The procedure was illustrated on a reactor with recycle. The selection of appropriate combinations of controlled and manipulated variables was relatively simple, since the interactions were limited. In this chapter the procedure will be applied to a distillation column. This is a unit operation with many interactions between the corrections that are made. Using a basic knowledge of the process dynamics, a basic control scheme is designed. Subsequently, two control schemes will be compared: a basic control scheme based on "material balance" control and a control scheme based on "energy balance" control. The distillation column can also be used to demonstrate the optimization of the control scheme. The principle is that the control scheme should be designed in such a way, that an objective function can be maximized.

34.1 Control Scheme for a Distillation Column

In this section, a control scheme will be developed, using Table 33.1 from the previous chapter.

34.1.1 Operation of the Process

Distillation is a much-used separation method in the chemical industry. Often, the separation process takes place in continuous distillation columns with one feed (F), one top product (D) and one bottom product (B), as shown in Fig. 34.1.

Separation is a result of counter-current contact between vapor and liquid on the trays. The driving force is the difference in relative volatility of the components. The vapor flow is generated in the reboiler, which is heated by steam or a hot liquid or a waste gas flow (H). If the feed is partly evaporated, the vapor flow is increased by the vapor part of the feed at the point where it enters the column.

The vapor flow from the top of the column is condensed in a condenser, which is cooled by water or air (C). It is also possible to transfer the heat of condensation to another process.

The distillate flows to an accumulator, from which part of the liquid is returned to the column as reflux (R). The other part of the liquid is top product (D). The internal reflux can be larger than the external reflux (R) if it is undercooled, thereby condensing some of the vapor.

Process Dynamics and Control: Modeling for Control and Prediction. Brian Roffel and Ben Betlem.

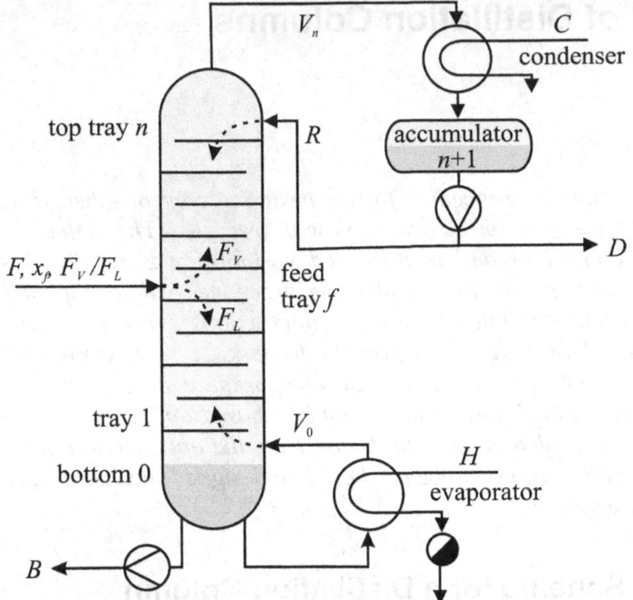

Fig. 34.1. Simple distillation column schematics.

The liquid flow in the column increases at the feed tray as a result of the liquid part of the feed. In the bottom the liquid flow is partially evaporated in the reboiler, the remainder is withdrawn as bottom product (B).

34.1.2 Goal of the Operation

The primary goal of the distillation process is the separation of one or more volatile components in the feed. Often there is a quality demand on the top product or bottom product. If the purity is less than the specification, the value of the product is often much less than when the purity requirements are met. The purity is sometimes expressed in the concentration of the main component, for example, in the case of technical solvents. Some of the other components are then more or less volatile than the main component. In a distillation column, the 'light ends' go to the top of the column and the 'heavy ends' to the bottom.

The purity can be expressed in the concentration of the key component: the 'heavy' key component in the top product and the 'light' key component in the bottom product. Let us assume that there is a strict quality requirement for the top product purity, which can be met by the right operation of the distillation column. There are no demands on the bottom quality; however, there is the potential that some valuable top product disappears with the bottom product. Consequently, we would like to control the bottom temperature, which is to some extent an indicator of the bottom purity.

In some distillation processes there are quality requirements for both products. This does not mean, however, that both products should exactly meet the quality requirements. It may be profitable to reduce the loss of a more valuable product in the less valuable product to the extent that the purity exceeds the specification requirement. In practical operation, it is not very common to install automatic quality control on both products. Automatic quality control of one product also gives an improvement of the quality of the other product.

34.1.3 System Boundaries and External Disturbances

The distillation column is considered without utilities. The feed enters the column without the presence of a buffer vessel, directly from another process unit. In that case, sources of disturbances are:

- feed flow
- feed composition
- feed enthalpy (vapor–liquid ratio and temperature)
- utilities: steam pressure and cooling water temperature

34.1.4 Selection of Controlled Variables

Process Conditions

The separation in a column section is determined by two degrees of freedom: the vapor flow and the liquid flow. In addition the pressure will affect the operation, it is assumed that it is set at the design pressure. For a good operation of the column, it is important the division of the feed over top and bottom product is adjusted properly, according to the mass and component balances. This is especially important in the case of a 'sharp' separation, which will be illustrated by means of an example.

Example

The feed of a distillation column consists of 40 tons/day 'light' component and 60 tons/day 'heavy' component, which has to be separated in products of 99% purity. If the flow of the top product would be 41 tons/day, then it has to contain at least 1 ton/day of the 'heavy component, since only 40 tons/day of the light component enters the column. This would mean that the purity would be less than 97.6%. A similar reasoning could be applied to the bottom product, if the top product flow would be 39 tons/day. The setting of the ratio D/B (or D/F, B/F, or another ratio between the same variables) is very critical and should be adjusted in varying circumstances, such as changes in feed composition.

Liquid Accumulations

The liquid levels on the trays of the distillation column are self-regulating. The level of the top accumulator, the level of the liquid storage tank of the condenser and the bottom level have to be controlled to keep them within boundaries.

Qualities

Good column operation is possible with automatic quality control of the most valuable product. Since the quality controller maintains the purity of one product, the purity of the other product can only vary within certain bounds. The fifth correcting variable can, if so desired, be used to realize optimal process operation.

In practice, a temperature controller is often used as a simple form of quality control. According to the phase rule of Gibbs, for binary mixtures the composition is fixed when the pressure and temperature are constant. Obviously, this does not hold for multi-component mixtures. One could also try to estimate the product quality from some tray temperature measurements and other easily measurable variables. This is called an inferential measurement.

In the following section, control of the top product quality (x_D) will be considered, while "control" of the bottom quality will be realized by control of the column bottom temperature T, which acts as an inferential measurement for the bottom quality.

For a favorable dynamic response, the measurement location should be located at the condenser and in front of the reboiler.

34.1.5 Throughput/Load

In this distillation example, it is assumed that the feed is determined by the previous process unit. This means that the load is fixed.

34.1.6 Selection of the Correcting Variables

If the feed is fed into the column without intermediate storage, it will not be possible to manipulate the vapor flows in both column sections independently. This is also true for the liquid flows. The number of correcting variables is then two: H or C and R (Fig. 34.1). On the vapor side, however, an extra degree of freedom can be used (C or H respectively), if it is used for pressure control. In the vapor line to the condenser and in the internal reboiler, no valves are permitted! These flows are determined by C and H.

On the liquid side, the tray contents are self-regulating, hence no additional degree of freedom is available.

Statically, there are no additional independent adjustable variables available at the top and bottom. Dynamically, there is a need for level controls at the top and bottom accumulator, hence D and B can be used as additional correcting variables.

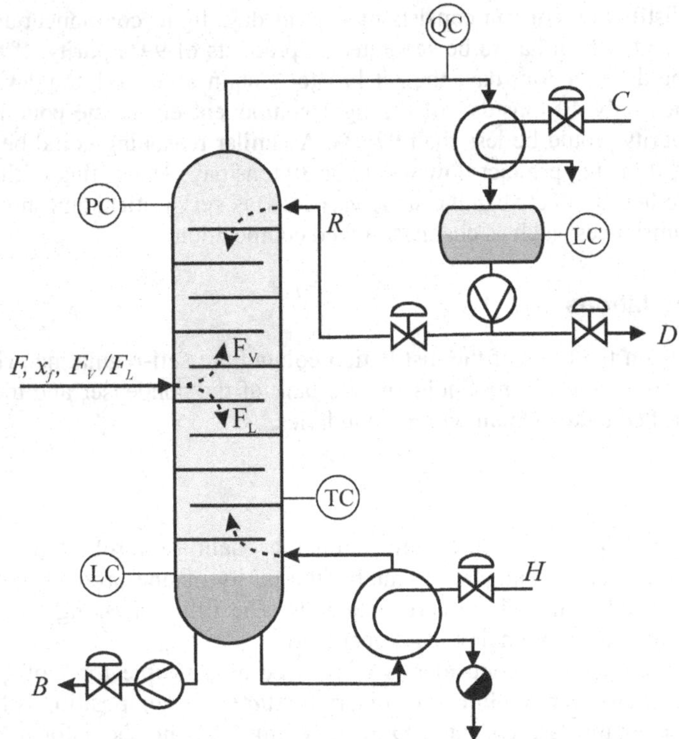

Fig. 34.2. Correcting variables of a distillation column with independent feed.

In total there are now five correcting variables (D, B, H, C, R) and five controlled variables (the level in the top and bottom accumulator (L_D en L_B); the pressure (P); and the top and bottom quality (x_D and T)). Figure 34. 2 shows the result.

34.1.7 Count of Degrees of Freedom

Since this was already discussed in the previous section, we will not address this further.

34.1.8 Power and Speed of Control

For every pair of correcting and controlled variables, the power and speed of control should be investigated. This results in a table, which can serve as a guide to develop a basic control scheme. The relationships shown in the sequel were developed by Rademaker *et al.* (1975) and discussed in Luyben *et al.* (1992).

Pressure Control

B and D should not be used for pressure control, owing to a lack of power of control. R cannot be used either, unless the degree of under-cooling is extremely large. C and H have a large power of control, since they affect the incoming and leaving vapor flow immediately.

The dynamic response of the pressure to variations in C and H can be analyzed with the help of Fig. 34.3, which shows part of the interactions in the distillation column. The response of the top vapor flow (V_c) on the cooling of the condenser (C) is usually relatively fast; the time constant is determined by the heat capacity of the pipes (J/K), divided by the sum of the heat transfer coefficients and corresponding areas (W/K) inside and outside the pipes. The evaporator (vapor flow V_H) will usually also show fast dynamic behavior. The difference between the vapor flow from the reboiler and to the condenser, yields after integration the response of the pressure (the pressure difference across the column is ignored for reasons of simplicity). The speed of pressure changes is limited by the total heat capacity of the trays, bottom and top.

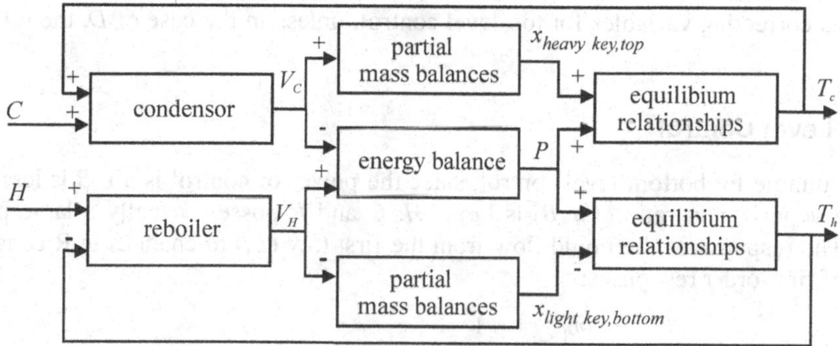

Fig. 34.3. Information flow diagram for temperature and pressure responses to changes in C and H.

The static value of the pressure response is determined by the so-called self-regulation in the reboiler and condensor (see Fig. 34.3): a higher pressure results in higher top and bottom temperatures, which result in improved heat transfer in the condensor and decreased heat transfer in the reboiler. Eventually, the equilibrium between V_C and V_H will be restored.

The self-regulation is, however, also affected by the composition responses (the so-called composition self-regulation): a step change in H results, via a higher pressure and higher top temperature, in an increased top vapor flow. This will result in an increase of less volatile component in the top, as a result of which the top temperature increases even more and also the top vapor flow increases even more. As a result of this positive feedback, the pressure might eventually end up at a lower value (see Fig. 34.4).

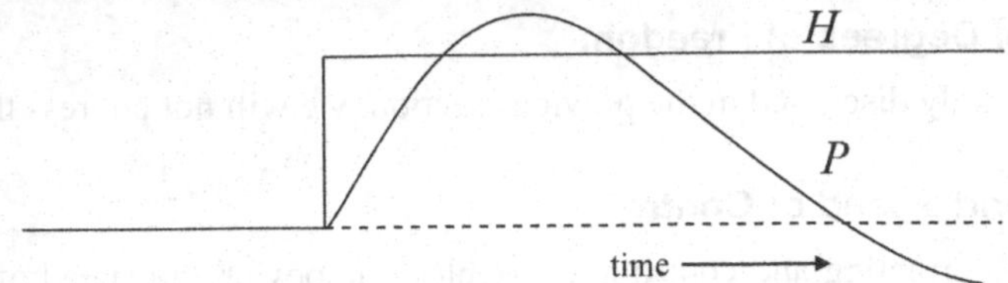

Fig. 34.4. Pressure repsonse to a step change in H.

It is evident that owing to this dynamic behavior, H is not suitable for pressure control. The self-regulation of the composition response in the bottom (the composition of the light key component in the bottom) yields however a negative feedback, as a result of which the pressure response to changes in C is not negatively affected.

The conclusion is that C and H are suited for pressure control (the speed of control is determined by small time constants!), unless, in the case of H, the 'composition self-regulation' in the condenser is too strong.

Top Level Control

B is not suitable for top-level control, since the power of control is nil. D is less suitable when the reflux ratio (R/D) is large. C and R possess usually a large power of control. The response of the level to changes in D, R and C are favorable for control: the most important dynamic element is an integrator. The response to variations in H is unfavorable: the condensate flow from the condenser will only increase if the pressure increases. The response can therefore be approximated by a cascade of pressure dynamics (large time constant) and level dynamics (integration). The conclusion is that C, R and D are candidates to serve as correcting variables for top-level control, unless in the case of D, the reflux ratio is large.

Bottom Level Control

D is not suitable for bottom level control, since the power of control is nil. B is less suitable when the vaporization ratio (V_H/B) is large. H, C and R possess usually a large power of control. The response of the liquid flow from the first tray (L_1) to changes in R consists of a cascade of first-order responses:

$$\frac{\delta L_1}{\delta R} = \frac{1}{(1+\tau_L s)^N} \tag{34.1}$$

in which τ_L = hydraulic time constant for each tray (including the downcomer) and N = number of trays.

Even if τ_L is small (for example 2 s.), a large number of trays (for example 40) will result in a considerable effective dead time (80 s.), which, together with the hydraulic dynamics of the reboiler and integration of the bottom level, will result in slow level control (oscillation period larger than 6 times 80 s).

The response of the bottom level to changes in H is dynamically unfavorable when the sensitivity of the liquid flow in the column for variations in the vapor flow has an unfavorable value. This is the case when the parameter

$$\lambda = \left(\frac{\partial L}{\partial V}\right)_{M_L} \tag{34.2}$$

in which L = liquid flow, V = vapor flow and M_L = amount of liquid on a tray (including downcomer),

is larger than 0.5 (see appendix 34.1 for a detailed analysis). Unfortunately, not much information on the value of λ as a function of the tray loading for different types of columns is available. Data that is available indicates that λ_L can be large at low tray loading (even > 1), at high tray loading values tend to be small or zero or can even become negative. (Betlem, 1998). The conclusion is that B is a favorable correcting variable, unless the value of V_h/B is large, and that H is a favorable correcting variable unless $\lambda > 0.5$.

Quality Control

The response of the key components can roughly be approximated by the response of the algebraic sum of the relative local flow changes, followed by a first order with large time constant (appendix 34.2):

$$\delta x_i = \left(\frac{\delta L_{i+1}}{L_{i+1}} - \frac{\delta V_{i-1}}{V_{i-1}}\right)\frac{K_x}{1+\tau_x s} \tag{34.3}$$

in which L_{i+1} = liquid flow to tray i, V_{i-1} = vapor flow to tray i and τ_x = large time constant (proportional to the square of the number of trays).

The response of the bottom quality to changes in the reflux is therefore unfavorable for automatic control for the same reason as was explained for the bottom level. This is not the case for the top quality or the quality on a tray close to the top, since the number of trays is small in that case. Similarly, the vapor flow (with pressure control on C or H respectively) is less suitable for bottom quality control when the parameter:

$$\lambda^* = \frac{V}{L}\left(\frac{\partial L}{\partial V}\right)_{M_L} \tag{34.4}$$

is larger than 0.5.

B and D have no direct impact on the product qualities. In conclusion, it can be said that R, C and H are suitable for control of the top quality. C and H are suitable for control of the bottom quality, unless $\lambda_L^* > 0.5$. There is then no alternative available.

Selection of the Basic Control Scheme

With the help of Table 34.1, which represents the conclusions and considerations of the previous sections, the different basic control schemes can be found. A double minus sign indicates an undesirable combination and a double plus sign a favorable combination.

Table 34.1. Interaction table between correcting and controlled variables.

	L_D	L_B	P	x_D	T
D	unless $R/D \geq 5$	—	—	—	—
B	—	unless $V_H/B \geq 5$	—	—	—
H	—	unless $\lambda > 0.5$	see 34.1.8 Pressure control	—	unless $\lambda^* > 0.5$
C	++	—	++	++	unless $\lambda^* > 0.5$
R	++	—	—	++	—

The control schemes have to be capable of accommodating variations in the feed flow and feed composition. For example: increased feed at the same feed composition means increased top and bottom product.

A more volatile feed at the same feed flow has to be reflected in more top product and less bottom product, etc. *D* and *B* can therefore not be omitted from the control scheme, they both have to be adjusted via controllers that are adapted to the changing circumstances.

A distillation column has five variables that can be adjusted:

- *D*, distillate flow
- *B*, bottom flow
- *H*, reboiler duty
- *C*, condenser duty (or vapor flow leaving the column)
- *R*, reflux ratio (internally or externally)

These five variables can be used to control the controlled variables. These five could be:

- L_D, liquid level in the reflux drum
- L_B, bottom level in the column
- *P*, pressure
- x_D, quality of the top product
- *T*, inferred quality of the bottom product

Combining five controlled variables with an equal number of correcting variables to *single-input–single-output* (SISO) control loops results in $5! = 120$ possibilities for developing the final control scheme.

In developing a basic control scheme, the criteria that should be considered are:

- power and speed of control of the control loops
- minimum interaction between the control loops
- minimum impact of the control loops on downstream process units

The most suitable control scheme is a compromise taking the control goals into account. A first scheme according to these criteria is shown in Fig. 34.5. If there are also stringent demands on the bottom quality, an additional quality controller can be installed that manipulates the setpoint of the column bottom temperature controller. In order to achieve disturbance rejection for the top quality, it would be advantageous to also install a temperature controller in the top of the distillation column and cascade the top quality controller to it, as shown in Fig. 34.5. Further disturbance rejection could be achieved by installing flow controllers on all the flows (*D*, *B*, *H*, *C*, *R*) and cascading the controllers shown in Fig. 34.5 to these flow controllers, i.e. cascade TC to FC on *R*, LC to FC on *D*, LC to FC on *B*, TC to FC on *H* and PC to FC on *C*. The advantage of cascade control was explained in detail in the previous chapter.

34.2 Material and Energy Balance Control

Often a goal of a control scheme for a distillation column is to maintain the quality of the top product on specification while also maintaining the material balance. Material balance control and energy balance control are two control schemes that can achieve this. Different criteria can be considered for the selection of a control scheme. Since the performance of both control schemes has a large impact on the profitable operation of the distillation column, this choice is not trivial.

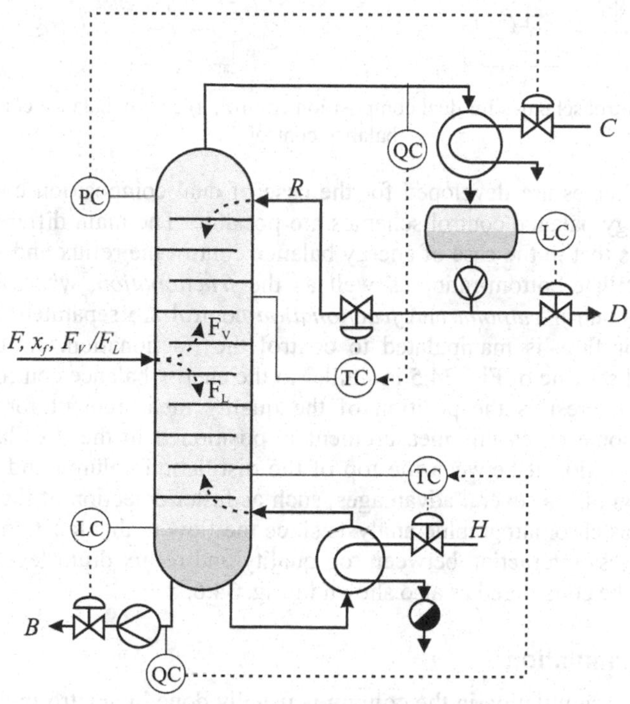

Fig. 34.5. Possibility for basic control scheme.

The name "material balance control" was introduced by Shinkey (1984). The different control schemes that the author developed were based on the concept of *relative gains* (= power of control) of the different input-output combinations. Speed of control was only considered as a secondary factor. A simple explanation is given by Ryskamp (1980). Also Van der Grinten (1970) presented a number of common control schemes for distillation columns. The latter author used behavioral models in the control scheme selection procedure. None of the mentioned references takes inverse responses into account when $\lambda > 0.5$. In the case of the more traditional approach, the energy balance control, the reflux ratio and/or vapor flow is used to control the top product quality, while the distillate and bottom flow are used to maintain the mass balance. In the case of the material balance control, one of the product flows is used to control product quality, while the other product flow maintains the material balance.

Figure 34.6 shows a schematic for energy balance control and material balance control.

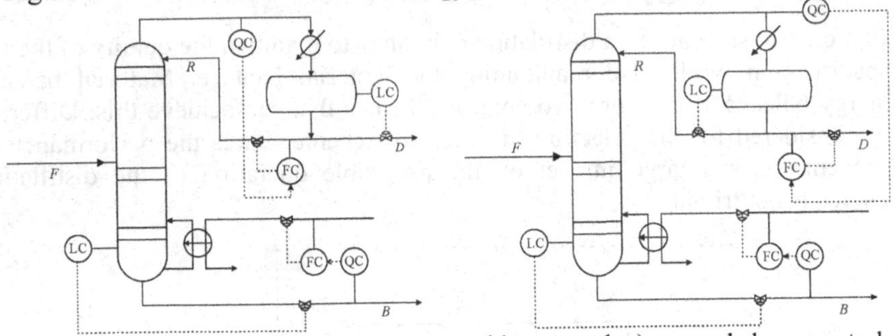

Fig. 34.6. Control schemes for dual composition control: a) energy balance control, b) material balance control.

The control schemes are developed for the case of dual composition control. Also other material and energy balance control schemes are possible. The main difference between the control schemes is that in the case of energy balance control the reflux and vapor flow affect the *cutpoint* (distillate-bottom-ratio) as well as the *fractionation*, whereas in the case of material balance control *cutpoint* and *fractionation* control are separated. Either the reflux ratio or the vapor flow is manipulated to control the fractionation. As can be seen, the developed control scheme of Fig. 34.5 is similar to the energy balance control scheme.

One point of interest is the position of the quality measurement for the top product composition. In some cases this measurement is positioned in the distillate flow, another possibility is to position it between the top of the distillation column and the reflux drum. The latter situation offers several advantages, such as faster detection of the top composition changes, easier gas chromatographic analysis since the flow at this location is already in the vapor phase and less interaction between top quality and reflux drum level. Therefore only this situation will be considered as also shown in Fig. 34.6.

Control of Accumulation

Control of liquid accumulation in the column is usually done by controlling the levels of the reflux drum and bottom. The level in the reflux drum can be controlled by the reflux flow (R), in which case the control scheme is called material balance control, or the top product draw-off (D), in which case it is called energy balance control. To make the right selection for reflux drum level control, the following factors should be considered:

- control goals for the level
- power of control of the level controller, with respect to changes in D en R
- the effect of control actions on downstream process units

As pointed out earlier, the power of control of D is limited when $R/D > 5$. If the reflux flow would stay constant during a change in vapor flow, a relatively large change has to be made in the top draw-off to maintain a constant reflux drum level. This might lead to a violation in a minimum or maximum draw-off constraint violation. Thus, at high reflux ratios (>5), manipulating the reflux flow would result in better controllability and sensitivity of the level of the reflux drum, the drawback will be a slow response of the top quality. The reverse is true for low values of R/D (<0.5). For reflux ratios between 0.5 and 5.0, either control scheme can be used.

If energy balance control is used in the case of $R/D > 5$, the small power of control can be increased by maintaining a constant R/D ratio. The product flow D is still used as correcting variable to maintain a constant reflux drum level. A higher level will then also result in an increased reflux flow, consequently, the level will respond faster.

The product flow of a distillation column that is controlled by an energy balance control scheme may fluctuate, since it is controlled by a level controller; consequently it will affect the downstream process unit. If a distillation column is controlled by a material balance control scheme, the distillate flow is affected by the slow quality control loop and disturbances will be smoothed and only partially propagated to downstream process units. Therefore, when the product flow is not allowed to fluctuate or if the one product flow is much larger than the other, a material balance control scheme is selected.

In the case of material balance control, the correcting actions of the quality controller on the product flow have no effect on the distillation column, i.e. the top product quality, until the level controller adjusts the reflux ratio. The controller should therefore be carefully tuned, such that the dynamics of the level control loop are reduced to a minimum. If material balance control is applied and the reflux drum is large, the power of control can be increased by keeping the ratio R/D constant with a flow ratio controller, which is adjusted by the level controller as a master controller. In that case the reflux is still the adjustable variable.

For most distillation columns, the bottom level is controlled by the bottom draw-off. Sometimes, the bottom draw-off is used for quality control. In that case, the bottom level can be controlled, by adjusting the load of the reboiler. Problems can originate when there is an inverse response. This happens when $\lambda > 0.5$.

The smallest product flow can only be kept constant if a flow controller is used for its control.

Interactions in the case of Dual Composition Control

When considering alternative control schemes, one should pay attention to interaction between control loops. Interaction means a mutual influence between the control loops, one-sided interaction is bothersome but does not affect the stability of the control system. An undesirable effect of interaction can be that the dynamic behavior of one control loop is strongly affected by another control loop. An extreme case of undesirable interaction is a situation in which a sign change takes effect, negative feedback changes to positive feedback and consequently the control loop becomes unstable.

Interaction could exist in case both product qualities have to be controlled. Interaction does not pose a problem, if the product quality of only one component is controlled. If both quality control loops are configured as single loop feedback controllers, the manipulation of the one control loop will affect the response of the other control loop and vice versa. For example, manipulation of the reboiler load to control the bottom product quality, will affect the top product quality. This interaction between the single loop controllers can lead to a considerable degradation in performance of the control system for the distillation column. Minimal interaction is required.

Dual Composition Energy Balance Control

Simulations were performed with a column separating toluene and *o*-xylene in a 30-tray column. The top product quality was controlled at 98% purity, the bottom quality at 97% purity. Holdup of the reflux drum was equal to ten times the average tray hold-up; hold-up of the bottom was equal to 15 times the average tray hold-up. The reflux to distillate flow ratio was equal to 1.6. As mentioned before, only the case where the top quality measuring device is located between the top of the column and reflux drum will be considered, because of its favorable dynamics.

The energy balance control scheme provided excellent control of the top quality, for a change in feed flow of +10% there was no noticeable change in top quality (<0.001%). The bottom quality controller responded as shown in Fig. 34.7.

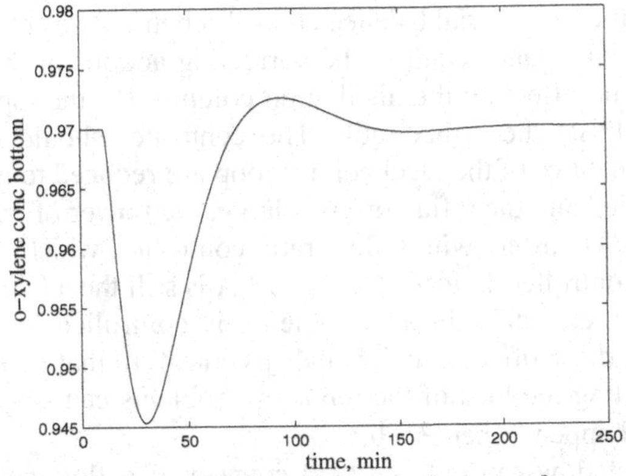

Fig. 34.7. *O*-xylene response to feed flow change in the case of energy balance control.

Dual Composition Material Balance Control

Tuning the top product quality controller proved to be more difficult in the case of a material balance control scheme. This can easily be understood from the block diagram of Fig. 34.8. The top half of Fig. 34.8 shows the energy balance control scheme. There is only one-sided interaction between the control loops, the quality control loop is an independent control loop while the level control loop is affected by the reflux changes. This is no problem for the level controller, since level control does not have to be tight.

The bottom half of the figure shows the material balance control scheme in which the top quality is controlled by the distillate draw-off. However, the draw-off does not affect the top quality but rather the level. The level in turn affects the reflux flow, which subsequently affects the top quality. This means that there is a severe degree of mutual interaction between the control loops. It was found that the control structure of Fig. 34.8b would result in an oscillatory behavior of the quality control loop for feed flow changes of +10%. Only addition of a feed-forward loop from distillate flow to reflux would stabilize the quality control loop. In that case, the distillate flow changes were subtracted from the reflux flow changes calculated by the level controller. After addition of feed-forward control, the response of the top and bottom qualities were similar to the responses of the energy balance control scheme for this situation with virtually no deviation from setpoint of the top quality and a bottom quality response similar to Fig. 34.7.

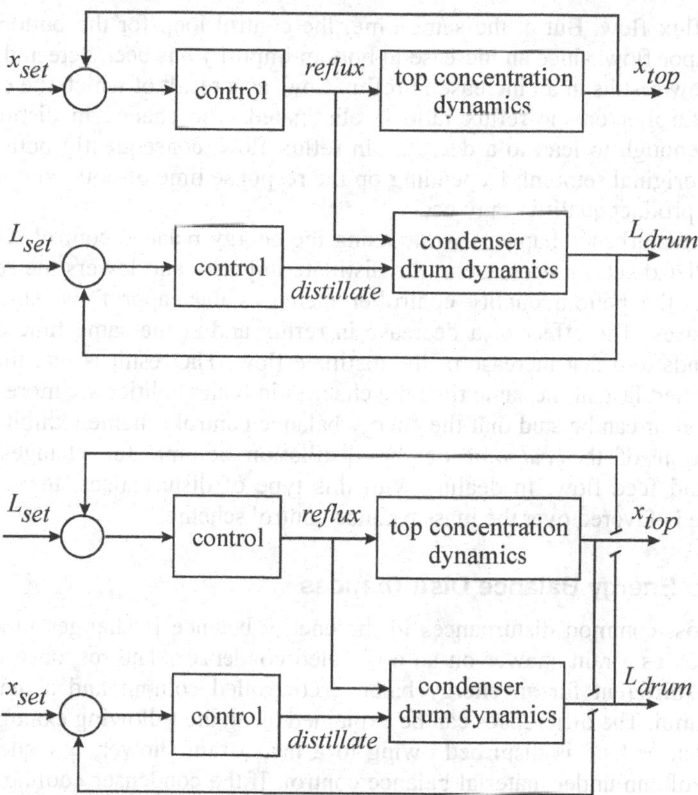

Fig. 34.8. Difference between energy balance and material balance control.

Sensitivity to Material Balance Disturbances

In many cases feed flow and composition disturbances are present and setpoint changes are normally not required. Process control schemes are designed to reduce the effect of disturbances on the operation of the process. In the case of a distillation column, in which the quality of both products is controlled, a material balance control scheme and an energy balance control scheme will bring the process back to a steady situation after a disturbance has taken place. However, different control schemes will show different dynamic responses to disturbances. In addition, there are control schemes that exhibit better self-regulation of the product quality to disturbances. In that case there will be a better response of the quality towards disturbances.

Energy balance control schemes use both product flows for level control. If the feed flow is partially evaporated, then the liquid part of the feed disappears with the bottom product and the vapor part of the feed with the top product. When the ratio between vapor and liquid flow approaches the D/B ratio, the effect of a changing feed flow on both product qualities becomes increasingly less. Hence, when the ratio between vapor and liquid flow in the column is approximately equal to D/B, the energy balance control scheme is less sensitive to feed disturbances than the material balance control scheme.

The difference between the two control schemes can also be demonstrated by looking at the responses of the quality control loops for a similar disturbance in the feed composition or feed flow. If the feed contains more of the light component than the heavy component, the top quality control loop will increase the distillate flow in the case of a material balance control scheme when a decrease in feed impurity is detected. Initially, this will lead to a decrease in reflux flow. But at the same time, the control loop for the bottom quality will increase the vapor flow, since an increase in bottom impurity has been detected. The increase of the vapor flow results in an increase in reflux flow, as a result of which the effect from the top quality controller on the reflux ratio is eliminated. The change in distillate flow will become large enough to lead to a decrease in reflux flow, consequently both qualities will return to their original setpoint. Depending on the response time of both control loops, large changes in the product qualities may occur.

If the same disturbance happens while using the energy balance control scheme, the top quality controller detects a decrease in the distillate impurity and lowers the reflux flow. At the same time, the bottom quality controller increases the vapor flow, since the bottom impurity increases. The effect of a decrease in reflux and at the same time an increase in vapor flow, leads to a fast increase of the distillate flow. The result is that the new desired *cutpoint* is reached fast, at the same time the changes in both qualities are more acceptable.

In conclusion, it can be said that the energy balance control scheme exhibits some degree of self-regulation of the *cutpoint* of the distillation column for changes in the feed composition and feed flow. In dealing with this type of disturbances, the energy balance control scheme is favored over the mass balance control scheme.

Sensitivity to Energy Balance Disturbances

One of the most common disturbances in the energy balance is changes in environmental conditions, such as a rain shower on an air-cooled condensor. The response to this type of disturbance is different for an energy balance controlled column and a material balance controlled column. The differences can be explained using the following example.

If the condenser load is disturbed owing to a heavy rain shower, a sequence of events happens in a column under material balance control. If the condenser cooling increases, the vapor flow in the column will increase (see Fig. 34.3). This will lead to an increase in the level of the reflux drum. Since the column is on material balance control, the liquid flow to the column will increase, as a result of which the ratio of vapor flow to liquid flow V/L remains more or less constant.

The *cutpoint* remains therefore the same and only the fractionation changes slightly. As a result, changes in both product qualities are small. Hence, a material balance control scheme will possess a certain degree of self-regulation for the internal reflux, this contributes to a minor effect on both product qualities.

In the case of an energy balance control scheme, in which the top product flow is adjusted by the reflux drum level controller, a similar change in condenser load may lead to a major shift in *cutpoint*, since the higher vapor flow is not compensated by a higher liquid flow. Consequently, the effect on the product qualities will be significant. Therefore a material balance control scheme is favored over an energy balance control scheme in the case of disturbances in energy flows. In case if energy balance control, the fractionation and the *cutpoint* can be maintained at setpoint by maintaining a constant vapor-reflux ratio, using a slave flow ration controller. The quality controller can act as the master controller, as explained in appendix 34.2.

Summary

Table 34.2 shows a summary of the guidelines for selection of an energy balance control scheme or a material balance control scheme. Interaction between the control loops has not been listed, since a simple decoupling action can often eliminate this undesired effect.

Table 34.2. Advantages and disadvantages of material and energy balance control schemes.

Factor	**Material balance control scheme**	**Energy balance control scheme**
Power of control of the product flow	for $R/D > 5$ ($V/B > 5$)	for $R/D < 0.5$ ($V/B < 0.5$)
Constant product quality requirement	0	0
Material balance disturbances (feed flow, feed composition)	—	++
Energy balance disturbances (cooling medium, steam, weather, etc.)	++	—

References

Betlem, B.H.L. (1998) Influence of tray hydraulics on tray column dynamics. *Chemical Engineering Science*, **53**, 3991–4003.
Grinten P.M.E.M. van der (1970), Procesregelingen, Prisma-Technica 140, Spectrum, Utrecht.
Luyben W.L. (ed.) (1992) *Practical Distillation Control*, Van Nostrand Reinhold, New York.
Shinskey F.G. (1984) *Distillation Control*, McGraw-Hill, New York.
Rademaker O., Rijnsdorp, J.E. and Maarlevelt, A. (1975) *Dynamics and Control of Continuous Distillation Units*, Elsevier Scientific Publishing Company, Amsterdam.
Ryskamp C.J. (1980) New strategy improves dual composition column control. *Hydrocarbon Processing*, June, 51–9.

Appendix 34.1 Impact of Vapor Flow Variations on Liquid Holdup

In chapter 16 the liquid dynamics for a distillation column were derived. In section 16.8 it was shown that the response of the bottom hold-up to vapor flow changes can be written as:

$$\frac{\delta M_0}{\delta V} = -\frac{1}{s}\left[1-\lambda+\frac{\lambda}{(1+\tau s)^n}\right] \tag{34.5}$$

Bottom level control using H is slowed down by the λ effect, if $\lambda > 0.5$, which is clear from the following approximation. The closed loop transfer function with proportional control is:

$$\phi_H \frac{K_L}{s}\left\{1-\lambda_1+\frac{\lambda_1}{(1+\tau_L s)^n}\right\} \tag{34.6}$$

in which K_L is the gain of the level controller, and ϕ_H contains the reboiler dynamics.

After approximation of the last term by a dead time $n\tau_L$, the oscillation period P_u and the closed loop gain $K_{L,u}$ on the limit of stability can be found. For simplicity reasons ϕ_H is assumed to be unity. If $\lambda = 0.5$ it is found that:

$$P_u = n\tau_L, \quad K_{L,u} = \infty \tag{34.7}$$

For $\lambda = 1$, Eqn. (34.6) reduces to the equation for bottom level control using the reflux, which is:

$$P_u = 4n\tau_L, \quad K_{L,u} = \frac{\pi}{2}\frac{1}{n\tau_L} \tag{34.8}$$

Appendix 34.2 Ratio Control for Liquid and Vapor Flow in the Column

For every component the partial mass balance on a tray can be written as:

$$\frac{d(M_{L,i}x_i)}{dt} = L_{i+1}x_{i+1} - L_i x_i + V_{i-1}y_{i-1} - V_i y_i \tag{34.9}$$

in which x = concentration if the liquid phase and y = concentration in the vapor phase, ignoring the vapor mass.

The total mass balance is:

$$\frac{dM_{L,i}}{dt} = L_{i+1} - L_i + V_{i-1} - V_i \tag{34.10}$$

Subtraction of x_i multiplied by the mass balance from the partial mass balance gives:

$$\frac{M_{L,i}}{L_{i+1}}\frac{dx_i}{dt} = (x_{i+1} - x_i) + \frac{V_{i-1}}{L_{i+1}}(y_{i-1} - y_i) + \frac{V_{i-1} - V_i}{L_{i+1}}(y_i - x_i) \tag{34.11}$$

The last term can be ignored since $V_{i-1} = V_i$.

Now assume that the column is initially at steady state, hence $dx_i/dt = 0$. If the ratio V_{i-1}/L_{i+1} is maintained constant, Eqn. (34.11) represents an equation for the concentration between successive trays. The only way in which the concentrations can change is through propagation of the concentration from tray to tray, which is a slow and gradually changing process. It is evident that the ratio V_{i-1}/L_{i+1} cannot be maintained constant on all trays, this is however possible at both column ends by controlling the ratios of the external flows.

35 Control of a Fluid Catalytic Cracker

In this chapter, control of a fluid catalytic cracker will be discussed. A simulator, based on an available model as discussed in the literature, has been developed, which allows the reader to simulate various control schemes and to see the effect of interaction between control loops. Also, the selection of control loops is discussed for this case study. In this particular case study, there are more controlled variables than manipulated variables, which creates the need to develop cascade control structures. Some problems that could occur are highlighted.

35.1 Introduction

A fluid catalytic cracker breaks down hydrocarbons with high boiling points into hydrocarbons with lower boiling points and higher added values. In the literature, modeling and control of catalytic crackers has been the subject of many papers (McFarlane *et al.*, 1993; Khandalekar and Briggs, 1995; Ansari and Tade, 2000). It does not happen very often that complete models of a fluid catalytic cracking unit are given; however, McFarlane *et al.* (1993) present a complete simplified model of such a processing unit. The model serves as a benchmark for non-linear multi-variable control problems. The model, called the Amoco model, will serve as the starting point for this chapter; a schematic diagram of the process is shown in Fig. 35.1.

The feed of a catalytic cracker is an in-line blending of multiple streams with different compositions originating from several refinery processes. In this typical case, gas oil is mixed with diesel oil (lighter than gas oil) and wash oil (heavier than gas oil) and serves as the feed F_{oil} to a furnace, where it is heated to about 670 °F (350 °C). The heated feed is mixed with the hot slurry recycle flow from the bottom of the fractionator. The side-stream feeds are on a fixed flow control setting; however, the total oil feed can be adjusted. The feed flow is injected into the reactor riser, where it is mixed with hot regenerated catalyst and evaporated totally. In the riser endothermic cracking takes place at a temperature and pressure of about 995 °F (535 °C) and 30 psia (2.0 bara) respectively. The gaseous reaction products are passed to a cooled main fractionator where the wet gas products (C_6 and lighter) are separated from the heavier liquid, which is recycled. After compression, the wet gas is separated in its subsequent fractions. The catalyst flows from the riser into the stripping section of the reactor. This is a fluidized bed in which steam is injected to remove entrained hydrocarbons.

Process Dynamics and Control: Modeling for Control and Prediction. Brian Roffel and Ben Betlem.

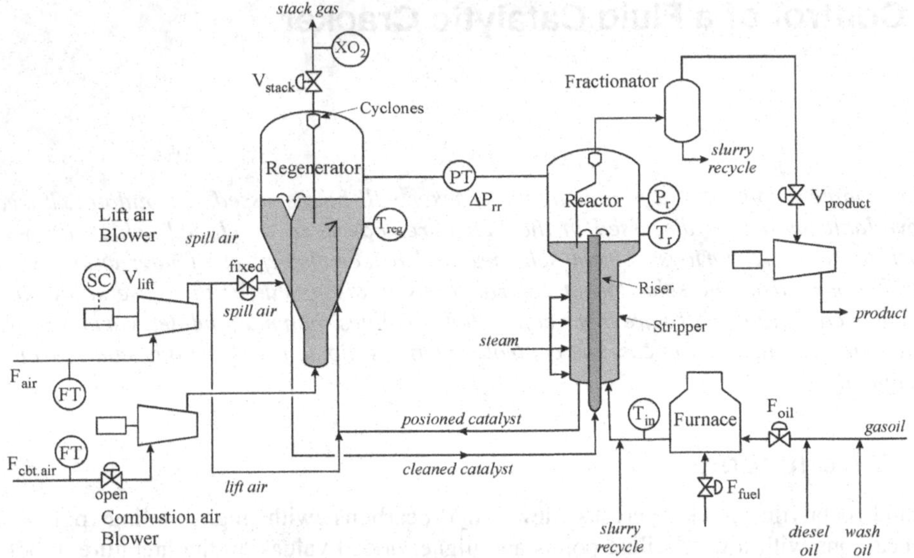

Fig. 35.1. Schematics of a fluid catalytic cracker.

As a result of the cracking reactions, coke is deposited on the catalyst, consequently the catalyst is poisoned and has to be regenerated. This exothermic regeneration process is carried out by circulating it to a fluidized bed regenerator, where under excess oxygen, the coke is burned off the catalyst at a temperature and pressure of about 1272 °F (690 °C) and 34 psia (2.3 bara) respectively. The process conditions should ensure that nearly all carbon monoxide produced in the bed is converted to carbon dioxide. The carbon monoxide concentration in the stack gas should meet the following constraint: $X_{CO} < 10^{-4}$ mol/mol.

The catalyst circulation rate takes care of the balance between coke formation and decoking and the exchange of heat between the endothermic cracking reaction and the exothermic regeneration. In this design, pressurized lift air is injected into the bottom of the lift pipe to ensure the catalyst feed to the regenerator. The regenerated catalyst flows over a weir into the regenerator standpipe and the head of the catalyst in this standpipe provides the driving force for the recycle flow. The head in the standpipe, the differential pressure between the reactor/regenerator combination together with the air lift flow rate determine the catalyst circulation rate. The pipelines of this circulation system should not include any valves.

Even though McFarlane *et al.* (1993) do not provide a model for the fractionator, the model for the fluid catalytic cracker is a complete model, accurate enough to show some of the dynamic effects and control interactions that are present.

The model essentially consists of the following sub-models for:

- feed system
- preheat system
- reactor

  The reactor model consists of various sub-models, such as:
  - o coke and wet gas yield models
  - o reactor mass balances
  - o reactor riser energy balance
  - o reactor riser pressure balance
  - o reactor and main fractionator pressure balances
- wet gas compressor

- regenerator
  The regenerator balances consist of:
  - energy balance
  - carbon balance
  - mass balance
  - equations for the volume fraction of catalyst
  - equations for catalyst entrainment
  - pressure balance
  - bed height and air lift calculations
- air blowers
  - combustion air blower
  - lift air blower
- catalyst circulation

35.2 Initial input–output Variable Selection

Hovd and Skogestad (1993) give a comprehensive treatment of the selection of a regulatory control structure for a catalytic cracking unit. This discussion includes the selection of controlled and manipulated variables. Based on this discussion and the discussion in Huq *et al.* (1995) and Khandelakar and Briggs (1993), the controlled and manipulated variables are selected, as shown in Table 35.1.

Table 35.1. Initial table of controlled and manipulated variables.

	F_{air}	P_r	ΔP_{rr}	T_{in}	X_{O_2}	T_r	T_{reg}
F_{oil}	Independent variable for optimization						
F_{fuel}							
V_{lift}							
$V_{product}$							
V_{stack}							

Some local controls are assumed to be in place, such as anti-surge control for the compressors. In addition, some valves are fully open or fixed, as shown in Fig. 35.1. The selection of controlled variables is strongly related to the goals that are set for operation of the fluid catalytic cracker. There are numerous constraints that all have to be met. Moreover, two main objectives should be met: maximization of the total feed rate and disturbance rejection.

On the horizontal axis in Table 35.1, the controlled variables are listed; the manipulated variables are listed vertically. The variables have the following meaning:

- F_{air}, air flow to regenerator
- P_r, reactor pressure, $P_r \leq 49.2$ psia
- ΔP_{rr}, pressure difference between reactor and regenerator, constraint values are: $-5.0 \leq \Delta P_{rr} \leq 2.0$ psia
- T_{in}, furnace outlet temperature, is equal to the temperature at the inlet of the reactor riser
- X_{O_2}, stack gas oxygen concentration, $X_{O_2} \geq 1.5\%$
- T_r, reactor riser temperature, $T_r \leq 1000$ °F
- T_{reg}, regenerator temperature, $T_{reg} \geq 1265$ °F
- F_{oil}, feed flow to furnace
- F_{fuel}, fuel flow

- V_{lift}, lift air blower steam valve, $0 \le V_{lift} \le 1.2$
- $V_{product}$, wet gas compressor suction valve position, $0 \le V_{product} \le 1.0$
- V_{stack}, stack gas valve position, $0 \le V_{stack} \le 1.0$

There can only be four manipulated variables: F_{fuel}, V_{lift}, $V_{product}$ and V_{stack} , since the oil flow is chosen as an independent variable. The model with these input and output variables is simulated in FCCU0.mdl. This program makes use of several subprograms: each subprogram simulates a particular piece of equipment. Step disturbances can be given in the manipulated variables to get an impression of the dynamics of the uncontrolled process. First it will be determined how these manipulated variables can be used to control the four controlled variables, which still have to be selected.

From Fig. 35.1 it is clear that the regenerator air supply can best be controlled by using V_{lift}. There will be no other manipulated variable that would give a more favorable and faster response. The valve in the spill air line is fixed to achieve a division of air between the spill air line and lift air line.

To operate the catalytic cracker at the required pressure, the reactor pressure P_r is chosen as a controlled variable. The pressure in the regenerator needs then to be controlled, to guarantee a reasonable pressure difference ΔP_{rr}, such that there is a guaranteed circulation of catalyst for regeneration. In this case we shall not control the pressure in the reactor and regenerator separately, but choose the reactor pressure and pressure difference between reactor and regenerator as controlled variables.

Now we need to make a choice how to control the pressure P_r and the pressure difference ΔP_{rr}. It is obvious that gas flows will be the proper choice to control pressures. Therefore the possible candidates are the stack gas flow valve V_{stack} and the wet gas compressor suction valve $V_{product}$. A step disturbance of –5% was given in $V_{product}$ and +5% in V_{stack}. The changes in ΔP_{rr} and P_r were monitored and the absolute value of the process gain was estimated from the tests. The results as shown in Table 35.2 were obtained (percent change divided by per cent change).

Table 35.2. Control variable selection for pressure and pressure difference control.

	ΔP_{rr}, pressure difference reactor and regenerator	P_r, reactor pressure
$V_{product}$, wetgas compressor suction valve	$K_p = 3.85$ Fast first-order response	$K_p = 0.46$ First order response with moderate time constant
V_{stack}, stack gas flow valve	$K_p = 17.88$ Fast first-order response with slow integrating tail	$K_p = 0.39$ Fast first-order response with slow integrating tail

As can be seen, the proper choice for controlling ΔP_{rr} is the stack gas flow valve V_{stack}. Since this gain is large, corrections will not be excessive and it is expected that interaction with reactor pressure control using the wet gas compressor suction valve $V_{product}$ will be limited.

There is one more choice that is relatively easy to make. Since T_{in} represents the temperature of the flow entering the reactor riser, it is crucial that it is controlled to provide the required energy for the cracking process. The obvious choice is to control it by using the fuel flow F_{fuel}.

Based on these choices, we have an initial basic control scheme, as shown in Table 35.3.

Table 35.3. Initial basic control scheme after first selection.

	F_{air}	P_r	ΔP_{rr}	T_{in}	X_{O_2}	T_r	T_{reg}
F_{oil}	Independent variable for optimization						
F_{fuel}				++			
V_{lift}	++						
$V_{product}$		++					
V_{stack}			++				

The process with this basic control scheme is shown in Fig. 35.2; it is simulated in program FCCU1A.mdl. The program can be used to study the tuning of the basic control scheme and the improved process operation compared with the uncontrolled process. Table 35.4 summarizes the controller settings.

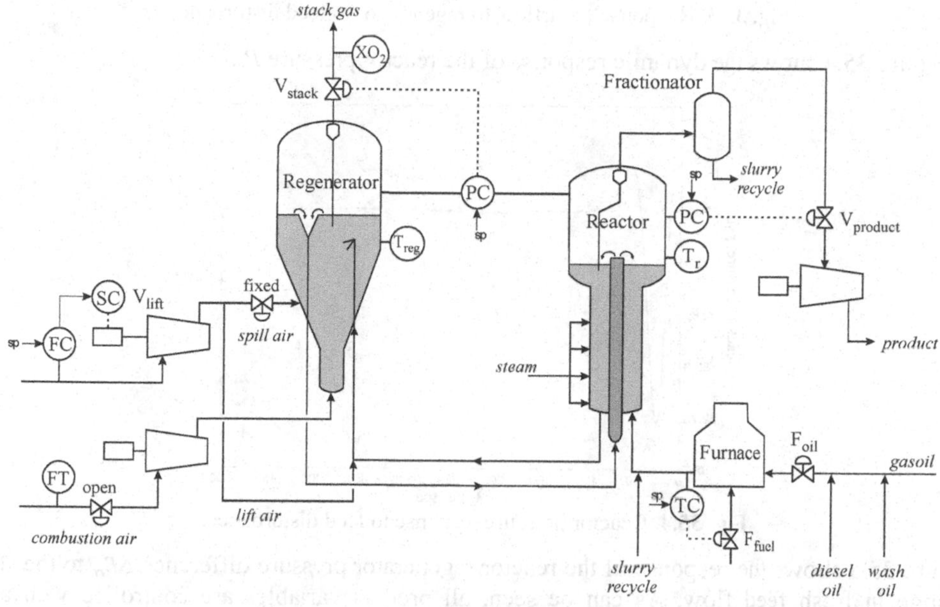

Fig. 35.2. Basic control scheme of a fluid catalytic cracker.

A step change of approximately –5% was given in one of the most important process variables, the fresh feed flow F_{oil} , which is a measure for the load of the process. The response of the controlled variables was monitored, Fig. 35.3 shows the response of the air flow. It can be seen that it is virtually constant.

Table 35.4. PID controller settings for basic control scheme.

	K_c	T_i	T_d
Air flow controller	5.0	2.0	0.0
Reactor pressure controller	0.2	10.0	0.0
Reactor/regenerator pressure difference controller	0.008	8.0	0.0
Furnace outlet temperature controller	0.03	6.5	10.0

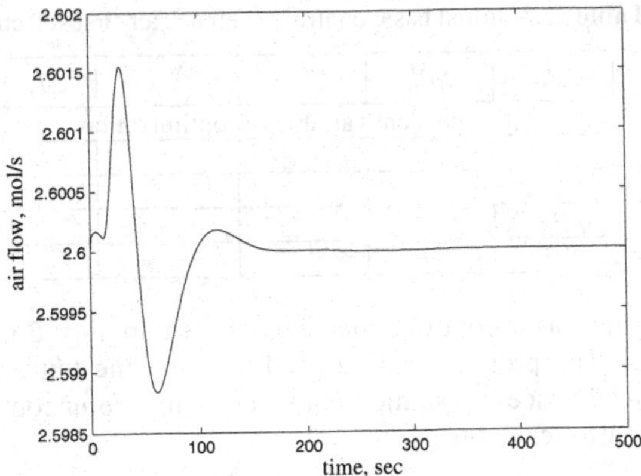

Fig. 35.3. Response of airflow to regenerator to feed disturbance.

Figure 35.4 shows the dynamic response of the reactor pressure P_r.

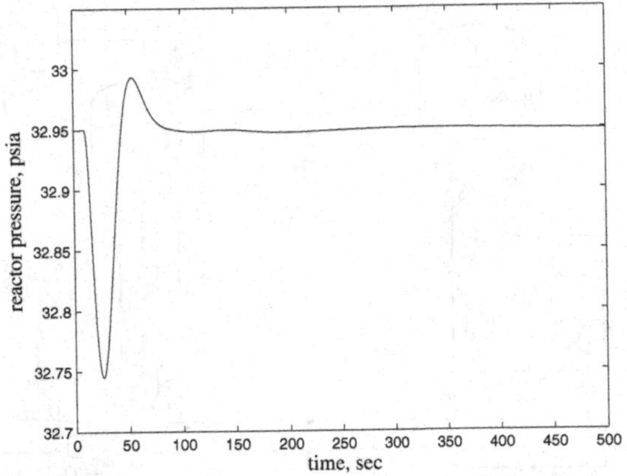

Fig. 35.4. Reactor pressure response to feed disturbance.

Figure 35.5 shows the response of the reactor/regenerator pressure difference ΔP_{rr} to the step change in fresh feed flow. As can be seen, all process variables are controlled well and deviations from setpoint are small.

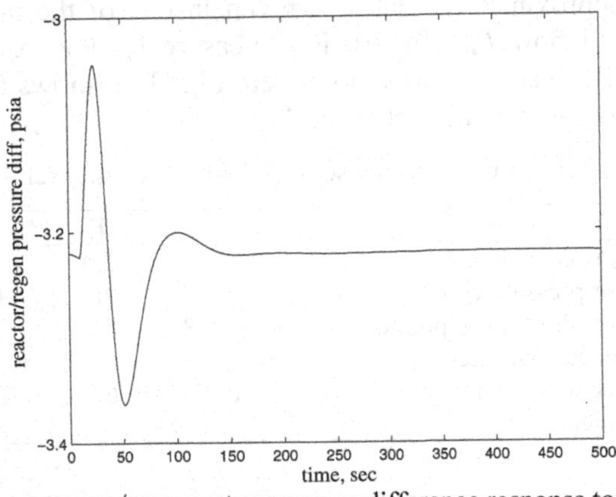

Fig. 35.5. Reactor pressure/regenerator pressure difference response to feed disturbance.

Figure 35.6 shows the response of the furnace outlet temperature, which is also the inlet temperature T_{in} of the reactor riser. It can be seen that there is an appreciable deviation (approximately 5 °F) from setpoint initially, however, the temperature soon stabilizes and reaches setpoint.

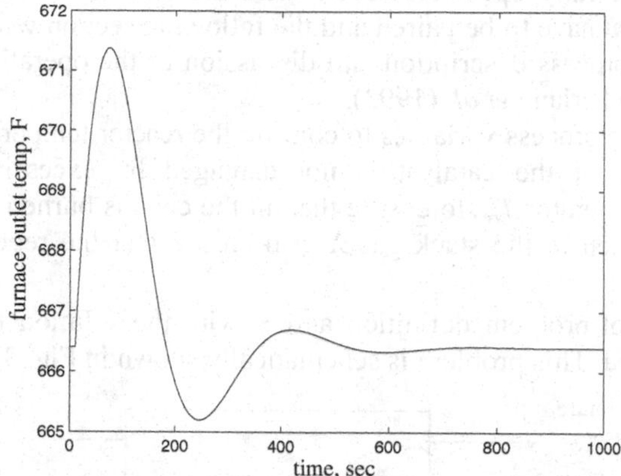

Fig. 35.6. Furnace outlet temperature response to feed disturbance.

35.3 Extension of the Basic Control Scheme

Of the four controlled variables, the furnace outlet temperature was the only one that gave a noticeable offset from setpoint, before stabilizing and returning to its setpoint. This response can be improved if a ratio controller is used that adjusts the fuel flow in ratio with the fresh feed flow. The temperature controller for furnace outlet temperature (T_{in}) can then be used as additional feedback. This is simulated in FCCU1B. The step response for a fresh feed change is shown in Fig. 35.7.

There is now only a maximum temperature offset of 0.8 °F instead of 5 °F, hence the ratio controller has improved the dynamic response considerably. The ratio setting is 0.34, which means that for every pound increase in fresh feed, the fuel flow is increased by 0.34 scf.

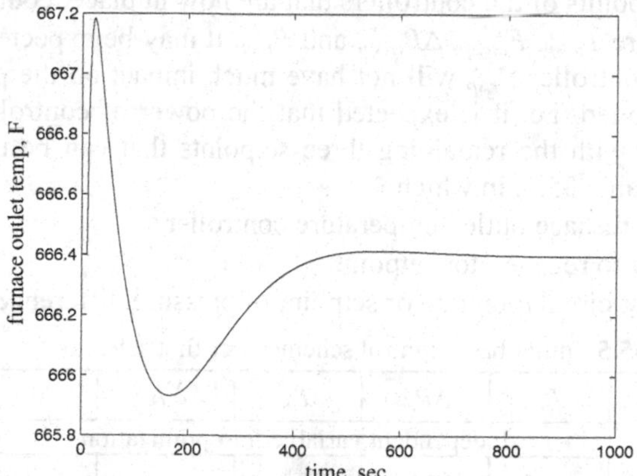

Fig. 35.7. Furnace outlet temperature response to feed disturbance using fresh feed/fuel flow ratio control.

The furnace outlet temperature controller has now slightly different controller settings: K_c = 0.02, T_i = 12.5 and T_d = 0.0.

35.4 Selection of the Final Control Scheme

Usually, catalytic cracking units are operated close to the constraint boundaries, which define an optimum for the oil feed throughput. The process is running at maximum throughput, hence some valves are fully open. However, there are still some measured variables and correcting variables that have to be paired and the following section will discuss this in more detail. For a detailed process description and discussion of the operation of the process the reader is referred to McFarlane *et al.* (1993).

There are three more process variables to control: the reactor temperature T_r to ensure that cracking takes place but the catalyst is not damaged by excessive temperatures, the temperature in the regenerator T_{reg} to ensure that all the coke is burned from the catalyst and the oxygen concentration in the stack gas X_{O_2} to ensure that the regenerator runs with an excess of oxygen.

This reduced control problem definition agrees with the solution that Khandalekar and Briggs (1995) presented. This problem is schematically shown in Fig. 35.8.

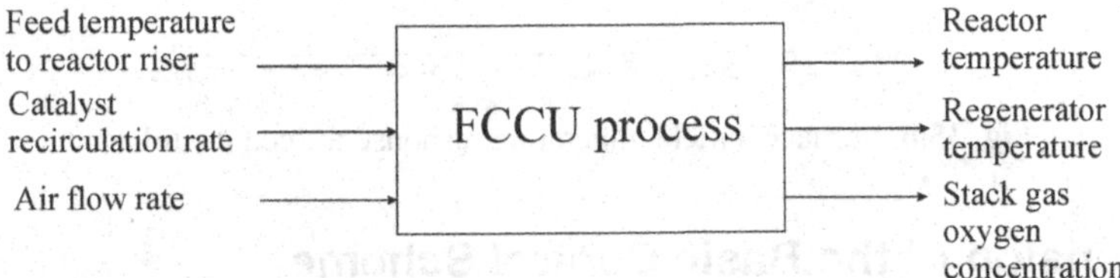

Fig. 35.8. Controlled and manipulated variables for non-linear model predictive control.

The authors used the Amoco model and compared it with a model developed by Lee and Groves (1985). They parameterized the Lee model to match the more complicated Amoco model by adjusting the heat of reaction and coke formation rate constant. They also studied non-linear multivariable control of process variables that showed much interaction.

The authors developed macroscopic non-linear approximate models to calculate the manipulated variables as shown in Fig. 35.8, and compared control performance to regulatory proportional-integral-derivative (PID) control. In this chapter the discussion will be limited to PI(D) control.

In principle the setpoints of the controllers that are now in place, could be used for master-slave control. These are $T_{in,sp}$, $F_{air,sp}$, $\Delta P_{rr,sp}$, and $P_{r,sp}$. It may be expected that the setpoint of the reactor pressure controller $P_{r,sp}$ will not have much impact on the process variables that still have to be controlled, i.e. it is expected that the power of control of $P_{r,sp}$ will be very small. This leaves us with the remaining three setpoints that can be used for control. The situation is given in Table 35.5, in which

- $T_{in,sp}$ is the setpoint furnace outlet temperature controller
- $F_{air,sp}$ is the air flow to regenerator setpoint
- $\Delta P_{r,rsp}$ is the catalyst circulation rate or setpoint of pressure difference controller

Table 35.5. Initial basic control scheme after first selection.

	F_{air}	P_r	ΔP_{rr}	T_{in}	X_{O_2}	T_r	T_{reg}
F_{oil}	Independent variable for optimization						
F_{fuel}				++			
V_{lift}	++						
$V_{product}$		++					
V_{stack}			++				
$T_{in,sp}$							
$F_{air,sp}$							
$\Delta P_{rr,sp}$							

It may not be immediately obvious that the value of ΔP_{rr} is proportional to the catalyst recirculation rate. This is shown in Fig. 35.9, which was obtained from the simulation, since the catalyst recirculation rate cannot be measured directly.

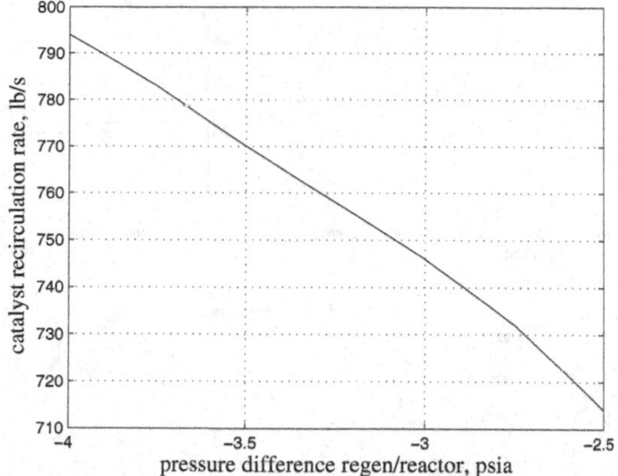

Fig. 35.9. Relationship between reactor/regenerator pressure difference and catalyst recirculation rate.

To determine which setpoint should be used to control which variable, step changes in the three setpoints were given: $\delta T_{in,sp}$ = + 10 °F, δF_{air} = +0.1 mol/s, $\delta\Delta P_{rr,sp}$ = –0.3 psia. The results are shown in Figs. 35.10, 35.11 and 35.12. It can easily be seen from Fig. 35.10 that the stack gas oxygen concentration can best be controlled by $\Delta P_{rr,sp}$. The response to the other two setpoints do not only give a major inverse response, the power of control is also very small compared with the power of control that is achieved for a change in $\Delta P_{rr,sp}$.

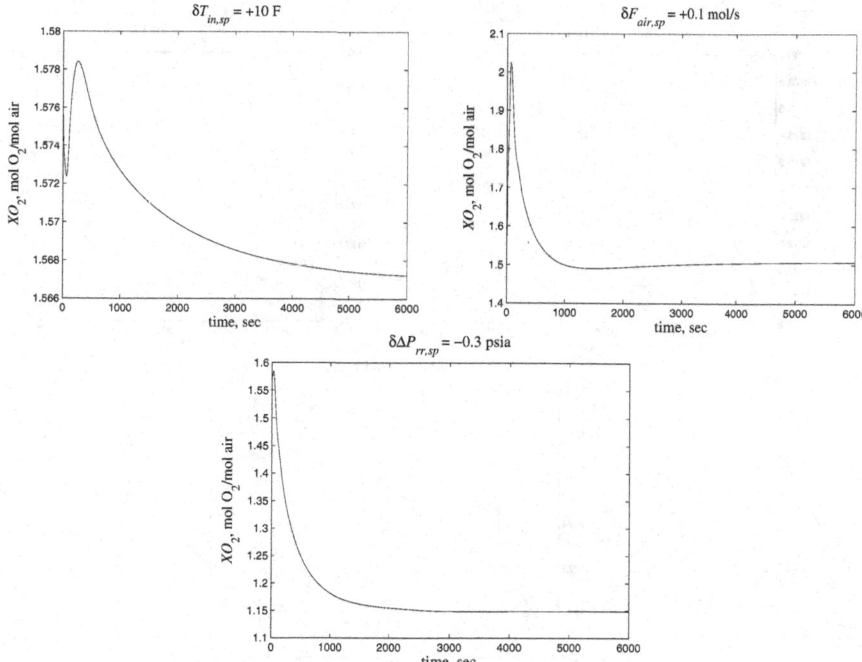

Fig. 35.10. Step responses in X_{O_2} for three remaining setpoints for control: $T_{in,sp}$, $F_{air,sp}$, $\Delta P_{rr,sp}$ (setpoint $X_{O_{2,sp}}$ = 1.577).

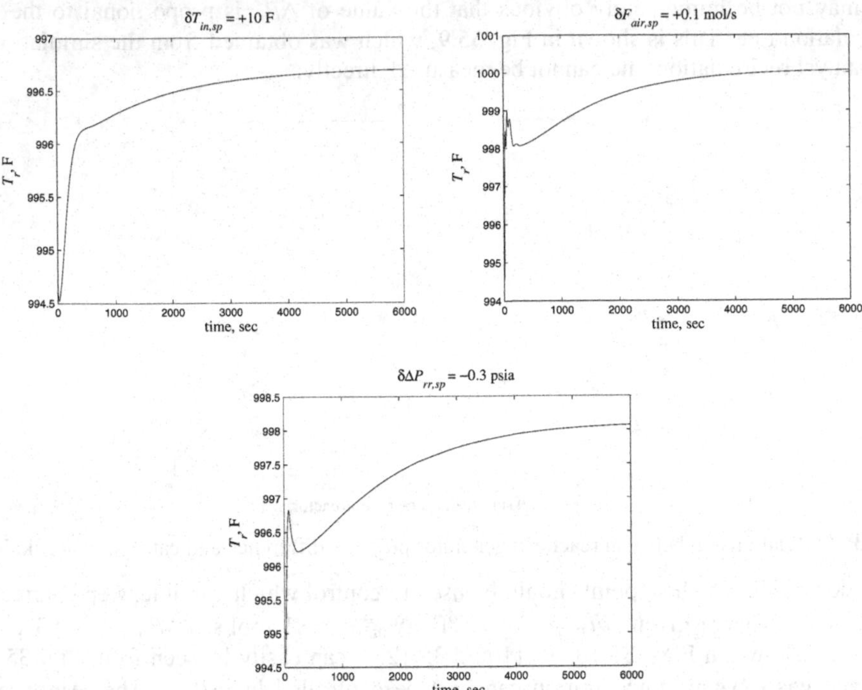

Fig. 35.11. Step responses in T_r for three remaining setpoints for control: $T_{in,sp}$, $F_{air,sp}$, $\Delta P_{rr,sp}$ (setpoint $T_{r,sp}$ = 994.6 °F).

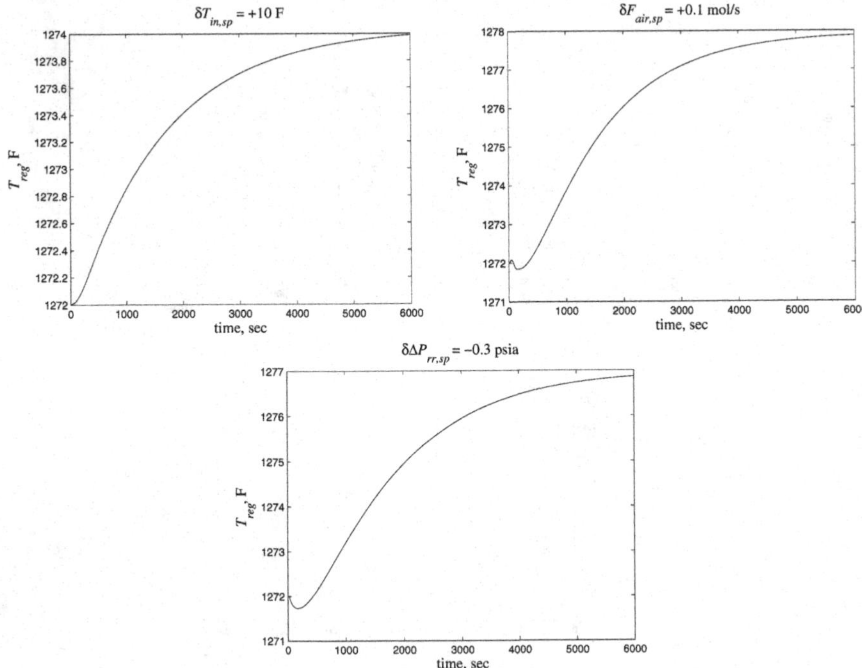

Fig. 35.12. Step responses in T_{reg} for three remaining setpoints for control: $T_{in,sp}$, $F_{air,sp}$, $\Delta P_{rr,sp}$ (setpoint $T_{reg,sp}$ = 1272 °F).

Figure 35.11 shows that the reactor riser temperature T_r can best be controlled by $T_{in,sp}$, since the step response for a change in air flow and $\Delta P_{rr,sp}$ show a large response initially, followed by an inverse response. This is not dynamically favorable for control.

Figure 35.12 shows that the regenerator temperature T_{reg} could be best controlled by $T_{in,sp}$. Also, $\Delta P_{rr,sp}$ and $F_{air,sp}$ could be used, although the response for these variables shows a minor inverse reaction initially.

Figure 35.13 shows the response for a feed change from 126 to 120 lb/s, with the two cascade controllers for T_r (controlled by $T_{in,sp}$) and X_{O_2} (controlled by $\Delta P_{rr,sp}$) installed. Controller settings are shown in Table 35.6. As can be seen, the controlled variables, the reactor temperature and oxygen stack gas concentration return to setpoint.

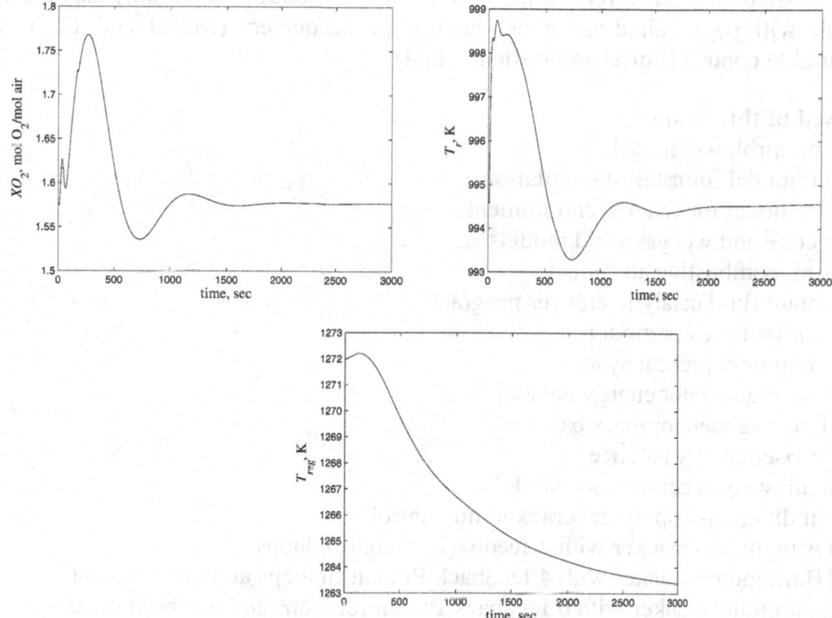

Fig. 35.13. Response of controlled reactor riser temperature, controlled stack gas oxygen concentration and uncontrolled regenerator temperature to step change in feed.

Table 35.6. Controller settings for two cascade controllers.

	K_c	T_i
$XO_2 - \Delta P_{rr}$ controller	0.3	25.0
$T_r - T_{in}$ controller	1.0	25.0

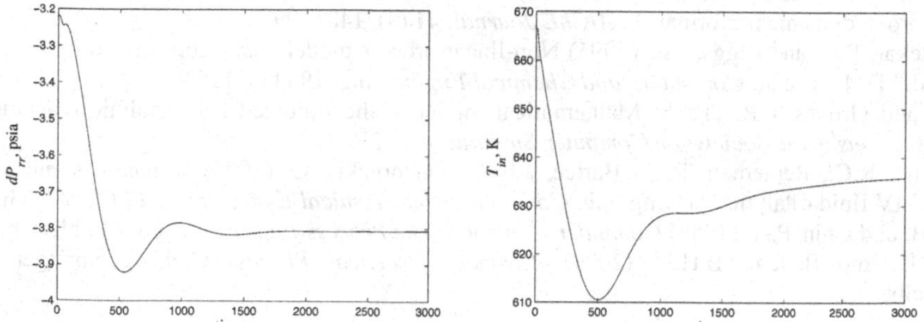

Fig. 35.14. Response of reactor-regenerator pressure difference and reactor inlet temperature to step change in feed.

The manipulated variables, shown in Fig. 35.14, have a nice response, although the reactor inlet temperature T_{in} takes a long time before it reaches steady state. The regenerator temperature is still uncontrolled and shows a considerable deviation from setpoint (about 9 °F).

There are two problems with the suggested control approach. The first one is that the feed decreased, hence one would expect that the recirculation rate of catalyst would be decreased. As Fig. 35.14 shows, $\Delta P_{rr,sp}$ is decreased, leading to a higher recirculation rate (Fig. 35.9). The second problem is that the regenerator temperature T_{reg} is uncontrolled and drops below its lower limit of 1265 °F. A third control loop will therefore be necessary. However, closing the third control loop creates so much mutual interaction between the control loops that they become unstable. If all three variables have to be controlled, a more advanced control approach will be required, such as the use of decouplers (Roffel and Chin, 1984) or multivariable control (Roffel and Betlem, 2004).

Files used in this chapter:
airblow.m: airblower model
catcirc.m: model for catalyst circulation
catentr.m: model for catalyst entrainment
coke.m: coke and wet gas yield models
combair.m: combustion air model
fccu.m: main fluid catalytic cracker program
liftair.m: lift air blower model
preheat.m: model preheat system
regeneb.m: regenerator energy balance
regenmb.m: regenerator mass balance
riseeb.m: riser energy balance
wgcomp.m: wetgas compressor model
FCCU0.mdl: open-loop model cracker, no control
FCCU1A.m: model cracker with 4 feedback PI control loops
FCCU1B.m: model cracker with 4 feedback PI control loops and one ratio controller
FCCU2.m: model cracker with 6 feedback PI control loops and one ratio controller

References

Ansari, R.M. and Tade, M.O. (2000) Constrained nonlinear multivariable control of a fluid catalytic cracking process. *Journal of Process Control*, **10**, 539–55.

Hovd, M. and Skogestad, S. (1993) Procedure for regulatory control structure selection with application to the FCC process. *AIChE Journal*, **39** (12), 1938–53.

Huq, I., Morari, M. and Sorensen, R.C. (1995) Modifications to model IV fluid catalytic cracking units to improve dynamic performance. *AIChE Journal*, **41** (4), 1481–99.

Khandalekar, P.D. and Riggs, J.B. (1995) Non-linear process model-based control and optimization of a model IV FCC unit. *Computers and Chemical Engineering*, **19** (11), 1153–68.

Lee, E. and Groves, F.R. (1985) Mathematical model of the fluidized bed catalytic cracking plant. *Transactions of the Society for Computer Simulation*, **2**, 219.

McFarlane, R.C., Reineman, R.C., Bartee, J.F. and Georgakis, C. (1993) Dynamic simulator for a model IV fluid catalytic cracking unit. *Computers and Chemical Engineering*, **17** (3), 275–300.

Roffel, B. and Chin, P.A. (1984) *Computer Control in The Process Industries*, Lewis Publishers.

Roffel, B. and Betlem, B.H.L. (2004) *Advanced Practical Process Control*, Springer Verlag, Heidelberg.

Appendix A. Modeling an Extraction Process

In this appendix an example will be given of the modeling of an extraction process. The exercise consists of five assignments: (i) defining the operational goal of the process and making an environmental diagram. In addition, simple control will be developed for the process, (ii) developing a physical dynamic process model, (iii) process linearization and development of transfer functions, (iv) simulation of the process and extension to multistage extraction and (v) control of the multistage processes.

A1: Problem Analysis

In an extraction vessel (Figure A1) an unbuffered wastewater flow F_A (the substrate) polluted with aniline with concentration $x_{A,in}$, is extracted by means of benzene with flow F_B (the solvent). The effective distribution coefficient α depends on the temperature. Steam with flow F_{steam} is added to raise the temperature to a required value of 50 °C. The vessel is will mixed by a stirrer with constant speed. Subsequently, in a phase separator the benzene (the extract) is separated from the water phase (the raffinate). In the separator the two levels can be measured. The fraction aniline in the wastewater should be decreased to x_A and increased in the benzene to x_B.

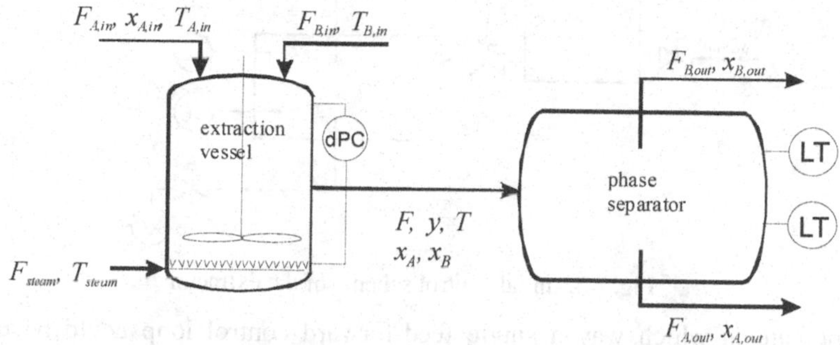

Fig. A1. Extraction process consisting of vessel and separator. y = weight fraction water, x_A = fraction aniline (ppm) in water, x_B = fraction aniline (ppm) in benzene.

Suppose the goal of the process is to produce a wastewater flow which meets the requirement.

a. **Q**: Design the environmental model for the goal mentioned. Indicate in the diagram: the manipulated variables, the external disturbances and the variables of the process, which should be controlled.
A: The output variables should be controlled, this can be achieved by manipulating the input variables $F_{B,in}$, $F_{A,out}$, $F_{B,out}$ F_{steam} and F.

Process Dynamics and Control: Modeling for Control and Prediction. Brian Roffel and Ben Betlem.

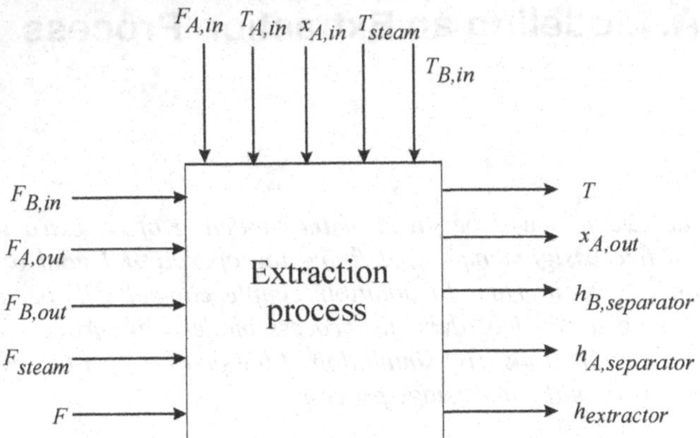

Fig. A2. Environmental diagram of the extractor.

b. **Q**: Sketch the process and add an initial control scheme consisting of single feedback control loops in agreement with the goal.
A: An initial control scheme could be the one shown in Fig. A3.

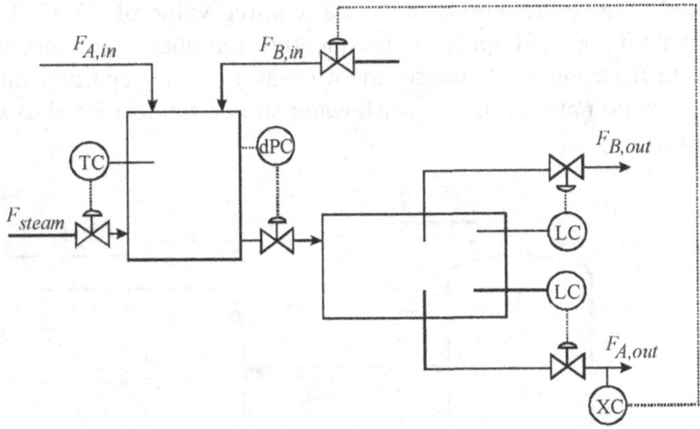

Fig. A3. Initial control scheme of the extractor.

c. **Q**: Indicate in which way a single feed-forward control loop could be used useful. Mention a disadvantage of this control.
A: Feed-forward extension. One possible extension could be to create a flow ratio controller *FrC* between F_A and F_B, this is a form of feed-forward control, the aniline outlet concentration could in that case reset the ratio. A general disadvantage of feed-forward control is that it is model-based (prone to modeling errors), in this case of ratio control the two models from F_A and F_B to the aniline outlet concentration are similar, which is an advantage. Another disadvantage is that variations in aniline concentration $x_{A,in}$ are not taken care of by the flow-ratio feed-forward controller.

A2: Dynamic Process Model Development

Process model

The following assumptions can be made for the dynamic model of the extraction process.

1. the fresh (regenerated) benzene contains no aniline
2. the distribution coefficient α depends only on the temperature and is independent of the concentration, according the relationship:

$$\alpha = A_{\alpha} e^{-E_{\alpha}/RT} \tag{A1}$$

3. no extraction occurs in the separator
4. water and benzene are ideal mixed in the extraction vessel
5. the liquid phases are at equilibrium
6. the steam enters at boiling point, hence the cooling down of the steam to the condensation point can be ignored
7. control of the contents of the extraction vessel is based on a pressure difference measurement manipulating the outflow of the extraction vessel, it can therefore be assumed that at ideal control, the total mass is constant.
8. control of the contents of the extraction vessel is ideal and can be considered as a part of the system
9. in the operation area, the density and the specific heat of water and benzene are nor a function of the temperature, neither a function of the aniline concentration. However, the density and the specific heat of the water-benzene mixture depends on the water-benzene ratio (see ρ and c_P values).
10. the heat capacities of the walls and mixer can be neglected.
11. the energy added by the stirrer can be neglected
12. the heat loss to the environment can be neglected.

Design the behavioral model, which is consistent with the environmental model, for the extraction vessel including the level controller (excluding the separator).). The inputs of the system are: F_A, $T_{A,in}$, x_A, $F_{B,in}$, $T_{B,in}$ F_{steam} T_{steam} and $M_{setpoint}$. The outputs are: the water fraction y, which can be expressed as a weight fraction and may be defined as the ratio of mass of water to the total mass, the aniline concentration in the water x_A and the temperature T. Please answer the following questions.

a. **Q**: Identify the state variables and determine the state equations. Indicate clearly where which assumption is used.
A: As state variables we can use (i) the water fraction y, (ii) the aniline fraction x_A in the water and (iii) the temperature T. The state equations will be derived under b.

b. **Q**: Transfer the state equations in such a way that the left-hand site contains only a derivative of one variable. Check whether the steady state component balance agrees with a one-stage extraction.
When you define an equation, describe clearly which equation it concerns (mass, component balance…). Indicate clearly where which assumption is used and indicate when additional assumptions were necessary.
A: Assuming ideal level control in the extraction vessel, the **overall mass balance** for the extraction vessel is:

$$\frac{dM}{dt} = F_{A,in} + F_{B,in} + F_{steam} - F = 0 \tag{A2}$$

in which F is the outlet flow. This equation can be written as:

$$F = F_{A,in} + F_{B,in} + F_{steam} \tag{A3}$$

The **water balance** for the extraction vessel is:

$$\begin{aligned} M\frac{dy}{dt} &= F_{A,in} + F_{steam} - yF \\ M_A &= yM, \quad M_B = M - M_A = (1-y)M \end{aligned} \tag{A4}$$

The **aniline component balance** is:

$$\begin{aligned} \frac{d(M_A x_A)}{dt} &= F_{A,in} x_{A,in} - yFx_A - J \\ \frac{d(M_B x_B)}{dt} &= J - (1-y)F \end{aligned} \tag{A5}$$

in which J = amount of aniline transferred from the water to the benzene.
Adding the foregoing two equations, together with $x_B = \alpha x_A$ and the two mass balances:

$$\begin{aligned} \frac{dM_A}{dt} &= F_{A,in} + F_{steam} - yF \\ \frac{dM_B}{dt} &= F_{B,in} - (1-y)F \end{aligned} \tag{A6}$$

gives:

$$(M_A + \alpha M_B)\frac{dx_A}{dt} = F_{A,in} x_{A,in} - (F_{A,in} + \alpha F_{B,in} + F_{steam})x_A \tag{A7}$$

Assuming the steam is at its boiling point, the **energy balance** for the extraction vessel is:

$$\begin{aligned} \frac{d(c_p MT)}{dt} &= M_{setpt}\frac{d(c_p T)}{dt} \\ &= c_{p,A} F_{A,in} T_{A,in} + c_{p,B} F_{B,in} T_{B,in} + F_{steam}(c_{p,A} T_{steam} + \Delta H) - \\ &\quad c_p FT \end{aligned} \tag{A8}$$

Using Eqn. (A3), (A4a) and the equation for the specific heat:

$$c_p = yc_{p,A} + (1-y)c_{p,B} \tag{A9}$$

the energy balance will become:

$$\begin{aligned} Mc_p\frac{dT}{dt} &= c_{p,A} F_{A,in}(T_{A,in} - T) + c_{p,B} F_{B,in}(T_{B,in} - T) + \\ &\quad F_{steam}(c_{p,A}(T_{steam} - T) + \Delta H_{evap}) \end{aligned} \tag{A10}$$

The state equations are (A4a), (A7) and (A10).

c. **Q**: Give the additional algebraic equations and additional assumptions, if required. Indicate clearly where which assumption is used and indicate whether additional assumptions are necessary.
A: Additional algebraic equations are eqns. (A3), (A4b) and (A9) and the equation for the distribution coefficient, eqn. (A1).

d. **Q**: Design the behavioral model for the extractor, which is consistent with the environmental model and ensure that one function produces one variable. Indicate clearly which equations are used in the diagram.

A: The environmental and detailed information flow diagrams are shown below:

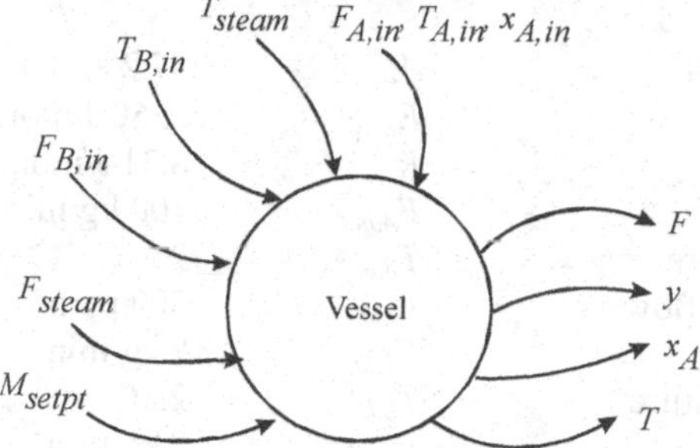

Fig. A4. Environmental diagram of the extractor.

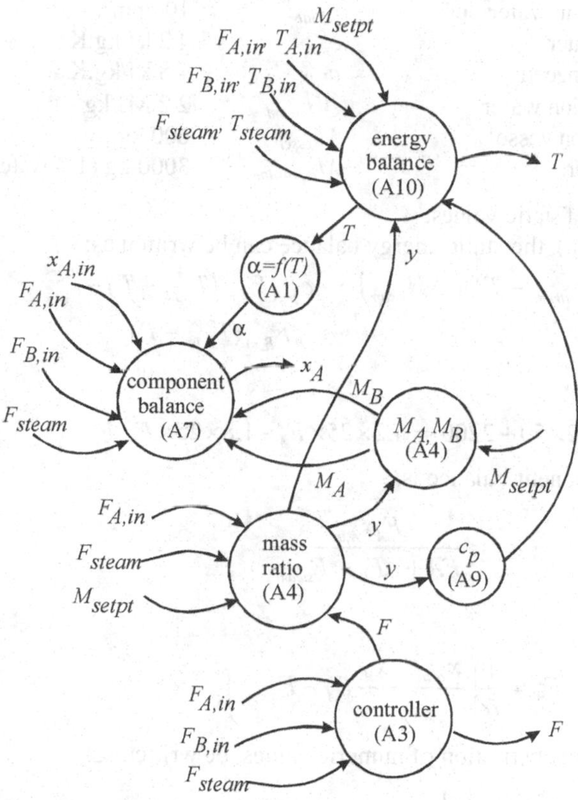

Fig. A5. Detailed information flow diagram of the extractor.

e. **Q**: Check the number of degrees of freedom.
When you suggest an equation, indicate clearly which equation you mean by giving it a name.
A: There are eight equations, three differential equations (for x_A, y and T) and five additional equations (for F, M_A, M_B, α and c_p). The following variables are used:

- eight unknowns: $x_A, y, T, F, M_A, M_B, \alpha$ and c_p
- eight known variables: $F_{A,in}, T_{A,in}, x_{A,in}, F_{B,in}, T_{B,in}, F_{steam}, T_{steam}, M_{setpt}$

f. **Q**: Calculate the benzene flow and steam flow for the steady-state conditions, using the following data:

distribution coefficient:	A_α	37.2 kg/kg
activation energy	E_α	2350 J/mol
gas constant	R	8.31 J/mol.K
waste-water flow	$F_{A,in}$	100 kg/min
wastewater temperature	$T_{A,in}$	25 °C
aniline in wastewater flow	$x_{A,in}$	300 ppm
benzene flow	$F_{B,in}$	? kg/min
benzene flow temperature	$T_{B,in}$	25 C
steam flow	$F_{steam,in}$	? kg/min
steam temperature	$T_{steam,in}$	100 °C
required vessel temperature	T_{vessel}	50 °C
required analine in water out	$x_{A,uit}$	10 ppm
heat capacity water	$c_{P,A}$	4.2 kJ/kg.K
heat capacity benzeen	$c_{P,B}$	1.8 kJ/kg.K
heat of evaporation water	ΔH_{steam}	2.2 MJ/kg
volume extraction vessel	M_{vessel}	600 kg
volume separator	$M_{separator}$	3000 kg (1/3 water, 2/3 benzene)

A: Calculation of static values.
Using Eqn. (A10), the static energy balance can be written as:

$$F_{steam}\left(c_{p,A}(T_{steam}-T)+\Delta H_{evap}\right)=-c_{p,A}F_{A,in}(T_{A,in}-T)- \\ c_{p,B}F_{B,in}(T_{B,in}-T) \tag{A11}$$

or

$$F_{steam}(4.2\times 50+2200)=4.2\times 25\times F_A+1.8\times 25\times F_B \tag{A12}$$

The static component balance is:

$$x_A=\frac{F_A x_{A,in}}{FA+\alpha F_B+F_{steam}} \tag{A13}$$

or

$$F_B=\frac{1}{\alpha}\left(\frac{x_{A,in}-x_A}{x_A}F_A-F_{steam}\right) \tag{A14}$$

which can, after substitution of numeric values, be written as:

$$F_B=\frac{1}{15.5}(29F_A-F_{steam}) \tag{A15}$$

Eqn. (A12) can be combined with Eqn. (A15) to give:
$F_{steam}=0.078F_A=0.078\times 100=7.8\ kg/min$
From eqn. (A14) F_B can then be calculated:
$F_B=1.87F_A=1.87\times 100=187\ kg/min$
The value of y can be calculated from Eqns. (A3) and (A4a):

$$y=\frac{F_A+F_{steam}}{F_A+F_B+F_{steam}}=\frac{100+7.8}{100+187+7.8}=0.366$$

A3 Dynamic Process Model Analysis

The assignment concerns the extractor of assignments A1 and A2. Only the extraction vessel is considered and some simplifications are made.

A wastewater flow F_A (the substrate) polluted with aniline $x_{A,in}$ is extracted by means of benzene with flow F_B (the solvent). The effective distribution coefficient α depends on the temperature. Steam with flow F_{steam} is added to raise the temperature. The vessel is well mixed by a stirrer with constant speed. The fraction aniline in the wastewater is reduced to x_A.

Assume that the process can be described by the following equations:
Component balance of aniline:

$$\left(M_A+\alpha M_B\right)\frac{dx_A}{dt}=F_A x_{A,in}-\left(F_A+\alpha F_B+F_{steam}\right)x_A \tag{A16}$$

Under the assumption that:

- the specific heat of component A and B are equal
- the steam enters at the condensation point
- $T_B < T$, $T_A < T$, $T_{steam} > T$

the energy balance can be written as:

$$\begin{aligned}c_p\left(M_A+M_B\right)\frac{dT}{dt}=c_pF_A\left(T_A-T\right)+c_pF_B\left(T_B-T\right)+\\ c_pF_{steam}\left(T_{steam}-T\right)+\Delta H_{steam}F_{steam}\end{aligned} \tag{A17}$$

The distribution coefficient can be given by eqn. (A1).
The steady-state distribution of M_A and M_B can be given by:

$$\frac{M_A}{M_B}=\frac{F_A+F_{steam}}{F_B} \tag{A18}$$

Define the residence time as:

$$\frac{M_A}{F_A+F_{steam}}=\frac{M_B}{F_B}=\tau \tag{A19}$$

Inputs:	F_A, F_B, F_{steam}	incoming flows A, B, steam	kg/s
	T_A, T_B, T_{steam}	temperatures of incoming flows	°C
	$x_{A,in}$	pollution fraction in F_A	ppm
Outputs:	T	temperature extraction vessel	°C
	x_A	pollution fraction in A in vessel	kg/kg
Internal variables:	α	distribution coefficient	kg/kg
	M_A, M_B	mass of A and B in vessel	kg
Constants:	c_p	specific heat	J/kg.°C
	ΔH_{steam}	heat of evaporation	J/kg

a. **Q**: Linearize the equations mentioned above and transfer these relationships to deviation variables. Indicate deviation variables with the symbol δ. Terms of the order equal to or higher than δ^2 may be ignored.
A: Linearization of the component balance gives:

$$\begin{aligned}(M_A+\alpha M_B)s\delta x_A = F_A\delta x_{A,in} + x_{A,in}\delta F_A - \\ x_A(\delta F_A + \delta F_{steam} + \alpha\delta F_B + F_B\delta\alpha) - \\ (F_A + F_{steam} + \alpha F_B)\delta x_A\end{aligned} \tag{A20}$$

which can be written as:

$$\begin{aligned}((M_A+\alpha M_B)s + F_A + F_{steam} + \alpha F_B)\delta x_A = \\ F_A\delta x_{A,in} + (x_{A,in} - x_A)\delta F_A \\ - x_A\delta F_{steam} - x_A\alpha\delta F_B - x_A F_B\delta\alpha\end{aligned} \tag{A21}$$

Linearization of the energy balance gives:

$$\begin{aligned}c_p(M_A+M_B)s\delta T = c_p(T_A - T)\delta F_A + c_p F_A(\delta T_A - \delta T) + \\ c_p(T_B - T)\delta F_B + c_p F_B(\delta T_B - \delta T) + \\ c_p(T_{steam} - T)\delta F_{steam} + c_p F_{steam}(\delta T_{steam} - \delta T) \\ + \Delta H_{steam}\delta F_{steam}\end{aligned} \tag{A22}$$

which can be written as:

$$\begin{aligned}((M_A+M_B)s + F_A + F_{steam} + F_B)\delta T = \\ F_A\delta T_A + F_{steam}\delta T_{steam} + F_B\delta T_B \\ (T_A - T)\delta F_A + (T_B - T)\delta F_B + \\ \left((T_{steam} - T) + \frac{\Delta H_{steam}}{c_p}\right)\delta F_{steam}\end{aligned} \tag{A23}$$

Linearization of the distribution coefficient gives:

$$\delta\alpha = \beta\delta T \tag{A24}$$

Using Eqn. (A18) and (A19), it can be shown that:

$$\frac{M_A + M_B}{F_A + F_{steam} + F_B} = \tau, \quad \frac{M_A + \alpha M_B}{F_A + F_{steam} + \alpha F_B} = \tau \tag{A25}$$

b. **Q**: Determine the transfer function between x_A and F_B.
Hint: use positive time constants to simplify the equations and calculations. Define clearly the meaning of the constants in the final result.
A: The transfer function $\delta x_A / \delta F_B$ can be determined as follows.
The component balance can be written as:

$$(\tau s + 1)\delta x_A = -K_1\delta F_B - K_2\delta T \tag{A26}$$

with

$$K_1 = \frac{x_A\alpha}{F_A + F_{steam} + \alpha F_B}, \quad K_2 = \frac{x_A\beta F_B}{F_A + F_{steam} + \alpha F_B} \tag{A27}$$

The energy balance can be written as:

$$(\tau s+1)\delta T=-K_3\delta F_B \tag{A28}$$

with

$$K_3=\frac{T-T_B}{F_A+F_{steam}+F_B} \tag{A29}$$

Substitution of the equation for δT into the component balance gives:

$$\frac{\delta x_A}{\delta F_B}=\frac{-K_1\tau s-(K_1-K_2K_3)}{(\tau s+1)^2} \tag{A30}$$

Using the numerical values from assignment A2, one gets:
$K_1 = 5.16\times10^{-8}$
$K_2 = 2.61\times10^{-8}$
$K_3 = 8.32\times10^{-3}$
$\tau = 2.04$ min
from which it can be seen that $K_2K_3 << K_1$.

c. **Q**: Is the system determined under b. stable? Can it oscillate? Under which conditions can an inverse response occur? Is this situation realistic? Try to explain how the system works physically.
A: The system is of second order, critically dampened. The two time constants, $\tau_{component}$ and τ_{energy} are equal, due to ideal mixing. The system does not oscillate and is not unstable. An inverse response could occur if:

$$K_1-K_2K_3<0\Rightarrow x_A\alpha-x_A\beta F_B(T-T_B)<0 \tag{A31}$$

or

$$\alpha<\frac{\partial\alpha}{\partial T}(T-T_B)F_B \tag{A32}$$

This is generally not the case, i.e. an inverse response is unlikely to happen.

d. **Q**: Determine the transfer function between x_A and F_{steam}.
A: The transfer function $\delta x_A/\delta F_{steam}$ can be determined as follows:
Relevant terms in the component balance are now:

$$((M_A+\alpha M_B)s+F_A+F_{steam}+\alpha F_B)\delta x_A=-x_A\delta F_{steam}-x_A\beta F_B\delta T \tag{A33}$$

from which

$$(\tau s+1)\delta x_A=-K_4\delta F_{steam}-K_2\delta T \tag{A34}$$

with

$$K_4=\frac{x_A}{F_A+F_{steam}+\alpha F_B} \tag{A35}$$

and K_2 as defined before.
Relevant terms in the energy balance are now:

$$((M_A + M_B)s + F_A + F_{steam} + F_B)\delta T = \left(T_{steam} - T + \frac{\Delta H_{steam}}{c_p}\right)\delta F_{steam} \tag{A36}$$

which can be written as:

$$(\tau s + 1)\delta T = K_5 \delta F_{steam} \tag{A37}$$

with:

$$K_5 = \frac{T_{steam} - T + \dfrac{\Delta H_{steam}}{c_p}}{F_A + F_{steam} + F_B} \tag{A38}$$

The required transfer function can now be written as:

$$\frac{\delta x_A}{\delta F_{steam}} = \frac{-K_4 \tau s - (K_4 + K_2 K_5)}{(\tau s + 1)^2} \tag{A39}$$

e. **Q**: Suppose that the influence of a change in F_B on a change in T is small and can be ignored. In addition, assume that the impact of a change in F_{steam} on a change in x_A is small compared to the impact of a change in T on a change in x_A. What is then the essential difference between the dynamics of x_A caused by changes in F_B and in F_{steam}? Which flow is more suitable for control of x_A?
A: The first assumption means that $K_3 = 0$. The second assumption means that $K_4 = 0$. The transfer functions can then be written as:

$$\frac{\delta x_A}{\delta F_{steam}} = \frac{-K_2 K_5}{(\tau s + 1)^2}$$
$$\frac{\delta x_A}{\delta F_B} = \frac{-K_1}{(\tau s + 1)} \tag{A40}$$

The dynamics of x_A for changes in F_B are faster than for changes in F_{steam}, first order compared with second order. Also, the gain K_1 is larger than K_2K_5 (as can be calculated from the numerical data), hence F_B is preferred for control of x_A.

A4 Dynamic Process Simulation

This assignment concerns the extractor of assignments A1, A2 and A3. Only the extraction vessel is considered and the steam heating is omitted. Three cases will be considered: a single extraction vessel, cross flow and counter-current flow in six extraction vessels. In all cases the target aniline concentration in the water is 10 ppm.

The wastewater flow F_A (the substrate) polluted with aniline $x_{A,in}$ is extracted by means of benzene F_B (the solvent). The vessel is well mixed by a stirrer with a constant speed. The fraction aniline in the wastewater is decreased to x_A.
Assume that the process can be described by the following equations.

The component balance of aniline for the water phase is:

$$(M_A + \alpha M_B)\frac{dx_A}{dt} = F_A x_{A,in} + F_B x_{B,in} - (F_A + \alpha F_B)\cdot x_A \tag{A41}$$

The component balance of aniline for the benzene phase is:

$$x_B = \alpha x_A \tag{A42}$$

The total mass is constant:

$$F = F_A + F_B \tag{A43}$$

The phase separation is ideal and is according the water benzene ratio:

$$y = \frac{M_A}{M_A + M_B} \tag{A44}$$

thus

$$\frac{dy}{dt} = \frac{1}{M_{tot}}(F_A - yF) \tag{A45}$$

Inputs:	F_A, F_B	incoming flows A, B	kg/s
	$x_{A,in}$	pollution fraction in F_A	ppm
Outputs:	x_A	pollution fraction in A in vessel	kg/kg
	y	water fraction, according to Eqn. (A44)	
Internal variables:	M_A, M_B	mass A and B in vessel	kg

Simulations using MATLAB Simulink:

a. **Q**: Simulate the response of x_A for a 25% step change in F_B for a single stage separation, when phase separation between A and B is instantaneous. Use the values mentioned below:

distribution coefficient:	α	15 kg/kg
wastewater flow	$F_{A,in}$	100 kg/min
aniline in wastewater flow	$x_{A,in}$	300 ppm
aniline in wastewater flow	$x_{A,out}$	10 ppm
benzene flow	$F_{B,in}$	? kg/min
aniline in fresh benzene	$x_{B,in}$	0 ppm
mass in extraction vessel	M_{tot}	600 kg

For additional information on how to set up your first Simulink model, see the information at the end of this appendix.

A: For a single-stage extraction, the following steady-state equation holds (see Eqn. A41) for the final aniline concentration:

$$x_{A,out} = \left(\frac{F_A}{F_A + \alpha F_B}\right) x_{A,in} \tag{A46}$$

Using the values given before, it can be calculated that $F_B = 193.33$ kg/min. The separation factor $F_A/(F_A + \alpha F_B) = 0.0333$.

Eqn. (A41), for the calculation of x_A, is shown in the lower part of the following diagram:

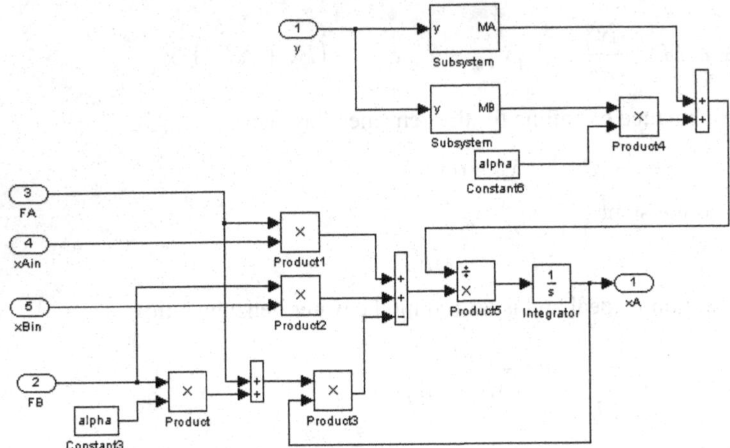

Fig. A6. Simulink diagram for calculation of x_A.

The calculation of M_A and M_B, eqn. (A44), noting that $M_{tot} = M_A + M_B$, is shown in the following diagram:

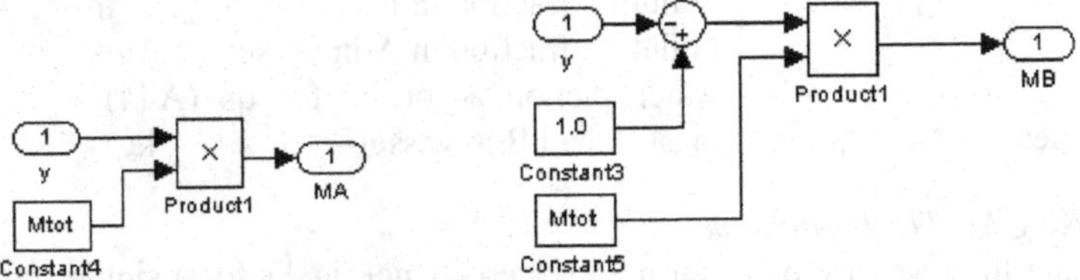

Fig. A7. Simulink diagram for calculation of M_A and M_B.

Equation (A45), for the calculation of y, is as constructed as follows:

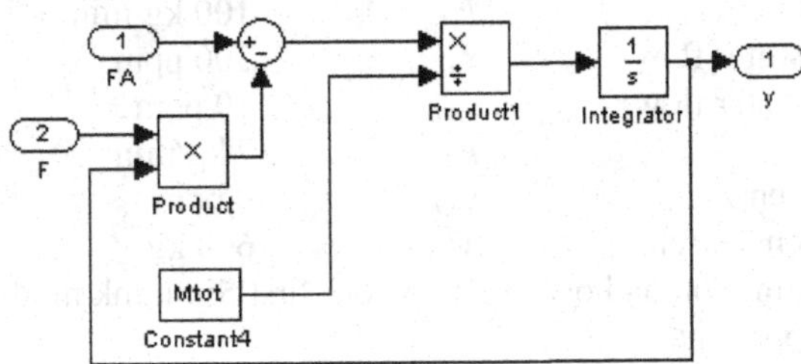

Fig. A8. Simulink diagram for the calculation of y.

The calculation of F from eqn. (A43) is a simple summation.

After running ExtractparSS, the response for a 25% step change in F_B can be calculated by running ExtrSS. The final aniline concentration in the water changes from 10 ppm to 8.05 ppm.

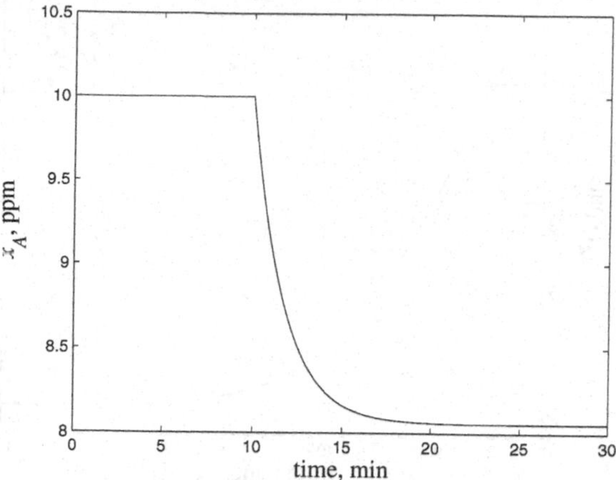

Fig. A9. Response of x_A to a step change in F_B of 25% for single-stage extractor.

b. **Q**: Suppose a multi-stage extraction is required, consisting of n consecutive cross flow stages. Sketch the response in $x_{A,n}$ (x_A in the nth vessel) due to a step disturbance in F_B. Make the single extraction a sub-model and take $n = 6$. What is the process behavior?
A: For cross flow extraction in six connected vessels, the following steady-state equation holds:

$$x_{A,out} = \left(\frac{F_A}{F_A + \alpha \dfrac{F_B}{n}} \right)^n x_{A,in} \tag{A47}$$

in which $n = 6$. Using the numerical values, it can be calculated that $F_B = 30.5$ kg/min. The separation factor $F_A/(F_A + \alpha F_B/n) = 0.5674$.

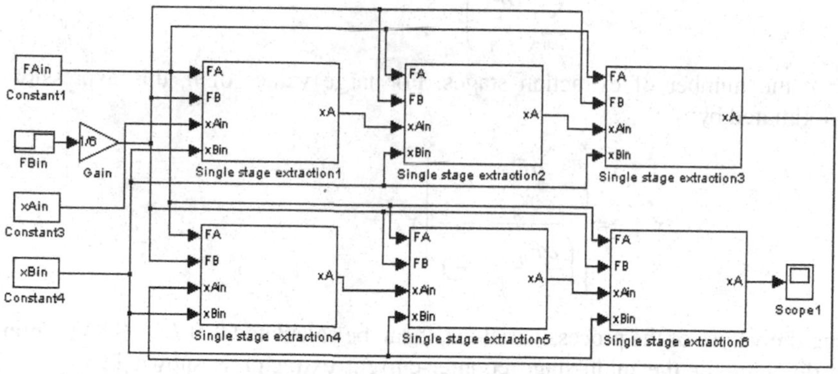

Fig. A10. Simulink diagram for a six stage cross flow extractor.

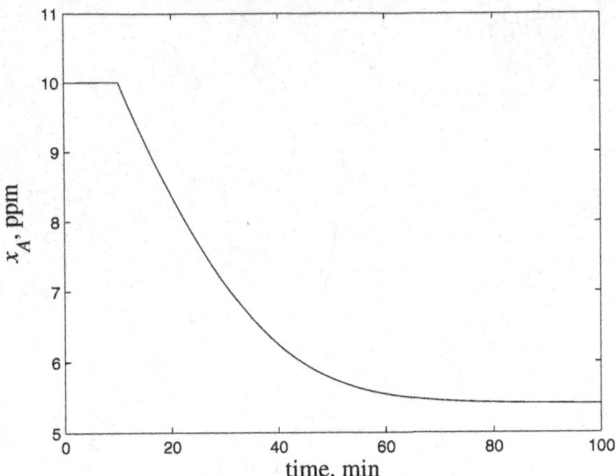

Fig. A11. Response of x_A to a step change in F_B of 25% for cross flow extractor.

The diagram for the multi-stage cross flow extractor is shown in Fig. A10. The benzene flow per stage is one sixth of the total benzene flow. After running ExtractparXF, the response for a 25% step change in F_B can be calculated by running ExtrXF. The response is shown in Fig. A11, the aniline concentration at the outlet of stage six changes from 10.0 to 5.40 ppm.

c. **Q**: Simulate a counter-current extraction of six stages. What is the process behavior and what is the fundamental difference with the single-stage and the cross flow system (dynamically and statically)?

A: For counter-current extraction, the following equation holds for the outlet aniline concentration:

$$x_{A,out} = \frac{1}{\sum_{i=0}^{n} \left(\frac{\alpha F_B}{F_A} \right)^n} x_{A,in} \tag{A48}$$

with n the number of extraction stages. For large values of n, this expression can be approximated by:

$$x_{A,out} = \left(\frac{\frac{\alpha F_B}{F_A} - 1}{\left(\frac{\alpha F_B}{F_A} \right)^{n+1} - 1} \right) x_{A,in} \tag{A49}$$

Using the values of the process variables, it can be calculated that $F_B = 9.84$ kg/min.
The diagram for the multi-stage counter-current extractor is shown in Fig. A12. After running ExtractparCC, the response for a 25% step change in F_B can be calculated by running ExtrCC.
The response for a step change of 25% in F_B is shown in Fig. A13. The aniline concentration at the outlet of the last stage changes from 10.0 to 3.54 ppm.
From the calculations and simulation, the following results can be summarized:

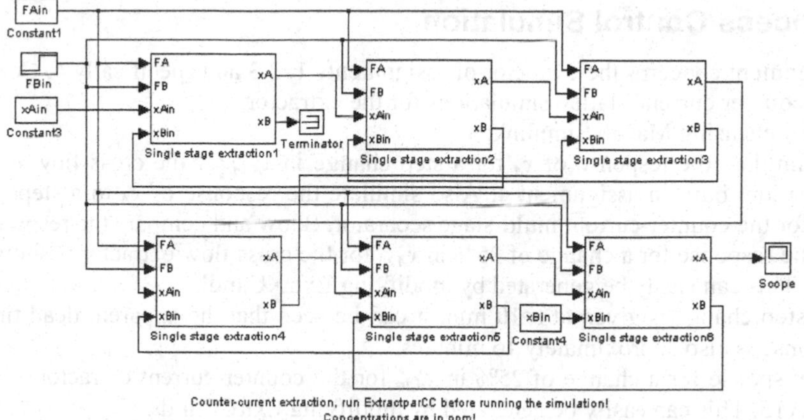

Fig. A12. Simulink diagram for a six stage counter current extractor.

	F_B, kg/min	Process gain $\Delta x_A/\Delta F_B$, ppm.min/kg	Time constant, min	Approximate time to steady state, min
Single stage	193.33	0.04	2.1	20.0
Cross flow	30.5	0.60	25.9	90.0
Counter-current	9.84	2.63	20.0	220.0

The counter-current extractor uses the least amount of benzene to lower the amount of aniline in the water from 300 to 10 ppm, the sensitivity for changes in the benzene flow is the largest, and the response to a change in the benzene flow is the slowest.

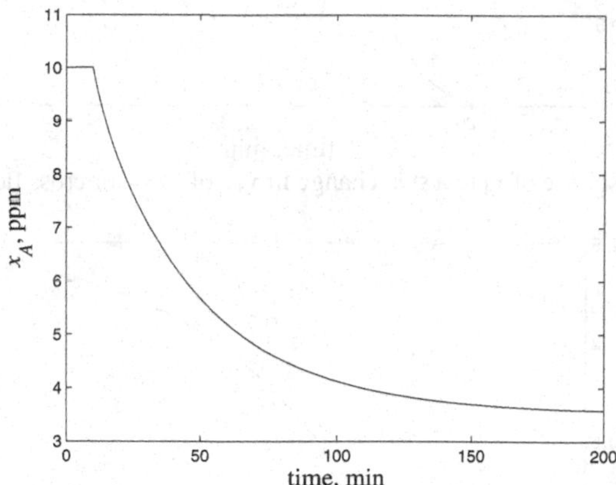

Fig. A13. Response of x_A to a step change in F_B of 25% for counter-current extractor.

d. **Q**: Write a report of the simulations with figures, details of the simulation and conclusions.
A: This has been done in the previous sections. One could make a final conclusion that the counter-current extraction is to be preferred statically and dynamically over a cross flow extractor.

A5: Process Control Simulation

This assignment concerns the extractor of assignments 1, 2 3 and specifically 4. Consider the flow and counter current Matlab simulations for the extractor.

Simulations using Matlab Simulink:

a. **Q**: Simulate the response of x_A for a step change in $x_{A,in}$ for the cross flow multi-stage separation, built in assignment 4. Also simulate the response of x_A to a step change in $x_{A,in}$ for the counter-current multi-stage separator. Show and compare the responses.
A: The response for a change of 25% in $x_{A,in}$ for the cross flow extractor is shown in Fig. A14. This can easily be generated by modifying ExtrXC.mdl.
The step change is given at t=10 min, it can be seen that the apparent dead time in the response is also approximately 16 minutes.
The response for a change of 25% in $x_{A,in}$ for the counter-current extractor is shown in Fig. A15. This can easily be generated by modifying ExtrCC.mdl.
The step change is given at t=10 min, it can be seen that the apparent dead time in the response is approximately 22 minutes in this case.

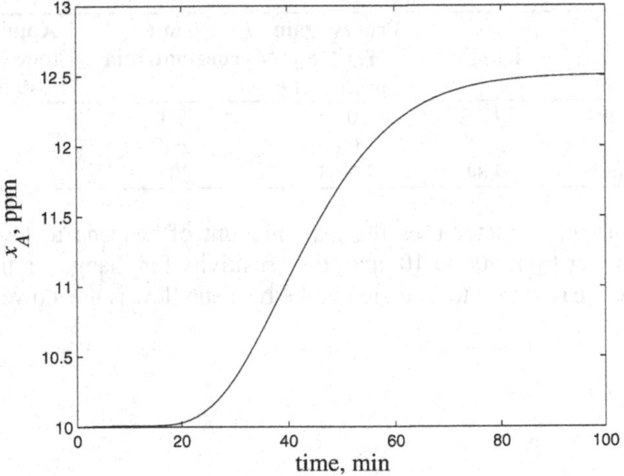

Fig. A14. Response of x_A to a step change in $x_{A,in}$ of 25% for cross flow extractor.

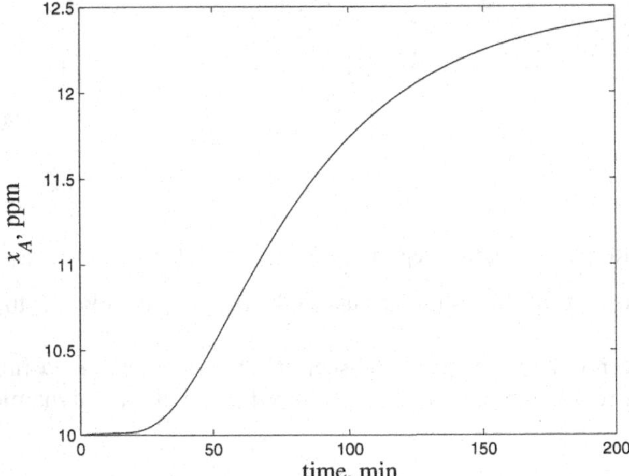

Fig. A15. Response of x_A to a step change in $x_{A,in}$ of 25% for counter-current extractor.

b. **Q**: Build a discrete PI controller which controls x_A of the cross flow extractor by means of F_B to the first stage and tune the controller for a setpoint change. Assume that the dynamics of the measurement and correction devices may be ignored. The control objective is to bring the controlled variable on target as soon as possible. The PI controller should have an execution interval of 1 minute. Show the setpoint change and controlled variable in one plot and the manipulated variable in another plot.
A: The Simulink diagram of a PI controller is shown in Fig. A16. The simulation model for this case is ExtrXFpifs.mdl. By running it, Fig. A17 is obtained.

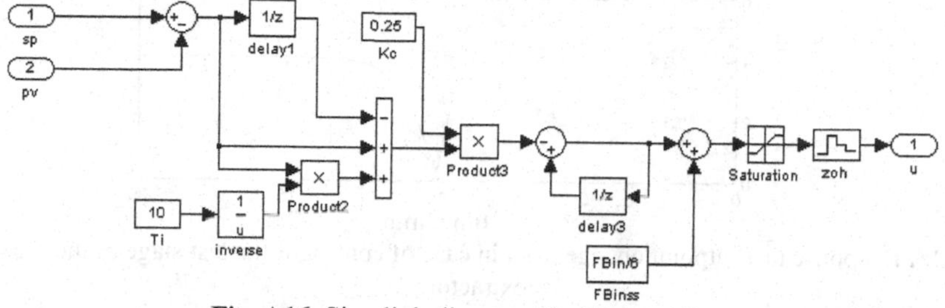

Fig. A16. Simulink diagram of a PI controller.

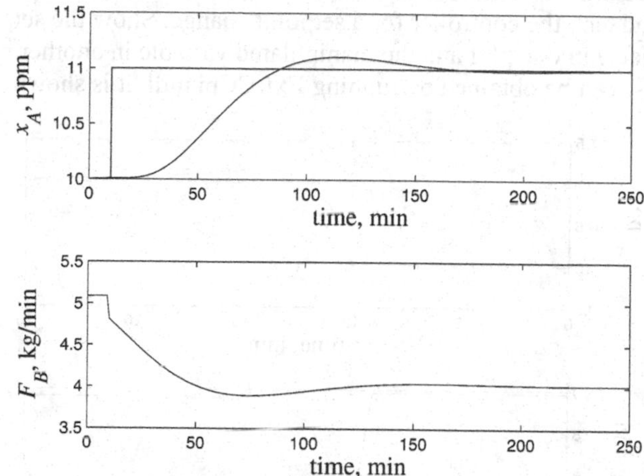

Fig. A17. Response to a setpoint change in x_A in case of control of the first stage of the cross flow extractor.

c. **Q**: Build a PI controller which shows control of x_A of the cross flow extractor by means of F_B to the last stage and tune the controller for a setpoint change. Show the setpoint change and controlled variable in one plot and the manipulated variable in another plot.
A: The response can be obtained by running ExtrXFpils.mdl, it is shown in Fig. A18.

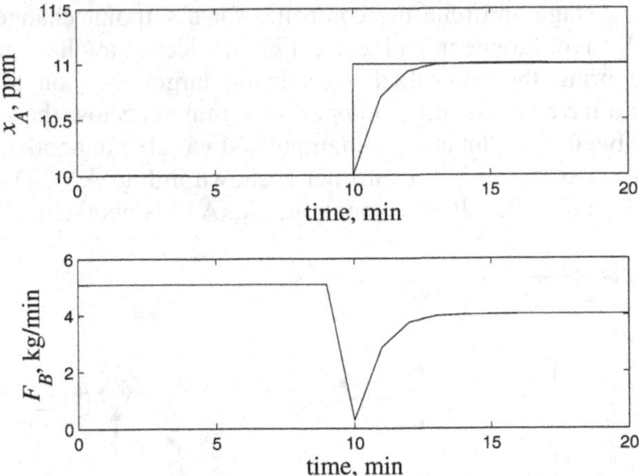

Fig. A18. Response to a setpoint change in x_A in case of control of the last stage of the cross flow extractor.

d. **Q**: Build a PI controller which shows control of x_A of the counter current extractor by means of F_B and tune the controller for a setpoint change. Show the setpoint change and controlled variable in one plot and the manipulated variable in another plot.
A: The response can be obtained by running ExtrCCpi.mdl, it is shown in Fig. A19.

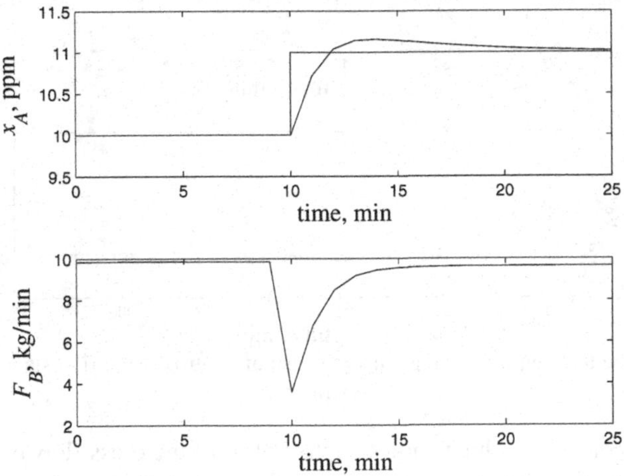

Fig. A19. Response to a setpoint change in x_A in case of control of counter current extractor.

e. **Q**: Make a table for the three control parameters of the three PI controllers. What are the differences between the controller settings and can the differences be explained from the process dynamics?
A: The PI settings are as follows:

Table A1. Controller settings for the three cases.

	K_c	T_i
Cross flow, first stage	0.25	10.0
Cross current, last stage	4.0	5.0
Counter current	5.0	4.0

The cross flow first stage controller has to be tuned very conservatively. Manipulation of the benzene flow affects the aniline concentration at the outlet of the first stage, this change has to be propagated to the outlet of the sixth stage, which introduces a transportation lag. This is not present in the other two cases. In case of manipulation of the benzene flow to the last stage of the cross flow setup, changes in benzene flow will have an immediate impact on the aniline concentration, hence control is fast. In case of counter-current flow, changes in benzene flow affect the aniline concentration with a relatively small time constant, hence control will also be fast.

Files used in this appendix:
ExtractparSS.m: file used to set single stage extractor parameters
ExtractparXF.m: file used to set cross flow extractor parameters
ExtractparCC.m: file used to set counter-current extractor parameters
ExtrSS.mdl: model of a single stage extractor
ExtrCC.mdl: model of a six stage counter current extractor
ExtrXF.mdl: model of a six stage cross flow extractor
ExtrXFpifs.mdl: model of control of the first stage of the cross flow extractor
ExtrXFpils.mdl: model of control of the last stage of the cross flow extractor
ExtrCCpi.mdl: model of control of the counter current extractor
ExtractPlot.m: file to plot uncontrolled extractor response
ControlPlot.m: file to plot controlled extractor response

Getting started in MATLAB Simulink

1) Make a subdirectory and create a file with a definition of the model parameters called ExtractDim.m. Any Simulink models and Matlab scripts you make should be saved in this subdirectory.
2) Open Matlab and set the current directory to your directory.

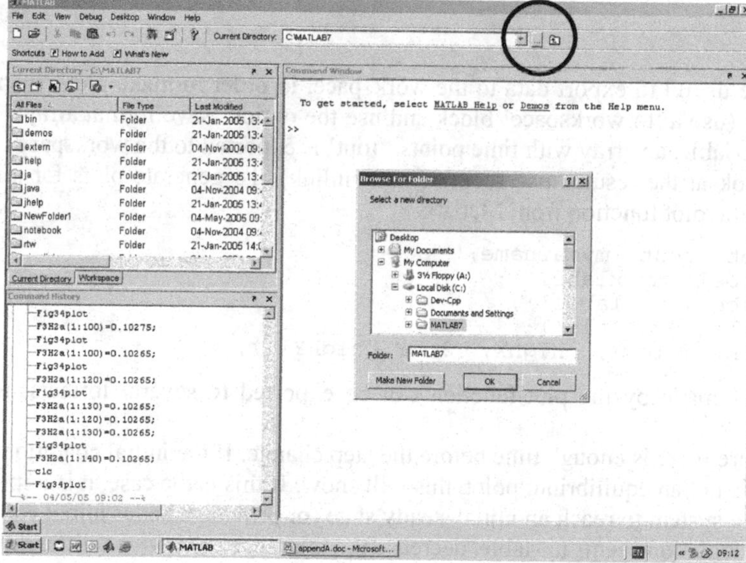

3) Run ExtractDim.m by typing **'ExtractDim'** in the command window. This will define constants needed. To change names or values, you can edit this file by typing **'open ExtractDim'** in the command window.
4) To see which variables are present in the workspace type **'whos'**
5) Open simulink by typing **'simulink'** in the command window

Hints

1) It will be much easier to find errors if the worksheet is well ordered.
   a) Try to avoid crossing lines.
   b) If the (sub)system does not seem to fit on a single screen, divide it into subsystems
   c) Use a subsystem per equation; insert a subsystem, open it and start with making the inputs/outputs and naming them.
2) Use the workspace

   Instead of providing a value as parameter, it is better to provide a variable, which is given a value in the workspace (or in a script). If there are many of instances of such a variable, it will be much easier to change it.

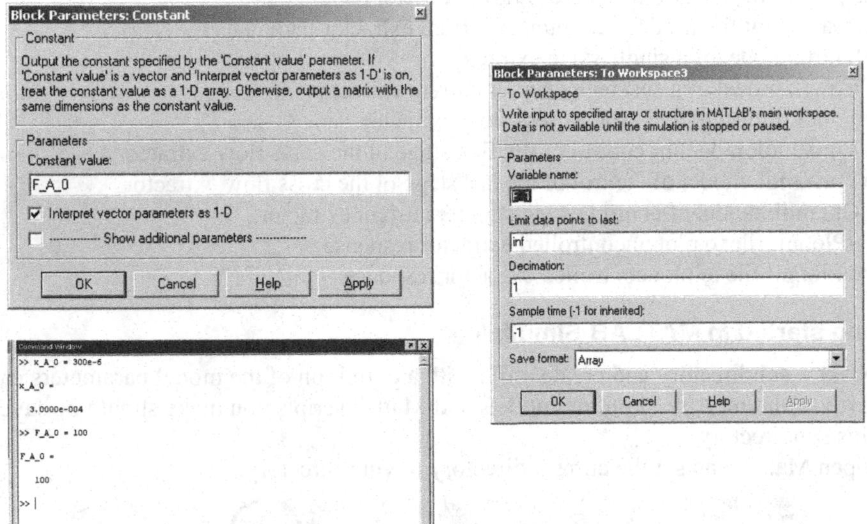

   It can be useful to export data to the workspace, in order to make plots, manipulate the data etc. (use a 'to workspace' block and use the option 'save format array'. In addition to the variable, an array with time points, 'tout' is exported to the workspace).

   To look at the results, use scopes (in Simulink). To generate plots for a report, you can use the plot function from Matlab:

```
plot( tout, myvarname,'-k' );
xlabel('my xlabel');
ylabel('y ylabel');
title('my title');
axis( [ lowX,  highX,  lowY, highY] );
```

   A plot made by the plot-function can be exported to several formats, for example jpg/eps.
3) Make sure there is enough time before the step change. If the initial condition of the system is not an equilibrium point, this will show. If this is the case, it is important to allow the system to reach an initial steady state (or provide a better initial condition).
4) If the calculation seems unstable, decrease the tolerance or choose a different solver. (Simulation --> Simulation parameters).
5) Make sure the system reaches steady state; if it does not, increase the simulation time. (Simulation --> Simulation parameters).

Index

Process Dynamics and Control: Modeling for Control and Prediction. Brian Roffel and Ben Betlem.